U0108368

人民幣國際化的
新進展

香港交易所的離岸金融產品創新

New Progress in RMB
Internationalisation

Innovations in HKEX's Offshore
Financial Products

人民幣國際化的新進展
香港交易所的離岸金融產品創新

主　　編：巴曙松

副 主 編：蔡秀清

編　　委：巴　晴　朱　曉

責任編輯：張宇程

封面設計：張　毅

出　　版：商務印書館（香港）有限公司
　　　　　香港筲箕灣耀興道 3 號東滙廣場 8 樓
　　　　　http://www.commercialpress.com.hk

發　　行：香港聯合書刊物流有限公司
　　　　　香港新界荃灣德士古道 220-248 號荃灣工業中心 16 樓

印　　刷：美雅印刷製本有限公司
　　　　　九龍觀塘榮業街 6 號海濱工業大廈 4 樓 A 室

版　　次：2023 年 12 月第 1 版第 4 次印刷
　　　　　© 2018 香港交易及結算所有限公司
　　　　　ISBN 978 962 07 6605 3
　　　　　Printed in Hong Kong
　　　　　版權所有　不得翻印

New Progress in RMB Internationalisation

Innovations in HKEX's Offshore Financial Products

Chief editor: BA Shusong

Deputy chief editor: Essie TSOI

Editors: Qing BA Jennifer ZHU

Executive editor: Chris CHEUNG

Cover design: Ann ZHANG

Publisher: The Commercial Press (H.K) Ltd.,
8/F, Eastern Central Plaza, 3 Yiu Hing Road,
Shau Kei Wan, Hong Kong

Distributor: The SUP Publishing Logistics (H.K.) Ltd.,
16/F, Tsuen Wan Industrial Center, 220-248 Texaco Road,
Tsuen Wan, N.T., Hong Kong

Printer: Elegance Printing and Book Binding Co. Ltd.
Block A, 4th Floor, Hoi Bun Building
6 Wing Yip Street, Kwun Tong, Kowloon, Hong Kong

© 2018 Hong Kong Exchanges and Clearing Limited
First edition, Fourth printing, December 2023

ISBN: 978 962 07 6605 3
Printed in Hong Kong

All rights reserved. No portion of this publication may be reproduced or transmitted in any form or by any means, electronic or mechanical, including photocopy, recording, or any information storage or retrieval system, without permission in writing from the copyright holder.

目　錄

Contents

序 言

為人民幣國際化
奠定堅實的產品和市場基礎

　　縱觀香港經濟金融體系發展的歷史，可以說是一部在東西方交匯與互動的過程中不斷探索自身獨特定位的歷史，香港也因此從荒蕪的漁村，一躍而成為繁榮的國際金融中心。一代又一代的香港人，為此付出了自己的努力與智慧。

　　在目前的全球格局變化中，香港應當如何找到自己在新的環境下的獨特定位？這一直是我在思考中的一個重要問題。

　　香港與紐約及倫敦並列為全球金融體系中十分活躍的國際金融中心，亦位處兩者之間的時區，更是輻射「一帶一路」沿線國家與地區的關鍵節點。歷史上，香港憑藉開放的市場環境、完善的法治、與國際接軌的市場制度、以及充足的專業人才成就其國際金融中心之地位，在全球金融中心的多種指數排名中，香港長期居於三甲之列。

　　當前，全球經濟金融體系正在經歷又一輪大規模的劇烈變化，一方面，中國在越來越深入地融入到全球金融體系中；與此同時，全球金融體系與中國市場的互動也日益頻繁。在這個過程中，香港作為國際金融中心，在「連接中國與世界」方面，將繼續發揮獨特的樞紐作用，「滬港通」、「深港通」和「債券通」的成功運行，更是強化了這一定位。

　　環顧全球金融體系，雖然中國的國內生產總值總量在迅速提升，但是人民幣資產融入到國際金融體系中的比重還處於起步階段。人民幣成功加入國際貨幣基金特別提款權（SDR）的貨幣籃子，佔比 10.92%，但是人民幣在外匯儲備、以及國際金融體系中的應用佔比，卻離這個比例還有相當的距離。我們判斷，未來一段時間，人民幣資產將有望加速融入到全球金融體系中、並且迅速提升其在全球金融體系中的影響力。在這個過程中，香港還能夠發揮重要的作用，特別是在人民幣「走出去」的過程中，香港離岸市場可充分利用資金進出自由的優勢，國際資金港的地位，進一步豐富人民幣計價的金融產品創新，成為推動人民幣走出去、實現國際人民幣跨境投融資、風險管理、金融創新的重要門戶。更為重要的是，「一國兩制」下的香港作為中國內地與國際市場的「超級聯繫人」，提供了根本的制度優勢。在「一國」的框架下，香港與中國內地的協作深度可以超越其他任何國家和地區，為全面開放體系提供無所比擬的可控性。「兩制」又使得香港擁有符合國際金融市場慣例的市場運行機制和制度框架，得以具備高度的靈活性來連接內地人民幣國際化過程中與國際市場之間的互動需求。

在中國構建全面開放新秩序的過程中，持續對外開放市場，實現中國資本全球化佈局、加快推進人民幣國際化將是持續發生且不可逆轉的重大轉變。中國未來的發展正在出現多個方面的重要轉變：第一，由以往的引進來為主，走向引進來與走出去並重，形成資金雙向流動、資源、資產雙向配置；第二，由以往的貿易、商品輸出為主，走向商品輸出與資本輸出相結合；第三，由以往的遊戲規則的遵循者，轉為既是遵循者也是遊戲規則的參與者和制定者。如何在資本項目未完全開放的情況下，平穩可控地實現上述目標？我相信這是當前中國經濟金融改革亟待解決又具有挑戰意義的重要課題。

如果在現有框架下，逐步開放境內市場和資本項目，推動資本、資源進一步雙向流動，固然可以實現中國經濟的全面開放，然而，鑒於內地與國際金融市場在規則體系、運作上的重大差異在相當長的時間內會存在，境內、外資金出現規模更大更頻繁的跨境流動也會帶來相當大的監管挑戰。因此，需要建立一個有序可控且不會對內地的宏觀經濟金融政策造成明顯衝擊的跨境平台，開發金融產品和發展金融生態圈，在既符合中國的金融開放的要求、又在國際認可的規則體系的基礎上，推動中國的資本、資產、資源與國際金融體系之間的良性平穩互動融合，提升中國資本在國際金融市場上的影響力和話語權。

這正是我們香港交易所近年來着力發展的「互聯互通」以及多項人民幣產品的着眼點：為中國提供一個既開放又風險可控的平台，將國際投資者和企業與中國企業和投資者連接起來，在推動金融產品創新的同時，充分保證交易便利性、資本暢通流動，無縫對接國內與國際市場，在中國與全球資源雙向配置中發揮獨特作用。

基於這些判斷，我們近年來在香港積極推動產品創新，我們認為有深度的、多元而且發達的金融產品市場，未來必定會在人民幣國際化的進程中成為重要的載體和支持力量。這本以香港人民幣離岸產品為主題的研究報告集，正是對我們近年來的一系列產品創新的跟蹤研究的成果。

這本文集嘗試把握國際視角和中國需求，集中探討了近年來香港交易所在股票、定息及貨幣產品、大宗商品等各個範疇開拓人民幣產品的最新進展，以及推動落實打造人民幣離岸產品中心的市場目標，為讀者展現了人民幣離岸產品發展的演進歷程，亦有助於引發讀者思考香港離岸金融創新在推動人民幣國際化和金融深化中所扮演的獨特角色和積極意義。

本書的第一部分介紹了香港交易所近年來在股票市場上推動的各項創新。股票通（又分為滬港通和深港通）是「互聯互通」模式在股票市場的先行先試，通過基礎設施和交易平台的跨境對接，實現香港和內地股票二級市場的互聯互通。建議中的「新股通」，即允許境內外投資者分別申購對方市場所發售的新股，實質是將股票市場的互聯互通模式從二級市場向一級市場的進一步延伸。容許有同股不同權架構的企業在港上市，特別是某些未錄得盈利、尚處於起步階段的新經濟、新能源、高端裝備製造等創新公司，有助於中國優質公司嫁接國際合作者、參與全球佈局和國際產業定價。本部分還對服務於內地和香港兩地跨境投資的指數及其相關產品作了介紹，為投資者投資內地 A 股和共同市場選擇有效工具提供借鑒。

本書的第二部分談及定息及貨幣產品，圍繞中國債券市場開放及債券通的制度創新等內容展開。債券通是互聯互通在債券及定息類產品市場上的具體體現，它不僅是一個實現人民幣跨境投資、雙向循環的載體，更可能成為發展多層次人民幣風險管理產品和金融工具的孵化器。通過債券通帶動國債期貨、美元兌人民幣（香港）期權合約、人民幣貨幣指數等衍生品的開發和運用，推動人民幣離岸金融產品生態圈發展，將提升香港人民幣定價能力和風險對沖能力，強化海外人民幣定價權。

本書的第三部分介紹了香港交易所首隻實物交收雙幣（以美元及人民幣定價

及結算）黃金期貨合約，以及以美元計價、現金結算的中國鐵礦石期貨合約。香港
交易所 2012 年收購倫敦金屬交易所後，積極推動在亞洲建立透明度高且流動性好
的大宗商品期貨市場，這不僅可滿足市場參與者對風險管理的需要，更促使香港
成為內地黃金、鐵礦石、乃至石油等重要商品連接國際市場的交易門戶，助推實
現中國商品市場的對外開放，強化中國在國際商品市場的定價權。

　　本書的第四部分結集了有關離岸人民幣中心的多篇文章，對離岸人民幣的流
動性、香港交易所人民幣產品的交易，以及為國內金融機構提供的場外結算方案
等話題進行了探討，顯示香港交易所致力成為便利全球投資者的離岸人民幣產品
交易及風險管理中心。

　　當前，中國金融市場在穩步推動雙向開放，越來越多的海外和內地投資者需
要參與香港市場，需要對國際金融產品和香港離岸人民幣市場有更多更深入的認
識。隨着香港交易所的「互聯互通」不斷延伸，越來越多的內地金融機構、企業，
以及個人投資者，都可能會利用香港交易所的平台進入國際金融市場。同時，隨
着內地經濟轉型推進、人民幣的進一步國際化，海外投資者亦越發希望加大人民
幣資產的配置，在內地金融市場還未完全開放之際，人民幣離岸產品以及「互聯互
通」平台也就提供了難得的便捷渠道。本書圍繞上述主題進行了較為系統的介紹，
並將一些貼近市場發展的第一手資料進行了梳理，使讀者不僅對香港人民幣離岸
市場和產品創新有一個全局的認識，也有利於支持內地投資者熟練運用這些工具、
平台和產品涉足海外市場投資，以及海外投資者投資內地市場，共同成為香港「互
聯互通」金融生態圈的活躍參與力量。

　　我相信，在人民幣國際化發展的新階段，香港可以通過「互聯互通」，成為連
接內地與全球的重要的門戶市場和離岸中心，為中國進一步融入國際金融體系、
提升國際市場的影響力帶來新動力。在這個過程中，我希望香港能湧現出更多類
型的產品，來推動這個激動人心的進程。

李小加

香港交易及結算所有限公司　集團行政總裁

Preface

Laying a solid product and market foundation for Renminbi internationalisation

The history of Hong Kong's economic and financial systems has been a history of perpetual exploration of self-positioning in the course of growing interactions between the East and the West. Along the way, Hong Kong has transformed from a barren fishing village into a thriving international financial centre, thanks to the efforts and wisdom of generations of the Hong Kong people.

How should Hong Kong find its unique position in the current changing global land-scape? This is a question I keep asking myself.

Hong Kong ranks alongside with New York and London as the world's most active international financial centres (IFC). Situated in a time zone in between New York and London, Hong Kong is at a key juncture to serve countries and regions along the Belt and Road. Thanks to its open market environment and rule of law, a market structure of international standards, and a large pool of professionals and experts, Hong Kong has lived up to its name as an IFC and for a long time ranked among the top three in various league tables ranking global financial centres.

Today's global economic and financial systems are in the midst of another round of large-scale drastic changes. China's integration with global finance is deepening. Inter-actions between global finance and the Chinese market are becoming frequent. In the process, Hong Kong, as an IFC, continues to play a unique pivotal role connecting China with the rest of the world, a role that has been enhanced by the success of Shanghai Connect, Shenzhen Connect and Bond Connect.

Globally, China's gross domestic product is rapidly rising, but the proportion of renmin-bi assets in international finance is still at a starting point. The renminbi is now included in the International Monetary Fund's Special Drawing Right (SDR) basket of currencies with a weighting of 10.92 per cent. However, the currency's share in foreign exchange reserves and its use in international finance are considerably lower than that percentage. We predict that, soon enough, the integration of renminbi assets into global finance will accelerate, and the currency's global influence will increase. In the process, Hong Kong can play a critical role, especially in the globalisation of the renminbi. Capitalising on the free flow of capital and its role as a port for international capital, the offshore market in Hong Kong can promote the innovation of renminbi financial products and become a crucial gateway for the currency's internationalisation and its application in cross-border investment and financing, risk management and financial innovation. More importantly, under "One Country, Two Systems", Hong Kong, as a "super-connector" between Mainland China and the international market, has fundamental institutional advantages. Under the framework of "One Country", Hong Kong can collaborate with the Mainland more intensively than any other country and region, giving the Mainland unparalleled control over its comprehensive market opening reforms. "Two Systems" preserves Hong Kong's market mechanisms and institutional frameworks that are in line with international financial market practices. This allows Hong Kong to have a high degree of flexibility to tackle the interactions between Mainland China and the international market during the course of the renminbi's internationalisation.

As China constructs a new order of comprehensive market opening, there are major, perpetuating and irreversible changes — the market's continuous opening up, global allocation of Mainland capital and accelerated internationalisation of the renminbi. China is heading towards significant changes in several aspects. Firstly, capital flow and asset and resources allocation will be two-way rather than one-way that focused on inflows only as in the past. Secondly, the dominance of trade and export of goods will move towards parallel export of both goods and capital. Thirdly, China will turn from a follower of rules into both a follower and a participant and developer of rules. How can these objectives be achieved smoothly and under control when the capital account is not yet fully opened up? This, I believe, is one of China's most important, challenging and urgent questions to answer as it proceeds with its economic and financial reforms.

Under the existing framework, gradual opening of the domestic market and capital account and promotion of further two-way flows of capital and resources will certainly lead to full-scale opening of China's economy. However, huge gaps between Mainland and international financial markets' institutional systems and operations will continue to exist for a long time. Regulatory challenges are expected to be considerable as cross-border flows of domestic and foreign capital become larger and more frequent. In the light of these, it is important to establish an orderly and controllable cross-border platform, that would not have significant impacts on the Mainland's macro economic and financial policies, for developing financial products and a financial ecosystem. Grounded on an internationally recognised regulatory structure while satisfying the requirements of China's financial opening, the platform will promote healthy and steady interactions of China's capital, assets and resources with the international financial system. The influential power of Chinese capital in the international financial market will thereby be enhanced.

This is precisely the focus of the mutual market access or connectivity programme and the renminbi products which HKEX has been developing with full efforts in recent years. The aim is to provide China with an open and controllable platform that connects international investors and enterprises with Chinese enterprises and investors. While promoting financial product innovation, the platform also fully guarantees the ease of trading, the smooth flow of capital and the seamless connection between the Mainland and international markets. This will play a unique role in the two-way allocation of resources between China and the rest of the world.

Based on these assessments, we have been actively promoting product innovation in Hong Kong in recent years. We believe that a deep, diversified and well-developed market for financial products will be an important vehicle and supporting force for the renminbi's internationalisation. This book of a collection of research reports on renminbi offshore products in Hong Kong is the product of a series of studies we have conducted on our latest product innovations.

This collection attempts to explore, from an international perspective and from the angle of China's demands, the latest developments of HKEX's renminbi products in asset classes ranging from stocks, fixed income and currency products to commodities, and the objective of building a renminbi offshore product centre in Hong Kong. It presents to readers the evolution of renminbi offshore products, and inspires readers to think about the unique role and significance of Hong Kong's offshore financial innovations in the renminbi's internationalisation process and in the deepening of the financial systems.

Part One of this book introduces HKEX's recent innovations in its stock market. Stock Connect (Shanghai Connect and Shenzhen Connect) is the first connectivity scheme with pilot implementation in the stock market. It aims at achieving mutual stock market access in the secondary market between Hong Kong and the Mainland through linking up the two markets' infrastructure and trading platforms. The proposed Primary Equity Connect, which allows Mainland and offshore investors to subscribe for newly issued shares in each other's market, is an extension of mutual stock market access from the secondary market to the primary market. Allowing the listing in Hong Kong of companies that adopt weighted voting rights, particularly pre-profit start-ups in the new economy, new energy and high-end equipment manufacturing sectors, will help quality Chinese companies connect with international partners, expand globally and participate in international pricing. This part also introduces Mainland-Hong Kong cross-border indices and other related products that provide benchmarks for investors investing in Mainland A shares and the mutual market.

Part Two discusses fixed income and currency products, examining the opening up of the Chinese domestic bond market and innovations in Bond Connect. Bond Connect is the manifestation of connectivity in bonds and fixed-income products. It is more than a vehicle for cross-border renminbi investment and two-way renminbi circulation. It could even become an incubator for developing multi-level renminbi risk management products and financial instruments. Bond Connect will stimulate the development and application of treasury bond futures, USD/CNH options, RMB currency index futures and other derivatives, thereby promoting the development of an ecosystem of renminbi offshore financial products. This will enhance Hong Kong's capability in renminbi pricing and risk hedging, and increase the pricing power of offshore renminbi.

Part Three introduces HKEX's first dual-currency (priced and settled in US dollar and renminbi) gold futures contract that is physically settled, and its USD-denominated and cash-settled iron ore China futures. After acquiring the London Metal Exchange in 2012, HKEX actively pushed for the establishment of a transparent and highly-liquid commodity futures market in Asia. Such a market will not only satisfy the risk management needs of market participants, but will also turn Hong Kong into a trading gateway connecting major Mainland commodities such as gold, iron ore and petroleum oil with the international market. This will help China's opening up of its commodity market and strengthen the country's pricing power in international commodities.

Several articles in Part Four on offshore renminbi centres discuss such topics as the liquidity of offshore renminbi, trading of renminbi products at HKEX and over-the-counter (OTC) clearing solutions for Mainland financial institutions. HKEX's commitment to serve global investors as a trading and risk management centre of offshore renminbi products is demonstrated.

Two-way opening of the Chinese financial market is in steady progress. An increasing number of overseas and Mainland investors need to participate in the Hong Kong market and need a wider and deeper understanding of international financial products and the offshore renminbi market in Hong Kong. As HKEX's connectivity programme continues to expand, more and more Mainland financial institutions, enterprises and individual investors may use HKEX's platform to enter the global financial market. With the Mainland's economic transformation and the renminbi's internationalisation under way, overseas investors increasingly want to increase their exposure to renminbi assets. Before the Mainland financial market is fully open, offshore renminbi products and the connectivity platform will continue to be valuable and convenient channels of access. This book provides a systematic introduction of the above themes and first-hand information on related market development. It will not only enable readers to have a full picture of the renminbi's offshore market and product innovations in Hong Kong, but will also empower, on the one hand, Mainland investors to utilise these instruments, platforms and products to invest in overseas markets and, on the other hand, overseas investors to participate in the Mainland market, thereby nurturing an active participant base in the Hong Kong ecosystem of financial connectivity.

In this new stage of the renminbi's internationalisation, I am confident that through mutual market access, Hong Kong can serve as an important gateway and offshore hub for the Mainland to connect with the global market and provide China with new impetus to integrate with international finance and increase its international market influence. In this process, I hope a greater variety of products will emerge in Hong Kong that will advance such an exciting journey.

Li Xiaojia, Charles

Chief Executive
Hong Kong Exchanges and Clearing Limited

第一部分

股　票

第 1 章

中華交易服務
中國 120 指數期貨

—— 跨境投資離岸對沖的好工具

2017 年 2 月 14 日

概　要

　　自滬港股票市場交易互聯互通機制試點（滬港通）於 2014 年 11 月正式開通，內地與香港市場之間的跨境股票投資活動揭開全新一頁。其後 2016 年 8 月 16 日宣佈設立深港股票市場交易互聯互通機制（深港通）並即時取消滬港通總額度之設立，及至 2016 年 12 月 5 日深港通正式開通，互通機制一再突破。滬港通與深港通合稱「互聯互通」機制。至此，滬深港三個市場之間一個「共同市場」的平台已然基本成形。這些發展無疑會促進跨境股票投資活動，同時亦產生越來越多對跨境股票投資組合的風險管理需求。

　　然而，環球市場中專為內地 A 股或共同市場跨境股票投資而設的指數期貨及期權等的相關風險管理工具卻十分稀缺。有衍生產品（期貨／期權）在內地以外的海外交易所掛牌買賣的中國相關指數，唯有中華交易服務中國 120 指數（中華 120）兼含內地 A 股及香港上市中資股票，其餘要不只是涵蓋內地 A 股（富時中國 A50 指數），要不就只包括境外（主要是香港）上市的中資股（包括富時中國 50 指數及 MSCI 中國外資自由投資指數）。相形之下，中華 120 的成份股覆蓋共同市場三家交易所（香港、上海及深圳），不少更是滬港通及深港通下的合資格股票，無論以上市交易所還是股票或行業的類別而言均有所及。此外，中華 120 與 A 股指數有很高的相關性。2016 年全年，中華 120 的回報率及波幅均高於其他的交易指數，股息收益率媲美內地藍籌的上證 50 指數。

　　基於中華 120 指數的這些特點，香港交易所衍生產品市場所提供的中華 120 期貨可以是投資者對沖其股票投資的有效風險管理工具，又或直接用作投資內地 A 股和共同市場的工具。若跟新加坡交易所的富時中國 A50 期貨比較，中華 120 期貨的交易所費用較低（按每合約名義金額計）、價位相對指數水平較小，持倉限額也較高，不失為投資 A 股和共同市場的一個方便兼具成本效益的離岸市場工具。參照歐元區斯托克 50 指數（EURO STOXX 50）及其衍生產品和結構性產品在服務歐洲共同市場的成功事例，中華 120 指數及其衍生產品和結構性產品應當可為內地和香港這個共同市場提供同樣的效益。

1 跨境投資的風險管理需要

1.1　滬港通及深港通促進跨境股票投資活動

　　滬港股票市場交易互聯互通機制試點（滬港通）於 2014 年 11 月正式推出，是中國內地與香港之間破天荒的市場互聯互通計劃的首個項目，為海外投資者投資內地股票市場及內地投資者投資香港股票市場開設嶄新的正式渠道。此前，境外人士參與內地證券市場的渠道只限於合格境外機構投資者（QFII）計劃及人民幣合格境外機構投資者（RQFII）計劃，境外散戶投資者只可透過 QFII 及 RQFII 提供的投資基金參與內地股市[1]。反向地，合格境內機構投資者（QDII）計劃及人民幣合格境內機構投資者（RQDII）計劃是內地參與海外證券市場的唯一全國性正式渠道[2]。

　　及至 2016 年 12 月 5 日，滬港通的延伸篇章 —— **深港股票市場交易互聯互通機制（深港通）**也正式開通，合資格證券範圍擴大。滬港通與深港通合稱「互聯互通」機制。互聯互通機制最重要的突破是自 2016 年 8 月 16 日中國證券監督管理委員會（中國證監會）與香港證券及期貨事務監察委員會（香港證監會）聯合宣佈設立深港通當日即時取消了總額度限制。深港通正式開通後，滬深港三個市場之間的「**共同市場**」模式基本成形；按此，三個市場上的各式金融產品，只要法規准許，內地及全球投資者均可跨境買賣。具體操作可分為南北兩個方向：內地投資者通過上海證券交易所（上交所）及深圳證券交易所（深交所）的交易平台落盤，指示買賣在香港上市的合資格產品（南向交易／港股通），以及國際投資者通過香港聯合交易所（聯交所）交易平台落盤，指示買賣在上交所及深交所上市的合資格產品（北向交易／滬股通及深股通）。**這樣一個沒有總額度限制的共同市場，讓投資者可從較長線的投資角度於內地與香港的市場中配置資產。**

　　在滬港通及深港通下，北向交易的合資格證券包括：上證 180 及上證 380 指

1　在 1992 年於上海及深圳交易所推出的 B 股市場（以外幣交易，與 A 股市場分開），是為內地首次嘗試將股票市場對外國投資者開放，但在市場開放的新浪潮下，B 股市場並不活躍。

2　地區政府有推出特殊計劃，如上海的「合格境內有限合夥人」（QDLP）計劃和深圳的「合格境內投資者境外投資試點」（QDIE）。但這些計劃只限於主要服務機構投資者及高端個人客戶的私募基金或投資工具，並不如 QDII 產品開放給一般投資者。

數的成份股；深證成份指數和深證中小創新指數成份股中所有市值不少於人民幣 60 億元的成份股；以及所有有相關 H 股在聯交所上市的上交所或深交所上市 A 股[3]。反向地，南向交易的合資格證券包括：恒生綜合大型股指數 (HSLI) 及恒生綜合中型股指數 (HSMI) 的成份股；市值 50 億港元或以上的恒生綜合小型股指數 (HSSI) 的成份股；以及所有有 A 股在內地市場 (上交所或深交所) 上市的 H 股[4]。 HSLI 及 HSMI 已佔恒生綜合指數 (HSCI) 總市值高達 95% 之多，而 HSCI 又涵蓋香港市場總市值最高的 95%[5]。在深港通正式開通當日 (2016 年 12 月 5 日)，南向交易合資格證券佔聯交所主板上市股票總市值的 87%，滬港通北向交易合資格證券佔上交所上市 A 股總市值的 81%，而深港通北向交易合資格證券佔深交所市場 (主板、中小板及創業板) 上市 A 股總市值的 71%[6]。

縱使現時合資格的證券只限於滬港通及深港通的指定股票範圍，滬港通及深港通已無形中打開了一個潛在的內地與香港股票共同市場，其股份總值 109,860 億美元 (2016 年 11 月底)、日均股份成交約 850 億美元 (2016 年截至 11 月)，**於全球交易所中按市值計排名第二 (僅次於紐約證券交易所)、按股份成交額計排名第二**[7]。

1.2 跨境股票投資的風險管理

共同市場平台設立後，跨境買賣活動增多可以預見，針對跨境股票組合的風險管理也隨之愈益重要。要對沖投資組合的風險，常用的風險管理工具包括上市股份及 / 或相關市場指數的期貨及期權。然而，市場上為內地投資者買賣港股提供的相關對沖工具卻乏善足陳 —— 內地的交易所至今未有提供任何香港指數 / 股票的期貨或期權產品。同樣地，香港亦欠缺 A 股指數期貨期權等的 A 股對沖工具。相關的衍生產品日後或會被納入共同市場模式，但在此之前，投資者可考慮利用

3　不包括不以人民幣交易的股份及被上交所或深交所實施風險警示的股份 (包括「ST 公司」及「*ST 公司」的股份以及須進行除牌程序的股份)。

4　不包括不以港元交易的股份及其相應 A 股被實施風險警示的 H 股。

5　資料來源：恒生指數有限公司網站。HSCI 的香港市場選股範疇包括所有於聯交所作第一上市的股份和房地產投資信託基金，但並不包括第二上市的證券、外國公司、優先股、債務證券、互惠基金及其他衍生產品。

6　資料來源：基於取自湯森路透及有關交易所網站的數據以及該等交易所網站所載的合資格證券名單。

7　國際證券交易所聯會 (WFE) 數據 (WFE 網站 2016 年 12 月 22 日資料)。日均成交額按 WFE 2016 年截至 11 月的數據的合併股份成交以及香港市場交易日總數 (225 日) 計算。排名按 2016 年截至 11 月的合併成交額計算。

其本地市場所提供的替代工具。

　　在香港買賣內地 A 股的環球投資者，可以使用替代性質的指數期貨作為對沖。滬港通和深港通下，資產配置可橫跨香港與內地（上海及深圳）市場，若有指數的成份股同時涵蓋這三個市場的上市股份，會是不錯的替代工具。由中華證券交易服務有限公司（「中華交易服務公司」，香港交易所合資公司）發展的中華交易服務中國 120 指數（中華 120）正好是這樣的一個跨境指數，並有相關的期貨合約在香港交易所買賣。事實上，在環球交易所當中，中華交易服務中國 120 指數期貨（中華 120 期貨）是唯一以同時追蹤中國內地 A 股及香港上市中資股票的指數作為相關指數的期貨合約。世界各地交易所所提供的其他中國相關指數期貨，要不只是涵蓋內地 A 股，要不就只涵蓋境外（主要在香港）上市的中資股。

　　以下第 2 節將綜覽這些指數及相關的衍生產品，第 3 節將中華 120 和中華 120 期貨與其他內地相關股票指數加以對照。

2　中國股票指數及其衍生產品

　　富時羅素（FTSE Russell）和 MSCI 是環球市場兩大指數供應商，各自編制的中國指數有 20 隻左右，有些僅覆蓋 A 股，有些單涵蓋非內地的境外上市中資股，也有些同時覆蓋內地上市及非內地上市中資股，但獲中國內地以外地方的交易所採用作為相關指數而提供期貨（及期權）產品的不多。

　　在內地，上交所和深交所共同成立的中證指數有限公司（中證）是最主要的指數供應商，既編制單一市場（上交所或深交所）指數，亦編制跨市場（上交所和深交所）指數。這些指數中，只有三個指數有期貨產品在中國金融期貨交易所（中金所）買賣。主要中國股票指數中，確認出有場內交易衍生產品者（下稱「交易指數」）摘要於下文表 1。

表 1：主要中國股票指數（有場內交易衍生產品者）				
指數	簡稱	成份股	衍生產品	上市交易所
涵蓋內地上市股份				
富時中國 A50 指數（FTSE China A50 Index）	富時A50	50隻在上交所及深交所上市的市值最大的A股	富時中國A50指數期貨	新加坡交易所（新交所）
滬深300指數	滬深300	300隻在上交所及深交所上市的市值最大及最具流通性的A股	滬深300指數期貨	中金所
滬深500指數	滬深500	500隻在上交所及深交所上市的中小型A股	滬深500指數期貨	中金所
上證50指數	上證50	50隻在上交所上市的市值最大及最具流通性的A股	上證50指數期貨	中金所
涵蓋香港上市中資股				
恒生中國企業指數	恒生國企指數	在聯交所上市的H股	H股指數期貨及期權	香港交易所
富時中國 50 指數（FTSE China 50 Index）	富時中國50	50隻在聯交所上市及買賣的市值最大及最具流通性中資股票（H股、紅籌股、民企股*）	E 小型富時中國 50 指數期貨（E-Mini FTSE China 50 In-dex Futures）	芝加哥商品交易所（CME）
			富時中國 50 指數期貨（FTSE China 50 Index Futures）	日本交易所集團（JPX）
涵蓋香港及海外上市中資股				
MSCI 中國外資自由投資指數（MSCI China Free Index）	MSCI中國自由指數	在內地以外地區上市、市值屬大中型的中資股，包括在聯交所上市的H股、紅籌股、民企股*及海外上市股票	MSCI中國指數期貨及期權	新交所
			MSCI中國外資自由投資指數期貨	歐洲期貨交易所（Eurex）
涵蓋內地及香港上市中資股				
中華交易服務中華120指數	中華120	80 隻在上交所及深交所上市的市值最大及最具流通性的 A 股及 40 隻在聯交所上市的市值最大及最具流通性內地公司股票（H 股、紅籌股及民企股 *）	中華120指數期貨	香港交易所

* H股是由中國內地註冊成立而於香港交易所上市的公司發行；紅籌股是由中國內地以外註冊成立而在香港上市的公司發行，該等公司由內地政府實透過直接或間接持股及/或派員擔任公司董事而控制；民企股亦是由中國內地以外註冊成立而在香港上市的公司發行，但該等公司在中國內地的業務是由民營企業人士運作。

資料來源：富時羅素、MSCI、中華交易服務、中證指數公司及相關交易所的網站。

　　現時已知在中國內地以外主要環球交易所有期貨產品買賣的主要中資股指數有五個（下稱「海外交易指數」），其中之一的富時中國 A50 指數（富時 A50）僅覆蓋內地上市 A 股，其期貨（「富時 A50 期貨」）在新加坡交易所（新交所）交易，另三個覆蓋香港上市中資股──富時中國 50 指數和恒生中國企業指數（恒生國企指數），或包括其他外國上市股票──MSCI 中國外資自由投資指數（MSCI 中國自由指數）；**只有中華交易服務中國 120 指數（中華 120）是同時覆蓋內地上市 A 股和香港上市中資股**。下圖 1 為具有場內交易衍生產品的中國股票指數按成份股劃分的分類圖。

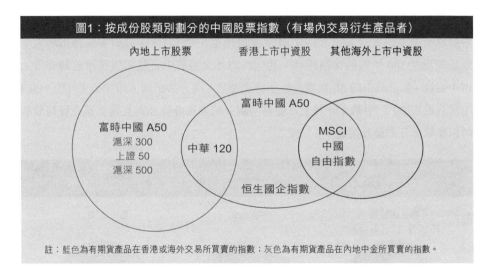

圖1：按成份股類別劃分的中國股票指數（有場內交易衍生產品者）

註：藍色為有期貨產品在香港或海外交易所買賣的指數；灰色為有期貨產品在內地中金所買賣的指數。

3　中華交易服務中國 120 指數及指數期貨

　　本節首先按市值和行業類別探討交易指數（有衍生產品在內地或海外衍生產品交易所買賣的指數）中於中港兩地上市成份股的組成，將不同指數的成份股組合作比較，並將之與互聯互通合資格股票及滬深港主要指數──上海：上證 A 股指數（上證 A）、深圳：深證 A 股指數（深證 A）及香港：恒生指數（恒指）──作比較。此外亦會研究指數間的相關性，及扼要比較各指數的表現。

最後是中華 120 期貨作為跨境投資風險管理工具的概述。

3.1　中華交易服務中國 120 指數
——A 股及共同市場指數的代表

就來自共同市場平台上三個市場（聯交所、上交所和深交所）的**成份股權重而言**，於 2016 年 11 月 30 日，中華 120 成份股中的聯交所和上交所股票權重幾乎相同，深交所股票的權重相對較輕（分別為 44%、43% 和 13%）。與現有中資股可交易指數比較，中華 120 更能代表共同市場：與 A 股指數比較，中華 120 成份股中的深股權重與富時 A50 相同，滬深 A 股權重比例（77:23）亦與滬深跨市場大型股指數滬深 300 指數的比例相近。其他在海外交易的中資股指數僅覆蓋聯交所上市中資股（如富時中國 50 指數及恒生國企指數）或另加美國上市中資股（如 MSCI 中國自由指數）。**中華 120 按上市交易所劃分的權重分佈亦與互聯互通合資格股票的相應權重分佈最為接近。**（見圖 2）

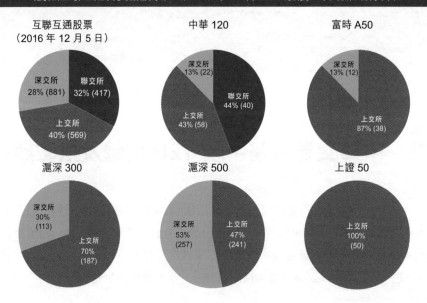

圖 2：中華 120 及其他中資股交易指數成份股的上市交易所權重佔比
（對照互聯互通合資格股票）（2016 年 11 月 30 日數據，另有説明除外）

互聯互通股票
（2016 年 12 月 5 日）

深交所 28% (881)　聯交所 32% (417)　上交所 40% (569)

中華 120

深交所 13% (22)　聯交所 44% (40)　上交所 43% (58)

富時 A50

深交所 13% (12)　上交所 87% (38)

滬深 300

深交所 30% (113)　上交所 70% (187)

滬深 500

深交所 53% (257)　上交所 47% (241)

上證 50

上交所 100% (50)

(續)

圖 2：中華 120 及其他中資股交易指數成份股的上市交易所權重佔比 （對照互聯互通合資格股票）（2016 年 11 月 30 日數據，另有說明除外）

富時中國 50

恒生國企指數

MSCI 中國自由指數 （2016 年 12 月 6 日）

註：不包括市值或權重數值並無數據的成份股。括號內為股票數目。

資料來源：互聯互通合資格證券名單來自香港交易所、上交所及深交所網站，市值資料來自香港交易所及湯森路透；中華 120 權重資料來自中華交易服務網站；MSCI 中國自由指數權重資料來自彭博；其他資料來自湯森路透。

此外，中華 **120** 的聯交所上市成份股包括 **H** 股、紅籌股及非 **H** 股內地民企（**P**股）[8]，不同股票類別的權重均類比互聯互通合資格股票的權重。（見圖 3）

圖 3：中華 120 成份股的股票類別權重佔比（對照互聯互通合資格股票）

互聯互通股票（2016 年 12 月 5 日）

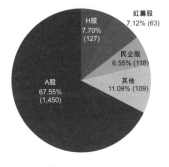

中華 120（2016 年 11 月 30 日）

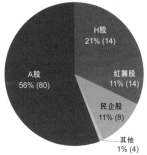

註：括號內為股票數目。

資料來源：互聯互通合資格證券名單來自香港交易所、上交所及深交所網站，市值資料來自香港交易所及湯森路透；中華 120 權重資料來自中華交易服務網站。

8　有關這些股票類別的定義，見表 1 的附註。

　　細看上述每個指數所**覆蓋的互聯互通合資格股票**，於深港通開通當日（2016年 12 月 5 日），就市值而言，中華 120 的覆蓋最為廣泛。以市值計，中華 120 覆蓋互聯互通合資格股票的 41.3%，其他海外交易指數中比例最高者的富時 A50 只覆蓋 22%，內地交易指數中比例最高者的滬深 300 指數則覆蓋 40.6%。以股票數目計，中華 120 的成份股代表了互聯互通合資格股票的 6.4%，比例本身不高，但已是海外交易指數中比例最高者（僅覆蓋聯交所上市股票的 MSCI 中國自由指數除外）。若按上市交易所分析所覆蓋的互聯互通股票，其他海外交易指數僅覆蓋內地A 股或聯交所上市股票，而並未同時覆蓋兩地股票。相反，**中華 120 的成份股分佈於聯交所、上交所和深交所三家交易所。**（見圖 4）

　　如按行業分析指數的組成，並與互聯互通股票比較，中華 120 中金融業權重較大（於 2016 年 11 月 30 日為 47%，互聯互通股票中的金融業權重為 25%），但其他行業的權重分佈則與互聯互通股票一樣均衡。中華 120 的行業分佈亦類似滬深跨市場藍籌指數滬深 300 指數。相反，若干其他海外交易指數較集中單一行業（金融業在富時 A50 及恒生國企指數的權重分別為 68% 及 71%；房地產在富時中國 50指數佔 72%）。特別的是，**中華 120 中的資訊技術及電訊服務成份股權重較為均衡，分別為 12% 及 6%**；相比之下，富時 A50 並無電訊服務股，資訊技術比重較低（1%）；而恒生國企指數則沒有資訊技術股，電訊股比重較低（2%）。（見圖 5）

　　上證 A 股和深證 A 股的行業組成相當不同，而**中華 120 由於覆蓋上交所和深交所股票，其行業組成較接近內地跨市場指數滬深 300 指數。**基於同一理由，其**行業覆蓋範圍亦較香港市場的恒指廣闊。**（見圖 5）

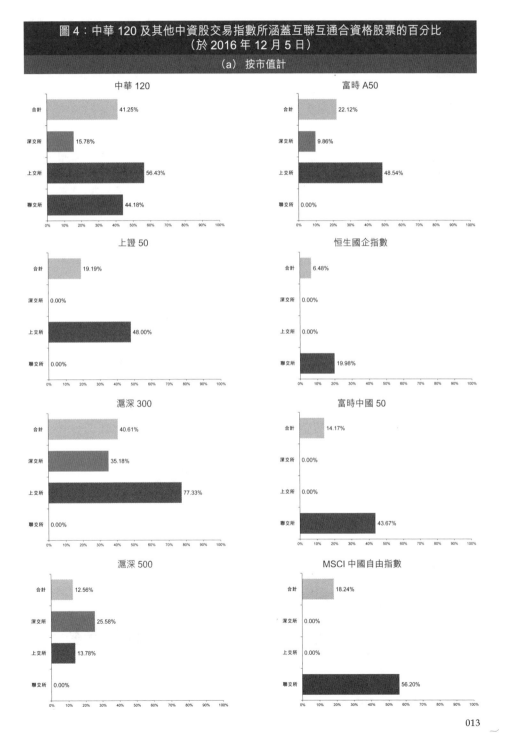

圖 4：中華 120 及其他中資股交易指數所涵蓋互聯互通合資格股票的百分比
（於 2016 年 12 月 5 日）
（a）按市值計

(續)

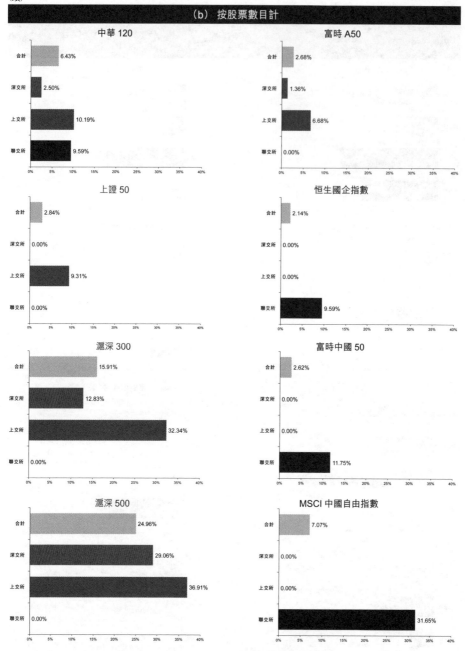

(b) 按股票數目計

中華 120

合計	6.43%
深交所	2.50%
上交所	10.19%
聯交所	9.59%

富時 A50

合計	2.68%
深交所	1.36%
上交所	6.68%
聯交所	0.00%

上證 50

合計	2.84%
深交所	0.00%
上交所	9.31%
聯交所	0.00%

恒生國企指數

合計	2.14%
深交所	0.00%
上交所	0.00%
聯交所	9.59%

滬深 300

合計	15.91%
深交所	12.83%
上交所	32.34%
聯交所	0.00%

富時中國 50

合計	2.62%
深交所	0.00%
上交所	0.00%
聯交所	11.75%

滬深 500

合計	24.96%
深交所	29.06%
上交所	36.91%
聯交所	0.00%

MSCI 中國自由指數

合計	7.07%
深交所	0.00%
上交所	0.00%
聯交所	31.65%

註：標題為藍色的指數為海外買賣的指數；標題為灰色的指數為內地買賣的指數。

資料來源：互聯互通合資格股票名單來自香港交易所、上交所及深交所網站；指數成份股名單來自中華交易
服務網站、湯森路透及彭博；市值資料來自香港交易所及湯森路透。

圖 5：中華 120 及其他中資股交易指數成份股的行業權重佔比
（對照互聯互通合資格股票和內地及香港主要市場指數）

（a）互聯互通股票（2016 年 12 月 5 日）

所有合資格股票

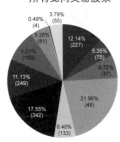

港股通股票

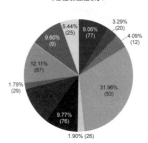

所有北向交易股票　　**滬股通股票**　　**深股通股票**

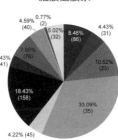

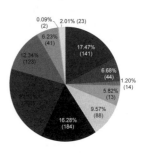

（b）交易指數（於 2016 年 11 月 30 日）

中華 120

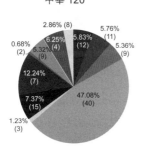

富時 A50

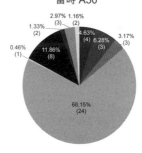

■ 非必需性消費品　　■ 必需性消費品　　■ 能源　　■ 金融

□ 醫藥衛生　　■ 工業　　■ 信息技術　　■ 原材料

■ 房地產　　■ 電信服務　　□ 公用事業

(續)

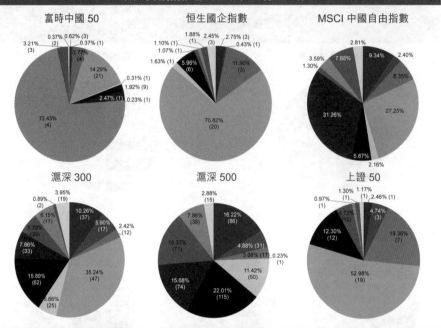

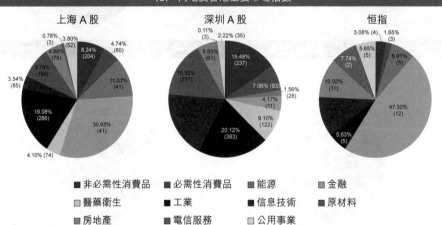

- ■ 非必需性消費品
- ■ 必需性消費品
- ■ 能源
- □ 金融
- □ 醫藥衛生
- ■ 工業
- ■ 信息技術
- ■ 原材料
- ■ 房地產
- ■ 電信服務
- □ 公用事業

註：不包括市值或權重數值並無數據的成份股。括號內為股票數目。標題為藍色的指數為海外買賣的指數；標題為灰色的指數為內地買賣的指數。由於四捨五入，數字總和未必等於 100%。

資料來源：互聯互通合資格股票名單來自香港交易所、上交所及深交所網站；中華 120 成份股名單來自中華交易服務網站；MSCI 中國自由指數行業組成資料來自 MSCI 網站；其他資料來自香港交易所及湯森路透。

　　中華 120 由於有具代表性的內地 A 股覆蓋率，因此能有效追蹤 A 股市場。如圖 6 所示，中華 120 的每日走勢緊隨上證 50 和滬深 300 指數。經進一步驗證，**中華 120 與 A 股指數有很高的相關性**。2011 年 1 月至 2016 年 11 月期間，中華 120 的日回報率與富時 A50、上證 50 及滬深 300 的相關系數約為 0.9（與富時 A50：0.904，與上證 50：0.905、與滬深 300：0.887），與上證 A 股的相關系數較高（0.869），縱使與深證 A 股的相關性則低很多（系數：0.671）。所有或大部分成份股均在香港上市的其他海外中資股交易指數（富時中國 50、恒生國企指數和 MSCI 中國自由指數）的日回報率，其彼此之間及與香港市場的恒指的相關性（期內相關系數為 0.944 或以上）均高於與 A 股指數的相關性（系數約為 0.6 或以下）。期內每一年的相關表現皆很相似。

　　（有關中華 120 與各指數的相關性，見圖 7；有關不同指數對的相關性，見附錄 1。）

圖 6：中華 120 及個別 A 股指數的每日收市值（2011 年 1 月至 2016 年 11 月）

── 中華 120　── 富時 A50　── 滬深 300　── 上證 50

資料來源：湯森路透。

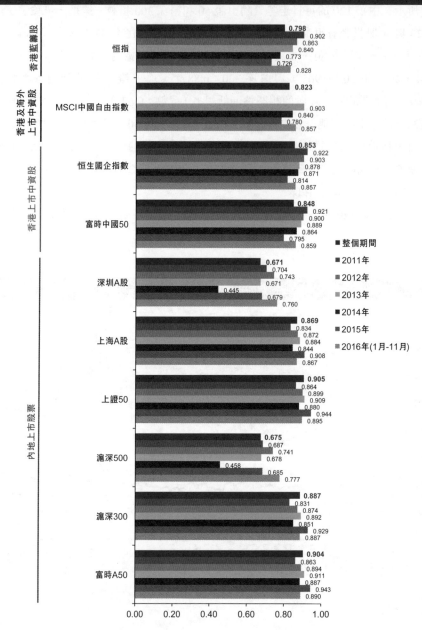

圖 7：中華 120 與個別指數每日回報率的相關系數 （2011 年 1 月至 2016 年 11 月）

資料來源：根據湯森路透每日指數收市值資料進行的分析。

　　中華 120 除了是追蹤 A 股市場的有效指數外，2015 年和 2016 年的股息率亦
與上證 50 指數相若，並高於滬深 300 指數、滬深 500 指數和富時 A50 指數（見圖
8）。此外，與這些交易指數相比，2015 年中華 120 在回報及波幅方面的表現均處
於中游位置——較低負回報率（-2.66%），波幅介乎其他指數之間；2016 年其表
現則優於所有這些指數——回報雖低，但為正數（0.17%），而所有其他指數均為
負回報，波幅亦為所有指數中最低者（見圖 9）。

圖 8：中華 120 及其他交易指數的股息收益率 （2015 年及 2016 年）

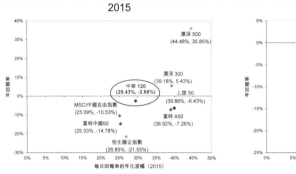

註：2016 年中華 120 的股息收益率是來自中華交易服務網站的 2016 年 9 月 30 日（而非年結日）的數
　　據。暫無 2015 年 MSCI 中國自由指數的數據。

資料來源：湯森路透、中華交易服務網站、MSCI 網站。

圖 9：中華 120 及其他交易指數的年回報率與每日回報率年化波幅 （2015 年及 2016 年）

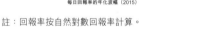

註：回報率按自然對數回報率計算。

資料來源：根據湯森路透的每日指數收市值計算。

總括而言，上述分析結果顯示中華 120 的特徵如下：

- 按上市交易所劃分的權重較任何其他海外交易指數更接近互聯互通合資格
  證券的相應權重；
- 所覆蓋的 A 股及聯交所上市的不同類別股票（H 股、紅籌股及 P 股）均與
  互聯互通證券相若；
- 在所有內地及海外交易指數中，包括及廣泛覆蓋聯交所、上交所和深交所
  三家交易所的成份股；
- 行業組成較香港市場的恒指更接近內地跨市場大型股指數滬深 300 指數，
  行業覆蓋亦更廣闊；
- 與 A 股指數有很高的相關性；及
- 2016 年回報率及波幅均優於其他交易指數，股息率與內地藍籌指數上證
  50 相若。

**作為共同市場藍籌指數，中華 120 可成為涉足內地與香港共同市場股票投資
及相關風險管理的投資產品的相關指數。**一如歐元區的領先跨市場藍籌指數歐洲
STOXX 50 指數（見下文），中華 120 可成為內地與香港共同市場的主要跨市場藍
籌指數。

歐洲 STOXX 50 指數於 1998 年 2 月 26 日由德意志銀行擁有的指數供應商
STOXX 推出，以迎接 1999 年 1 月 1 日歐元正式啟用所誕生的歐元區。該指數旨
在「為歐元區提供一個可代表超級行業先導者的指數」。該指數由 11 個歐元區國
家（1999 年加入歐元區的 11 個歐盟創始成員國奧地利、比利時、芬蘭、法國、德
國、愛爾蘭、意大利、盧森堡、荷蘭、葡萄牙和西班牙）的 19 個超級行業中的 50
家公眾持股市值最高的大型公司組成。該指數佔歐洲 STOXX 全市場指數（TMI）
公眾持股市值約 60%，後者覆蓋相關國家公眾持股市值約 95%[9]。

金融機構獲授權使用歐洲 STOXX 50 指數作為全球多種投資產品（如交易所
買賣基金（ETF）、期貨及期權，以及結構性產品）的相關指數。德意志銀行經營
的電子市場 —— 歐洲期貨交易所（Eurex）—— 於 1998 年 6 月推出的歐洲 STOXX
50 指數期貨，被譽為全歐洲流動性最高的衍生工具，2016 年在 Eurex 的年成交合

9　資料來源：歐洲期貨交易所（Eurex）網站和維基百科。

約總數為 3.74 億張，平均每日成交合約 146 萬張 [10]。

借鑒為歐洲共同市場提供服務的歐洲 STOXX 50 指數及其衍生產品和結構性產品，中華 120 指數及其衍生產品和結構性產品應能夠滿足內地與香港共同市場的同樣需要。

3.2　中華 120 期貨 —— 有助於投資 A 股及共同市場

中華 120 期貨合約於 2013 年 8 月 12 日在香港交易所衍生產品市場香港期貨交易所（期交所）推出，**是全球市場唯一以同時追蹤內地上市 A 股和香港上市中資股的指數作其相關指數的期貨合約**。其他在中國內地以外地方買賣的中國相關指數期貨或僅以 A 股為基礎，或僅以海外上市中資股（主要是香港股票）為基礎（見第 2 節表 1）。在其他指數期貨中，只有新交所的富時 A50 期貨合約提供涉足內地 A 股的機會。下表 2 是中華 120 期貨與新交所富時 A50 期貨的簡明對照。

表 2：香港交易所中華 120 期貨與新交所富時 A50 期貨的比較		
特點	中華 120 期貨	富時 A50 期貨
相關指數	內地 A 股及在聯交所上市的中資股（包括 H 股、紅籌股及民企股）	內地 A 股
合約金額	每個指數點 50 港元（於 2016 年 11 月 30 日約為 263,844 港元或 34,014 美元）	每個指數點 1 美元（於 2016 年 11 月 30 日約為 10,500 美元）
合約月份	現月、下月及之後的兩個季月	最近兩個連續月及一年內的 3 月、6 月、9 月及 12 月
大手交易	最低 100 張合約	最低 50 手
持倉限額	300,000 張合約	15,000 張合約
最低上落價位（價位波動）	0.5 個指數點（25 港元或約 3 美元）	2.5 個指數點（2.50 美元）
最低上落價位相對指數	0.0095%	0.0237%
按金	（2016 年 12 月 1 日起生效）初始：16,450 港元（約 2,121 美元）維持：13,170 港元（約 1,698 美元）	（於 2016 年 11 月 4 日）初始：495 美元維持：450 美元
按金佔名義價值百分比（%）	初始：6.23%維持：4.99%合計：11.22%	初始：4.7%維持：4.3%合計：9%

10　資料來源：Eurex 網站 2016 年 3 月的「EURO STOXX 50 Index Quanto Futures」；Eurex 網站 2016 年 12 月的「Eurex Monthly Short Statistics」。

(續)

表 2：香港交易所中華 120 期貨與新交所富時 A50 期貨的比較		
特點	中華 120 期貨	富時 A50 期貨
交易所費用	10 港元（2017 年 12 月 31 日前折收 5 港元）（約 1.3 美元，折收 0.6 美元）	0.80 美元（結算費）
交易所費用佔名義價值百分比（%）	0.0038%（按折扣後費用計為 0.0019%）	0.0076%

註：根據 2016 年 11 月 30 日指數收市值計算——中華 120 為 5276.87 點，富時 A50 為 10537.38 點。
　　匯率為 2016 年 11 月底香港金融管理局（香港金管局）網站所報匯率 1 美元兌 7.757 港元。
資料來源：香港交易所及新交所網站。

　　如上文第 3.1 節所述，中華 120 的成份股為跨境共同市場平台上的內地 A 股和香港上市中資股，指數表現與 A 股市場有很高的相關性。因具有這些特徵，中華 120 期貨可成為投資者對沖內地 A 股市場持倉的一個有效風險管理工具，有關持倉包括以富時 A50 和滬深 300 兩隻 A 股指數為基礎的交易所買賣基金（ETF）以及透過在共同市場下互聯互通買賣的持倉。中華 120 期貨亦可作為投資者同時涉足內地及香港股市的交易工具。

　　中華 120 期貨作為 A 股對沖或投資的工具，相對富時 A50 期貨而言，其在按每張合約名義價值計的交易費用較低（見上文表 2），縱使其保證金稍高。與富時 A50 期貨比較，中華 120 期貨擁有相對於相關指數水平較高的持倉限額及較小的最低上落價位，因此可提供具成本效益的另類投資交易機會。

　　此外，就內地與香港共同市場的投資組合進行風險管理，中華 120 期貨亦是一個方便而又具成本效益的離岸市場工具。同時擁有內地 A 股和香港中資股持倉的投資者，可以根據中華 120 與其投資組合持倉的關聯而考慮利用中華 120 期貨對沖其在共同市場的持倉。

附錄一

中華 120 指數與
個別指數的每日回報率的相關性

（2016 年 1 月至 2016 年 11 月）

註：各指數的數據從期內首個有數據提供之日起計算。
　　每日回報率 = LN（當日指數收市值 / 上日指數收市值）

指數	首個日回報率的日期
中華120	03/01/2011
富時A50	04/01/2011
滬深300	04/01/2011
滬深500	04/01/2011
上證50	04/01/2011
上海A股	04/01/2011
深圳A股	04/01/2011
富時中國50	03/01/2011
恒生國企指數	03/01/2011
MSCI 中國自由指數	11/01/2013
恒指	03/01/2011

Pearson 相關系數
（所有系數在統計學上於 0.1% 水平具顯著相關性）

（個案數目）

期間	指數	內地上市股票						香港上市中資股		香港及海外上市中資股	香港主要指數
		富時A50	滬深300	滬深500	上證50	上海A股	深圳A股	富時中國50	恒生國企指數	MSCI中國自由指數	恒指
整個期間	中華120	0.904 (1,438)	0.887 (1,436)	0.675 (1,436)	0.905 (1,436)	0.869 (1,436)	0.671 (1,436)	0.848 (1,427)	0.853 (1,423)	0.823 (952)	0.798 (1,423)
	富時A50		0.939 (1,436)	0.663 (1,436)	0.993 (1,436)	0.910 (1,436)	0.658 (1,436)	0.590 (1,401)	0.613 (1,397)	0.588 (946)	0.519 (1,397)
	滬深300			0.863 (1,436)	0.948 (1,436)	0.984 (1,436)	0.856 (1,436)	0.586 (1,398)	0.605 (1,394)	0.608 (943)	0.522 (1,394)
	滬深500				0.682 (1,436)	0.885 (1,436)	0.988 (1,436)	0.470 (1,398)	0.480 (1,394)	0.507 (943)	0.428 (1,394)
	上證50					0.921 (1,436)	0.676 (1,436)	0.593 (1,398)	0.615 (1,394)	0.590 (943)	0.523 (1,394)
	上海A股						0.872 (1,436)	0.580 (1,398)	0.598 (1,394)	0.602 (943)	0.518 (1,394)
	深圳A股							0.468 (1,398)	0.476 (1,394)	0.501 (943)	0.427 (1,394)
	富時中國50								0.986 (1,456)	0.982 (958)	0.963 (1,456)
	恒生國企指數									0.962 (955)	0.944 (1,456)
	MSCI中國自由指數										0.957 (955)

（續）　　　　　　　　　　　　　　　　　　　　　　　　　　　　　（個案數目）

期間	指數	內地上市股票						香港上市中資股		香港及海外上市中資股	香港主要指數
		富時A50	滬深300	滬深500	上證50	上海A股	深圳A股	富時中國50	恒生國企指數	MSCI中國自由指數	恒指
2011 年	中華 120	0.863 (244)	0.831 (244)	0.687 (244)	0.864 (244)	0.834 (244)	0.704 (244)	0.921 (247)	0.922 (246)	. (0)	0.902 (246)
	富時 A50		0.963 (244)	0.800 (244)	0.995 (244)	0.955 (244)	0.825 (244)	0.622 (237)	0.629 (236)	. (0)	0.598 (236)
	滬深 300			0.916 (244)	0.965 (244)	0.990 (244)	0.934 (244)	0.579 (237)	0.587 (236)	. (0)	0.559 (236)
	滬深 500				0.803 (244)	0.922 (244)	0.990 (244)	0.460 (237)	0.470 (236)	. (0)	0.445 (236)
	上證 50					0.957 (244)	0.825 (244)	0.624 (237)	0.631 (236)	. (0)	0.602 (236)
	上海 A 股						0.931 (244)	0.592 (237)	0.598 (236)	. (0)	0.573 (236)
	深圳 A 股							0.471 (237)	0.481 (236)	. (0)	0.454 (236)
	富時中國 50								0.993 (246)	. (0)	0.974 (246)
	恒生國企指數									. (0)	0.970 (246)
	MSCI 中國自由指數										. (0)
2012 年	中華 120	0.894 (243)	0.874 (243)	0.741 (243)	0.899 (243)	0.872 (243)	0.743 (243)	0.900 (247)	0.903 (247)	. (0)	0.863 (247)
	富時 A50		0.968 (243)	0.827 (243)	0.991 (243)	0.962 (243)	0.833 (243)	0.642 (237)	0.660 (237)	. (0)	0.588 (237)
	滬深 300			0.925 (243)	0.975 (243)	0.990 (243)	0.930 (243)	0.606 (237)	0.624 (237)	. (0)	0.550 (237)
	滬深 500				0.841 (243)	0.928 (243)	0.991 (243)	0.486 (237)	0.502 (237)	. (0)	0.432 (237)
	上證 50					0.973 (243)	0.842 (243)	0.649 (237)	0.668 (237)	. (0)	0.594 (237)
	上海 A 股						0.924 (243)	0.614 (237)	0.631 (237)	. (0)	0.557 (237)
	深圳 A 股							0.487 (237)	0.500 (237)	. (0)	0.436 (237)
	富時中國 50								0.987 (247)	. (0)	0.962 (247)
	恒生國企指數									. (0)	0.951 (247)
	MSCI 中國自由指數										. (0)

（續）　　　　　　　　　　　　　　　　　　　　　　　　　　（個案數目）

期間	指數	內地上市股票						香港上市中資股		香港及海外上市中資股	香港主要指數
		富時A50	滬深300	滬深500	上證50	上海A股	深圳A股	富時中國50	恒生國企指數	MSCI中國自由指數	恒指
2013年	中華120	0.911 (238)	0.892 (238)	0.678 (238)	0.909 (238)	0.884 (238)	0.671 (238)	0.889 (233)	0.878 (232)	0.903 (234)	0.840 (232)
	富時A50		0.958 (238)	0.691 (238)	0.995 (238)	0.941 (238)	0.680 (238)	0.683 (231)	0.682 (230)	0.683 (232)	0.611 (230)
	滬深300			0.853 (238)	0.963 (238)	0.986 (238)	0.842 (238)	0.668 (231)	0.668 (230)	0.680 (232)	0.609 (230)
	滬深500				0.705 (238)	0.862 (238)	0.986 (238)	0.513 (231)	0.516 (230)	0.546 (232)	0.485 (230)
	上證50					0.946 (238)	0.692 (238)	0.681 (231)	0.680 (230)	0.681 (232)	0.611 (230)
	上海A股						0.840 (238)	0.665 (231)	0.667 (230)	0.677 (232)	0.610 (230)
	深圳A股							0.512 (231)	0.513 (230)	0.543 (232)	0.485 (230)
	富時中國50								0.988 (244)	0.986 (237)	0.960 (244)
	恒生國企指數									0.977 (236)	0.946 (244)
	MSCI中國自由指數										0.964 (236)
2014年	中華120	0.887 (245)	0.851 (245)	0.458 (245)	0.880 (245)	0.844 (245)	0.445 (245)	0.864 (238)	0.871 (238)	0.840 (245)	0.773 (238)
	富時A50		0.938 (245)	0.454 (245)	0.993 (245)	0.911 (245)	0.454 (245)	0.580 (238)	0.628 (238)	0.539 (245)	0.462 (238)
	滬深300			0.707 (245)	0.947 (245)	0.973 (245)	0.700 (245)	0.563 (238)	0.599 (238)	0.533 (245)	0.446 (238)
	滬深500				0.479 (245)	0.725 (245)	0.981 (245)	0.337 (238)	0.323 (238)	0.346 (245)	0.278 (238)
	上證50					0.918 (245)	0.476 (245)	0.573 (238)	0.619 (238)	0.530 (245)	0.448 (238)
	上海A股						0.707 (245)	0.585 (238)	0.619 (238)	0.557 (245)	0.474 (238)
	深圳A股							0.316 (238)	0.304 (238)	0.330 (245)	0.260 (238)
	富時中國50								0.969 (247)	0.980 (247)	0.928 (247)
	恒生國企指數									0.942 (247)	0.871 (247)
	MSCI中國自由指數										0.954 (247)

(續)　　　　　　　　　　　　　　　　　　　　　　　　　　　　　　　　（個案數目）

期間	指數	內地上市股票						香港上市中資股		香港及海外上市中資股	香港主要指數
		富時A50	滬深300	滬深500	上證50	上海A股	深圳A股	富時中國50	恒生國企指數	MSCI中國自由指數	恒指
2015年	中華120	0.943 (244)	0.929 (244)	0.685 (244)	0.944 (244)	0.908 (244)	0.679 (244)	0.795 (237)	0.814 (237)	0.780 (244)	0.726 (237)
	富時A50		0.928 (244)	0.614 (244)	0.994 (244)	0.902 (244)	0.606 (244)	0.609 (237)	0.646 (237)	0.587 (244)	0.517 (237)
	滬深300			0.848 (244)	0.936 (244)	0.986 (244)	0.838 (244)	0.635 (237)	0.667 (237)	0.627 (244)	0.553 (237)
	滬深500				0.632 (244)	0.880 (244)	0.988 (244)	0.532 (237)	0.554 (237)	0.544 (244)	0.479 (237)
	上證50					0.909 (244)	0.624 (244)	0.612 (237)	0.650 (237)	0.591 (244)	0.522 (237)
	上海A股						0.865 (244)	0.632 (237)	0.663 (237)	0.625 (244)	0.549 (237)
	深圳A股							0.534 (237)	0.553 (237)	0.545 (244)	0.484 (237)
	富時中國50								0.982 (247)	0.990 (247)	0.960 (247)
	恒生國企指數									0.967 (247)	0.922 (247)
	MSCI中國自由指數										0.959 (247)
2016年 1月-11月	中華120	0.890 (224)	0.887 (222)	0.777 (222)	0.895 (222)	0.867 (222)	0.760 (222)	0.859 (225)	0.857 (223)	0.857 (229)	0.828 (223)
	富時A50		0.946 (222)	0.792 (222)	0.988 (222)	0.920 (222)	0.773 (222)	0.574 (221)	0.592 (219)	0.587 (225)	0.525 (219)
	滬深300			0.933 (222)	0.957 (222)	0.991 (222)	0.920 (222)	0.573 (218)	0.585 (216)	0.589 (222)	0.516 (216)
	滬深500				0.804 (222)	0.959 (222)	0.994 (222)	0.515 (218)	0.517 (216)	0.530 (222)	0.456 (216)
	上證50					0.932 (222)	0.785 (222)	0.573 (218)	0.591 (216)	0.585 (222)	0.525 (216)
	上海A股						0.945 (222)	0.564 (218)	0.573 (216)	0.577 (222)	0.505 (216)
	深圳A股							0.500 (218)	0.502 (216)	0.514 (222)	0.443 (216)
	富時中國50								0.991 (225)	0.973 (227)	0.979 (225)
	恒生國企指數									0.960 (225)	0.965 (225)
	MSCI中國自由指數										0.956 (225)

附錄二

中華交易服務中國 120 指數成份股

（於 2016 年 11 月 30 日）

編號	成份股代號	成份股名稱	上市交易所	股份類別	權重(%)
1	135	昆侖能源有限公司	聯交所	紅籌股	0.16
2	144	招商局港口控股有限公司	聯交所	紅籌股	0.22
3	151	中國旺旺控股有限公司	聯交所	其他	0.34
4	267	中國中信股份有限公司	聯交所	紅籌股	0.62
5	270	粵海投資有限公司	聯交所	紅籌股	0.30
6	322	康師傅控股有限公司	聯交所	其他	0.18
7	384	中國燃氣控股有限公司	聯交所	民企股	0.18
8	386	中國石油化工股份有限公司	聯交所	H股	1.23
9	392	北京控股有限公司	聯交所	紅籌股	0.16
10	656	復星國際有限公司	聯交所	民企股	0.26
11	688	中國海外發展有限公司	聯交所	紅籌股	0.87
12	700	騰訊控股有限公司	聯交所	民企股	9.72
13	728	中國電信集團公司	聯交所	H股	0.46
14	762	中國聯合網絡通訊集團有限公司	聯交所	紅籌股	0.60
15	836	華潤電力控股有限公司	聯交所	紅籌股	0.22
16	857	中國石油股份天然氣股份有限公司	聯交所	H股	0.99
17	883	中國海洋石油有限公司	聯交所	紅籌股	1.55
18	939	建設銀行股份有限公司	聯交所	H 股	4.94
19	941	中國移動通信集團公司	聯交所	紅籌股	4.62
20	960	龍湖地產有限公司	聯交所	民企股	0.15
21	966	中國太平保險控股有限公司	聯交所	紅籌股	0.28
22	992	聯想集團有限公司	聯交所	紅籌股	0.33
23	998	中信銀行股份有限公司	聯交所	H 股	0.40
24	1044	恒安國際有限公司	聯交所	民企股	0.47
25	1109	華潤置地有限公司	聯交所	紅籌股	0.46
26	1288	農業銀行股份有限公司	聯交所	H 股	0.71
27	1398	工商銀行股份有限公司	聯交所	H 股	3.66
28	1880	百麗國際控股有限公司	聯交所	其他	0.20

(續)

編號	成份股代號	成份股名稱	上市交易所	股份類別	權重(%)
29	2007	碧桂園控股有限公司	聯交所	民企股	0.35
30	2318	中國平安保險 (集團) 股份有限公司	聯交所	H 股	1.99
31	2319	中國蒙牛乳業有限公司	聯交所	紅籌股	0.39
32	2328	中國人民財產保險股份有限公司	聯交所	H 股	0.53
33	2601	中國太平洋保險 (集團) 股份有限公司	聯交所	H 股	0.74
34	2628	中國人壽保險股份有限公司	聯交所	H 股	1.49
35	3328	交通銀行股份有限公司	聯交所	H 股	0.56
36	3333	中國恒大集團	聯交所	民企股	0.19
37	3799	達利食品集團有限公司	聯交所	民企股	0.08
38	3968	招商銀行股份有限公司	聯交所	H 股	0.78
39	3988	中國銀行股份有限公司	聯交所	H 股	2.62
40	6808	高鑫零售有限公司	聯交所	其他	0.17
41	600000	上海浦東發展銀行股份有限公司	上交所	A 股	1.49
42	600011	華能國際電力股份有限公司	上交所	A 股	0.32
43	600015	華夏銀行股份有限公司	上交所	A 股	0.62
44	600016	中國民生銀行股份有限公司	上交所	A 股	2.25
45	600018	上海國際港務 (集團) 股份有限公司	上交所	A 股	0.17
46	600019	寶山鋼鐵股份有限公司	上交所	A 股	0.32
47	600023	浙江浙能電力股份有限公司	上交所	A 股	0.23
48	600028	中國石油化工股份有限公司	上交所	A 股	0.54
49	600030	中信證券股份有限公司	上交所	A 股	1.40
50	600036	招商銀行股份有限公司	上交所	A 股	1.92
51	600048	保利房地產 (集團) 股份有限公司	上交所	A 股	0.70
52	600050	中國聯合網絡通信股份有限公司	上交所	A 股	0.57
53	600104	上海汽車集團股份有限公司	上交所	A 股	0.84
54	600276	江蘇恒瑞醫藥股份有限公司	上交所	A 股	0.66
55	600485	北京信威科技集團股份有限公司	上交所	A 股	0.20
56	600519	貴州茅台酒股份有限公司	上交所	A 股	1.60
57	600585	安徽海螺水泥股份有限公司	上交所	A 股	0.36
58	600606	綠地控股集團股份有限公司	上交所	A 股	0.04
59	600637	上海東方明珠新媒體股份有限公司	上交所	A 股	0.33
60	600690	青島海爾股份有限公司	上交所	A 股	0.33
61	600795	國電電力發展股份有限公司	上交所	A 股	0.39

(續)

編號	成份股代號	成份股名稱	上市交易所	股份類別	權重(%)
62	600837	海通證券股份有限公司	上交所	A股	1.36
63	600871	中石化石油工程技術服務股份有限公司	上交所	A股	0.07
64	600887	內蒙古伊利實業集團股份有限公司	上交所	A股	1.20
65	600893	中航動力股份有限公司	上交所	A股	0.27
66	600900	中國長江電力股份有限公司	上交所	A股	0.88
67	600958	東方證券股份有限公司	上交所	A股	0.45
68	600999	招商證券股份有限公司	上交所	A股	0.55
69	601006	大秦鐵路股份有限公司	上交所	A股	0.44
70	601088	中國神華能源股份有限公司	上交所	A股	0.34
71	601166	興業銀行股份有限公司	上交所	A股	2.25
72	601169	北京銀行股份有限公司	上交所	A股	1.23
73	601186	中國鐵建股份有限公司	上交所	A股	0.43
74	601211	國泰君安證券股份有限公司	上交所	A股	0.30
75	601288	中國農業銀行股份有限公司	上交所	A股	1.23
76	601318	中國平安保險（集團）股份有限公司	上交所	A股	3.94
77	601328	交通銀行股份有限公司	上交所	A股	1.61
78	601336	新華人壽保險股份有限公司	上交所	A股	0.39
79	601390	中國中鐵股份有限公司	上交所	A股	0.55
80	601398	中國工商銀行股份有限公司	上交所	A股	1.10
81	601601	中國太平洋保險（集團）股份有限公司	上交所	A股	0.95
82	601628	中國人壽保險股份有限公司	上交所	A股	0.43
83	601633	長城汽車股份有限公司	上交所	A股	0.13
84	601668	中國建築股份有限公司	上交所	A股	1.65
85	601669	中國電力建設股份有限公司	上交所	A股	0.33
86	601688	華泰證券股份有限公司	上交所	A股	0.65
87	601727	上海電氣集團股份有限公司	上交所	A股	0.26
88	601766	中國中車股份有限公司	上交所	A股	1.10
89	601788	光大證券股份有限公司	上交所	A股	0.21
90	601800	中國交通建設股份有限公司	上交所	A股	0.25
91	601818	中國光大銀行股份有限公司	上交所	A股	0.65
92	601857	中國石油天然氣股份有限公司	上交所	A股	0.37
93	601898	中國中煤能源股份有限公司	上交所	A股	0.11
94	601901	方正證券股份有限公司	上交所	A股	0.35
95	601985	中國核能電力股份有限公司	上交所	A股	0.34

(續)

編號	成份股代號	成份股名稱	上市交易所	股份類別	權重（%）
96	601988	中國銀行股份有限公司	上交所	A 股	0.74
97	601989	中國船舶重工股份有限公司	上交所	A 股	0.66
98	601998	中信銀行股份有限公司	上交所	A 股	0.21
99	1	平安銀行股份有限公司	深交所	A 股	0.66
100	2	萬科企業股份有限公司	深交所	A 股	2.10
101	69	深圳華僑城股份有限公司	深交所	A 股	0.24
102	166	申萬宏源集團股份有限公司	深交所	A 股	0.42
103	333	美的集團股份有限公司	深交所	A 股	0.97
104	538	雲南白藥集團股份有限公司	深交所	A 股	0.36
105	625	重慶長安汽車股份有限公司	深交所	A 股	0.37
106	651	珠海格力電器股份有限公司	深交所	A 股	1.37
107	725	京東方科技集團股份有限公司	深交所	A 股	0.69
108	776	廣發證券股份有限公司	深交所	A 股	0.58
109	858	宜賓五糧液股份有限公司	深交所	A 股	0.68
110	895	河南雙匯投資發展股份有限公司	深交所	A 股	0.22
111	1979	招商局蛇口工業區控股股份有限公司	深交所	A 股	0.46
112	2024	蘇寧雲商集團股份有限公司	深交所	A 股	0.44
113	2252	上海萊士血液製品股份有限公司	深交所	A 股	0.21
114	2304	江蘇洋河酒廠股份有限公司	深交所	A 股	0.43
115	2415	杭州海康威視數字技術股份有限公司	深交所	A 股	0.46
116	2594	比亞迪股份有限公司	深交所	A 股	0.30
117	2736	國信證券股份有限公司	深交所	A 股	0.44
118	2739	萬達電影院線股份有限公司	深交所	A 股	0.31
119	300059	東方財富信息股份有限公司	深交所	A 股	0.46
120	300104	樂視網信息技術（北京）股份有限公司	深交所	A 股	0.38

註：H 股是由中國內地註冊成立而於香港交易所上市的公司發行；紅籌股是由中國內地以外註冊成立而在香港交易所上市的公司發行，該等公司由內地政府實體透過直接或間接持股及／或派員擔任公司董事而控制；民企股是由中國內地以外註冊成立但在香港交易所上市的民營企業發行。

資料來源：中華交易服務網站。

第2章

滬港通與深港通下的互聯互通

—— 內地及全球投資者的「共同市場」

2017 年 3 月 23 日

摘 要

　　滬港股票市場交易互聯互通機制試點（滬港通）於 2014 年 11 月 17 日正式推出，是中國內地與香港之間破天荒的市場互聯互通計劃，為海外投資者投資內地股票市場及內地投資者投資香港股票市場提供嶄新的正式渠道。此渠道實現全程封閉兼有序的人民幣跨境資金流，減低對內地市場的潛在金融風險。深港股票市場交易互聯互通機制（深港通）亦已於 2016 年 12 月 5 日推出，為滬港通的延伸篇章，進一步擴充合資格股票範圍。滬深港之間的「共同市場」模式已然基本形成。中國內地與香港之間的共同市場模式是中國內地資本賬開放進程中極具象徵意義的突破，為內地投資者開啟越來越多的環球投資機會，同時也為全球投資者打開更多的內地投資機會。

　　滬港通的經驗顯示，內地投資者對於透過港股通投資港股的需求正與日俱增。他們的投資不止於大型藍籌股，亦涉及不同行業的中小型股。滬股通的交投顯示全球投資者對內地多元化行業的中小型股亦興趣日濃。深港通涵蓋更多小型股，以迎合內地及全球投資者的需要。兩地監管機構已就交易所買賣基金納入互聯互通標的達成共識，具體時間將另行公告。若獲監管機構批准，日後亦可能納入債券及其他證券、商品及衍生產品。透過「共同市場」模式下的南向交易，內地投資者面向全球資產配置的機遇，或許能獲得更佳的潛在回報，以及利用較內地市場更為多元化的投資及風險管理工具。在國際市場的交易經驗亦有助內地投資者（特別是散戶投資者）趨向成熟。

　　隨着跨境投資活動日增，可能有越來越大的需求將人民幣股票衍生產品、人民幣利率及貨幣衍生產品等相關跨境組合對沖工具納入共同市場模式中。

1 股票市場交易互聯互通機制試點
—— 破天荒連通內地股票市場，邁向「共同市場」

　　股票市場交易互聯互通機制試點於 2014 年 11 月推出首項計劃 —— 滬港股票市場交易互聯互通機制試點（滬港通）。儘管交易範圍有限制，這是接通內地證券市場與海外市場股票交易的破天荒機制。此前，境外人士參與內地證券市場的渠道只限於合格境外機構投資者（QFII）計劃及人民幣合格境外機構投資者（RQFII）計劃，境外散戶投資者只可透過 QFII 及 RQFII 提供的投資基金參與內地股市[1]。反向地，合格境內機構投資者（QDII）計劃及人民幣合格境內機構投資者（RQDII）計劃是內地參與海外證券市場的唯一全國性正式渠道[2]。繼滬港通順利開通後，深港股票市場交易互聯互通機制（深港通）亦於 2016 年 8 月宣佈將會啟動，並已於同年 12 月推出。滬港通與深港通於下文合稱「互聯互通」機制。

　　互聯互通機制是中國內地資本賬開放的重要里程碑。在實施每日額度及跨境資金流全程封閉下，跨境資金投資活動可在受密切監控下進行及有序發展，減低對內地股票市場的潛在金融風險。此機制日後可因應內地市場的開放進程，在規模、範圍及市場領域等方面擴容，目標是為內地及全球投資者建立中國內地與香港的「共同市場」。

　　以下分節簡介兩項股票市場交易互聯互通計劃。第 2 節載列互聯互通自推出後的表現。第 3 節則探討「共同市場」模式可提供的機遇。

1.1　滬港通

　　中國證券監督管理委員會（中國證監會）與香港證券及期貨事務監察委員會（香港證監會）於 2014 年 4 月發出聯合公告，宣佈開展中國內地與香港的股票市場

1　在 1992 年於上海及深圳交易所推出的 B 股市場（以外幣交易，與 A 股市場分開），為內地首次嘗試將股票市場對外國投資者開放，但在市場開放的新浪潮下，B 股市場交易並不活躍。

2　地區政府有推出特殊計劃，如上海的「合格境內有限合夥人」（QDLP）計劃和深圳的「合格境內投資者境外投資試點」（QDIE）。但這些計劃只限主要服務機構投資者及高端個人客戶的私募基金或投資工具，並不如 QDII 產品開放給一般投資者。

交易互聯互通機制試點 —— 滬港通。香港交易所全資附屬公司香港聯合交易所有限公司(聯交所)與上海證券交易所(上交所)聯手建立跨境買賣盤傳遞及相關技術基礎設施(交易通)。香港交易所另一全資附屬公司香港中央結算有限公司(香港結算)與中國內地的證券結算所中國證券登記結算有限責任公司(中國結算)則共同建立結算及交收基礎設施(結算通)。經過多月的市場準備及系統測試後,滬港通於 2014 年 11 月 17 日正式開通。機制旨在於設定的合資格範圍內,容許香港及海外投資者在內地市場買賣上交所上市股票(「滬股通」或「滬港通北向交易」)及內地投資者在香港市場買賣聯交所上市股票(「滬港通下的港股通」或「滬港通南向交易」)。

在初期階段,滬股通合資格證券包括以下在上交所上市的 A 股(「滬股通股票」):

- 上證 180 指數及上證 380 指數的成份股;以及
- 有 H 股同時在聯交所上市的上交所上市 A 股;

但不包括不以人民幣交易的滬股及被實施風險警示的滬股[3]。

滬港通下的港股通合資格證券包括以下在聯交所主板上市的股票(「港股通股票」):

- 恒生綜合大型股指數(HSLI)成份股;
- 恒生綜合中型股指數(HSMI)成份股;以及
- 有相關 A 股在上交所上市的 H 股;

但不包括不以港幣交易的港股及其相應 A 股被實施風險警示的 H 股。

在滬股通股票中,上證 180 指數成份股是上交所最具市場代表性的 180 隻 A 股,而上證 380 指數則是由 380 家規模中型的公司組成,綜合反映上交所在上證 180 指數以外一批新興藍籌公司的表現[4]。因此,上證 180 指數的滬股通股票視為對應於港股通 HSLI 股票的「大型」股,而上證 380 指數的滬股通股票則為與港股通 HSMI 股票相應的「中型」股。

於 2017 年 2 月底,合資格證券範圍中共有 715 隻上交所上市滬股通股票(包

---

3 指相關股份被上交所實施「風險警示」,包括 ST 公司及 *ST 公司的股份以及須根據上交所規則進行除牌程序的股份。

4 資料來源:上交所網站。

括 139 隻僅合資格出售的股票 [5]) 及 317 隻聯交所上市港股通股票 [6]。

投資者資格方面，所有香港及海外投資者均可參與滬股通交易，但只有內地機構投資者及擁有證券賬戶及資金賬戶餘額合計不低於人民幣 50 萬元的個人投資者方可參與港股通。

滬股通（北向）交易方面，在香港的投資者透過香港經紀進行買賣，交易則在上交所平台執行。港股通（南向）交易方面，內地投資者透過內地經紀進行買賣，交易則在聯交所平台執行。滬股通及港股通跟隨交易執行平台各自的市場規則。具體而言，內地 A 股市場不可進行即日回轉交易，但香港市場則容許。滬股通股票僅以人民幣進行買賣及交收，港股通股票則以港元進行買賣，內地投資者再與中國結算或其結算參與人以人民幣進行交收。

滬港通下的交易受制於投資額度，最初設有跨境投資價值上限的總額度及每日額度。滬股通及港股通的總額度分別為人民幣 3,000 億元及人民幣 2,500 億元。其後在宣佈建立深港通當日（2016 年 8 月 16 日）已取消總額度之設。每日額度現時仍然適用，按「淨買盤」計算的滬股通股票每日上限為人民幣 130 億元，港股通股票則為人民幣 105 億元（於 2016 年底折合約 117 億港元）。

自 2014 年 11 月滬港通推出後，投資者對滬股通及港股通的興趣時有不同。推出後大部分時間均見滬股通交易較港股通交易更為活躍，但自 2015 年後期以來則見港股通交易額日漸增長並有見超越滬股通。2016 年 12 月深港通推出，該月的北向交易顯著提升，數據顯示有近三成歸功於深股通。（詳見第 2 節。）

1.2　深港通

中國證監會與香港證監會於 2016 年 8 月 16 日聯合宣佈建立深港股票市場交易互聯互通機制（深港通），是基於滬港通推出以來平穩運行的基礎而建立的股票市場交易互聯互通機制試點的延伸項目。深圳證券交易所（深交所）、聯交所、中國結算及香港結算按類似滬港通的形式建立深港通，其後於 2016 年 12 月 5 日開通。含深港通的互聯互通機制現包括以下交易通：

5　原為合資格滬股，但根據預設規定其後不再合資格者。
6　資料來源：香港交易所及上交所網站 2017 年 2 月 28 日資料。

股票市場交易互聯互通機制

		滬港通	深港通
交易類別	北向	滬股交易通	深股交易通
	南向	滬港通下的港股交易通	深港通下的港股交易通

深股通合資格證券包括：

- 深證成份指數和深證中小創新指數成份股中所有市值不少於人民幣 60 億元的成份股；以及
- 有相關 H 股在聯交所上市的所有深交所上市 A 股；

但不包括不以人民幣交易的深股及被實施風險警示的深股[7]。

除滬港通下的港股通合資格證券外，深港通下的港股通合資格證券範圍擴展至包括：

- 所有市值 50 億港元或以上的恒生綜合小型股指數 (HSSI) 成份股；以及
- 所有聯交所上市公司中同時有 A 股在深交所上市的 H 股；

但不包括不以港幣交易的港股及其相應 A 股被實施風險警示的 H 股。

深港通下的港股通合資格內地投資者與滬港通相同，但透過深港通買賣深交所創業板上市股票的合資格投資者初期只限於相關香港規則及規例所界定的機構專業投資者[8]。

深港通亦沿用滬港通的每日額度，同時不設總額度。

下表 1 概述滬港通及深港通的主要異同。

---

[7] 指相關股份被深交所實施「風險警示」，包括 ST 公司及 *ST 公司的股份以及須根據深交所規則進行除牌程序的股份。

[8] 有關「機構專業投資者」的定義見《證券及期貨條例》下的《證券及期貨 (專業投資者) 規則》。

表 1：滬港通及深港通主要特點		
特點	滬港通	深港通
滬／深股通合資格證券	• 上證 180 指數的成份股及上證 380 指數的成份股 • 有 H 股同時在聯交所上市的上交所上市 A 股	• 深證成份指數和深證中小創新指數成份股中市值在人民幣 60 億元或以上的成份股 • 有相關 H 股在聯交所上市的深交所上市 A 股
	• 不包括被實施風險警示的 A 股及不以人民幣交易的 A 股	
	• 合資格可買可賣的股票共 576 隻（於 2017 年 2 月 28 日）	• 合資格可買可賣的股票共 904 隻（於 2017 年 2 月 28 日）
港股通合資格證券	• 恒生綜合大型股指數（HSLI）成份股 • 恒生綜合中型股指數（HSMI）成份股	
	• 有相關 A 股在上交所上市的 H 股	• 市值 50 億港元或以上的恒生綜合小型股指數（HSSI）的成份股 • 有 A 股在上交所或深交所上市的 H 股
	• 不包括其相應 A 股被實施風險警示的 H 股及不以港幣交易的港股	
	• 317 隻股票（於 2017 年 2 月 28 日）	• 417 隻股票（較滬港通下的港股通股票多出 100 隻）（於 2017 年 2 月 28 日）
滬／深股通合資格投資者	• 所有香港及海外投資者（個人及機構）	• 創業板合資格股票：初期僅限於機構專業投資者 • 其他合資格股票：所有香港及海外投資者（個人及機構）
港股通合資格投資者	• 內地機構投資者及擁有證券賬戶及資金賬戶餘額合計≧人民幣 50 萬元的個人投資者	
每日額度	• 北向：人民幣 130 億元 • 南向：人民幣 105 億元	
總額度	• 沒有	
北向交易、結算及交收	• 按照上交所及中國結算在上海市場的慣例	• 按照深交所及中國結算在深圳市場的慣例
南向交易、結算及交收	• 按照聯交所及香港結算的市場慣例	

2 互聯互通迄今的表現
（截至 2016 年底）

2.1 北向及南向交易的整體表現

　　北向及南向交易的成交量隨着市場氣氛的轉變而時有不同。然而，北向交易的日均成交額佔內地 A 股市場總體日均成交額一直維持在 1% 至 2% 相對窄幅的水平。反觀南向交易，在滬港通推出後首九個月，南向交易成交量漲跌互現，但自 2015 年第四季，南向交易額在聯交所主板市場總成交的佔比出現強勁上升趨勢，從 2015 年 9 月佔主板日均成交額的 2.1% 升至 2016 年 9 月的 10.8%；縱使其後有所回落，但上升勢頭持續。南向交易日均成交額於 2016 年 6 月再度超過北向交易，是 2015 年 4 月以來的首次，其後月度亦屢次超過。（見下圖 1 及 2）

　　值得注意的是，在 2016 年 12 月 5 日深港通開通日起至 2016 年底的 17 個北向交易日中，深港通北向交易佔互聯互通北向交易總成交額的 27%，佔北向買盤總額的 40%，顯示國際投資者對深股有相當興趣。

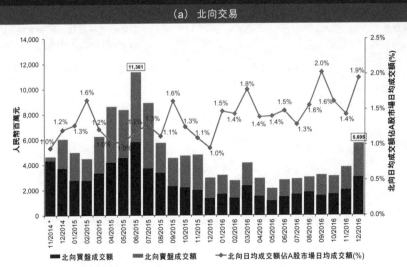

圖 1：互聯互通平均每日成交額 （2014 年 11 月至 2016 年 12 月）

(a) 北向交易

■北向買盤成交額　■北向賣盤成交額　◆北向日均成交額佔A股市場日均成交額(%)

* 自 2014 年 11 月 17 日滬港通開通之日起計。

註：自 2016 年 12 月 5 日起包括於當日推出的深港通數據；作為基礎比較的 A 股市場數據自該日起包括深交所 A 股市場。

（續）

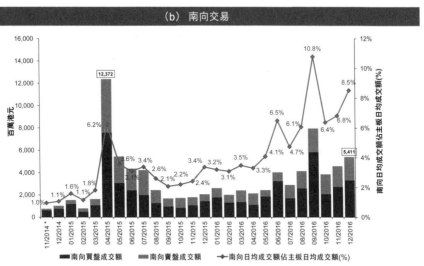

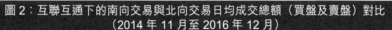

* 自 2014 年 11 月 17 日滬港通開通之日起計。

註：在計算佔市場日均成交總額的百分比時，北向交易／南向交易的交易額為雙邊成交額（買盤及賣盤分別計
　　算在內），而市場總成交額為單邊成交額（買盤及賣盤以單一交易額計算）。自 2016 年 12 月 5 日起包括
　　於當日推出的深港通數據。

資料來源：香港交易所。

圖 2：互聯互通下的南向交易與北向交易日均成交總額（買盤及賣盤）對比
（2014 年 11 月至 2016 年 12 月）

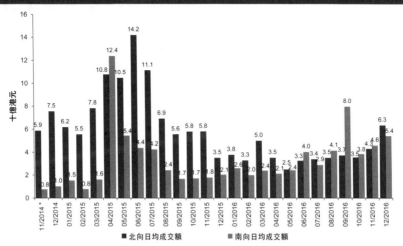

* 自 2014 年 11 月 17 日滬港通開通之日起計。

註：北向交易成交額按月末匯率（來自香港金融管理局網站）轉換為港元。自 2016 年 12 月 5 日起包括於當
　　日推出的深港通數據。

資料來源：香港交易所。

此外，自 2015 年後期開始，南向交易的平均每日買盤淨額均遠高於北向交易。自互聯互通推出至 2016 年底，南向交易只有兩個月錄得淨賣盤，相比之下北向交易則曾錄得六個月的淨賣盤（見圖 3）。其間 485 個南向交易日中，有 86% 的時間出現淨買盤，相對於 494 個北向交易日中有 56% 的日子出現淨買盤。然而，北向交易及南向交易兩者按淨買盤基礎計算的每日額度使用量一直不高 —— 滬港通下僅有 18% 的北向交易日及 20% 的南向交易日的每日額度用量曾超過 10%，以及有 6% 的北向交易日及 6% 的南向交易日的每日額度用量曾超過 20%[9]。在深港通下，17 個交易日中有 4 日（即 24%）的北向交易日的每日額度用量超過 10% 及只有 1 日（即 6%）超過 20%，而南向交易額度則從未超過 10%。（見下圖 4 及表 2）

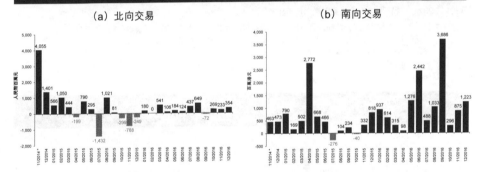

圖 3：互聯互通下的北向交易與南向交易日均淨買盤／賣盤成交額
（2014 年 11 月至 2016 年 12 月）

（a）北向交易　　　　　　　　　　（b）南向交易

* 自 2014 年 11 月 17 日滬港通開通之日起計。

註：自 2016 年 12 月 5 日起包括於當日推出的深港通數據。

資料來源：香港交易所。

9　就北向交易而言，每日額度用量超過 10% 意指淨買盤超過北向交易每日額度人民幣 130 億元的 10%，即淨買盤超過人民幣 13 億元。南向交易方面亦同理，只是以南向交易淨買盤每日額度人民幣 105 億元為計算基準，貨幣轉換按湯森路透每日人民幣兌港幣的匯率計算。

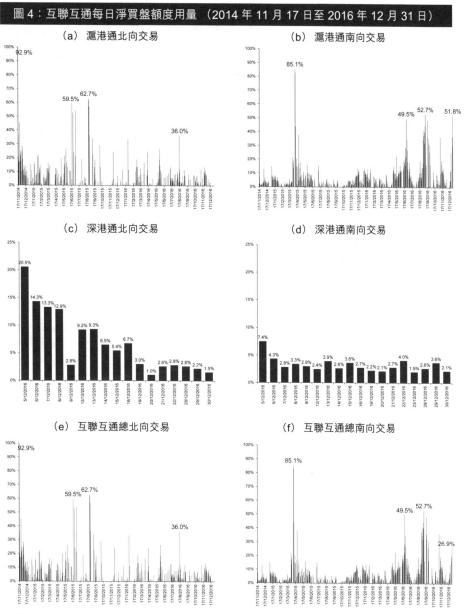

圖 4：互聯互通每日淨買盤額度用量（2014 年 11 月 17 日至 2016 年 12 月 31 日）

(a) 滬港通北向交易

(b) 滬港通南向交易

(c) 深港通北向交易

(d) 深港通南向交易

(e) 互聯互通總北向交易

(f) 互聯互通總南向交易

註：自 2016 年 12 月 5 日起，互聯互通總北向交易及總南向交易的每日總額度是滬股通與深股通之和。

資料來源：香港交易所；用以計算南向額度用量的每日成交額按湯森路透的每日匯率轉換為人民幣。

表2：互聯互通每日額度使用量 （2014年11月17日至2016年12月31日）				
	滬港通		深港通	
	北向交易	南向交易	北向交易	南向交易
交易日總數	494	485	17	18
錄得淨買盤日數佔比	55%	85%	100%	100%
每日額度使用量範圍	北向交易 （日數／相對總日數佔比）	南向交易 （日數／相對總日數佔比）	北向交易 （日數／相對總日數佔比）	南向交易 （日數／相對總日數佔比）
>0% - 10%	186 ／ 37.7%	317 ／ 65.4%	13 ／ 76.5%	18 ／ 100%
>10% - 20%	58 ／ 11.7%	66 ／ 13.6%	3 ／ 17.6%	0 ／ 0%
>20% - 30%	17 ／ 3.4%	10 ／ 2.1%	1 ／ 5.9%	0 ／ 0%
>30% - 40%	5 ／ 1.0%	11 ／ 2.3%	0 ／ 0%	0 ／ 0%
>40% - 50%	1 ／ 0.2%	6 ／ 1.2%	0 ／ 0%	0 ／ 0%
>50% - 60%	4 ／ 0.8%	2 ／ 0.4%	0 ／ 0%	0 ／ 0%
>60% - 70%	2 ／ 0.4%	0 ／ 0%	0 ／ 0%	0 ／ 0%
>70% - 80%	0 ／ 0%	0 ／ 0%	0 ／ 0%	0 ／ 0%
>80% - 90%	0 ／ 0%	2 ／ 0.4%	0 ／ 0%	0 ／ 0%
>90% - 100%	1 ／ 0.2%	0 ／ 0%	0 ／ 0%	0 ／ 0%
互聯互通總和	北向交易		南向交易	
交易日總數	494		485	
錄得淨買盤日數佔比	56%		86%	
每日額度使用量範圍	北向交易		南向交易	
>0% - 10%	191 ／ 38.7%		324 ／ 66.8%	
>10% - 20%	58 ／ 11.7%		67 ／ 13.8%	
>20% - 30%	17 ／ 3.4%		9 ／ 1.9%	
>30% - 40%	5 ／ 1.0%		10 ／ 2.1%	
>40% - 50%	1 ／ 0.2%		6 ／ 1.2%	
>50% - 60%	4 ／ 0.8%		1 ／ 0.2%	
>60% - 70%	2 ／ 0.4%		0 ／ 0%	
>70% - 80%	0 ／ 0%		0 ／ 0%	
>80% - 90%	0 ／ 0%		2 ／ 0.4%	
>90% - 100%	1 ／ 0.2%		0 ／ 0%	

2.2　環球投資者對互聯互通北向股票的興趣

在滬港通推出初期，環球投資者北向買賣及持有滬股通股票主要涉及市值龐大的上證 180 指數成份股（佔 2014 年成交金額的 94% 及 2014 年底持股金額的 96%）。北向交易中，買賣中型的上證 380 指數成份股的比例由 2014 年佔 6% 逐漸增至 2016 年的 23%。經滬股通持有此等中型股的比例曾於 2015 年年底升至 22%，又於 2016 年底回落至 17%，但仍遠高於 2014 年底的 4%。不過，環球投資者對滬股通股票的興趣始終以內地大型藍籌股為主。（見圖 5a 及 5b）

2016 年 12 月 5 日推出的深港通方面，環球投資者北向買賣及持有的股票亦主要是深證成份指數的藍籌股 —— 佔 2016 年交易金額的 90% 及年底持股金額的 93%。（見圖 5c 及 5d）

圖 5：互聯互通 —— 北向交易金額及投資者持股金額按股票類別的分佈（2014 年 11 月至 2016 年 12 月）

（a）滬股通交易金額

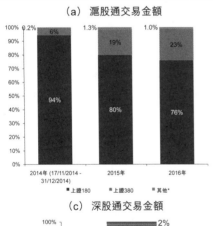

（b）滬股通持股金額

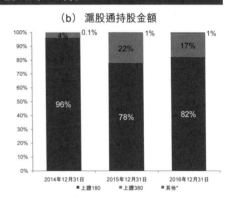

（c）深股通交易金額

（d）深股通持股金額

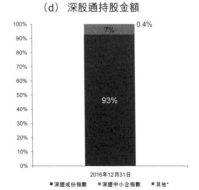

(續)

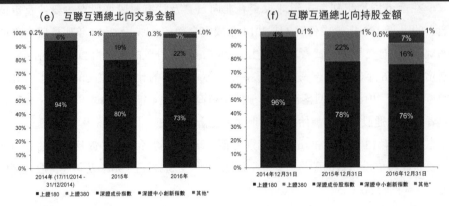

圖 5：互聯互通 —— 北向交易金額及投資者持股金額按股票類別的分佈
（2014 年 11 月至 2016 年 12 月）

* 「其他」包括非為合資格指數成份股但有 H 股同時在聯交所上市的 A 股以及期內從合資格股票名單中剔除的股票（只可賣出）。

註：深港通數據是由 2016 年 12 月 5 日深港通推出之日起計算。因四捨五入關係，百分比的總和或不等於 100%。

資料來源：香港交易所。

　　滬股通下環球投資者對內地消費板塊（非必需性消費品及必需性消費品）股票的興趣保持平穩，自滬港通推出以來，該板塊股票佔北向交易金額略增至 2016 年的 20%，持股佔比亦有所上升。深股通下的消費板塊佔更重要的比重 —— 2016 年交易金額的 47% 及年底持股金額的 58%。滬股通與深股通合計，環球投資者於 2016 年底持有的內地消費板塊高達 38%。

　　滬股通下的內地工業板塊股票亦頗具吸引力 —— 該板塊股票佔 2016 年北向交易金額及期末持股金額達 17%。滬股通下的金融板塊股票（均為市值大的上證 180 指數成份股）的較大佔比則漸降 —— 由 2014 年分別佔北向交易金額及期末持股金額的 51% 及 43% 降至 2016 年底的 31% 及 20%。深股通下的信息科技股票對環球投資者亦相當吸引，2016 年其交易額佔 16% 及年底持股額佔 15%。深股通的推出促使內地信息科技股在互聯互通下的北向交易佔比的進一步提升。（見圖 6）

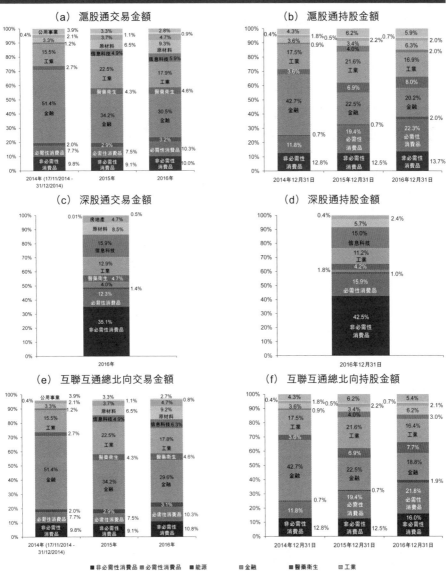

圖 6：互聯互通 —— 北向交易金額及投資者持股金額按行業類別的分佈
（2014 年 11 月至 2016 年 12 月）

註：深港通數據是由 2016 年 12 月 5 日深港通推出之日起計算。因四捨五入關係，百分比的總和或不等於
　　100%。

資料來源：香港交易所；股份分類採用來自彭博或湯森路透的環球指數分類標準（GICS）。

　　在涉及大型的上證 180 指數股份的北向交易中，金融股的佔比最大，而涉及中型的上證 380 指數股份的北向交易中，工業板塊股票佔相當高比重。消費板塊股票亦佔大型上證 180 指數股份北向交易的相當大比重，而消費及信息科技板塊股票佔中型上證 380 指數股份北向交易的比重呈顯著上升趨勢。誠然，縱使中型上證 380 指數股份中沒有金融類股票，其成份股中的信息科技、原材料及醫藥衛生板塊的股票較大型上證 180 指數股份更能吸引多元化的投資。（見圖 7 及 8）

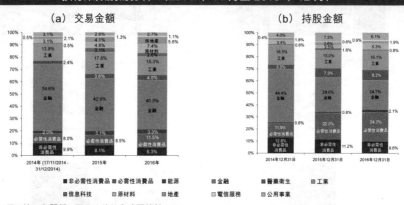

圖 7：滬港通 —— 上證 180 指數成份股的北向交易金額及投資者持股金額按行業類別的分佈（2014 年 11 月至 2016 年 12 月）

註：因四捨五入關係，百分比的總和或不等於 100%。

資料來源：香港交易所；股份分類採用來自彭博或湯森路透的環球指數分類標準（GICS）。

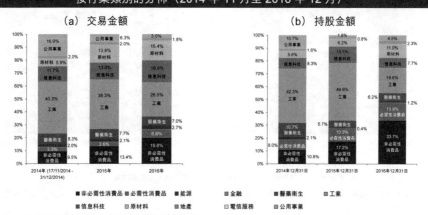

圖 8：滬港通 —— 上證 380 指數成份股的北向交易金額及投資者持股金額按行業類別的分佈（2014 年 11 月至 2016 年 12 月）

註：因四捨五入關係，百分比的總和或不等於 100%。

資料來源：香港交易所；股份分類採用來自彭博或湯森路透的環球指數分類標準（GICS）。

　　至於深股通方面，屬藍籌股指數的深證成份指數的北向交易中，消費板塊股票的交易及持股金額佔比相當高，信息科技股的佔比亦頗高；而涉及深證中小創新指數股份的北向交易中，工業、信息科技以及原材料板塊股票均有相當大的比重，非必需性消費品板塊股票則在持股比重方面佔優。（見圖 9 及 10）

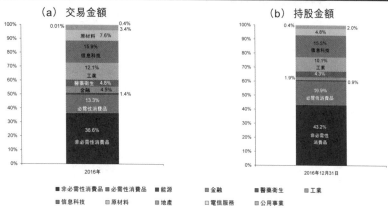

圖 9：深港通 —— 深證成份指數成份股的北向交易金額及投資者持股金額按行業類別的分佈（2016 年 12 月）

註：數據自 2016 年 12 月 5 日起深港通推出之日起計。因四捨五入關係，百分比的總和或不等於 100%。
資料來源：香港交易所；股份分類採用來自彭博或湯森路透的環球指數分類標準（GICS）。

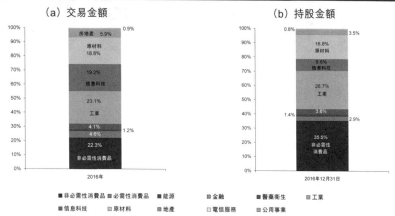

圖 10：深港通 —— 深證中小創新指數成份股的北向交易金額及投資者持股金額按行業類別的分佈（2016 年 12 月）

註：數據自 2016 年 12 月 5 日起深港通推出之日起計。因四捨五入關係，百分比的總和或不等於 100%。
資料來源：香港交易所；股份分類採用來自彭博或湯森路透的環球指數分類標準（GICS）。

2.3　內地投資者對港股通股票的興趣

　　圖11顯示各類合資格港股通股票的交易及投資者持股情況。與北向的滬股通投資相較之下，南向的港股通交易及持股金額在2014年互聯互通推出之時多集中於恒生綜合中型股指數成份股。到2016年期間，某程度上已轉為以恒生綜合大型股指數成份股為主。儘管如此，恒生綜合中型股指數股份仍佔港股通2016年全年的交易金額及期末持股金額的相當大比重（~40%）。

　　除了恒生綜合大型股指數及中型股指數成份股外，在深港通下的合資格南向交易股票還包括恒生綜合小型股指數成份股，後者所佔深港通2016年18個交易日的南向交易的比重相當高，與中型股的比重同為42%，小型股在深港通下的持股比重較中型股更高（46%對比42%）。然而互聯互通總體南向持股分佈方面，恒生綜合小型股的比重因其低資產價值定義的內在性質關係仍相對甚低（4%）。

圖11：互聯互通 ── 南向交易金額及投資者持股金額按股票類別的分佈（2014年11月至2016年12月）

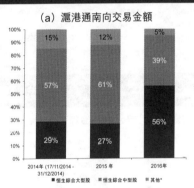

(a) 滬港通南向交易金額

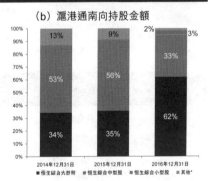

(b) 滬港通南向持股金額

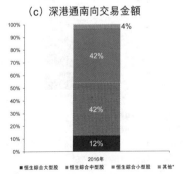

(c) 深港通南向交易金額

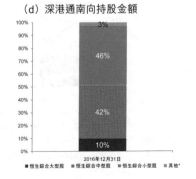

(d) 深港通南向持股金額

（續）

圖 11：互聯互通 —— 南向交易金額及投資者持股金額按股票類別的分佈
（2014 年 11 月至 2016 年 12 月）

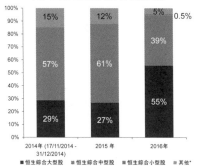

（e）互聯互通總南向交易金額

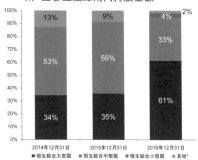

（f）互聯互通總南向持股金額

* 「其他」包括非為合資格指數成份股但有 A 股同時在上交所或深交所（2016 年 12 月 5 日起計）上市的 H 股
　以及國內從合資格股票名單中剔除的股票（只供賣出）。

註：深港通數據是由 2016 年 12 月 5 日深港通推出之日起計算。持股類別為期末之狀態，交易期間之股票類
　　別可能有變；因股票類別變更，滬港通下於期末會持有恒生綜合小型股，但其買賣則計算入「其他」類別。
　　因四捨五入關係，百分比的總和或不等於 100%。

資料來源：交易數據來自香港交易所；持股數據來自 Webb-site Who's Who 數據庫。股份分類採用恒生指數
　　　　　有限公司的分類。

　　從行業來看，投資者對金融股的興趣日濃，互聯互通下的金融股於 2016 年獨
佔鰲頭，成為交易及持股最多的行業。其他較受歡迎的行業還有消費品製造業及
地產建築業。深港通下的南向交易和持股則並未集中於金融股，而有相當比重分
佈於消費品類、地產建築、資訊科技及工業類股票。（見圖 12）

圖 12：互聯互通 —— 南向交易金額及投資者持股金額按行業類別的分佈
（2014 年 11 月至 2016 年 12 月）

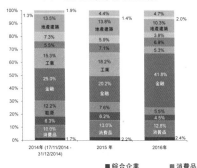

（a）滬港通南向交易金額

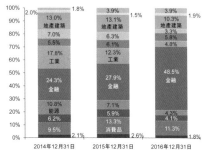

（b）滬港通南向持股金額

(續)

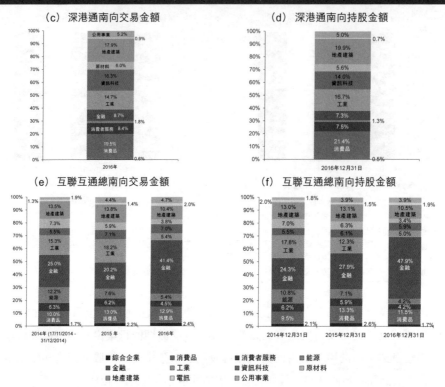

圖 12：互聯互通 —— 南向交易金額及投資者持股金額按行業類別的分佈
（2014 年 11 月至 2016 年 12 月）

註：深港通數據是由 2016 年 12 月 5 日深港通推出之日起計算。因四捨五入關係，百分比的總和或不等於
100%。

資料來源：交易數據來自香港交易所；持股數據來自 Webb-site Who's Who 數據庫。股份分類採用恒生指數
有限公司的分類。

　　但是，南向的港股通投資以金融股為主的情況主要見於恒生綜合大型股指數
股份方面。在恒生綜合中型股指數股份方面，內地投資者經港股通買賣及持有的
行業類別更為多元化。2016 年，南向交易及持有的恒生綜合中型股指數成份股
中有相當大的比例是消費品製造業股份（約 25-27%，若計及消費品服務業更超過
30%）及地產建築業股份（約 16-18%）。金融股只列位第三，與其於恒生綜合大型
股指數成份股的港股通交易及持股中的主力地位顯然有別。在深港通下恒生綜合
大型及中型股指數成份股的南向交易金額及持股金額按行業類別的分佈同樣有相

若的顯著分別。而深港通下的恒生綜合小型股指數成份股中的資訊科技股明顯能吸引頗高的交易與持股比重。（見圖 13 至 15）

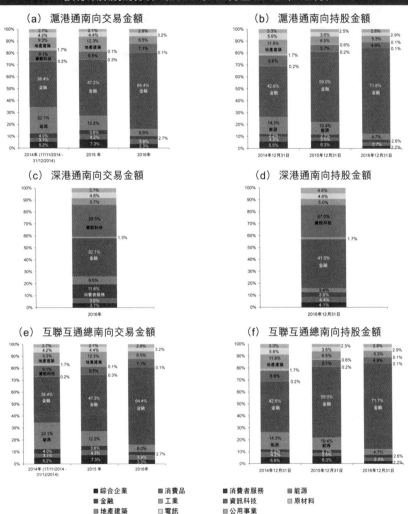

圖 13：互聯互通 —— 恒生綜合大型股指數成份股的南向交易金額及投資者持股金額按行業類別的分佈（2014 年 11 月至 2016 年 12 月）

註：深港通數據是由 2016 年 12 月 5 日深港通推出之日起計算。持股類別為期末之狀態，交易期間之股票類別可能有變。因四捨五入關係，百分比的總和或不等於 100%。

資料來源：交易數據來自香港交易所；持股數據來自 Webb-site Who's Who 數據庫。股份分類採用恒生指數有限公司的分類。

圖 14：互聯互通 —— 恒生綜合中型股指數成份股的南向交易金額及投資者持股金額按行業類別的分佈（2014 年 11 月至 2016 年 12 月）

(a) 滬港通南向交易金額

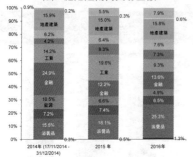

(b) 滬港通南向持股金額

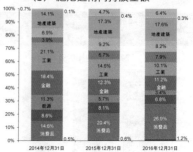

(c) 深港通南向交易金額

(d) 深港通南向持股金額

(e) 互聯互通總南向交易金額

(f) 互聯互通總南向持股金額

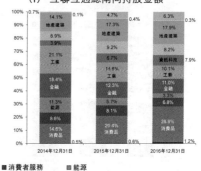

■ 綜合企業　■ 消費品　■ 消費者服務　■ 能源
■ 金融　■ 工業　■ 資訊科技　■ 原材料
■ 地產建築　□ 電訊　□ 公用事業

註：深港通數據是由 2016 年 12 月 5 日深港通推出之日起計算。持股類別為期末之狀態，交易期間之股票類別可能有變。因四捨五入關係，百分比的總和或不等於 100%。

資料來源：交易數據來自香港交易所；持股數據來自 Webb-site Who's Who 數據庫。股份分類採用恒生指數有限公司的分類。

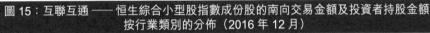

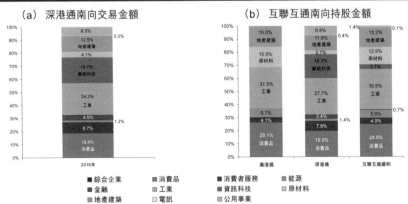

圖 15：互聯互通 —— 恒生綜合小型股指數成份股的南向交易金額及投資者持股金額
按行業類別的分佈（2016 年 12 月）

（a）深港通南向交易金額　　　　（b）互聯互通南向持股金額

註：深港通數據是由 2016 年 12 月 5 日深港通推出之日起計算。持股類別為期末之狀態，交易期間之股票類
別可能有變；因股票類別變更，滬港通下於期末會持有恒生綜合小型股，但其買賣則計算入「其他」類別。
因四捨五入關係，百分比的總和或不等於 100%。

資料來源：交易數據來自香港交易所；持股數據來自 Webb-site Who's Who 數據庫。股份分類採用恒生指數
有限公司的分類。

　　換言之，在互聯互通機制下，內地投資者對多種不同類型行業的中型港股通
股票有相當大興趣。相對大型股中金融股佔比相當重的情況，**較小型的股份反為
內地投資者提供更多元的股份行業投資選擇**。

3 「共同市場」模式
—— 給予內地及全球投資者的機遇

　　在深港通推出後，縱使在指定合資格股票範圍內運作，滬深港三地的「共同
市場」模式已然基本形成。由於股票市場交易互聯互通機制可以擴容，這無形中打
開了一個潛在的內地與香港股票共同市場，其股份總值 105,140 億美元（2016 年
底）、日均股份成交約 843 億美元（2016 年），於全球交易所中按市值計排名第二

（僅次於紐約證券交易所）、按股份成交額計排名第二[10]。「共同市場」模式更可擴展至股票以外的多個範疇。按中國證監會與香港證監會於 2016 年 8 月 16 日原則上批准建立就深港通的聯合公告所披露，兩家監管機構已就將交易所買賣基金納入機制內合資格股票達成共識，將於深港通運行一段時間及滿足相關條件後再宣佈推出日期。此外，中國證監會與香港證監會將共同研究及推出其他金融產品，便利及迎合內地與環球投資者對於管理對方股票市場的價格風險的需要。

在「共同市場」模式下，可以向內地投資者提供各式各樣的海外產品，亦可以向全球投資者提供不同的內地產品。南向交易為內地投資者（個人及機構）打開投資海外資產的規範化渠道。此渠道全程封閉、每日額度的使用受審慎監控，而又在無總額度下提供相當的靈活度。在無總額度限制下，投資者可較以往更自由地配置跨境資產組合投資，這等同為內地投資者提供了全球資產配置機會。按此機制操作，資金全程封閉：在機制下人民幣先轉成港元用於購買海外資產，他日出售海外資產時再轉換成人民幣匯回中國內地，實質上完全避免長遠的資金外流問題。這樣的模式等同擴展了內地投資者可投資資產的種類。在此環境下，互聯互通渠道補足了內地可投資資產相對短缺的問題，內地資金可以投資於海外，或者能獲得較內地市場更佳的潛在回報。有見及此，中國保險監督管理委員會於 2016 年 9 月初發出政策文件[11]，容許保險資金參與滬港通下的港股通交易。深港通的合資格內地投資者與滬港通相同，「共同市場」在深港通擴闊投資範圍下會為這些內地投資者提供更為多元化的南向投資選擇。

此外，港股通對內地投資者而言實際上是投資外幣（與美元掛鈎的港元）。在人民幣貶值預期下，南向投資提供了從幣值角度看的另類投資選擇。

基於「共同市場」模式的合資格工具可予擴充，相信向內地投資者提供的投資工具範圍將會日趨多元化，儘管短期內或只能提供現貨市場證券，包括股票及有可能納入的交易所買賣基金。深港通推出後，港股通的合資格股票除了 HSLI 及 HSMI 成份股外，還包括市值 50 億港元或以上的 HSSI 成份股，以及有 A 股在內地市場上市的所有 H 股（不只限於上交所上市 A 股）。HSLI 及 HSMI 已涵蓋恒生

10　國際證券交易所聯會（WFE）數據（WFE 網站 2017 年 1 月 20 日取得的市值資料及 2017 年 3 月 1 日取得的成交資料）。日均成交額按 WFE 2016 年數據的合併成交額以及內地市場交易日總數 244 日計算。排名按 2016 年截至 12 月的合併成交額計算。

11　《關於保險資金參與滬港通試點的監管口徑》，2016 年 9 月 9 日。

綜合指數（HSCI）總市值 95%，佔香港市場總市值 95%[12]。故此，合資格股票名單添加了 100 多隻股票。更重要的是，擴充後的股票範圍加入了許多不同行業，包括新經濟行業如資訊科技和消費品及服務。

　　再者，香港市場的主要參與者為國際專業機構投資者，這樣的國際證券市場所給予的交易經驗對內地投資者（特別是散戶投資者）而言具一定價值。成熟市場的專業投資策略一般基於股票的基本面以及經濟及行業因素，這有助於平衡內地部份投資者的短期投機交易行為。因此港股通的交易經驗預期將有助內地投資者基礎日趨成熟。

　　除二級市場的交易外，共同市場的模式下亦可推出集資市場／一級市場（即首次公開招股市場）的互聯互通（新股通）（須獲監管機構批准），讓兩邊市場的投資者可認購對方市場的首次公開招股的股份。模式下所涵蓋的產品日後（須獲監管機構批准）還可以延伸至債券、商品及風險管理工具，包括股票衍生產品、人民幣利率及貨幣衍生產品。事實上，鑒於股票市場交易互聯互通計劃已順利實施，如何切合投資者對沖其跨境股票組合的需要會是當前急切的問題：現在內地投資者可買賣港股，但卻沒有香港指數／股票期貨及期權作為對沖，同樣地香港亦欠缺 A 股指數期貨期權等 A 股對沖工具。相關的衍生產品日後或會被納入共同市場模式。

　　「共同市場」模式實際上是中國內地資本賬開放進程中極具象徵意義的突破，在內地資本市場可能作全面開放前，這是個可長線提供極為多元化的投資及風險管理工具配套的市場模式。按此，內地投資者可受惠於更佳的資產配置及投資組合管理，全球投資者亦獲得開放渠道，能在有相關風險管理工具可用的情況下捕捉更多內地投資機會。

12　HSCI 所涉及的香港市場整體包含所有在聯交所作主要上市的股票及房地產投資信託基金（REIT），不包括第二上市的證券、境外公司、優先股、債務證券、互惠基金及其他衍生產品。（資料來源：恒生指數有限公司網站）

第3章

新股通

—— 內地與香港市場互聯互通及
人民幣國際化的突破性契機

2017 年 8 月 28 日

摘 要

　　隨着滬港通及深港通（合稱「滬深港通」）先後於 2014 年 11 月及 2016 年 12 月開通，內地與香港市場互聯互通平台基本形成。然而，這個平台現階段只限於股票交易市場上的買賣，兩地市場的投資者仍未能參與對方的新股市場，以致投資者無法從另一方的首次公開招股中尋求投資機遇，這亦實質上妨礙了共同市場要實現匯集資金、支援發行人進行集資的本意。

　　香港交易所《戰略規劃 2016-2018》提出了新股通計劃[1]，為市場提供突破性機遇，以協助完成共同市場在股市部分的版圖。新股通的概念是提供機制容許內地投資者在香港市場認購首次公開招股（南向），以及允許香港的全球投資者認購內地市場的首次公開招股（北向）。所發行的首次公開招股股份上市後，另一方市場的投資者可透過現有滬深港通機制進行買賣。按此聯通模式，循新股通認購股份及經滬深港通買賣股份將有效實現一個密封式機制。

　　對內地市場及香港市場各自而言，在國際化層面上的發展同樣面臨瓶頸，新股通預期可令兩者共同受惠。目前內地股票市場的國際化程度在多個方面仍相對偏低，這包括投資者基礎、發行人基礎及制度結構等。合格境外機構投資者（QFII）和人民幣合格境外機構投資者（RQFII）在內地股市的參與度仍然偏低（總持股價值佔滬深證券交易所總流通市值不足 0.3%）；境外公司尚未獲准在內地交易所上市；內地的市場慣例也尚未與國際慣例接軌。至於另一端的香港，參與股票市場的投資者已高度國際化，但上市發行人方面則仍不然。面對這些短板，新股通可助內地市場及香港市場同時拓展自身的國際版圖，以至整個共同市場的國際化程度也得以提升。

　　內地與香港共同市場的國際化本身並不是一個終極目標，而是中國達致更平衡經濟、更有效市場開放及最終使人民幣更高度國際化的戰略的其中一環。共同市場模式下的新股通計劃正好提供了改善現狀的機會。

1　須經監管部門批准。

對內地而言，新股通可（1）為內地投資者提供全球資產配置的新機遇，從而改善國家的資產負債表；（2）以較低成本利便市場雙向開放；（3）在人民幣資本項目自由兌換方面向前邁進一步；（4）支持市場發展國際投資者基礎；（5）為內地企業帶來更多上市機會；以及（6）協助培育內地投資者基礎；而又能同時以適當監控措施減低潛在風險。對香港而言，新股通將有助吸引海外公司來港上市、增加投資者參與度而進一步激活市場，並為市場中介機構帶來更多商機。

實施新股通較滬深港通計劃涉及更多包括監管及營運方面的議題。不過，相信只要機制模式設計合宜，符合內地與香港共同市場的最佳利益，這些問題大致上都可迎刃而解，最終會惠及中國國民經濟賬目以至人民幣國際化這大戰略。

1 欠缺新股互通的共同市場

1.1　交易市場互聯互通 —— 滬深港通

內地與香港股票市場交易互聯互通機制試點計劃（「試點計劃」）於 2014 年 11 月 17 日由滬港通揭開序幕，內地與香港市場的跨境股票投資自此接通，到 2016 年 12 月 5 日更進一步推出深港通。於 2016 年 8 月 16 日正式宣佈深港通時，更即時取消試點計劃最初採用的總額度安排。（本文內，滬港通及深港通合稱「滬深港通」。）至此，滬深港三地的共同市場平台基本形成，打開了一個潛在的內地與香港股票共同市場，其股票總值 105,140 億美元（2016 年底），日均股份成交約 843 億美元（2016 年），於全球交易所中市值排名第二（僅次於紐約證券交易所），按股份成交額計也排名第二[2]。此外，共同市場模式或可由股票擴展至交易所買賣基金（ETF）等合資格交易工具[3]。

滬深港通計劃讓香港及海外投資者在計劃的合資格範圍內買賣在上海證券交易所（上交所）或深圳證券交易所（深交所）上市的內地市場證券（滬港通及深港通下各自的滬股通及深股通），以及讓內地投資者在計劃的合資格範圍內買賣在香港聯合交易所（聯交所）上市的香港市場證券（滬港通及深港通下各自的港股通）。

滬股通的合資格證券涵蓋在上交所上市的上證 180 指數及上證 380 指數的成份股，以及有相關 H 股同時在聯交所上市的上交所上市 A 股，但不包括不以人民幣交易的滬股及被實施風險警示[4]的滬股。深股通合資格證券涵蓋在深交所上市的深證成份指數和深證中小創新指數成份股中所有市值不少於人民幣 60 億元的成份股，以及有相關 H 股在聯交所上市的所有深交所上市 A 股，但不包括不以人民

2　國際證券交易所聯會（WFE）數據（WFE 網站 2017 年 1 月 20 日取得的市值資料及 2017 年 3 月 1 日取得的成交資料）。日均成交額按 WFE 2016 年數據的合併成交額以及內地市場交易日總數 244 日計算。排名按 2016 年截至 12 月的合併成交額計算。

3　按中國證券監督管理委員會（中國證監會）與香港的證券及期貨事務監察委員會（香港證監會）於 2016 年 8 月 16 日原則上批准建立深港通的聯合公告所披露，兩家監管機構已就將交易所買賣基金納入機制內合資格股票達成共識，將於深港通運行一段時間及滿足相關條件後再宣佈推出日期。

4　指相關股份被上交所實施「風險警示」，包括「ST 公司」及「*ST 公司」的股份以及須根據上交所規則進行除牌程序的股份。

幣交易的深股及被深交所實施風險警示的深股。滬港通下的港股通合資格證券涵蓋在聯交所主板上市的恒生綜合大型股指數 (HSLI) 成份股及恒生綜合中型股指數 (HSMI) 成份股，以及有相關 A 股在上交所上市的 H 股，但不包括不以港幣交易的港股及其相應 A 股被實施風險警示的 H 股。除滬港通下的港股通合資格證券外，深港通下的港股通合資格證券亦包括所有市值 50 億港元或以上的恒生綜合小型股指數 (HSSI) 成份股，以及所有聯交所上市公司中同時有 A 股在深交所上市的 H 股，但不包括不以港幣交易的港股及其相應 A 股被實施風險警示的 H 股。

於 2017 年 6 月 28 日，滬港通下合資格可買可賣的滬股通股票共 574 隻、港股通合資格股票共 310 隻；深港通下合資格可買可賣的深股通股票共 901 隻、港股通合資格股票共 418 隻。換言之，上交所及深交所上市 A 股分別約 44% 及 45% 以及聯交所主板約 24% 屬滬深港通合資格證券[5]。至 2016 年底，北向交易的日均成交額佔內地 A 股市場成交總額約 2%，而南向交易的日均成交額則佔聯交所主板成交總額約 8%[6]。

1.2　新股市場商機

滬深港通計劃是接通內地證券市場與海外市場的破天荒機制。然而，目前這機制只限於二級股票交易市場上的買賣，兩個市場的投資者尚未可參與對方市場的一級股票集資市場。投資者無法參與另一方市場新上市公司首次公開招股所提供的投資機會。根據國際證券交易所聯會 (WFE) 的數據，過去兩年香港、上海及深圳均位列全球首次公開招股集資額排名的前十名 (見圖 1)。過去八年中，香港有五年均高踞首次公開招股集資額榜首 (見圖 2)。

5　資料來源：香港交易所、上交所及深交所網站。

6　見本書第 2 章〈滬港通與深港通下的互聯互通 —— 內地及全球投資者的「共同市場」〉。

圖1：首次公開招股集資額最高的十大交易所（2015年及2016年）

(a) 2015年

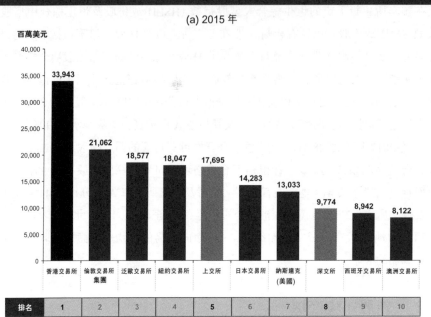

(b) 2016年

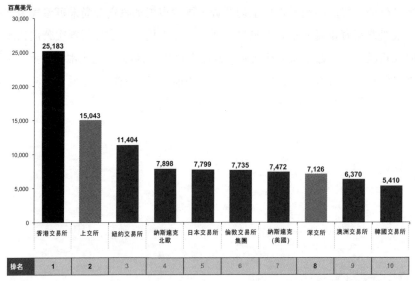

資料來源：WFE 網站。

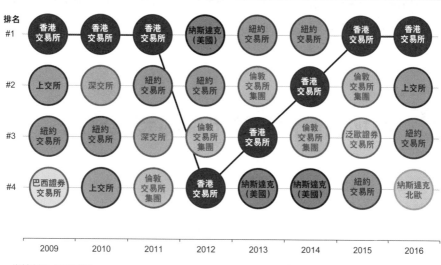

圖 2：按首次公開招股集資額排名的全球四大交易所（2009 年至 2016 年）

排名
#1 香港交易所　香港交易所　香港交易所　納斯達克（美國）　紐約交易所　紐約交易所　香港交易所　香港交易所

#2 上交所　深交所　紐約交易所　紐約交易所　倫敦交易所集團　香港交易所　倫敦交易所集團　上交所

#3 紐約交易所　紐約交易所　深交所　倫敦交易所集團　香港交易所　倫敦交易所集團　泛歐證券　紐約交易所

#4 巴西證券交易所　上交所　倫敦交易所集團　香港交易所　納斯達克（美國）　納斯達克（美國）　紐約交易所　納斯達克北歐

2009　2010　2011　2012　2013　2014　2015　2016

資料來源：WFE 網站。

　　股票市場方面，滬深港通僅僅接通二級交易市場，但尚未打通集資市場，內地與香港的共同市場始終未臻圓滿。事實上，欠缺集資市場互通或會損害投資者在交易市場聯通方面的權益，造成市場不公。以較近期的交通銀行股份有限公司（交通銀行）分拆交銀國際控股有限公司（交銀國際）作例子，交通銀行 H 股在香港上市，其 A 股在上交所上市。交通銀行於 2016 年 8 月宣佈建議分拆交銀國際在聯交所上市，並為現有股東提供交銀國際發行新股的保證配額。然而，受法律及政策體制所限，交通銀行只可向當時的 H 股股東而未能向 A 股股東提供保證配額[7]。部分市場人士[8]亦建議，接通新股市場應有助解決市場這類不公平現象。

　　香港交易所於其《戰略規劃 2016-2018》中，提出了深港通以及新股通作為進一步拓展市場互聯互通的計劃。在深港通於 2016 年 12 月開通後，新股通計劃預期可為內地投資者的環球資產配置進一步提供新機遇。計劃亦會支持內地股票市場進一步國際化及協助完善互聯互通機制，從而促使共同市場匯聚的資金能更有

7　交通銀行 2016 年 9 月 12 日的公告對此有所解釋，指出的原因包括滬港通並無為 A 股股東設立認購香港市場新發行股份的機制。

8　見《東方日報》及《香港經濟日報》2017 年 1 月 18 日的報導。

效發揮支援發行人集資及投資者買賣股票的功能。就更廣泛的層面而言，共同市場若能加添新股通以至其他進一步的聯通計劃，還可助中國推進經濟發展及人民幣國際化等更宏大的藍圖。以下就此一一闡釋。

2 內地及香港股票市場的發展瓶頸

市場進一步開放及國際化一直是內地資本市場發展的主要政策方向。中國國民經濟和社會發展第十三個五年規劃綱要（2016-2020）概述為市場全方位開放建立新模式的措施，包括擴大金融業雙向開放及資本市場開放。尤其是，將上海發展成國際金融中心一直是中央政策。為達到這個目標，中國人民銀行（人行）與其他政府部門於 2015 年聯合發表的規劃 [9] 中，提出多項措施加快上海國際金融中心的發展。國務院於 2017 年 3 月進一步發表規劃 [10]，推進中國（上海）自由貿易試驗區（上海自貿區）金融改革及市場開放措施，包括進一步深化區內開放創新及有序推進資本項目開放及金融慣例國際化等試驗計劃。

這邊廂，香港一直是知名的國際金融中心 [11]，其資本市場奉行符合國際標準的市場慣例，國際投資者活躍其中。香港資本市場一直利用自身優勢支持內地資本市場開放及國際化，例如透過讓中國企業來港上市、經滬深港通接通內地二級市場交易等。**內地市場全然國際化固然需要時間，但香港市場在國際化層面方面也仍有不足。**兩邊市場均需要在若干方面創新突破，互惠互利。以下分節論述兩個市場的發展瓶頸狀況，有關創新突破於第 4 節作討論。

9　《進一步推進中國（上海）自由貿易試驗區金融開放創新試點加快上海國際金融中心建設方案》，2015 年 10 月 30 日。

10　《全面深化中國（上海）自由貿易試驗區改革開放方案》，2017 年 3 月 31 日。

11　根據英國智庫 Z/Yen 集團與中國（深圳）綜合開發研究院共同編制的全球金融中心指數（2017 年 3 月），香港名列全球第四大金融中心。上海、北京及深圳分別排第 13、16 及 22 位。

2.1　內地股市的「國際化」

在投資者方面，於滬深港通推出前，合資格投資內地股票市場的外地投資者只有合格境外機構投資者（**QFII**）及人民幣合格境外機構投資者（**RQFII**）。2017年 4 月 26 日數據顯示，國家外匯管理局向 281 家 QFII 及 183 家 RQFII 分別批出 907.65 億美元（約人民幣 6,266.38 億元）及人民幣 5,420.04 億元的總投資額度[12]。當中，香港註冊的機構獲批最多名額及最高投資額度 —— 佔 QFII 額度的 23% 及 RQFII 額度的 49%（見圖 3）。

然而，所有獲批的 QFII 及 RQFII 投資額度合計，也只佔上海及深圳股市總流通市值不足 3%[13]。再者，並非全部 QFII 及 RQFII 投資額度均會投資於股市。

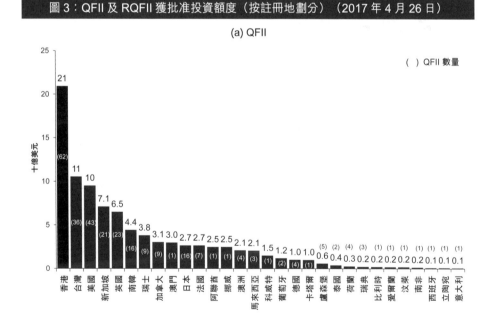

圖 3：QFII 及 RQFII 獲批准投資額度（按註冊地劃分）（2017 年 4 月 26 日）

(a) QFII

12　資料來源：國家外匯管理局網站。

13　根據 2016 年 4 月 26 日的總批准額度佔內地交易所 2017 年 3 月底的股本證券總流通市值人民幣 41.405 萬億元 —— 上交所（人民幣 254,101.07 億元）及深交所（人民幣 159,953.68 億元）—— 計算（資料來源：上交所及深交所網站各自的每月統計數據）。

（續）

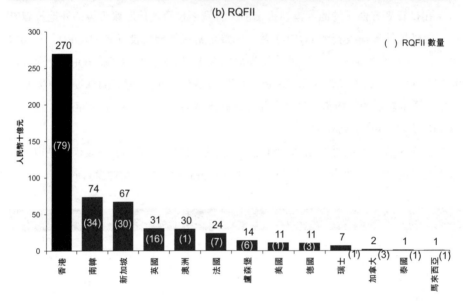

圖 3：QFII 及 RQFII 獲批准投資額度（按註冊地劃分）（2017 年 4 月 26 日）

(b) RQFII

（ ）RQFII 數量

資料來源：國家外匯管理局網站。

　　至 2016 年底，上海及深圳股市的中央結算所 —— 中國證券登記結算有限責任公司（中國結算）—— 共有 32.6 萬個「法人」投資者戶口，但當中只有 1,088 個 QFII 戶口（上海 543 個，深圳 545 個）及 1,078 個 RQFII 戶口（上海 534 個，深圳 544 個），即合共佔戶口總數少於 1%[14]。

　　就持股市值而言，2017 年 3 月底 QFII 投資於內地股市的總金額為人民幣 1,144.4 億元，佔上交所及深交所總流通市值少於 0.3%[15]。滬深港通已然為外地投資者開拓另一途徑投資於內地市場，惟其參與規模仍相當低 —— 北向交易只佔內地 A 股市場交易額的 2% 或以下（見上文第 1.1 節）。

　　在發行人方面，外資公司現時仍未可在內地股票市場上市。上海政府 2009 年

14　資料來源：中國結算 2016 年 12 月統計數據，中國結算網站。註：不同的 QFII 及 RQFII 基金產品會使用不同的投資者戶口。

15　有關 QFII 投資的資料來源：西南證券有關 2017 年第一季 QFII 持股狀況的研究報告，2017 年 5 月 1 日。

提出在上交所設立國際板的方案 [16]，獲中央政府在 2011 年公佈的中國國民經濟和社會發展第十二個五年規劃綱要中支持探討其可行性，但至今未見太大進展。

在股票以外的現貨產品方面，涉外產品只有少數追蹤境外資產的 ETF。於 2016 年底，上交所共有 75 隻 ETF，當中 6 隻是跨境 ETF（8%）；而深交所則有 48 隻 ETF，當中只有 2 隻追蹤境外產品（分別追蹤恒生指數及納斯達克 100 指數）[17]。

在市場結構方面，內地股市主要以散戶為主。市場常規、規則及法規均為滿足內地市場發展進程的特別需要而設，未必貼近國際市場常規。

總括而言，內地股市在投資者基礎、發行人基礎及制度架構等多個範疇上的國際化程度仍相對偏低。提高外資參與度將有助上海邁向成為國際金融中心這目標。

2.2　香港股市的「國際化」

以投資者參與度而言，香港股市是一個非常國際化的市場，外地投資者於香港交易所現貨市場交易的比重較本地投資者的交易比重為高（2016 年分別為 40% 及 36%）（見圖 4），他們來自世界各地 —— 亞洲區的外地投資者報稱來自 18 個地區，亞洲及歐美以外地區的外地投資者報稱來自 53 個地區 [18]。

16　上海政府發表的《貫徹國務院關於推進〈兩個中心〉建設實施意見》，2009 年 5 月 11 日。
17　資料來源：上交所及深交所網站。
18　資料來源：香港交易所現貨市場交易研究調查 2016。

圖 4：香港交易所現貨市場交易金額的分佈（按投資者來源地劃分）（2016 年）

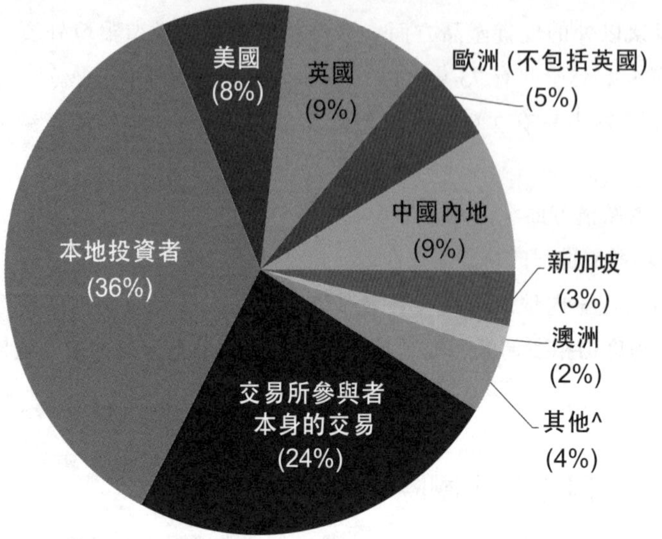

^「其他」涵蓋來自日本、台灣、亞洲其他地區及世界其他地區。

註：由於四捨五入，數字的總和未必相等於 100%。

資料來源：香港交易所現貨市場交易研究調查 2016。

由 2010 年至 2017 年第一季止，香港市場幾乎所有首次公開招股活動均有向國際投資者作公開發售，期內首次公開招股集資額超過 80% 來自國際發售部分 [19]（見圖 5）。換言之，在香港，不論是首次公開招股的一級市場認購活動還是二級市場交易活動，國際投資者同樣活躍。

---

19 參與國際發售的國際投資者包括香港投資者、內地投資者及其他海外投資者，而香港許多機構投資者均源自海外。

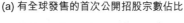

圖 5：香港首次公開招股中全球發售的佔比（2007 年至 2017 年首季）

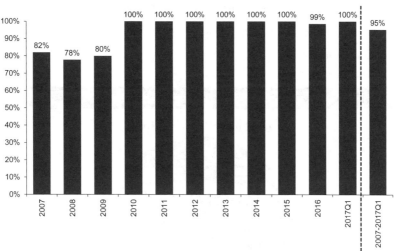

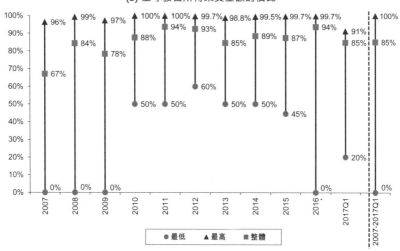

資料來源：香港交易所。

　　在上市發行人方面，情況恰好相反，在香港交易所整體市場（主板及創業板）上市的，以市值及成交額計大部分是內地企業——H 股、紅籌股及內地民營企業——按市值計共佔 64%（2017 年 6 月底），按成交額計共佔 74%（2017 年 1 月至 6 月）（見圖 6）。

過去 10 年（2008 年至 2017 年第一季）的新上市公司中，僅 8% 來自香港及中國內地以外地區，共佔首次公開招股集資總額的 20%。以公司數目計，內地民營企業佔比最高（47%）；以首次公開招股集資額計，H 股公司佔比最高（48%）（見圖 7）。換言之，**香港一級市場的服務對象主要是香港及內地公司，在服務國際公司方面尚未盡展所長。**

圖 6：聯交所主板上市公司的分佈

(a) 按市值計（2017 年 6 月底）

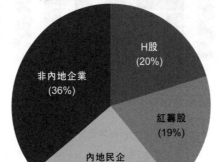

(b) 按成交金額計（2017 年 1 月至 6 月）

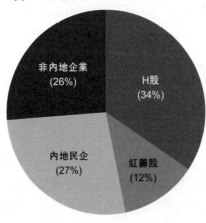

資料來源：香港交易所。

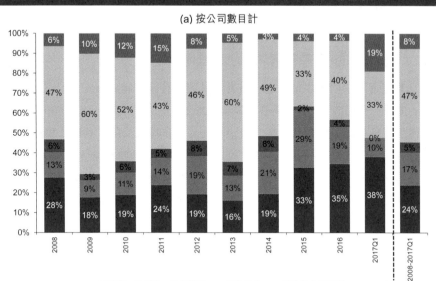

圖 7：聯交所主板新上市公司的分佈（2008 年至 2017 年首季）

(a) 按公司數目計

圖例：■ 香港　■ H股　■ 紅籌股　□ 內地民企　■ 其他海外來源地

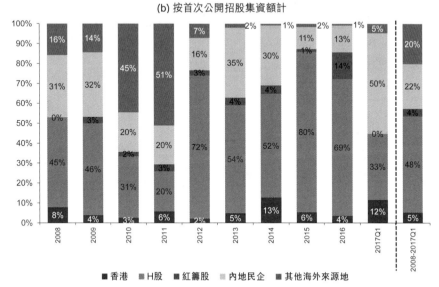

(b) 按首次公開招股集資額計

圖例：■ 香港　■ H股　■ 紅籌股　□ 內地民企　■ 其他海外來源地

資料來源：香港交易所。

　　總括而言，以投資者參與程度觀之，香港股市是極為國際化的市場，但以上市發行人類別計則國際化程度偏低。

2.3　內地與香港共同市場的「國際化」

　　內地及香港共同市場的概念，主要就是透過特設的聯通模式，向國際及內地投資者開放一個渠道，投資於結合內地與國際元素的大型市場。就股本市場而言，滬深港通現行的聯通模式只限於二級市場交易，自然也局限了共同市場的參與者（包括發行人及投資者）可得享的裨益。股票市場的主要功能是讓發行人籌集資金。要達到這個目的，發行人在新上市時發售股份是基礎所在，一方面令私營企業可籌集發展業務所需的資金，另一方面亦為投資者提供多元化的投資機遇。以聯交所主板為例，首次公開招股集資額佔股份集資總額的比重於 2011 年曾高見 54%，最近 2016 年亦達 40%；過去 10 年，股份發售也是公司上市後集資的最主要採用方式（見圖 8）。

圖 8：在聯交所主板的股份集資金額 —— 首次公開招股與上市後發行集資金額的佔比以及發售股份集資金額的佔比（2007 年至 2016 年）

(a) 首次公開招股及上市後集資金額　　　　(b) 上市後集資金額（按集資方式）

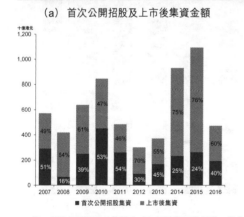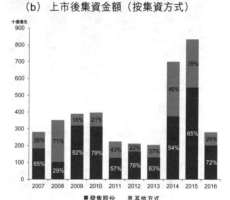

註：首次公開招股的集資方式包括發售以供認購、發售現有證券及配售。上市後發售股份集資的方式包括配售、供股及公開招股；其他方式包括代價發行、行使權證及購股權計劃。

資料來源：香港交易所。

內地與香港共同股票市場的國際化，意味着市場擁有國際投資者基礎及國際發行人基礎。香港股市具有國際投資者基礎，只是國際發行人基礎較薄弱，而內地股市則兩方面均有待更大進展。二級市場聯通後，這個共同市場的國際化發展已踏前一步，但如上文所述，一級市場一日未聯通，共同市場的聯通機制仍未完善，會窒礙市場國際化發展的進程。

3　中國宏圖：國民經濟賬目、市場開放及人民幣國際化

內地與香港共同市場的國際化本身並不是一個終極目標，而是中國達致更平衡經濟、更有效市場開放及最終使人民幣更高度國際化的戰略的其中一環。

3.1　改善國家資產負債表的可能性

中國的國家資產負債表[20]上，在資產項目下，2007 年至 2013 年間的對外直接投資及其他境外資產（不包括國際儲備）不斷上升 —— 對外直接投資由 2007 年的人民幣 7,180 億元上升至 2013 年的人民幣 61,470 億元；後者則由人民幣 38,660 億元上升至人民幣 61,610 億元。若以其佔金融資產總值的百分比計算，對外直接投資的佔比日益增加，2013 年升至 1.7%，後者的佔比相若，但多少呈下降趨勢。相比之下，國際儲備在 2013 年國家資產負債表中佔金融資產的比重較高（7.5%）。國際儲備的金額後來雖從 2014 年的近期新高（38,430 億美元）回落，但在 2016 年仍維持於相對較高的水平（30,105 億美元）[21]。至於表上負債項目下，2007 年至 2013 年間外國直接投資及其他對外負債的金額亦不斷增加，兩者佔負債總額的百分比近年保持平穩，分別約 4% 及 1.5%（見圖 9）。

20　資料來源：Wind 資訊（一手資料來源：中國社會科學院）。

21　資料來源：中國國家統計局。

圖9：中國對外直接投資／外國直接投資及其他對外資產／負債（2007年至2013年）

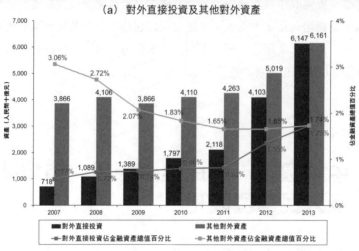

(a) 對外直接投資及其他對外資產

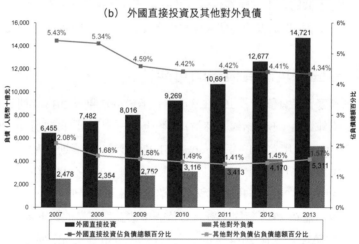

(b) 外國直接投資及其他對外負債

資料來源：Wind 資訊。

註：所有對外負債均為金融負債。

　　儘管對外投資不斷增長，經常賬的數據顯示中國 2010 年至 2014 年間的投資
收入為負數（見圖 10），與美國的國民經濟賬目所見明顯不同。圖 11 清楚顯示，
美國 2007 年至 2016 年的國際投資頭寸淨額負數愈來愈大，而投資收入則為正數
並漸增。美國大部分投資收入均來自組合投資（見圖 12）。

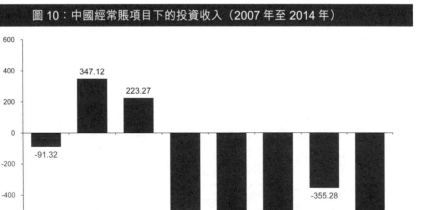

圖 10：中國經常賬項目下的投資收入（2007 年至 2014 年）

資料來源：Wind 資訊。

圖 11：美國的國際投資頭寸淨額及投資收入（2007 年至 2016 年）

資料來源：美國經濟分析局。

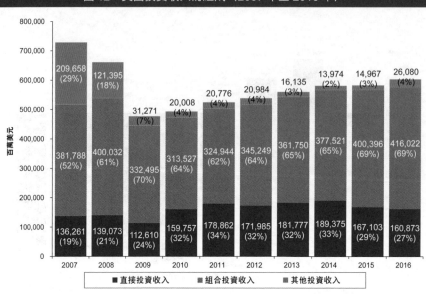

圖 12：美國投資收入的組成（2007 年至 2016 年）

資料來源：美國經濟分析局。

事實上，美國自 90 年代起二十多年來的收入淨額均錄得正數，而國際投資頭寸淨額則一直惡化。這勾起學者們要破解這個長久以來的謎團 —— 為何美國面向全球各國明明是個淨借貸國，但其對外頭寸卻能錄得淨收入？他們發現了兩大因素：（1）美國的對外股本結餘淨額是正數，對外債務結餘淨額是負數，而股本收益率高於債務收益率；及（2）美國對外直接投資資產所獲得的收益一直高於外國人在美國直接投資賺取的收益 [22]。諾貝爾獎得主保羅‧克魯曼（Paul Krugman）評論道：「美國資產通常是美國公司在外地的附屬公司⋯⋯ 當許多境外資金都購買（美國）國債時，⋯⋯美國資產的回報率要比美國負債所付出的回報高⋯⋯ 因此，美國在外地擁有的資產的收入一直高於外國在美國擁有的資產的付酬」[23]。還有一項研究發現，美國擁有的外國資產所賺取的收入與要向外國擁有的美國資產支付的收

22 Alexandra Heath 著作 "What explains the US net income balance?"〈為甚麼美國可有淨收入？〉，國際清算銀行工作文件第 223 號（2007 年 1 月）。

23 資料來源：保羅‧克魯曼於《紐約時報》的專欄 "The Conscience of a Liberal"〈自由主義者的良知〉中 2011 年 12 月 31 日的文章 "US Net Investment Income"〈美國投資收入淨額〉（https://krugman.blogs.nytimes.com）。

入迥異，除卻一些次要因素外，可以歸因於稅制的不同（美國稅率一般較美國擁有外國資產的所在國家為高），以及美國所投資的外國國家有較高風險（使得調整風險後的收益高於外國擁有的美國資產）[24]。同一研究亦發現，在股票及債務組合方面，申索與負債的平均收益幾乎相等（1990 年至 2010 年期間）。

　　美國的情況對中國極具參考價值，中國的情況好比是美國的反照：境外資產淨額正數，其投資收入淨額卻負數。中國有龐大的國際儲備，大部分投資於收益率極低的美國國債[25]。另一方面，中國以優惠政策吸引外國直接投資，付出了相對較高的成本，其對外直接投資又飽受當地一定程度的反對及監管障礙。此外，中國在歐美等發達國家的對外直接投資，其風險調整收益會不及在發展中國家（如中國）的外國直接投資。**美國的情況點出了一個可以改善中國國家資產負債表的方法，就是放寬對外組合投資，以期增進國際投資收入淨額。朝此方向，發展這一富於國際元素的內地與香港共同市場平台，應當極具前景**（見下文第 4 節）。

3.2　金融市場的進一步開放

　　正如上文第 2 節所論述，進一步開放金融市場是十三五規劃的一大政策線，國家亦投入了大量政策支持發展上海為國際金融中心。回望 2009 年，在上交所設立國際板的建議亦獲得政府的政策支持，好讓外國公司在當地交易所掛牌上市。此政策主張及對對外直接投資的政策支持，揭示了中國的經濟發展政策的進展 —— 首先是自 90 年代起讓中國企業走出去集資（及讓外國資金流入），進而讓國內資金通過對外直接投資及組合投資渠道如合格境內機構投資者（QDII）（2006年起）及滬深港通（2014 年起）等流出去，再而**日後可能容許外國企業進入內地市場集資**。

　　對開放市場的憧憬是雙向（對內及對外）的全面開放金融市場，讓外國資金投資於中國內地本土的金融產品，也讓內資金投資於外國金融產品。基於現時內地的金融市場體系及慣例尚與國際成熟市場有別，再加上內地部門對國內金融穩

24　Stephanie E. Curcuru 及 Charles P. Thomas 著作 "The return on US direct investment at home and aboard" 〈美國本土及對外直接投資的回報〉，聯邦儲備系統委員會國際金融討論文件第 1057 號（2012 年 10 月）。

25　中國外匯儲備的分佈並無正式對外披露。部分資料來源按中國的經濟數據估計約七成是美元資產。（資料來源：維基百科〈揭秘：中國 3 萬億美元外匯儲備是如何配置〉，http://finance.sina.com.cn/）。

定性有所顧慮,中國要從現時的有限度開放終至一天全面開放市場並非易事,亦不是朝夕可成。內地與香港共同市場平台體現封閉式系統,覆蓋範圍又可逐步擴容,可對上述開放市場的漫長進程提供助力(見下文第 4 節)。

3.3　邁向資本項目全面可兌換

有序實現人民幣資本項目可兌換是國家在十三五規劃訂明的目標,而開放金融市場正正是實現此目標的關鍵。有分析指,在國際貨幣基金組織分類為資本項目的 40 個子項中,中國只有數項仍不可兌換 [26],主要涉及非居民境內發行股票、貨幣市場工具、衍生工具及其他工具等。為使人民幣成為可兌換及自由使用的貨幣,正如人行 2015 年年報所載,進一步開放金融市場的主張可包括:

- 進一步放寬及促進 (1) 境內居民投資海外金融市場及 (2) 外國投資者投資內地金融市場;以及
- **讓合資格外國公司在內地市場發行股份。**

為此,具集資市場互通功能的內地與香港共同市場平台當可助人民幣資本項目可兌換更進一步(見下文第 4 節)。

4　新股通 —— 突破性機遇

內地和香港股市在國際化發展上碰到的瓶頸,包含集資市場互聯互通(新股通)及交易市場互聯互通(滬深港通)的共同市場互聯互通模式很可能會是一解決之道。圖 13 為此模式的概念圖。

26　其餘項目為可兌換或基本可兌換或部分可兌換。資料來源:〈人民幣資本項目開放的現狀評估及趨勢展望〉,載《第一財經》(2016 年 4 月)(http://www.yicai.com/)。

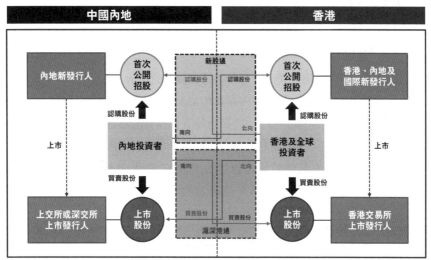

圖 13：包含集資及交易市場互聯互通的內地與香港共同市場互聯互通模式的概念圖

註：新股通及滬深港通實現全程封閉的資金流系統。

　　新股通的概念是允許內地投資者在香港市場認購首次公開招股（南向），以及允許香港的國際投資者認購內地市場的首次公開招股（北向）。所發行的首次公開招股股份上市後，另一方市場的投資者可透過現有滬深港通機制進行買賣。按此聯通模式，**循新股通認購股份及經滬深港通買賣股份將有效實現一個密封式機制**。

　　相較於 QFII 及 RQFII 機制，通過滬深港通投資內地股市的外地投資者毋須向內地監管機構申請特別資格或投資額度或在內地券商及託管商開立戶口。經滬深港通買賣內地股票沿用其於香港股市交易的慣用方式。同樣地，內地投資者經滬深港通買賣港股沿用其內地市場慣用的交易方式。再者，滬深港通讓內地個人投資者直接參與買賣港股而毋須透過 QDII 的產品作投資渠道。源於滬深港通所提供的此等交易效率與便利，若服務一級集資市場新股認購的新股通與服務二級市場股票買賣的滬深港通作無縫對接，對投資者會特別有利。

　　實施新股通將有助改善內地市場的國際投資者參與度，以及香港市場的國際發行人基礎，使兩個市場更國際化。此外，新股通也會為內地投資者帶來更多國內市場短期內未能提供的投資機會。更重要的是，正如前文第 3 節所述，內地與香港共同市場模式有助達成中國更遠大的目標：改善國家資產負債表、進一步雙

向開放市場以及人民幣國際化。加入新股通將共同市場的聯通範圍擴大，相信將
會是一大突破。新股通的潛在裨益論述如下。

對內地市場及投資者的潛在裨益

(1) 帶來新的環球資產配置渠道，可改善國家資產負債表

2006 年至 2015 年十年間，國內存款增加至約人民幣 337,080 億元（約 51,150
億美元），複合年增長率達 14%[27]。同期對外直接投資的複合年增長率為 32%，即
使 2006 年推出了 QDII 計劃，對外證券投資的複合年增長率卻錄得負數 (-0.2%)[28]。
對外證券投資佔境內存款總額的比例從 2011 年起維持於 5% 左右的相對偏低水
平，但對外直接投資則大升至 2015 年底約佔 21%（見圖 14）。

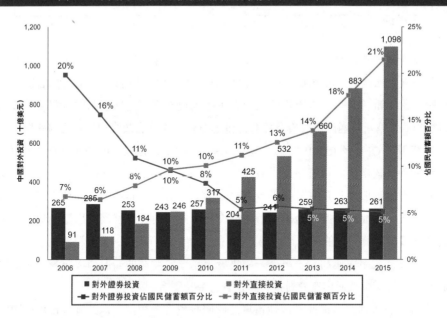

圖 14：中國內地對外直接投資及證券投資（2006 年至 2015 年）

資料來源：Wind 資訊。

---

27 資料來源：Wind 資訊。
28 複合年增長率乃按 Wind 資訊的年度數據計算得出。

　　此外，人民幣自 2015 年 8 月起呈貶值之勢 [29]（見圖 15），市場恐怕跌勢會持續一段時間。內地投資者（包括政府機關及企業的投資部門）紛紛尋求非人民幣資產，以分散投資組合，也可於人民幣繼續貶值時作對沖。

圖 15：人民幣指數及人民幣兌美元匯價每日收報（2014 年至 2017 年 5 月）

註：RXY 全球在岸人民幣指數是香港交易所與湯森路透合作推出的指數，用於計量在岸人民幣兌一籃子主要
　　國際貨幣的表現。CFETS 人民幣匯率指數是中國外匯交易中心（CFETS）計量人民幣兌一籃子主要國際
　　貨幣的官方指數。
資料來源：湯森路透。

　　相對於國內存款的充裕，境內投資產品的供應卻不足以應付人民對投資保值及為資產增值的需要。就如一學術研究所載，這引致中國內地自 1990 年以來的資產短缺情況，導致資產泡沫這樣的後果，其現象表現於地產市場、股市甚至消費品的價格飆升 [30]。該研究解釋資產短缺程度是以國內儲蓄（即產資需求）減去資產供給所佔國內儲蓄的百分比來衡量；而資產供給是債券、股票、借出款、短期存款變化和持有的非本幣資產的總額。作為一參考點，2015 年於滬、深交易所的股本

29　觸發點為 2015 年 8 月 11 日人行改革人民幣兌美元匯率中間價報價機制，使之更加市場化。
30　楊勝剛、梁粲，《中國資產短缺問題研究》，2015（http://www.sinoss.net）。

集資總金額約為人民幣 15,400 億元 [31]，少於 2015 年底國內儲蓄額的 5%，與 2015
年對外證券投資的水平相若（見圖 14）。

　　由此可見，內地投資者對全球資產配置有迫切需求。再者，要改善國家資產
負債表，亦需增加海外投資的收入。「一帶一路」策略會促進對發展中國家的投資，
固然可增加海外直接投資的收入 [32]，但加大海外投資組合將可帶來潛在的龐大收入。

　　除滬深港通的南向交易外，現行的 QDII 計劃是海外證券投資的唯一合法渠
道，但有額度限制，亦須經審批。事實上，QDII 的投資額度自 2015 年 3 月以來
並無增長，至 2017 年 6 月仍維持於 899.93 億美元的水平（見圖 16），與 2015 年
底國內存款相比只佔不足 2%。

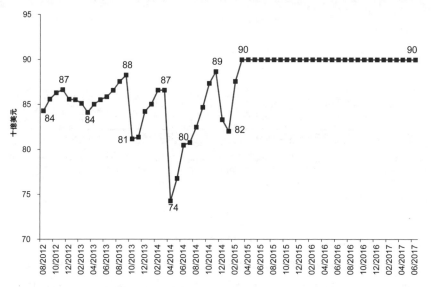

圖 16：已批准的 QDII 投資額度（2012 年 8 月至 2017 年 6 月，月底數字）

資料來源：國家外匯管理局網站。

31　2015 年，於上交所的股票發行集資總額為人民幣 871,082 百萬元，於深交所的（包括主板、中小企業板、創業板）
　　則為人民幣 668,902 百萬元（資料來源：滬交所、深交所網站上的每月市場統計數據）。2015 年底的國內儲蓄額
　　為人民幣 337,080 億元（資料來源：Wind 資訊）。

32　「一帶一路」計劃包括「絲綢之路經濟帶」及「21 世紀海上絲綢之路」計劃。「絲綢之路經濟帶」連接中亞、西亞、
　　中東至歐洲，伸延至南亞及東南亞。「21 世紀海上絲綢之路」則橫跨東南亞、大洋洲及北非。中國國際貿易研究中
　　心在 2015 年 8 月發佈的《「一帶一路」沿線國家產業合作報告》中，列出了 65 個會參與一帶一路計劃的沿線國家。

　　除 QDII 外，國家外匯管理局亦特設「綠色通道」，若干國內投資者 (主要是內地基礎投資者) 可經此渠道向外管局申請特別許可，以認購將於香港上市的內地企業的首次公開發售股份。這是中國郵政儲蓄銀行首次公開招股時開始實行的做法，自此曾應用於多家內地企業首次上市，但始終屬特事特辦，特別許可亦只適用於特定企業，也須符合國家外匯管理局的特別規定[33]。

　　南向新股通將使內地投資者能認購將於香港上市的國際公司的新股，等同為國內投資者開闢多一個正式的環球資產配置渠道，更可能較現有渠道廣闊。與現時只有交易市場聯通相比，南向新股通將可在更大程度上改善內地資金的海外投資組合。

(2)　促進雙向市場開放

　　能否成功開放內地本土市場供國際公司集資，取決於內地市場對潛在發行人在集資需要及資金成本考量上有多大吸引力。一些有意在中國擴展業務的大型國際公司可能會為了國內上市地位的品牌效益而感興趣，但這類的發行人未必很多，容易無以為繼。再者，內地股市的監管框架與國際發達市場大相徑庭，即使設立了國際板，合規成本也可能很高，大部分有意上市的外國發行人恐怕會望而卻步。

　　相形之下，透過內地與香港共同市場模式下的新股通，外國發行人向內地投資者發售股份時可遵從其更熟悉、更國際化的香港股市規則及標準。正如下圖 17 所示，共同市場的聯通平台[34] 能讓內地境內資金及金融產品流出，反之亦能讓國際資金及金融產品流入中國，任何一邊的共同市場投資者及發行人都不用強為適應另一邊市場的做法。

33　據市場參與者所述，在「綠色通道」許可下，國家外匯管理局要求發行人在境外上市募集資金結束後要調回募集所得資金，並要求境內基礎投資者在出售認購所得股份後調回部分 (若非全部) 資金至國內。一般認為這是個苛刻的「雙重結匯」規定。
34　某些項目需要經過監管部門的批准。

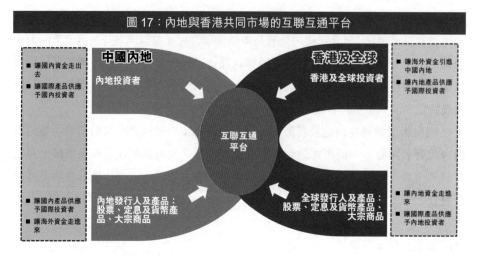

圖 17：內地與香港共同市場的互聯互通平台

(3) 人民幣資本項目可兌換更進一步

共同市場的南向新股通將等同在資本項目下實行多一項的人民幣可兌換，即讓非居民於境內發行股票或股本類證券。日後，共同市場大可續推其他合適的聯通計劃（例如金融衍生產品的發售），在其餘資本項目實現人民幣可兌換。

儘管長遠而言，中國總有其他方法實現人民幣資本項目全面可兌換，但透過新股通及共同市場的其他聯通計劃，則有助以可控方式（見下文第 (7) 點）加快這個進程。

(4) 發展國際投資者基礎

正如上文第 2.1 節所述，外國人在內地股市的持股量佔上交所及深交所流通市值總額不足 0.3%，而外國投資者在香港股市的交易佔比則達 40%。**北向新股通有助開拓內地的國際投資者基礎**，那是股市國際化不可或缺的要素。

通過北向新股通的首次公開招股將會對所有國際投資者開放。相較於只開放於合格外國投資者（QFII 及 RQFII）的國內首次公開招股而言，北向新股通的首次公開招股預料會涉及發行人和證券業（國內及國際業界）更大的國際市場推廣力度，國際投資者將能獲得大量發行人資訊，可能較針對國內投資者的國內招股更全面，從而更清楚了解投資價值。再者，外地投資者可以經新股通於一級集資市場獲取相對較大持股量而不會像在二級市場買入大量股份般對股價造成影響。

(5)　為內地企業帶來更多上市機會

在中國內地相對較快的經濟發展下，國內存款不斷增長，另一方面現有及新成立企業亦不斷需要資金作發展及擴充之用，兩邊似乎正可各取所需。然而，內地股市現時的首次公開招股步伐未必能適時應付大量的企業申請。據彭博報導[35]，截至 2016 年 9 月，仍待中國證監會批准在上交所或深交所上市的首次公開招股申請人逾 830 名。中國證監會努力加快審批[36]，至 2017 年 2 月已將輪候隊伍縮短至約 700 名，但申請人一般仍需輪候 18 個月或以上[37]。按地方媒體報導，還有待省級審批方可輪候上市的公司申請更超過 700 宗[38]。官方交易所數據顯示上交所及深交所於 2016 年的新上市總數分別有 103 及 124 宗，2017 年截至 5 月則分別有 103 及 108 宗。按 2017 年已加快處理新股上市申請的速度（約每年 500 宗）估計，現有的申請者隊伍（包括省級輪候申請）可於三年內消化，但這並未計中國經濟發展而不斷有新申請者加入，更不用說國家在熊市時可能會實施暫緩新股上市的政策（有先例可援）。

對於在內地輪候上市的內地企業來說，到香港首次公開招股無疑是另一選擇，但這意味有關企業選擇面向主要是香港及全球投資者，很大程度上要捨卻內地投資者。若落實南向新股通，在港上市的內地企業亦可向內地投資者發售股份。如此，在香港首次公開上市會成為以內地投資者客源為目標的內地企業的另一個實際選擇，同時亦讓他們有機會接觸全球投資者。故此，**新股通有助舒緩內地市場發展現階段對首次公開招股的制約**。

(6)　培育內地投資者基礎

香港市場的國際監管框架及市場慣例，對內地投資者而言是一個極具價值的訓練場。2006 年推出的 QDII 計劃使內地機構投資者受惠，亦間接透過 QDII 產品惠及內地散戶投資者。2014 年 11 月推出滬港通後，更進一步為內地散戶投資者

35　"Few in China's IPO queue likely to benefit from fast-track reform"〈中國加快處理 IPO 申請受惠的輪候者寥寥可數〉，《彭博》，2016 年 9 月 12 日。

36　2016 年 9 月，中國證監會讓在任何全國 592 個窮困地區註冊的公司插隊，但似乎只為財富再分配，而非舒緩首次公開招股的瓶頸（《彭博》，2016 年 9 月 12 日）；2017 年 2 月，據報中國證監會考慮為全國最大科技公司提供捷徑插隊。

37　"China to let big tech firms jump IPO queue"〈中國讓科技大企插隊上市〉，《海峽時報》，2017 年 2 月 25 日。

38　"Few in China's IPO queue likely to benefit from fast-track reform"〈中國加快處理 IPO 申請受惠的輪候者寥寥可數〉，《彭博》，2016 年 9 月 12 日。

直接提供在交易市場買賣方面的練習場。至於在集資市場認購新股，內地的市場慣例與國際迥異。部分出於供求失衡，內地投資者可能都習慣了新股在內地上市時股價例升。**南向新股通可讓內地投資者有機會接觸國際市場的做法，在新股認購及上市後股價變動方面（可升亦可跌）獲得國際經驗，有助培育內地投資者趨向成熟。**

(7)　資本外流的風險可控

與滬深港通的交易市場模式設計相同，新股通涉及的資金流也是在封閉的系統內完成，使出售認購股份時（假定透過滬深港通），資金會回流申購股份的投資者的原有市場。如此便可舒緩內地市場資金外流的憂慮。

新股通對香港市場及全球投資者的潛在裨益

(1)　吸引國際公司上市

南向新股通等同打開了內地投資者的資金池，可吸引全球有意的發行人在港上市集資。內地有龐大的境內存款基礎，且業務機會處處，對於想進行大規模首次公開招股的國際公司及有意在中國發展業務的公司來說，透過南向新股通在香港上市相信極具吸引力。此外，有內地投資者作為潛在認購者，較小型的國際公司也會更有信心可於香港成功上市。

(2)　有更多投資者參與使市場交投活躍

為使共同市場的股份招股及交易循環更為完備，運作上集資市場的新股通預期會與交易市場的滬深港通相連，使兩邊市場的投資者透過新股通認購的股份可透過滬深港通買賣。在新股通的帶動下，香港市場預期會有更多國際公司上市，也因為內地投資者的參與而擴大了流動資金池，市場活動更見活躍。全球投資者都會樂見集資市場及交易市場的流動資金增加。

(3)　活躍於集資及交易市場活動的市場中介機構同樣受益

集資及交易市場活動增加，香港的市場中介機構（包括投資銀行、律師事務所、會計及審計師行以及證券經紀）就有更多業務機會。

新股通對內地與香港共同市場的潛在裨益

實行新股通有助完善內地與香港共同市場的股票市場生態系統，是聯繫國際投資者與內地公司及內地投資者與國際公司的突破性機會，有助共同市場進一步國際化。**內地與香港共同市場可以擴容，自然能成為內地與國際投資者的金融超市，為他們提供世界各地各式各樣的金融產品。**

5 開通新股通的考量

正如滬深港通機制，新股通亦是一全新概念，在實務問題及市場影響方面均需要審慎考慮，但只要模式設計得當，各種挑戰性難題大都可迎刃而解，而裨益將會超過所需成本。各種挑戰詳述如下。

(1)　市場競爭或會增加

如境外投資者透過北向新股通增加參與內地首次公開招股，可能會被視為在內地新股市場股切內需之上再增動力，市場或憂慮原已相當熾熱的內地新股市場會火上加油。但其實正好相反，只要模式設計得當，**新股通將為內地一級市場帶來更公平的定價及更健康的發展。**根據市場現行的新股配發流程，QFII 及 RQFII 在內地首次公開招股市場中往往只認購到少量股份。內地市場的新股配發機制與國際慣例相當不同，或限制了外地投資者於其一級集資市場的參與度。北向新股通的設計可着力提高新股認購及分配以及定價流程等方面的外資參與度，這將有助內地一級市場逐步貼近國際常規。

第二，內地市場或會憂慮新股通會加劇香港爭取內地發行人上市的競爭，令發行人選擇香港而不在內地市場上市。但這不是大問題，原因是新股通不會影響中國企業 (特別是 H 股公司) 在海外上市的現行監管規定。新股通對發行人的額外吸引力會是打開內地投資者的股份認購需求。**事實上，內地有大批申請人輪候上市，新股通反有助舒緩內地處理首次公開招股的壓力。**

第三，市場或會關注香港市場與內地正在籌劃的國際板在爭取海外公司上市方面可能存在競爭。就此而言，內地國際板尚未確定發行人上市資格要求及運作詳情，**構思中的新股通相信不會對其有任何影響**。相反，市場人士認為，啟動南向新股通後，香港爭取國際公司來港上市其實是對海外證券交易所構成更多競爭。事實上，內地股市要達至國際化，需要先行先試多樣的措施，再適時配合市場發展作出適當調整。**提供新股通的共同市場就有如中國的一種離岸國際板，可與其在岸國際板同時運行。**

至於香港市場，市場一般認為南向新股通所增加的流通量於新股認購方面不會構成競爭，反而有利。

(2) 監管及投資者保障

市場監管是開通新股通前需要審慎考量的重要範疇，當中包括發行人及投資者資格，以及共同市場雙方的證券交易所、市場監管機構、中介機構、發行人與投資者的責任及義務。**最終目的是提供一個公平的市場環境，具備充分的投資者保障及風險管控。**

內地與香港股市的一級市場制度（包括新上市資格、上市要求及招股章程的披露要求等）不盡相同。透過南向新股通機制進行以內地投資者為對象的首次公開招股，或要施加額外的發行人資格及披露要求 [39]。透過北向新股通機制進行以香港本地及環球投資者為目標的首次公開招股，亦可能要有額外披露要求。此外，對於從事戰略產業的內地企業，中央政府或有若干外資擁有權限制，故北向新股通的實施或會涉及若干有關首次公開招股的特定資格要求或外資認購的特定限制。

至於新股通下若發行人出現合規問題而令投資者利益受損，應採用哪個市場的投資者保障機制、雙方市場監管機構的角色以及發行人、證券交易所及中介機構的責任等問題，也需要一一釐清。此外，市場中介機構亦關注跨境投資者的信貸風險。然而，只要制定明確穩健的監管框架及操作設計得當，上述問題都應該可以解決。

---

39 舉例：發行人向跨境（例如英美市場）投資者發售新股，須遵守進行跨境發售的市場當地的若干監管規定（包括資料披露要求）。

此外，內地與香港股市須同時進行**大量投資者教育**，令兩地投資者都了解對方的一級市場常規和投資行為，保障其本身的投資。

(3)　操作事宜

不論是北向還是南向，新股通都會涉及將首次公開招股的認購程序延伸至共同市場的另一方。有別於滬深港通下的二級市場交易毋須修訂上市股份的主場交易、結算及交收常規，新股通下首次公開招股的本土市場股份認購及分配的常規必須審慎設計，以容納跨境投資者。

有待解決的問題包括但不止於：(1) 跨境散戶投資者可否經新股通認購股份，抑或新股通是否只開放予跨境機構投資者？ (2) 新股通股份如何配發予跨境投資者？ (3) 跨境投資者會是一個獨立的股份認購組別，還是與主場投資者的申購一併處理？ (4) 跨境投資者認購股份應遵守不同的市場規則還是沿用首次公開招股所屬市場的規則？ (5) 跨境投資者由首次公開招股所屬市場的中介機構提供服務，還是由投資者所屬市場的中介機構提供服務？ (6) 為跨境投資者提供服務的中介機構是否要遵守不同的監管規定 (例如：「認識您的客戶」規則及配售指引)？

基於香港與內地的首次公開招股程序及市場慣例差別相當大，**新股通的操作模式須兼顧發行人及投資者的利益，以及保持市場公平、公開、公正的基本原則，預計在設計上會相當費力。**

首次公開招股後的股份買賣有必要**無縫銜接滬深港通**，故監管及操作框架須為此作好充足準備。在香港，券商一向有為認購首次公開招股股份的投資者提供融資的做法，有市場人士要求滬深港通容許大宗交易以利便該等融資支援服務。原因在於若無大宗交易設施，投資者如在二級市場大量出售已認購股份來回付券商的融資，會影響股價。實行新股通而未有於滬深港通容許大宗交易，對須獲券商融資的跨境投資者而言會是一絆腳石。在銜接新股通與滬深港通時，此等操作細節必須顧及。

6 總結

滬深港通啟動，內地與香港的股票二級市場交易正式互聯互通，但若一級（新股）市場尚未聯通，內地與香港共同市場終究不完整。推出新股通可完善這個共同市場的基本功能，就是讓共同市場中的投資者可買賣股票、發行人也可集資。新股通亦可優化內地與香港股市以至整個共同市場現時尚未真正國際化的層面。新股通啟動後，共同市場可更有效地支持中國的宏觀藍圖發展，達致經濟更平衡、市場更開放而最終人民幣更國際化。

新股通對共同市場兩邊都有益處。對內地而言，新股通可（1）為內地投資者提供全球資產配置的新機遇，從而改善國家的資產負債表；（2）以較低成本利便市場雙向開放；（3）在人民幣資本項目自由兌換方面向前邁進一步；（4）支持市場發展國際投資者基礎；（5）為內地企業帶來更多上市機會；以及（6）協助培育內地投資者基礎；而又能同時以適當監控措施減低潛在風險。對香港而言，新股通有助吸引海外公司來港上市、增加投資者參與度而進一步激活市場，並為市場中介機構帶來更多商機。

實施新股通較滬深港通計劃涉及更多包括監管及營運方面的議題。不過，相信只要機制模式設計合宜，符合內地與香港共同市場的最佳利益，這些問題大致上都可迎刃而解，最終會惠及中國國民經濟賬目以至人民幣國際化這大戰略。

英文縮略詞

ETF　　交易所買賣基金（Exchange-traded fund）

QDII　　合格境內機構投資者（Qualified Domestic Institutional Investor）

QFII　　合格境外機構投資者（Qualified Foreign Institutional Investor）

RQFII　人民幣合格境外機構投資者（Renminbi Qualified Foreign Institutional Investor）

備註

本研究報告的內容有參考香港一級市場參與者（包括投資銀行、券商、律師行、會計師事務所及基金經理）對新股通概念的意見及反饋。

第4章

中華交易服務
港股通精選 100 指數

—— 互聯互通機制下港股通投資的
重要指標

2017 年 11 月 24 日

摘 要

滬港通及深港通（合稱「滬深港通」）先後於 2014 年 11 月及 2016 年 12 月開通，標誌着內地與香港證券共同市場基本形成，內地投資者今後投資海外市場不但機會比從前多，限制亦較以往透過合格境內機構投資者計劃（QDII）等渠道為少。統計數字顯示內地投資者對透過滬深港通的港股通買賣香港上市股票的興趣日濃。

隨着投資者對共同市場的興趣日增，開發相關指數服務支援這個市場的持續增長及發展亦是理所當然。指數不但可追蹤本地／區域／全球市場、市場個別板塊或跨市場表現，更日益用作被動投資工具如交易所買賣基金（ETF）或指數期貨及期權等衍生產品的相關資產，投資者可藉此參與個別市場或對沖個別市場中的投資。

中華交易服務港股通精選 100 指數（中華港股通精選 100）是滬深港通相關指數中專門追蹤港股通合資格股票（港股通股票）的指數，具有以下特色：

(1) 以市值及成交額計，於港股通股票的覆蓋率相對較高；

(2) 追蹤純香港概念的投資，就非同時在內地上市的香港上市股票而言有很高的代表性，因此純反映內地以外地區的投資機會，與國內證券市場的相關性僅屬溫和；

(3) 港股通股票中增長型板塊股票的覆蓋率較高，例如內地民企及新經濟行業的股票；

(4) 由於其成份股分佈的關係，指數自推出以來大部分時間的市盈率均較恒指及恒生國企指數為高，但股息率卻較低，回報率的波幅亦較香港及內地主要指數低。

基於這個指數能高度代表港股通股票及新經濟增長型企業股票，其成份股的潛在投資機會仍有待透過港股通獲進一步發掘。朝此方向，中華港股通精選 100 可能會是開發作港股通投資的 ETF 等被動型投資工具的一個有用基準指標。

1 內地投資者在內地與香港證券共同市場的投資機會

1.1　內地與香港證券共同市場的形成 [1]

互聯互通機制試點計劃（「試點計劃」）於 2014 年 11 月啟動，首先推出**滬港股票市場交易互聯互通機制（滬港通）**，為海外投資者投資內地股市及內地投資者投資海外資產開闢了全新的渠道。透過滬港通，在計劃容許的範圍內，通過連接內地上海證券交易所（上交所）與香港聯合交易所有限公司（聯交所）及相關結算所的交易及結算基礎設施，香港及海外投資者可買賣在上交所上市的證券（「**滬股通**」或「**北向交易**」），內地投資者可買賣在聯交所上市的證券（「**港股通**」或「**南向交易**」）。

滬港通成功推出及運作暢順後，**深港股票市場交易互聯互通機制（深港通）**於 2016 年 12 月開通。就像滬港通，透過連接聯交所與深圳證券交易所（深交所）以及其結算公司的基礎設施，深港通容許「北向交易」，讓香港及海外投資者可買賣深交所上市證券，及「南向交易」，讓內地投資者可買賣聯交所上市證券。除滬港通批准的「大型」及「中型」股票外，深港通的合資格證券範圍亦包括小型股票 [2]。（下文滬港通及深港通統稱「滬深港通」。）

在深港通推出後，縱使只限指定合資格股票範圍內運作，滬深港三地的「共同市場」模式基本成形。由於股票市場交易互聯互通機制可以擴容，這無形中打開了一個潛在的內地與香港股票共同市場，其股份總值達 105,140 億美元（2016 年底）、日均股份成交約 843 億美元（2016 年），於全球交易所中按市值計排名第二（僅次於紐約證券交易所）、按股份成交額計排名第二 [3]。

1　亦可見本書第 2 章〈滬港通與深港通下的互聯互通 —— 內地及全球投資者的「共同市場」〉。

2　有關滬深港通計劃的合資格證券，見香港交易所、上交所及深交所的網站。

3　國際證券交易所聯會（WFE）數據（WFE 網站 2017 年 1 月 20 日取得的市值資料及 2017 年 3 月 1 日取得的成交資料）。日均成交額按 WFE 2016 年數據的合併成交額以及內地市場交易日總數 244 日計算。排名按 2016 年截至 12 月的合併成交額計算。

1.2　對內地投資者的價值

　　滬深港通計劃推出前，內地投資者要投資海外市場，合格境內機構投資者
(QDII) 計劃是唯一正式渠道，這須受國家外匯管理局 (外管局) 通過的投資額度
規限。內地合資格的機構投資者在獲得 QDII 資格許可及認可額度後可直接投資
海外，但內地個人投資者擬投資海外市場，則要透過既有的 QDII 產品進行。不
過，QDII 的投資額度總額自 2015 年 3 月起至 2017 年 10 月一直不曾增加，維持
於 899.93 億美元不變[4]。

　　滬深港通下的港股通為內地個人及機構投資者打開了直接投資海外資產的一
個全新規範化渠道。有別於 QDII 計劃，滬深港通不設總額，只有適用於淨買盤
的每日額度[5]。此渠道全程封閉、每日額度的使用受審慎實時監控，而又在無總額
下提供相當的靈活度。由於每日額度按淨買盤計算，而且每日重訂一次 (交易時段
內若觸及額度，當日餘下時間將只接受合資格證券的跨境賣盤)，整個交易日基本
上不會有股票成交額限制。滬深港通下的共同市場因此等同擴展了內地投資者可
投資資產的覆蓋面，能為他們提供潛在回報不俗的投資機會。

　　滬深港通下的港股通交易近一年來不斷穩步增長，突顯**內地投資者透過此渠
道買賣香港上市股票的興趣強烈且有增無減**。這將令市場對相關服務的需求不斷
增加，包括指數服務及相關投資工具。(有關滬深港通下的港股通成交資料，見附
錄一。)

2　適用於投資共同市場的指數服務

2.1　內地與香港股票共同市場的指數編制

　　為向內地與香港共同市場提供更佳的相關市場服務，香港交易及結算所有限

4　資料來源：外管局網站。

5　根據每日額度，滬股通及深股通股票每日「淨買盤」上限為人民幣 130 億元，滬港通及深港通的港股通股票則各
　　為人民幣105億元。

公司（香港交易所）、上交所和深交所於 2012 年共同成立**中華證券交易服務有限公司（中華交易服務）**。中華交易服務最先開展的，是覆蓋滬深港市場的跨境指數，為開發惠及內地及全球投資者投資共同市場的可交易指數產品奠下基礎。下圖 1 顯示中華交易服務現時的指數系列。

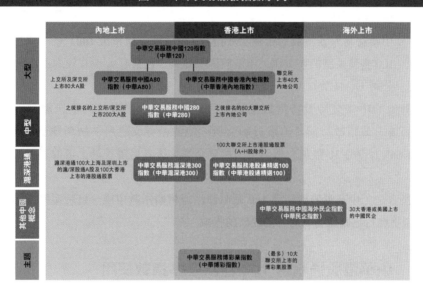

圖 1：中華交易服務指數系列

資料來源：中華交易服務網站。

服務內地與香港共同市場的中華交易服務指數有兩大類：

(1) **中華交易服務中國指數系列**——由內地及 / 或香港上市的內地企業股組成

- **中華交易服務中國 120 指數（中華 120）**——成份股為 120 家大型企業，其中 80 家為內地交易所上市的 A 股公司，40 家為聯交所上市的內地公司

- **中華交易服務中國 A80 指數（中華 A80）**——成份股為中華 120 內的 80 家 A 股公司

- **中華交易服務中國香港內地指數（中華香港內地指數）**——成份股為中華 120 內的 40 家內地公司

- **中華交易服務中國 280 指數（中華 280）**——成份股為 280 家中型企業，

規模位於中華 120 的 120 家大型企業之後，包括 200 隻內地交易所上市的 A 股及 80 家聯交所上市的內地公司

(2) 中華交易服務滬深港通指數系列 —— 由內地及／或香港上市的滬深港通合資格股票組成

- **中華交易服務滬深港 300 指數（中華滬深港 300）** —— 成份股為上交所及深交所各自首 100 隻最大型的滬深股通 A 股及香港上市首 100 隻最大型的港股通股票
- **中華交易服務港股通精選 100 指數（中華港股通精選 100）** —— 成份股為 100 隻最大型的聯交所上市港股通股票，不包括有 A 股在內地上市的 H 股公司（A+H 股）

其他的中華交易服務指數包括中華交易服務中國海外民企指數及中華交易服務博彩業指數，前者追蹤於聯交所、紐約證券交易所、納斯達克或 NYSE American[6] 上市的 30 隻最大型的中國民營企業，後者追蹤香港上市博彩股的整體表現。

換言之，**中華港股通精選 100 是滬深港通相關指數中唯一追蹤滬深港通計劃下合資格進行港股通交易的香港股票的指數。**

2.2 中華港股通精選 100 作為基準指數使用

股票指數（或股市指數）以個別股票市場的上市交易股票的價格變動為基礎，用以計量該市場或其市場板塊的表現。股票指數一向被投資者廣泛用作基準，將本身或財務管理人的投資組合表現與股票指數所計量的整體市場或市場板塊的表現作一比較。

股票指數因應市場需求而創設、制定、計算、管理及發佈。舉凡投資者有興趣投資某個別市場或市場板塊又或作跨市場投資，對股票指數就會有需求，以作為評估其投資表現的基準。在 19 世紀創設、用以計量紐約股市表現的道瓊斯工業平均指數（道指）據知是首隻股票指數。如今，市場上已有不同類型的股票指數涵蓋不同的市場範疇（見表 1）。

6　早年稱美國證券交易所（American Stock Exchange），比較近期稱 NYSE MKT。

表 1：股票指數的主要類別		
股票指數類別	性質	例子
「國家」指數	計量個別國家或地區的主要股市（或主要證券交易所）的表現	**美國**：道指及納斯達克 100 指數； **英國**：富時 100 指數； **日本**：日經 225 指數； **香港**：恒生指數（恒指）； **上海**：上證綜合指數； **深圳**：深證綜合指數；
「板塊」指數	計量個別交易所市場特定板塊的表現	**恒生綜合規模指數**：恒生綜合大型股指數、恒生綜合中型股指數、恒生綜合小型股指數； **深交所**：中小板綜合指數、創業板綜合指數
「區域」指數	計量按地理或工業化／收入水準界定的特定地區的股市表現	歐盟 STOXX50 指數 [7]、MSCI 新興市場指數 [8]
「行業」指數	計量特定行業股票的表現，可以是同一市場或不同市場的股票	恒生行業分類指數 [9]、STOXX 亞太 600 行業指數 [10]
「全球」指數	計量全球多個地區的股市表現	MSCI 世界指數 [11]、標普全球 100 指數 [12]
「主題」及「策略」指數	根據若干投資主題或策略而設立的指數	追蹤香港上市中資企業的恒生中國企業指數（恒生國企指數）[13]、恒指波幅指數

　　指數按計算方法會有多種版本，其分別可以在於成份股權重的釐定方法及股息的計算方法。最常見的包括只考慮成份股價格的**價格回報指數**，和計及將股息再投資的**總回報指數**。另一分別是在權重方式 —— 可以只按價格、或按全市值加權、或按自由流通量調整的市值加權。道指是**價格加權**指數的表表者 [14]；恒指則採

7　歐盟 STOXX 50 指數是歐元區的藍籌股指數，涵蓋 11 個歐元區國家的 50 隻股票（資料來源：STOXX 網站 https://www.stoxx.com/）。

8　MSCI 新興市場指數涵蓋 24 個新興市場國家的大中型企業，成份股 843 隻（資料來源：MSCI 網站 https://www.msci.com）。

9　恒生行業分類指數包括金融、公用事業、地產以及工商業分類指數。

10　STOXX 亞太 600 行業指數系列分別涵蓋 19 個行業，成份股來自日本、香港、澳洲和新加坡等多個亞太市場（資料來源：STOXX 網站 https://www.stoxx.com/）。

11　MSCI 世界指數涵蓋 23 個發達市場國家的大中型企業，成份股 1,654 隻（資料來源：MSCI 網站 https://www.msci.com）。

12　標普全球 100 指數計量全球股票市場中最重要的跨國藍籌公司的表現，100 隻成份股均具高流動性（資料來源：標普道瓊斯指數網站 https://us.spindices.com）。

13　恒生國企指數原是用來追蹤香港上市 H 股公司。2017 年 8 月，編算該指數的恒生指數有限公司公佈，指數將於 2018 年 3 月至 2019 年 3 月分五階段加入紅籌股（內地以外地方註冊成立的國家控股企業）及內地民營企業（內地民企或 P 股）。

14　這很大程度是因為歷史原因，因道指推出時尚未有電腦作指數的自動計算。

用自由流通市值加權法，個別證券的權重上限為 10%[15]，這是股票市場中最常見的指數計算方法。

中華港股通精選 100 被視為「主題式」基準指數，所計量的是滬深港通下港股通的投資表現。

除可計量股市表現外，股票指數（或用於股票以外的其他資產，如地產及商品市場的一般指數）更日益用作被動投資工具如交易所買賣基金（ETF）或指數期貨及期權等衍生產品的相關資產，投資者可藉此參與個別市場或對沖個別市場中的投資。ETF 近年尤其愈來愈普及 —— 根據 ETFGI[16] 的資料，2016 年底有 3.548 萬億美元投資於 6,630 隻分佈全球的 ETF 或交易所買賣產品（ETP），2016 年淨流入為 3,893.4 億美元，是連續 35 個月錄得淨流入[17]。根據貝萊德的資料，2017 年首季全球的 ETF 市場錄得淨流入 1,891 億美元，創下歷史最高紀錄，當中有 1,091 億美元流入股票 ETF[18]。

ETF 是追蹤指數、商品、債券或一籃子資產的有價證券，如同普通股一樣在證券交易所交易。投資 ETF 等同投資其相關資產。若 ETF 的相關資產是股票指數，相關資產組合會與有關指數中成份股的相對權重相同。ETF 日益普及，主因是其能提供一簡單途徑讓投資者可投資幾乎任何資產類別、地區或行業，而所涉的成本比積極管理的基金為低[19]。

因此，針對個別資產組別的市場或其可投資的市場板塊編算合適的指數，不僅是計量市場表現所需，還可通過設立以該指數為基礎的 ETF 去協助促進或利便投資該市場或資產組別，增加流動性。要透過 ETF 成功達到後者的目的，必須符合兩大條件：（1）指數必須可予投資，即指數的組成部分必須能夠在自由、公開的市場交易；（2）毋須負擔高昂的交易成本或產生市場影響而能夠按指數成份的各自所佔權重購買所有該等成份資產。

投資者對內地與香港股票共同市場的興趣日增，開發相關指數服務及指數相關的投資產品以支援這個市場的持續增長及發展亦是理所當然。

15　資料來源：恒生指數網站 http://www.hsi.com.hk。

16　ETFGI 是全獨立的研究及諮詢公司，服務客戶包括全球領先的機構及專業投資者、全球 ETF 及 ETP 的業界以及其監管機構和顧問（http://etfgi.com）。

17　資料來源：〈ETF industry grew faster than hedge funds in 2016 — ETFGI〉，載於《International Adviser》雜誌，2017 年 3 月 6 日。

18　資料來源：〈貝萊德：首季全球 ETF 吸 1.4 萬億新高〉，載於《香港經濟日報》，2017 年 5 月 4 日。

19　參考見〈The evolution of the ETF industry〉，載於《Pensions & Investments》雜誌，2017 年 1 月 31 日。

2.3　共同市場的指數掛鈎投資工具

　　首個專為內地與香港股票共同市場而開發的跨境指數掛鈎投資產品 [20] 於 2013 年推出，就是香港交易所衍生產品市場於 2013 年 8 月 12 日推出的中華 120 指數期貨，同年並有三隻相關的指數 ETF 於香港交易所證券市場上市（見表 2）。由於相關指數都屬跨境指數系列，該等產品被視為「共同市場概念」的指數掛鈎產品，有別於共同市場概念以外獨立創設而覆蓋香港股票或內地 A 股又或香港或海外上市的內地公司的指數 ETF。

表 2：香港的共同市場概念指數掛鈎產品（2017 年 8 月底）			
期貨產品			推出日期
中華交易服務中國 120 指數期貨			12/08/2013
交易所買賣基金（雙櫃台股票）	港元櫃台	人民幣櫃台	上市日期
華夏中華交易服務中國 A80 指數 ETF *	3180	83180	26/08/2013
南方東英中華 A80 ETF	3137	83137	23/09/2013
易方達中華交易服務中國 120 指數 ETF	3120	83120	21/10/2013

註：該 ETF 已於 2017 年 11 月 10 日除牌。
資料來源：香港交易所。

　　香港遠在滬深港通推出之前已有跨境指數掛鈎產品供香港及海外投資者投資內地證券市場。首隻加入內地上市股票作為相關資產並於聯交所上市的 ETF 是 2001 年 11 月 28 日上市的 iShares 安碩 MSCI 中國指數 ETF [21]，第二隻是 2004 年 11 月 18 日上市的 iShares 安碩富時 A50 中國指數 ETF。於 2017 年 8 月底，聯交所 93 隻上市實物股票指數 ETF 中，約三成的相關資產有內地上市股票在內 [22]，相關指數包括 MSCI 中國指數、富時中國 A50 指數、上證 50 指數、滬深指數、中華交易

---

20 「跨境」是指內地與香港的跨境，而跨境投資產品是指讓內地投資者投資香港資產及 / 或香港投資者投資內地資產的產品。

21 MSCI 中國指數涵蓋中華人民共和國註冊成立並在上交所、深交所或聯交所上市的公司。成份股包括內地上市的 B 股、香港上市的 H 股、P 股及外地上市股票（如美國預託證券）。MSCI 於 2017 年 6 月宣佈會於 2018 年 6 月開始將中國 A 股納入其新興市場指數及全球指數系列，包括 MSCI 中國指數。

22 數字不包括雙櫃台的複計。資料來源：香港交易所網站。

服務指數及其他。所有這些聯交所上市的 ETF 佔 2017 年首八個月 ETF 成交總額的 28%[23]。

至於內地，2017 年 8 月底共有 85 隻 ETF 於上交所上市及 53 隻 ETF 於深交所上市，當中上交所上市的只有三隻跨境 ETF，以恒指及恒生國企指數為相關指數；深交所只有一隻以恒指為相關指數的跨境 ETF。該隻以恒指為相關指數的 ETF 於 2017 年 1 月至 8 月的成交金額在所有深交所上市 ETF 中排名第五，佔期內深交所上市 ETF 成交總額的 1.6%（期內成交金額最高的三隻 ETF 已佔深交所上市 ETF 成交總額的 87%）[24]。

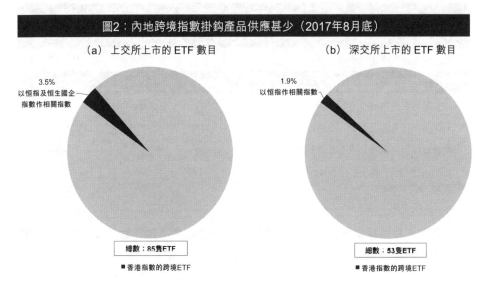

圖2：內地跨境指數掛鈎產品供應甚少（2017年8月底）

資料來源：上交所及深交所網站上的每月統計數字。

縱使預料中潛在需求強勁，但能照顧共同市場方面投資需要的跨境指數掛鈎產品，在內地明顯供應稀缺。

---

23 資料來源：按載於香港交易所網站的 2017 年 8 月「交易所買賣基金成交統計」的數據作計算。
24 資料來源：載於深交所網站的每月統計數字。

3　中華港股通精選 100 指數
── 可供南向投資買賣的指數

中華港股通精選 100 於 2014 年 12 月 15 日推出，以 2008 年 12 月 31 日為基日，基點為 2000 點，成份股為港股通股票中市值排名首 100 隻的聯交所上市股票（不包括在內地與香港兩邊上市的 A+H 股）。指數編算是常用的自由流通市值加權法，權重上限為 10%。指數的實時數據每五秒發佈一次。

此指數的特別之處，在於其為唯一的滬深港通相關指數，所追蹤的是可透過港股通買賣的港股。此指數對南向交易的用處，在於其能**代表南向合資格股份，以及其組合成份有別於其他現有中港市場指數，可作另類投資選擇**。詳細說明見以下分節。

3.1　大型南向股票的覆蓋率

中華港股通精選 100 按市值排名挑選成份股，入選的都是港股通股票中流通量大的最大型股票，儘管只佔港股通股票總數的 24%（2017 年 8 月底），按市值及市場成交額 [25] 計的佔比卻分別為 68% 及 53%（見圖 3）。除 H 股外（注意指數並不包括 A+H 股），指數於其他各類股票都有高度代表性（見 3.2 節）。

---

25 「市場成交額」指股票市場總成交金額，不單指南向交易的成交。

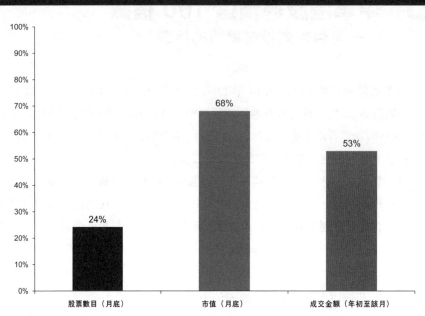

圖 3：中華港股通精選 100 股票佔所有港股通股票百分比（2017 年 8 月）

資料來源：股票清單來自香港交易所、上交所、深交所及中華交易服務網站；市場數據來自香港交易所。

3.2　純香港概念，有高度代表性

中華港股通精選 100 包含只在香港上市的股票，不包括有 A 股在內地市場上市的 A+H 股。因此，此指數計算純粹對香港上市股票的投資，這給予內地投資者除投資國內股票以外的一個特有機會，來透過滬深港通進行投資。

相對整體港股通股票而言，中華港股通精選 100 的成份股於非 H 股中的代表性更高：按市值計，港股佔比為 33% 對 25%、內地民企為 31% 對 26%、紅籌股為 25% 對 20%；按成交金額計（2017 年截至 8 月），港股佔比為 25% 對 16%、內地民企為 38% 對 27%、紅籌股為 21% 對 13%（見圖 4）。特別要說明的是，內地民企是現時中國轉型至新經濟模式的主要增長板槐。

圖 4：中華港股通精選 100 成份股按股票類別對照所有港股通股票（2017 年 8 月底）

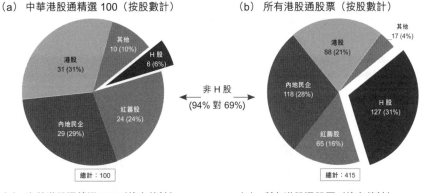

（a）中華港股通精選 100（按股數計）

（b）所有港股通股票（按股數計）

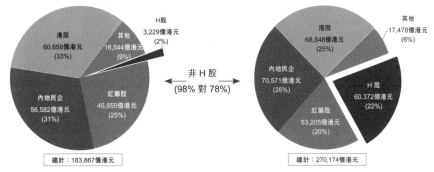

（c）中華港股通精選 100（按市值計）

（d）所有港股通股票（按市值計）

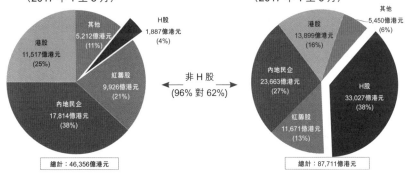

（e）中華港股通精選 100（按成交金額計）
（2017 年 1 至 8 月）

（f）所有港股通股票（按成交金額計）
（2017 年 1 至 8 月）

註：以上是香港交易所所作的股票分類，分類會考慮上市公司是否來自中國，對 H 股、紅籌股及內地民企以
　　外的其他公司，會看其成立來源地及註冊成立地。

資料來源：股票名單來自香港交易所、上交所、深交所及中華交易服務網站；市場數據來自香港交易所。

若不計算 H 股，中華港股通精選 100 所有其他類別的成份股所包含的港股通股票的佔比都很高：按市值計為 80% 至 95%，按市場總成交金額計為 75% 至 96%，雖則佔南向成交金額比例略低（62% 至 91%）（見圖 5）。

図5：中華港股通精選 100 成份股佔所有港股通股票的比率（按股票類別）（2017 年 8 月）

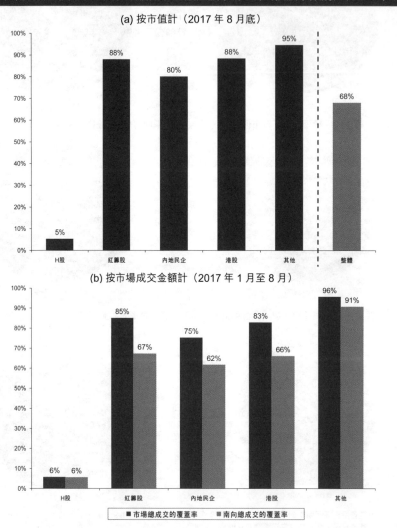

(a) 按市值計（2017 年 8 月底）

(b) 按市場成交金額計（2017 年 1 月至 8 月）

■ 市場總成交的覆蓋率　■ 南向總成交的覆蓋率

註：以上是香港交易所所作的股票分類，分類會考慮上市公司是否來自中國，對 H 股、紅籌股及內地民企以外的其他公司，會看其成立來源地及註冊成立地。

資料來源：股票名單來自香港交易所、上交所、深交所及中華交易服務網站；市場數據來自香港交易所。

3.3　行業分佈互有相似，但有更多投資新經濟的機會

　　按行業板塊劃分，中華港股通精選 100 的股票組合與所有港股通股票同樣分散，但金融類股份較少（按市值計為 22% 對 31%，按市場成交金額計為 20% 對 31%），反而來自被視為新經濟中高速增長的板塊的資訊科技股份，以市值及成交金額計佔比都較高（見圖 6）。

圖 6：中華港股通精選 100 成份股按行業類別對照所有港股通股票（2017 年 8 月底）

（a）中華港股通精選 100（按股數計）

（b）所有港股通股票（按股數計）

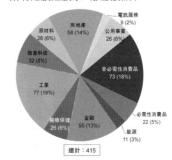

（c）中華港股通精選 100（按市值計）

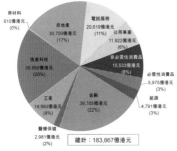

（d）所有港股通股票（按市值計）

（e）中華港股通精選 100（按成交金額計）（2017 年 1 月至 8 月）

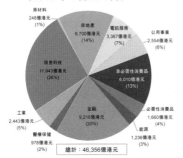

（f）所有南向股票港股通股票（按成交金額計）（2017 年 1 月至 8 月）

資料來源：股票名單來自香港交易所、上交所、深交所及中華交易服務網站；市場數據來自香港交易所。

如圖 7 所示，中華港股通精選 100 包含相對較多新經濟板塊中的港股通股
票 26：

- 資訊科技 —— 按市值計佔 90%，按市場成交金額計佔 82%；
- 必需性消費品 —— 按市值計佔 75%，按市場成交金額計佔 77%；
- 非必需性消費品 —— 按市值計佔 68%，按市場成交金額計佔 57%；
- 醫療保健 —— 按市值計佔 56%，按市場成交金額計佔 50%。

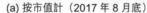

圖7：中華港股通精選100股票相對所有港股通股票佔比（按行業板塊分類）（2017 年 8 月）

(a) 按市值計（2017 年 8 月底）

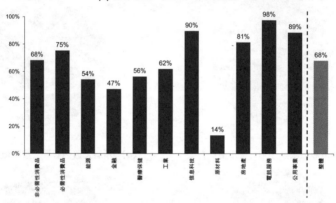

(b) 按市場成交金額計（2017 年 1 月至 8 月）

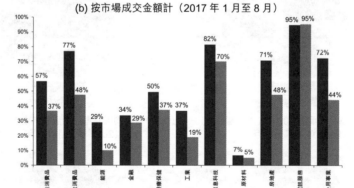

資料來源：股票名單來自香港交易所、上交所、深交所及中華交易服務網站；市場數據來自香港交易所。

---

26 新經濟板塊參照中證邁格中國新經濟指數所涵蓋的板塊（資料來源：中證指數有限公司網站）。

3.4　對照香港及內地主要指數的表現

　　圖 8 對比了中華港股通精選 100 與香港、上海及深圳股市主要股票指數的每日走勢。中華港股通精選 100 從基日（2008 年 12 月 31 日）至 2017 年 8 月底止這八年多以來，其累計回報率達 102%，大幅拋離香港兩大指數恒指（66%）及恒生國企指數（36%），僅次於上證 380 指數（118%）、深證 A 股指數（125%）及深圳中小板綜合指數（144%）（見表 3）。

圖 8：中華港股通精選 100 與香港及內地主要指數每日收市
（重整基日：2008 年 12 月 31 日）（2008 年 12 月 31 日至 2017 年 8 月 31 日）

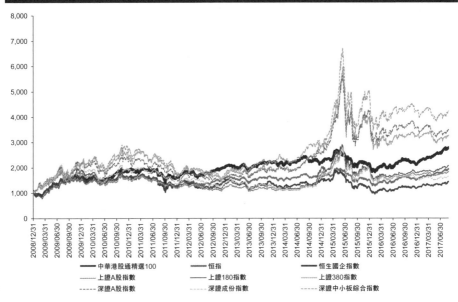

資料來源：中華港股通精選 100 的數據來自中華交易服務，其他指數的數據來自湯森路透。

表 3：中華港股通精選 100 與香港及內地主要指數累計回報率
（2008 年 12 月 31 日至 2017 年 8 月 31 日）

指數	累計回報率
中華港股通精選 100	102.23%
恒指	66.48%
恒生國企指數	35.86%
上證 A 股指數	61.03%
上證 180 指數	72.52%
上證 380 指數	118.58%
深證 A 股指數	125.23%
深證成份指數	51.15%
深證中小板綜合指數	144.08%
深交所創業板綜合指數	74.67%

註：回報率為自然對數回報率。

資料來源：按指數每日收市計算 —— 中華港股通精選 100 來自中華交易服務，其他指數來自湯森路透。

　　數據顯示中華港股通精選 100 的歷史回報率，不論短期還是長期均跑贏香港主要指數。另一方面，內地指數回報率較為波動，尤其深交所指數的回報率在過去幾年顯著波動，近年甚至處於較低（或負數）水準，而中華港股通精選 100 則有正數的回報率。事實上，中華港股通精選 100 自推出以來，其每日回報率的年化波幅大部分時間均低於香港及內地的主要指數。（見表 4 及圖 9）

表 4：中華港股通精選 100 與香港及內地主要指數的各期間回報率
（截至 2017 年 8 月 31 日）

指數	1 年	3 年	5 年	7 年
中華港股通精選100	21.41%	15.26%	46.11%	54.55%
恒指	19.67%	12.26%	36.16%	30.89%
恒生國企指數	16.87%	2.98%	19.65%	-0.95%
上證A股指數	8.59%	41.64%	49.56%	24.15%
上證180指數	15.79%	49.27%	56.52%	31.00%
上證380指數	0.01%	41.02%	68.42%	36.73%
深證A股指數	-4.44%	47.44%	84.14%	50.93%
深證成份指數	0.54%	32.16%	27.56%	-4.71%
深證中小板綜合指數	-2.35%	48.21%	88.19%	54.82%
深交所創業板綜合指數	-16.09%	45.57%	121.22%	83.82%

註：回報率為自然對數回報率。

資料來源：按指數每日收市計算 —— 中華港股通精選 100 來自中華交易服務，其他指數來自湯森路透。

圖 9：中華港股通精選 100 與香港及內地主要指數的回報率及波幅 （2009 年至 2017 年 8 月）

(a) 年度 / 期間回報率

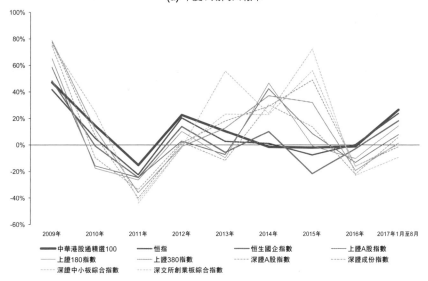

(b) 每日回報率的年化波幅

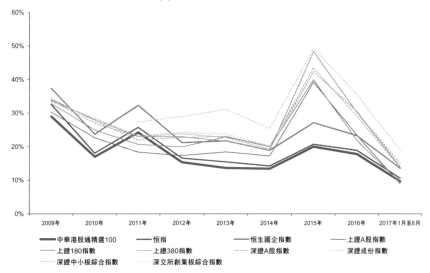

註：回報率為自然對數回報率。年化波幅為期內每日回報率的年度化標準差。
資料來源：按指數每日收市計算 —— 中華港股通精選 100 來自中華交易服務，其他指數來自湯森路透。

儘管指數回報率各有不同，但中華港股通精選 100 的每日回報率與香港主要指數恒指及恒生國企指數相關性頗高 —— 於 2009 年 1 月至 2017 年 8 月期間，中華港股通精選 100 與恒指及恒生國企指數的相關系數分別為 0.978 及 0.899，與內地主要指數的相關性僅處於中低水平 —— 相關系數為 0.5 或以下（見表 5）。

表 5：中華港股通精選 100 與香港及內地主要指數的每日回報率相關系數 (2009 年 1 月至 2017 年 8 月)	
與指數的相關性	相關系數
恒指	0.978
恒生國企指數	0.899
上證 A 股指數	0.507
上證 180 指數	0.509
上證 380 指數	0.435
深證A股指數	0.437
深證成份指數	0.461
深證中小板綜合指數	0.407
深交所創業板綜合指數	0.341

註：相關系數為 Pearson 相關系數；所有系數在統計學上有顯著相關的水平為 0.1%。
資料來源：按指數每日收市計算 —— 中華港股通精選 100 來自中華交易服務，其他指數來自湯森路透。

2014 年 12 月至 2017 年 8 月期間，在僅追蹤香港股票的指數當中，中華港股通精選 100 的月底市盈率一直高於恒生國企指數，大部分時間亦高於恒指，儘管中華港股涌精選 100 的股息率在同期大部分時間均低於該兩隻指數[27]（見圖 10）。這或反映出中華港股通精選 100 成份股主要來自目前經濟發展階段中的增長行業（見上文第 3.3 節），這些行業會將盈利再投資產生資本收益而非分派作股息，而恒指及恒生國企指數成份股則多為已處於較成熟發展階段的傳統經濟公司。

此外，相對於深圳股市（普遍被視為由增長型股份組成），中華港股通精選 100 於 2017 年 8 月底的股息率（3.08%）高於深證成份指數（1.04%）、中小板指數（0.84%）及深交所創業板綜合指數（0.69%）[28]。

---

27 指數市盈率及股息率為指數中各成份股的市盈率及股息率的加權平均數。
28 深交所指數股息率資料來源：國證指數 2017 年 8 月的指數運行月報（http://index.cninfo.com.cn/）。

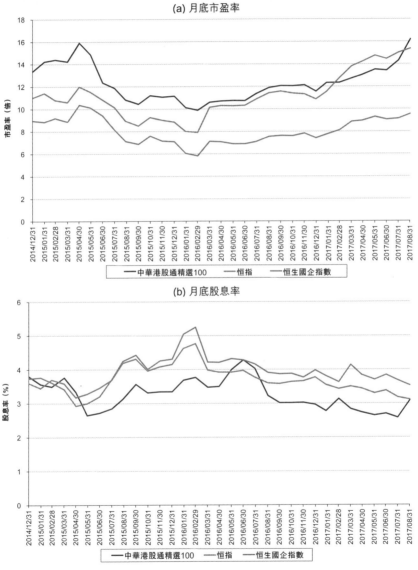

圖 10：中華港股通精選 100 及香港主要指數的市盈率及股息率
（2014 年 12 月至 2017 年 8 月）

(a) 月底市盈率

(b) 月底股息率

註：有關數字為各指數成份股月底股息率的加權平均數。

資料來源：中華港股通精選 100 來自中華交易服務，恒指及恒生國企指數來自恒生指數公司網站。

3.5　中華港股通精選 100 帶來的港股通投資機遇

總括而言，中華港股通精選 100 具有以下特點，可追蹤滬深港通南向交易的投資：

(1) 以市值及成交額計，於港股通股票的覆蓋率相對較高；

(2) 追蹤純香港概念的投資，就非同時在內地上市的香港上市股票而言有很高的代表性，因此純反映內地以外地區的投資機會，與國內證券市場的相關性僅屬溫和；

(3) 港股通股票中增長型板塊股票的覆蓋率較高，例如內地民企及新經濟行業的股票；

(4) 由於其成份股分佈的關係，指數自推出以來大部分時間的市盈率均較恒指及恒生國企指數為高，但股息率卻較低，回報率的波幅亦較香港及內地主要指數低。

有趣的是，儘管有以上實證，今年截至 8 月為止，中華港股通精選 100 成份股的港股通成交額佔這些股份的總市場成交額的百分比（10%）卻少於全部港股通股份的港股通成交額佔其總市場成交額的百分比（14%）。不論按股份類別還是行業劃分，都發現有相同現象（見圖 11）。

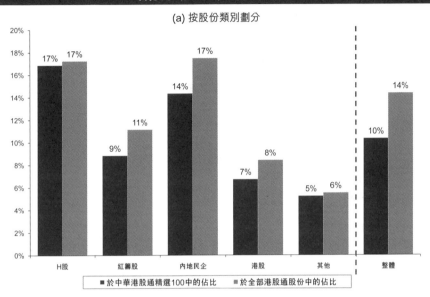

圖 11：中華港股通精選 100 及全部港股通股份的港股通交易佔其總市場成交額的百分比（2017 年 1 月至 8 月）

(a) 按股份類別劃分

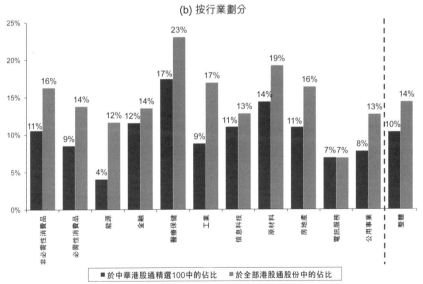

(b) 按行業劃分

資料來源：證券名單來自香港交易所、上交所、深交所及中華交易服務網站；市場數據來自香港交易所。

換言之，**中華港股通精選 100 成份股的潛在投資機遇仍待透過港股通交易獲進一步發掘**。中華港股通精選 100 滿足了應有的先決條件，可支援開發作港股通投資的 ETF 等被動型投資工具 —— 該指數成份股可在公開市場自由買賣，而成份股的買賣預期不會對市場有影響，因為該等股份按市值計是合資格經港股通買賣的首百大股份（見上文第 2.2 節有關 ETF 成功的必要條件）。因此，中華港股通精選 100 是開發 ETF 等被動型投資工具的有用基準指標。

4 總結

滬深港通啟動後，內地與香港建立起共同市場，內地投資者今後投資海外市場不但機會比從前多，限制亦較以往透過 QDII 等渠道為少。統計數字顯示內地投資者對透過滬深港通的港股通買賣香港上市股票的興趣日濃。

中華港股通精選 100 是滬深港通相關指數中專門追蹤港股通股票的指數，純反映中國內地以外地區的投資機遇。基於這個指數能高度代表港股通股票及新經濟增長型企業，其成份股的潛在投資機會仍有待透過港股通獲進一步發掘。朝此方向，中華港股通精選 100 可能會是開發作港股通投資的 ETF 等被動型投資工具的一個有用基準指標。

附錄一

滬深港通的南向交易活動

　　自滬港通開通後，北向交易在 2016 年之前按日均成交金額計，大部分時間都是遠多於南向交易。但由 2015 年底開始，南向交易漸呈上升趨勢，起初增幅緩慢，及至 2016 年 12 月深港通推出後開始急速增長（見圖 A1）。南向交易在聯交所主板市場總成交的佔比由 2015 年 9 月佔主板日均成交額[29]1.0% 躍升至 2017 年 9 月的峰位 6.1%。相較之下，北向交易日均成交額只佔內地 A 股市場總成交約 1%（見圖 A2）。此外，期內月份的南向交易日均成交額多次超過北向交易。

圖 A1：滬深港通下北向交易及南向交易的日均成交額（買盤及賣盤）
（2014 年 11 月至 2017 年 9 月）

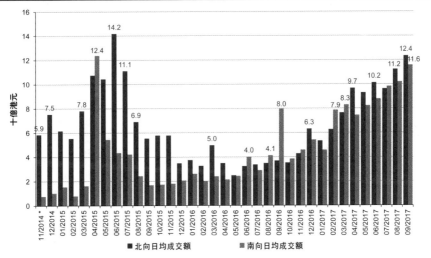

* 自 2014 年 11 月 17 日滬港通開通之日起計。

註：滬深港通的交易總額含買盤及賣盤。北向交易的成交額按月末匯率（來自香港金融管理局網站）轉換為港
　　元。自 2016 年 12 月 5 日起包括於當日推出的深港通數據。

資料來源：香港交易所。

---

29 為與單邊計算的聯交所主板成交總額作較恰當的比較，計算方式是將雙邊計算（含買盤及賣盤）的南向交易總額除
　　以二得出單邊數字來計算比率。在計算北向交易佔內地 A 股市場成交額的比率採用同一方式。

圖 A2：滬深港通日均成交額及其佔市場總成交的比重（2014 年 11 月至 2017 年 9 月）

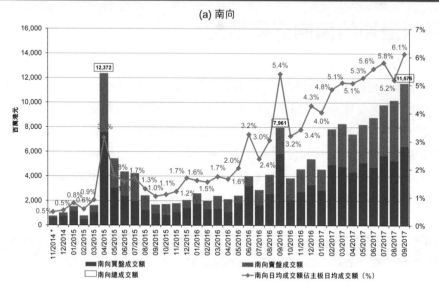

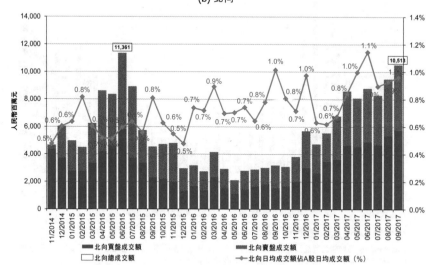

* 自 2014 年 11 月 17 日滬港通開通之日起計。

註：在計算佔市場日均成交總額的百分比，是將北向交易／南向交易的交易總額（買盤及賣盤）除以二，獲得
　　的單邊數字再與單邊市場總成交額（買盤及賣盤以單一交易額計算）來計算比率。自 2016 年 12 月 5 日
　　起包括於當日推出的深港通數據。內地 A 股市場的基本參考數據包括深交所自深港通開通當日起的 A 股
　　市場數據。

資料來源：香港交易所。

此外，自 2015 年後期開始的大部份時間，南向交易的平均每日買盤淨額均遠高於北向交易。由推出至 2017 年 9 月底，南向交易只有兩個月錄得淨賣盤，相比之下北向交易則曾錄得六個月的淨賣盤（見圖 A3）。截至 2017 年 9 月，內地投資者透過南向交易買入港股的累計買盤淨額達 6,040 億港元，而環球投資者透過北向交易買入內地股票的累計買盤淨額則為人民幣 3,145 億元（約 3,720 億港元）。

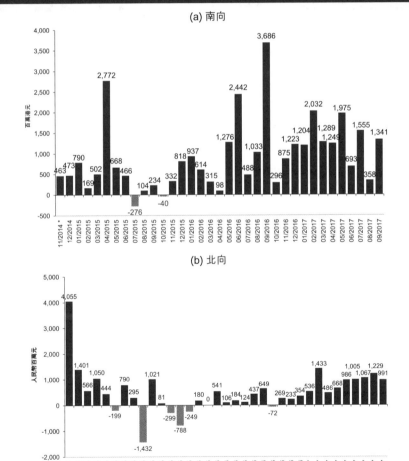

圖 A3：滬深港通南向交易與北向交易的日均淨買盤 / 賣盤成交額（2014 年 11 月至 2017 年 9 月）

(a) 南向

(b) 北向

* 自 2014 年 11 月 17 日滬港通開通之日起計。

註：自 2016 年 12 月 5 日起包括於當日推出的深港通數據。

資料來源：香港交易所。

附錄二

中華港股通精選 100 成份股名單
（2017 年 9 月底）

代號	股份名稱	權重（%）
700	騰訊控股有限公司	11.27
5	滙豐控股有限公司	10.21
1299	友邦保險有限公司	7.89
941	中國移動有限公司	5.52
1	長江和記實業有限公司	3.06
388	香港交易及結算所有限公司	2.93
2888	渣打集團有限公司	2.87
16	新鴻基地產發展有限公司	2.09
883	中國海洋石油有限公司	2.04
1113	長江實業集團有限公司	1.91
2388	中銀香港（控股）有限公司	1.82
11	恒生銀行有限公司	1.65
2	中電控股有限公司	1.61
27	銀河娛樂集團有限公司	1.60
2007	碧桂園控股有限公司	1.50
3	香港中華煤氣有限公司	1.40
175	吉利汽車控股有限公司	1.34
688	中國海外發展有限公司	1.26
3333	中國恒大集團	1.21
6	電能實業有限公司	1.15
1928	金沙中國有限公司	1.12
2018	瑞聲科技控股有限公司	1.10
2382	舜宇光學科技（集團）有限公司	1.08
1658	中國郵政儲蓄銀行股份有限公司	1.01
4	九龍倉集團有限公司	0.96
66	香港鐵路有限公司	0.92
762	中國聯合網絡通信（香港）股份有限公司	0.89
267	中國中信股份有限公司	0.76
17	新世界發展有限公司	0.75

(續)

代號	股份名稱	權重（%）
1109	華潤置地有限公司	0.75
2328	中國人民財產保險股份有限公司	0.72
12	恒基兆業地產有限公司	0.71
1114	華晨中國汽車控股有限公司	0.71
669	創科實業有限公司	0.69
2319	中國蒙牛乳業有限公司	0.68
728	中國電信股份有限公司	0.63
1093	石藥集團有限公司	0.63
1038	長江基建集團有限公司	0.61
1044	恒安國際集團有限公司	0.61
288	萬洲國際有限公司	0.55
384	中國燃氣控股有限公司	0.53
23	東亞銀行有限公司	0.52
2313	申洲國際集團控股有限公司	0.52
20	會德豐有限公司	0.51
83	信和置業有限公司	0.49
1177	中國生物製藥有限公司	0.49
2688	新奧能源控股有限公司	0.49
656	復星國際有限公司	0.48
19	太古股份有限公司	0.47
101	恒隆地產有限公司	0.47
151	中國旺旺控股有限公司	0.47
1099	國藥控股股份有限公司	0.47
1359	中國信達資產管理股份有限公司	0.44
144	招商局港口控股有限公司	0.43
966	中國太平保險控股有限公司	0.43
270	粵海投資有限公司	0.41
291	華潤啤酒（控股）有限公司	0.39
960	龍湖地產有限公司	0.39
992	聯想集團有限公司	0.38
371	北控水務集團有限公司	0.37
522	ASM 太平洋科技有限公司	0.37
1128	永利澳門有限公司	0.37
1357	美圖公司	0.37
1972	太古地產有限公司	0.35
2689	玖龍紙業（控股）有限公司	0.33
425	敏實集團有限公司	0.32

(續)

代號	股份名稱	權重（%）
2282	美高梅中國控股有限公司	0.32
586	中國海螺創業控股有限公司	0.31
836	華潤電力控股有限公司	0.31
6808	高鑫零售有限公司	0.31
10	恒隆集團有限公司	0.30
257	中國光大國際有限公司	0.30
322	康師傅控股有限公司	0.30
607	豐盛控股有限公司	0.30
1169	海爾電器集團有限公司	0.30
2020	安踏體育用品有限公司	0.30
69	香格里拉（亞洲）有限公司	0.29
135	昆侖能源有限公司	0.28
551	裕元工業（集團）有限公司	0.28
981	中芯國際集成電路製造有限公司	0.28
1193	華潤燃氣控股有限公司	0.28
659	新創建集團有限公司	0.27
683	嘉裡建設有限公司	0.27
1816	中國廣核電力股份有限公司	0.27
14	希慎興業有限公司	0.26
494	利豐有限公司	0.26
813	世茂房地產控股有限公司	0.26
3311	中國建築國際集團有限公司	0.26
392	北京控股有限公司	0.24
867	康哲藥業控股有限公司	0.23
8	電訊盈科有限公司	0.22
1060	阿裡巴巴影業集團有限公司	0.22
3377	遠洋集團控股有限公司	0.22
165	中國光大控股有限公司	0.21
3320	華潤三九醫藥股份有限公司	0.20
880	澳門博彩控股有限公司	0.18
293	國泰航空有限公司	0.16
241	阿裡健康資訊技術有限公司	0.13
3799	達利食品集團有限公司	0.13
1929	周大福珠寶集團有限公司	0.12

第 5 章

新經濟產業的崛起

—— 融資需求及香港的新角色

2017 年 12 月 19 日

摘 要

當傳統增長動力逐步減弱，投資和出口等舊經濟部門出現停滯時，中國以新產業、新業態、新商業模式為特徵的新經濟板塊開始快速發展，逐步成為推動經濟結構轉型、新舊動能轉換的重要推手。然而不容忽視的是，多元化、多層次的投融資機制不健全，對中國新興產業發展形成一定制約。傳統的銀行信貸、創投私募基金，未能充分滿足新經濟產業發展的融資需求，而以首次公開招股形式上市融資也未能妥善解決創始人的股權架構問題。如何進一步完善資本市場融資功能，幫助新興科創企業融資將是推動中國科創企業發展，實現新經濟板塊快速增長的關鍵。

打造一個真正開放、多元創新的融資市場，可助推提升中國新興企業的國際競爭力，另一方面，中國經濟轉型和創新產業的不斷崛起將為香港注入新動力，使香港在中國企業實現國際化佈局和經濟開放中找到新的角色和定位。

1 新經濟產業崛起，為中國經濟增長注入新動力

當前中國產業結構正在發生前所未有的轉變，經濟增長引擎正逐步脫離出口和投資拉動的舊有模式，向經濟轉型和企業創新方向轉變。

1.1 傳統增長動力逐步減弱，投資和出口等舊經濟部門出現停滯

過去近 40 年發展，中國以高投資、高出口為導向，以投資密集型、能源密集型和勞力密集型的經濟結構為依託，經歷了全球經濟領域內最大、歷時最久也最廣泛的經濟貿易增長。從 1978 年至 2016 年國內生產總值年均增長 9.6%，人均國內生產總值從人民幣 385 元提高到 53,980 元，成為世界第二大經濟體（見圖 1）。

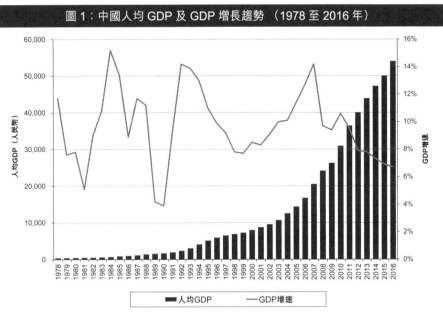

圖 1：中國人均 GDP 及 GDP 增長趨勢（1978 至 2016 年）

資料來源：Wind 資訊。

然而在 2008 年全球金融危機後,世界經濟持續低迷,國際經濟環境發生重大變化,中國經濟亦出現「三期疊加」的階段性下行特徵,經濟增長遭遇瓶頸。投資方面,投資邊際效益持續下降,對經濟增長的帶動作用不斷降低。2006 年至 2016 年,固定資產投資同比增速從 24.33% 下降至 8.1%,投資對國內生產總值增長的拉動作用從高峰時期的 8.0% 下降到 2016 年的 2.8%。貿易方面,全球貿易量增長連續五年低於 3% 水平,中國 2015 年對外貿易總額下滑近 8%,2016 年更低於 2015 年水平[1]。在當前去全球化趨勢的情況下,中國的貿易增速已經難以回復過去高於全球貿易平均增速的發展態勢,2009 年以後,對外貿易對國內生產總值增長貢獻為負數(見圖 2)。

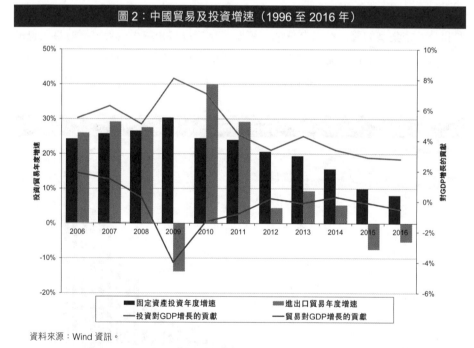

圖 2:中國貿易及投資增速(1996 至 2016 年)

資料來源:Wind 資訊。

1　資料來源:國家統計局。

1.2　新經濟板塊崛起，得益於技術和政策層面的雙重推進

以新產業、新業態、新商業模式為特徵的新經濟板塊開始快速發展，逐步成為推動中國經濟結構轉型，新舊動能轉換的重要推手。

新經濟板塊的崛起主要得益於兩方面因素：

一方面，科技創新逐步成為商業模式變革的主要推動力。具體體現在五大運用領域：

第一，**互聯網**已由最初的一種改善溝通的工具變成支撐整個商業運作的核心基礎設施。數字化的商業運用，引領商業模式創新和消費者行為變化。

第二，**數據**成為一種新的資源，以知識創新、雲計算、人工智能，大數據開發為代表的技術成為新一輪產業革命的核心驅動力。

第三，**共享經濟**以更低成本和更高效率實現剩餘資源的供需匹配。從住宿（如AirBnB）到交通運輸（如滴滴打車），共享經濟涵蓋多個領域，發展出以協同消費、協作經濟、點對點經濟為特徵的全新商業模式。

第四，**金融科技（Fintech）**的運用。金融科技囊括了支付清算、電子貨幣、網絡借貸、區塊鏈、智能投顧、智慧合同等多個領域，通過將金融與科技相結合，創造出新的業務模式，顛覆了金融市場的傳統服務方式。麥肯錫研究表明，隨着數字化銀行的廣泛使用，平均 40% 的傳統銀行業務收入將發生改變 [2]。

第五，**實現製造業全產業升級**。新能源、新材料、高端裝備製造產業成為全球新一輪產業發展方向，為全球結構性經濟復蘇提供堅實支撐。

另一方面，2008 年全球金融危機之後各國出台了一系列政策導向和戰略規劃，為新經濟產業的發展提供了方向指引和制度保障。

從全球範圍來看，國際金融危機以後，發達國家重新重視實體經濟發展，紛紛制定以重振製造業為核心的再工業化戰略，對生物製藥、電子設備、智能技術、材料科技、清潔能源等多個新經濟行業推出鼓勵政策，培育新的經濟增長點。

美國政府率先積極發展先進製造技術、智能製造、新能源、生物技術、信息等新興產業，促進高端製造回流，以創造更多高附加值、高技術含量的本土崗位，從而重新擁有具備強大競爭力的新工業體系。2009 年 12 月，美國公佈了《重振美

2　資料來源：《中國銀行業的明天在哪裏》，麥肯錫研究，2016 年。

國製造業框架》，2011 年 6 月和 2012 年 2 月，相繼啟動《先進製造業夥伴計劃》和《先進製造業國家戰略計劃》，實施「再工業化」。隨着美國「再工業化」戰略推進，其製造業佔全球比重穩步上升至 2014 年的 16.6%。美國製造業就業人數持續上升，截至 2015 年底累計新增 70 萬人，製造業回流效果初見端倪[3]。

　　歐洲各國政府也制定並實施了重振工業或打造未來工業的政策。其中，德國提出的「工業 4.0」國家戰略計劃瞄準新興產業，謀求新的產業鏈升級和競爭優勢，在全世界引起了極大反響。德國「工業 4.0」戰略是將代表着互聯網時代，如雲計算、大數據、3D 打印、網絡安全等技術整合到工業製造領域，旨在提升製造業的智能化水平，建立具有適應性、資源效率的智慧工廠。自德國工業 4.0 項目推出以來，該國已有 47% 公司參加到工業 4.0 戰略中，12% 的公司已把工業 4.0 戰略付諸實踐[4]，成為當地乃至歐洲經濟轉型的重要推手。

　　中國於 2010 年出台的《國務院關於加快培育和發展戰略性新興產業的決定》，2015 年又出台了《中國製造 2025》戰略規劃，圍繞實現製造強國的戰略目標，明確了 9 項戰略任務和重點[5]。以全球視野和戰略思維，實現技術革命對全行業的滲透，打造中國製造業的競爭新優勢。這一政策舉措不僅為中國搶佔未來新經濟產業和高科技發展制高點提供了重要的戰略指引，對於中國實現經濟增長動能轉換，推動中國由製造業大國向製造業強國轉變具有重要的戰略意義。

3　資料來源：美國勞動統計局。

4　資料來源：德國 3 大協會（德國機械及製造商協會、信息技術和通信與新媒體協會、電子電氣製造商協會）調查結果。

5　9 項戰略任務和重點分別為新一代信息技術產業、高檔數控機床和機器人、航空航天裝備、海洋工程裝備及高技術船舶、先進軌道交通裝備、節能與新能源汽車、電力裝備、農機裝備、新材料、生物醫藥及高性能醫療器械以及與之配套的生產性服務業。

表 1：美、德、中等國支持高科技製造業發展的主要政策比較		
國家	推出時間	主要政策
美國	2009	《重振美國製造業框架》
	2011	《先進製造業夥伴計劃》
	2012	《先進製造業國家戰略計劃》
	2013	《從互聯網到機器人 —— 美國機器人路線圖》
	2015	《美國創新新戰略》
德國	2010	《德國 2020 高科技戰略》
歐盟	2006	《創建創新型歐洲》
	2010	《歐洲 2020 戰略》
	2014	《歐盟「地平線 2020」計劃》
日本	2007	《日本創新戰略 2025》
韓國	2009	《新增長動力規劃及發展戰略》
中國	2010	《國務院關於加快培育和發展戰略性新興產業的決定》
	2015	《中國製造2025》戰略規劃

資料來源：各政府網站及新聞。

1.3　中國新經濟行業已迎來重要戰略發展期

　　一方面，全球經濟分工調整和技術變革為中國新經濟行業帶來重大歷史機遇，另一方面，中國也需要通過創新驅動來引導經濟向更高質量、更有效率、更可持續的方向發展。新經濟替代舊經濟，並實際轉化為未來中國經濟增長和生產率提升的關鍵力量，是中國走出「L 型」經濟趨勢，成功實現經濟轉型的重要標誌。

　　根據中國的「十三五」規劃，到 2020 年戰略性新興產業相關產值規模將超過 60 萬億元，佔 GDP 比重超過 15%。根據麥肯錫研究預測，到 2025 年僅移動互聯網、雲計算、先進機器人與新一代基因組等 12 項重大技術的突破，每年就將產生 14 萬億至 33 萬億美元的直接經濟價值[6]。可以預見，未來 10 年將是中國高端裝備製造業、新一代信息技術與信息服務產業、新材料、新能源、節能與環保等新興行業的重要戰略發展期。

　　隨着中國的經濟結構優化加快，新興產業[7]已經出現快於經濟總體的發展態

6　資料來源：〈麥肯錫發佈 12 大顛覆技術物聯網、雲、機器人、自動汽車在列〉，新華網頁，2015 年 9 月 29 日。
7　中國對新經濟沒有給予具體定義。在本文中，新興產業與新經濟範疇類似。

勢。以上市公司為樣本，截止 2016 年底，A 股上市公司中共有 1,152 家戰略性新興產業企業，佔上市公司總數的 38%。2016 年戰略性新興產業上市公司實現營業收入 3.25 萬億元人民幣，同比增速為 17.7%，連續 4 年保持增長；利潤增速為 22.3%，較上年高出 8.5 個百分點，亦高出上市公司總體平均水平 16.2 個百分點，相對優勢進一步擴大。企業平均銷售毛利潤率為 25.1%，高出 A 股平均水平 7.5 個百分點，效益持續領航 A 股市場。根據財新中國新經濟指數估算，於 2017 年 4 月新經濟投入佔整個經濟投入的比重為 31.8%[8]，表明目前中國新舊經濟結構已經出現重新調整，以新產業、新業態、新商業模式為代表的新經濟產業已初步成為支撐中國企業績效增長的主要力量。

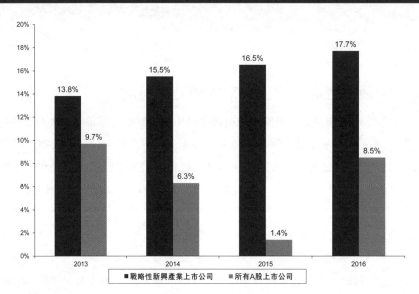

圖 3：戰略性新興產業上市公司與所有 A 股上市公司的營收年度增速對比（2013 年至 2016 年）

資料來源：國家信息中心。

8　資料來源：財新智庫網頁（http://pmi.caixin.com/2017-05-02/101085054.html）。

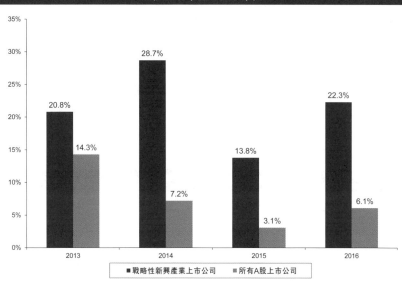

圖 4：戰略性新興產業上市公司與所有 A 股上市公司的年度利潤增速對比
（2013 年至 2016 年）

資料來源：國家信息中心。

2 現有渠道未能充分滿足新經濟產業發展的融資需求

　　科技產業在成長過程的不同階段具有不同的風險特徵和資金需求。早期的創新型企業一般具有核心技術和知識產權，但往往面臨資本投入少、創業週期長，風險和不確定高的難題，需要多種金融工具和孵化環境支持新興技術的創新和產業化進程。

　　不容忽視的是，中國多元化、多層次的投融資機制尚不健全，對中國新經濟產業發展形成一定制約。2016 年，中國以民營企業佔據主導地位的深圳證券交易所中小企業板和創業板全年籌資額為 7,068 億元，相比之下，全國社會融資規模同年新增 17.8 萬億元，當中企業債券融資規模為 2.9 萬億元[9]，而非金融企業股票融資規模為 1.7 萬億元，中小企業板和創業板體量只佔中國整個融資體系的一小部分（見圖 5）。另有研究顯示，在大中型非國有企業中，85% 以上依靠自有資金投資；剩餘來自外部融資，其中三分之二以銀行信貸方式解決資金需求，三分之一通過資本市場融資（見圖 6），說明目前科創企業通過銀行貸款、債券以及股票等資本市場融資的佔比較低，現有渠道較難完全滿足新經濟產業的融資需求。**如何完善資本市場融資功能，將是推動中國科創企業發展，實現新經濟板塊快速增長的關鍵。**

圖 5：中小企業板和創業板融資、債券融資、股票融資規模（2006 年至 2016 年）

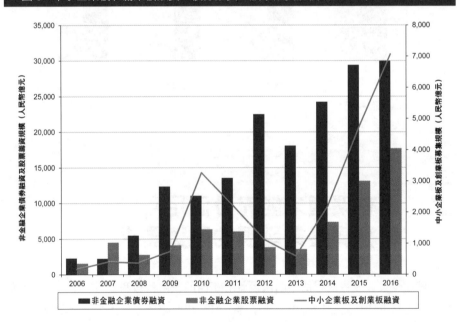

資料來源：Wind 資訊。

9　資料來源：Wind 資訊。

圖 6：大中型非國有企業依靠不同來源資金的比例（2016 年）

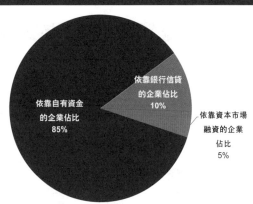

依靠自有資金
的企業佔比
85%

依靠銀行信貸
的企業佔比
10%

依靠資本市場
融資的企業
佔比
5%

資料來源：中國民營企業發展研究報告，《民銀智庫研究》第 57 期。

2.1　銀行信貸

新經濟企業在發展初期，由於企業特殊性往往較難獲得銀行體系的信貸支持。其主要原因在於以下幾點。

第一，新興技術及其科技創新模式往往存在較大的不確定性和特殊性，並不適用於傳統的銀行估計模式。第二，與傳統公司相比，新經濟公司（特別是互聯網業務）往往具有「輕資產」的特性，這也意味着進入壁壘低於傳統企業，因此在這些公司發展的早期階段（在實現盈利之前），需要投入大量資源爭奪市場份額。在競爭激烈的環境下，這些企業在起步階段往往沒有盈利記錄或缺乏良好信用歷史，又難以找到合適的擔保人，盈利模式和現金流量皆不穩定，銀行出於控制風險考慮而不願提供信貸支持。第三，現有法律規定金融機構對高新、科創企業的信貸支持只能要求固定利率，即收益是固定的，無法實現與高風險相對等的高收益，從而降低了金融機構對科創企業的信貸積極性。

因此，當前中國商業銀行融資體系支持新經濟產業及高科技企業的規模明顯不足。截至 2015 年末，新經濟及高科技行業[10] 獲得的銀行貸款餘額為 5,627 億元人民幣，佔整體銀行貸款總量比重不到 1%，顯著低於戰略性新興產業在國內生產

10　包括信息傳輸、計算機服務和軟件業及科學研究、技術服務和地質勘查業等行業。

總值中的規模佔比（見圖 7）。

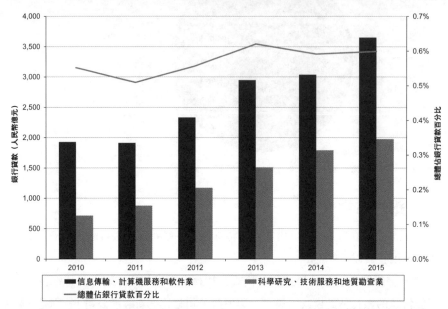

圖 7：新經濟產業及高科技企業從銀行體系獲得融資規模及佔比（2010 年至 2015 年）

註： 新經濟產業及高科技企業包括以下行業板塊 ——「信息傳輸、計算機服務和軟件業」及「科學研究、技術
服務和地質勘查業」。

資料來源：Wind 資訊。

2.2　創業投資和私募股權投資基金

目前創業投資（Venture Capital,“VC”）及私募股權投資（Private Equity,“PE”）
基金是新經濟科創企業獲得融資的最主要來源，而私募股權投資基金規模更大，
對新經濟企業發展更重要。隨着全球對新經濟投入資本持續增加，全球私募投資
規模從 2012 年的 452 億美元逐年增加到 2015 年的 1,285 億美元（見圖 8）。

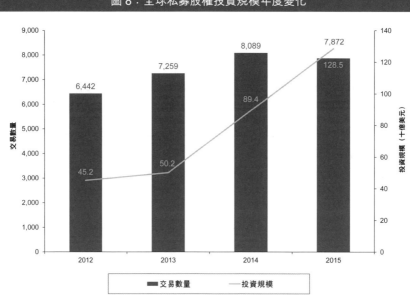

圖 8：全球私募股權投資規模年度變化

資料來源：Venture Plus Q4 2015, KPMG 和 CB Insights。

　　一般而言，從私募股權投資基金獲得融資取代上市，對科創企業的早期成長帶來不少好處。

(1) 從公司戰略發展角度來看，私募階段能讓公司專注與長遠戰略而不是被短期收益所局限，減少管理層花在與外界股東打交道上的時間和資源，能在不透露商業細節的同時保持競爭優勢。

(2) 私募股權投資基金除了為科創企業提供融資之外，還會提供一系列的增值服務。基於資本增值的考慮，私募機構通常會協助企業完善管理制度和治理機構，調動資源幫助企業發展，比如在人力資源、市場營銷、企業戰略等方面為科創企業尋找更好的培育環境和平台。此外，私募還會為企業提供豐富的行業、供應商和客戶等外部信息，幫助企業及時了解市場，促進內外部資源整合，提升競爭優勢。

(3) 隨着全球寬鬆貨幣流動性環境持續，私募基金已經可以為科創企業提供足夠的長期限資金，越來越多的公司通過一級市場的融資就能實現超過10 億美元的估值。

　　這就是為甚麼大量科創企業不斷推遲上市時間主要原因。2014 年美國科技公司上市時的平均年齡為 11 年，而在 1999 年美國科技公司上市時的平均年齡為 4 年。隨着上市準備時間的延長，許多風險投資人要花上更長的時間才能將投資回報變現 [11]。

　　但是從投資人角度來看，普通投資人通常希望科創企業的運作有更多的透明度和信息披露，私募方式運作難以滿足這些要求。而私募投資人通常要求合夥人承諾在 7 至 10 年的時間內取得收益。上市或併購是實現投資回報的最主要途徑，這也是私募投資人催促公司上市的主要原因。

2.3　IPO 上市

　　IPO 不僅是實現私募資本退出的最主要途徑，更可幫助公司從股票市場獲取大量資本，提升公司信譽，實現商業佈局和戰略增長，為公司長期發展帶來正向積極作用。

　　由於新經濟公司的科技創新及商業模式具有不同於其他行業的特殊性，新經濟公司在考慮上市地點時，除了資本容量、流動性等金融市場基本要素外，會更加關注與科創公司戰略發展相配合相適應的其他因素。比如，同行企業的上市地點選擇是否有利於形成產業集聚效應，是否有足夠分析師對新經濟公司的特殊業務模式進行深入專業分析，上市後能否與當地市場的科技能力結合形成新的創新動力等等。

　　另一個不容忽視的問題是，一旦 IPO 上市，隨着股權融資規模的不斷擴張，企業創始人的股權會逐漸稀釋，企業控制權自然會受到影響，這對以創始人為基本模式的科創企業尤為不利。一方面是融資所帶來的機遇和資產增值，另一面則是從創始人控股管理向民主透明的上市治理機制轉型可能帶來衝突。如何在企業成長與控制權之間權衡，是已經上市或準備上市的科創企業家們必須考慮的重要問題。

　　目前主要流行的做法是採用不同投票權的雙層股權制度滿足企業，特別是科技公司的私募投資者或創始人的要求，通過對資本治理結構的靈活設計，使得創

11　資料來源："Grow fast or die slow: Why unicorns are staying private"，麥肯錫研究，2016 年 5 月。

始人在公司進行 IPO 後，在不持有大多數股份的前提下仍能保持對公司的控制權。以百度在納斯達克上市為例，百度借鑒了谷歌（Google）的不同投票權股權設置，將股票分為 A、B 兩類，在美國股市新發行股票為 A 類股，而所有原始股為 B 類股。向外部投資人公開發行的 A 類股，每股只有 1 票的投票權；上市前原始股為 B 類股，每股有 10 票的投票權。B 類股可以按照 1：1 的比例轉換成 A 類股，而 A 股卻不能轉換為 B 類股。通過這種不同投票權股權設置，百度創始股東僅握有小比例但高投票權的股票，可有效控制上市後的企業。百度既獲得了需要的資金，又使創辦人的控制權得到了保障。

3 香港將在中國新興企業國際化佈局中承擔新的角色

　　一直以來，香港以其特殊的地理區位優勢和高度專業化、國際化的投資環境，成為中國內地與全球市場的聯繫紐帶與橋樑。在中國經濟轉型和創新產業不斷崛起的大背景下，香港市場如何與中國新興企業共融共生，為之提供最合適的國際化平台？這又將如何推動香港科創生態圈發展，成為香港經濟轉型的新動力？

3.1　具備「立足中國本土，實現國際慣例」的「主場優勢」，可成為新經濟企業的首選的資本運作市場

　　中國企業在上市和融資市場的選擇上，除了考慮市場規模與交易量、監督透明度與法制環境、估值水平、融資效率等主要因素外，該資本市場對中國企業所具有的熟悉度和契合度可能是更為重要的考慮因素（稱之為「本土效應」（home-country effect））。多年來，香港一直是眾多內地企業國際化的第一站。這與香港市場在全球資本市場中能夠為內地企業引入最多的國際投資者，且對內地企業最為熟悉不無關係。

　　在中國企業三十年不斷成長過程中，香港始終發揮了連接國際資本與內地企

業的樞紐作用，承擔着為內地企業籌集資金，改制轉型等多重功能。從 1993 年內地註冊企業 (H 股公司) 在香港上市開始，香港逐漸成為內地企業上市的融資門戶，而於海外註冊成立的紅籌股公司更早於上世紀八十年代已見於香港市場。目前香港已經成為內地企業 IPO 的最主要海外資本市場，截至 2017 年 11 月底，已有 1,041 家內地企業 (包括 H 股、紅籌股和民營企業) 在港上市，佔所有上市公司數量的 50%，佔港股市值的 66%[12]。這些數據表明，香港在中國企業海外上市中具備獨特「主場」優勢，而這些優勢在中國新經濟企業的戰略發展和各個成長階段，能夠發揮出優於美國等其他市場的價值發現和估值功能，繼而為科創企業的資本運作、轉型整合、併購擴張等一系列戰略發展提供最合適的國際化平台。通過香港資本市場的轉化，可為中國培育具有社會轉型創新意義的、優秀的科技公司。

3.2 憑藉開放的國際制度環境，可助力新經濟企業構建區域總部和進行國際化佈局

香港是公認的國際金融中心，香港的開放市場、法治環境、與國際接軌的監管制度與市場體系、專業人才和中英雙語環境是成就國際融資平台必不可少的條件。即使內地金融市場近年來已有了長足發展，但香港作為國際金融中心的獨特優勢依然十分突出，對內地企業進行海外投資及尋找合作夥伴具有重要的支持作用。

近年來，中國越來越多的高科技和新經濟企業走出國門，擴大對外投資，收購國外先進技術、研發能力和優勢項目，在海外建立區域總部。在助推中國企業「走出去」方面，香港除了提供低成本融資渠道外，還可以利用豐富的管理資源和通訊便利等優勢，幫助高科技企業突破貿易壁壘、監管障礙，為全球擴展制定有效的發展策略。例如，2015 年光大控股收購 Lampmaster 這家全球領先的高精密工業設備公司，通過光大香港進行運作，不僅以香港為平台設立了全球併購基金，籌集外部資金，享受香港平台提供的法律和人才優勢。再例如，金風科通過收購德國風機製造商 Vensys 公司的直驅永磁技術，加速產業技術經驗的積累，在美國企業核心技術上實現了跨越式升級，成為中國領先的風機製造企業。金風科技在

---

12 資料來源：香港交易所。

歐洲、澳大利亞、南非、拉美等國均有運營。在海外運營中，金風科技聘用香港工作的管理人員，更貼近海外市場和客戶需求。

未來中國科技創新企業在國際市場開展離岸投融資業務，他們均可以利用香港市場靈活的金融工具和融資支持體系，充分享受香港資本市場制度靈活性帶來的便利。香港可為他們在不同資本市場之間進行合作安排，孵化成長，為中國企業收購國際同行與先進技術，及尋找海外科技開發夥伴、進行技術合作和全球化佈局搭建平台。

3.3　對接中國新經濟資產和全球資本，成為全球投資人佈局中國新經濟資產的主要平台

目前中國的新經濟企業出現的一個顯著特徵在於，在引進或借鑒了發達國家的技術或創新後，經過中國市場培育後又催生了創新和變革，進而形成了世界性的影響力。CB Insights 數據庫顯示，全球 183 家獨角獸公司 [13]，43 家來自中國，其中估值在 100-600 億美元的有螞蟻金服、小米、滴滴出行、陸金所、眾安保險 [14]、大疆科技、美團點評等，分佈在互聯網金融、電子商務、信息傳媒等新經濟公司最為活躍的領域。這些公司一旦上市，有可能成為市場上的真正巨鯨，但每間公司的成長又異常複雜和漫長，投資人面臨巨大風險和不確定性。

香港的優勢在於，它不僅是中國企業走出去的首要平台，也是國際投資者佈局中國資產的最為熟悉的平台。中國各種各樣的新經濟資產規模已具雛形，根據中國科技部火炬中心聯合長城企業戰略研究所發佈的「獨角獸」企業榜單，中國有 131 家獨角獸企業 [15] 估值已達到 4,876 億美元，企業平均估值 37.2 億美元，中國已成為僅次於美國的全球第二大獨角獸聚集地。而香港市場，較其他市場更熟悉中國企業的經營狀況、盈利模式，投資者也更容易與中國科創企業管理團隊溝通並形成外部約束。隨着新經濟公司在此聚集，就可以形成圍繞不同新經濟行業發展的產業生態圈和投資者群體，以豐富的產品、完善的專業團隊和市場功能，在中國新經濟資產和全球資本之間搭建一個有效連接，實現全球金融資源與中國資產

13　「獨角獸」企業是指成立 10 年以內、估值超過 10 億美元、獲得過私募投資且尚未上市的企業。
14　眾安保險已於 2017 年 9 月 28 日在香港交易所正式掛牌上市。
15　該數據調查所涵蓋的獨角獸企業樣本與 CB Insights 的有一定區別。

的對接和整合，將新經濟科創企業的成果真正轉化成中國經濟增長動力。同時，對中國市場感興趣的國際新經濟公司，也可以將香港作為踏板，將在港上市作為涉足中國市場的第一步，繼而擴展國際企業在中國的業務。

3.4 科創企業及相關資本聚集於香港，將大大加快香港「科技生態圈」建設步伐，推動香港與中國內地新經濟企業共融共生、協同發展

目前香港本地的科技創新步伐正在加快，香港創新及科技局、香港科學院接連成立，2016-17 年《財政預算案》中提及多項創新科技措施及未來科技基建的籌備，2015 及 2016 年《施政報告》也提出多項創科政策，涉及政府開支金額超過182 億港元，為香港科創企業發展注入強心劑 [16]。**一旦將中國及全球新經濟企業引入，更可在科技創新行業的上、中、下游不斷豐富業態佈局，形成區域創新核心體系，增強產業集聚內企業的比較競爭優勢。同時配合港府近年來提出的「智慧城市、健康老齡化和機械人技術」的應用發展計劃，引入新經濟公司將加速香港「智能城市應用生態圈」的塑造，推動香港作為國際大都市、綠色生態城市的可持續發展。**

通過打造一個真正開放、多元創新的多層次金融市場，將為香港市場注入新的科技創新和城市轉型動力，香港也可助推提升中國新經濟企業的國際競爭力，在中國企業成就國際化和經濟體系開放中承擔新的角色和定位。新經濟產業崛起，為中國經濟增長注入新動力。

---

16 資料來源：香港創新科技署網頁。

第二部分

定息及貨幣產品

第6章

湯森路透 / 香港交易所
人民幣貨幣指數

2016 年 10 月

摘 要

湯森路透／香港交易所人民幣貨幣指數（RXY 指數）為人民幣兌中國最重要貿易夥伴的一籃子貨幣匯率提供獨立、透明和公允的基準。

RXY 指數採取可靠清晰的計算方法，以 WM／路透[1] 同日即期匯率為基礎計算，並嚴格遵循國際證監會組織（IOSCO）對金融市場基準的原則。這樣的計算方式令 RXY 指數比市場參與者內部自行開發而多不公開的人民幣估值模式更富優勢，相信能成為市場廣泛使用的人民幣基準指數。

RXY 指數亦能與內地央行為制定外匯政策而推出的 CFETS 人民幣匯率指數相輔相成，既與後者有高度相關性，亦會每小時發佈相關指數值，為所有市場參與者提供了高透明度和高使用度的指標。環顧整個市場，RXY 指數或許是現時唯一公開及可作買賣的人民幣指數系列，適合為期貨、期權及交易所買賣基金（ETF）等金融工具提供參考基準。隨着人民幣國際化不斷推進及內地金融市場日漸開放，有意增加人民幣投資及使用更多不同對沖工具的市場參與者將可望受益。

1　WM／路透指 World Markets Company/Reuters。

1　滿足市場對可交易人民幣指數的需求

1.1　人民幣日漸國際化

2015 年 12 月 11 日，內地央行中國人民銀行（人行）在中國外匯交易中心（CFETS）推出三隻新的作政策性用途的人民幣匯率指數，作為反映人民幣兌一籃子主要國際貨幣匯率表現的基準指數。這分別是 CFETS 人民幣匯率指數、BIS 貨幣籃子人民幣匯率指數及 SDR 貨幣籃子人民幣匯率指數[2]。

新政策對推動國際投資者轉變觀察人民幣匯率視角具重要意義，即由單一兌美元匯價轉變為兌多種貨幣匯價。此舉旨在減低人民幣兌美元匯率屏幕上予人的波動幅度的觀感，須知人民幣兌美元匯率大幅波動有時僅是反映美國國內的經濟事件，而與人民幣的國際價值無甚關聯。

將觀察人民幣匯率的視角轉為其兌一籃子貨幣的表現，主要由於中國今天無疑已成為國際貿易大國，國際貿易及金融活動使用人民幣已愈趨廣泛，單單一個美元兌在岸人民幣（USD/CNY）的雙邊匯率已無法反映中國與全球多個國家的貿易及金融關係。

近年與中國進行貿易的國家數目日增，中國已成為許多國家的主要貿易夥伴。世界貿易組織（WTO）的統計數據[3] 顯示，中國於 2004 年已取代日本，於 2007 年及 2009 年亦分別超越美國及德國，成為全球最大出口國。2015 年中國的商品出口額達 2.27 萬億美元，保持世界首位。中國貿易夥伴的數目及類別不斷增長，增進了對人民幣的需求，同時亦突顯出以雙邊匯率反映人民幣表現的局限性。（見圖 1）

2　CFETS 人民幣指數參考 CFETS 貨幣籃子，包括在 CFETS 掛牌的各人民幣對外匯交易幣種。BIS 貨幣籃子人民幣匯率指數參考國際結算銀行（BIS）貨幣籃子。SDR 貨幣籃子人民幣匯率指數參考國際貨幣基金組織的特別提款權貨幣籃子。

3　WTO《2015-2016 世界貿易》（World trade in 2015-2016），2016 年 4 月。

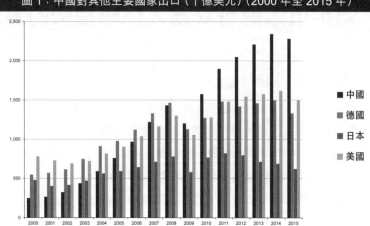

圖1：中國對其他主要國家出口（十億美元）（2000年至2015年）

資料來源：聯合國商品貿易數據庫。

　　除了中國在國際貿易的強勢地位，人民幣作為結算及投資貨幣的使用量亦不斷增長。根據 SWIFT[4] 數據顯示，2016年7月人民幣作為國際支付貨幣的市場份額排名第五，在短短兩年間跳升兩級。（見圖2）

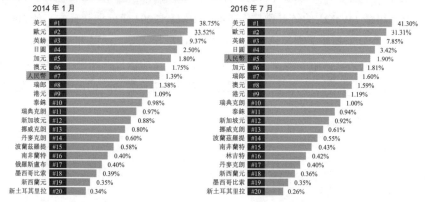

圖2：人民幣作為國際支付貨幣的市場份額（2014年1月及2016年7月）

資料來源：2016年7月 SWIFT, RMB tracker。

4　SWIFT《人民幣追蹤》（RMB tracker），2016年8月。

中國國家主席習近平於 2013 年提出「一帶一路」倡議，隨之啟動的大規模項
目預計將進一步增加對人民幣的需求。實施這項發展戰略會透過提升現有貿易基
礎設施並建構新設施，從陸路「絲綢之路經濟帶」及海路「21 世紀海上絲綢之路」，
將中國與亞洲、東亞、歐洲及非洲連接起來。中國已承諾投入 400 億美元到中國
絲路基金 [5]、1,000 億美元到亞洲基礎設施投資銀行（亞投行）[6] 及 500 億美元到新開發
銀行 [7]，支持「一帶一路」戰略中的項目。這勢將開拓人民幣國際化的版圖，提升與
「一帶一路」相關的貿易及金融交易對人民幣的需求。

1.2　人民幣市場自由化

最近 20 年，中國經濟在匯率機制的支持下實現了巨大飛躍。在中國的全球化
發展初期，人民幣緊盯美元匯率。2005 年，中國將與美元掛鈎的匯率制度轉為有
管理的浮動匯率制度，每日允許人民幣在兌美元 ±0.3% 的區間內波動 [8]，2014 年 3
月更將此區間放寬至 ±2% [9]。真正的里程碑是 2015 年中國匯率制度的改革，為人
民幣國際化奠定了基石：人行於 2015 年 8 月 11 日對人民幣兌美元的定價機制引
入重大改動，「做市商在每日銀行間外匯市場開盤前，參考上日銀行間外匯市場收
盤匯率，綜合考慮外匯供求情況以及國際主要貨幣匯率變化向中國外匯交易中心
提供中間價報價」[10]。

中國採用新的以市場定價為基礎的人民幣匯率機制，顯示出政府已做好準備，
允許人民幣匯率可在相當程度上參考主要貨幣及由市場力量推動。中國政府朝着
人民幣國際化而採取的措施獲國際貨幣基金組織（IMF）認同，2015 年 11 月 IMF
宣佈將人民幣納入特別提款權（SDR）籃子貨幣，於 2016 年 10 月生效。

外匯機制的市場化改革為使用人民幣從事更自由的跨境金融活動打下堅實基
礎。其後中國又出台進一步的放寬措施，包括放寬合格境外機構投資者（QFII）
投資額度及簡化其申請手續、開放境外機構投資者直接投資銀行間債券市場

5　"China's Silk Road dream falls into place with US$40b fund"，《南華早報》，2015 年 2 月 17 日。
6　〈中國批准亞洲基礎設施投資銀行協議〉，《經濟時報》，2015 年 11 月 4 日。
7　新開發銀行網址：http://www.ndb.int/brics-bank-to-begin-funding-of-projects-from-april-kamath.php。
8　《有管理的浮動匯率制度是既定政策》，人行，2016 年 7 月 15 日。
9　《中國人民銀行公告 [2014 年]5 號》，人行，2014 年 3 月 17 日。
10　"China defends new currency regime"，《金融時報》，2015 年 8 月 13 日。

（CIBM）。此等舉措使全球市場參與者及決策者對清晰的人民幣基準需求日高，以便可用以分析人民幣波動，更好地掌握趨勢及走向以權衡其全球人民幣風險敞口，以及進一步平衡有關人民幣的政策措施。

人行實施新的人民幣兑美元定價機制後不久，2015 年 12 月再推出 CFETS 人民幣匯率指數，將中國 13 個主要貿易夥伴的貨幣納入其中。同時發佈的還有 BIS 貨幣籃子人民幣匯率指數和 SDR 貨幣籃子人民幣匯率指數。這三隻指數為全球經濟活動所涉及的人民幣匯率波動提供新的基準指標。

表 1：利便人民幣國際化以至晉身全球儲備貨幣的近期政策	
日期	政策
2015 年 8 月 11 日	人行改革人民幣兑美元匯率中間價報價機制
2015 年 11 月 30 日	IMF 宣佈將把人民幣納入特別提款權（SDR）
2015 年 12 月 11 日	推出中國外匯交易中心（CFETS）人民幣匯率指數
2016 年 2 月 4 日	國家外匯管理局放寬合資格境外機構投資者（QFII）額度及簡化申請程序
2016 年 2 月 24 日	人行宣佈容許境外機構投資者直接投資中國銀行間債券市場（CIBM）
2016 年 5 月 27 日	人行及國家外匯管理局宣佈銀行間債券市場直接准入方案實施細則
2016 年 10 月 1 日	IMF 將人民幣納入 SDR

除出於監控人民幣波動之需要外，全球市場參與者無論是為管理人民幣投資組合又或通過看升／看跌人民幣匯率謀取投資收益，都需要對沖人民幣匯率風險，故來自這方面的需求也將不斷增加。要滿足這些需求，市場極需一隻具透明度兼可予買賣的人民幣指數及相關匯率工具。而這正是湯森路透／香港交易所人民幣貨幣指數（RXY指數或RXY指數系列）推出的背景。整個指數系列是由香港交易所與湯森路透聯合開發，於2016年6月23日正式推出，相信可切合現有人民幣指數無法滿足的市場需求（詳見下文第2及3節）。

2　匯價指數的國際經驗

2.1　央行指數

外匯市場近代史顯示不少國家央行均設有貨幣指數作為比較本國與其主要貿易夥伴貨幣匯價的經濟指標。1973 年美國聯邦儲備局推出的貿易加權美元指數即為一例，該指數現涵蓋 26 種貨幣，每年調整貨幣權重一次。貿易加權美元指數的主要功能是作為政策的宏觀經濟指標，地位超然，但儘管如此，卻並未發展成為金融交易工具。

在歐元區，歐洲央行由 1999 年開始發佈歐元兩種有效匯率，一種是兌歐元區主要貿易夥伴 19 種貨幣的匯價，另一種覆蓋更廣，反映與 38 個國家貨幣的貿易關係。類似貿易加權美元指數，覆蓋較廣的歐元有效匯率的貨幣權重亦是每年調整一次，能有力顯示歐元幣值。兩種歐元有效匯率主要用作評估外部經濟狀況以及國際價格和成本競爭力的重要指標，二者的表現也是歐洲央行評估歐元區內貨幣狀況和制定歐盟貨幣政策戰略的重要元素。

央行外匯指數近期最顯著的發展，應是 2015 年 11 月推出的 CFETS 人民幣匯率指數。雖然該指數的組成方法並非完全透明、公開，但能反映人民幣相對中國重要貿易夥伴貨幣的幣值。該指數參照中國外匯交易中心直接與人民幣買賣的一籃子 13 種貨幣計算。指數中每種貨幣的權重採用考慮轉口貿易因素的貿易權重法計算 [11]。CFETS 人民幣匯率指數推出後，在引導外匯市場參與者衡量人民幣走勢方面無疑發揮了重要作用，成功將注意力從雙邊人民幣兌美元匯價轉為參照一籃子貨幣。

2.2　可作交易的指數

有別於央行的政策貨幣指數，金融市場參與者亦有制定其他貨幣基準指標，這些指標符合金融市場要求及更適合用作交易用途。當中最成功的例子之一是由

11　資料來源：中國外匯交易中心網站（http://www.chinamoney.com.cn）。

ICE 美國期貨交易所（ICE Futures U.S.）創立、只覆蓋加元、瑞士法郎、歐元、英鎊、日圓和瑞典克朗六種貨幣的美元指數（USDX）。USDX 期貨合約其後於 1985 年 11 月在 ICE 美國期貨交易所上市。自推出以來，USDX 的貨幣權重只改動過一次，就是 1999 年以歐元取代數個歐洲貨幣。儘管近 10 年國際經濟形勢轉向中國，令現時 USDX 的一籃子貨幣並不完全反映美國最新經貿關係，但 USDX 期貨已成為全世界最獲廣泛認同的交易貨幣指數期貨，2015 年合約成交 1,200 萬張 [12]。不少資產值數以百萬計的交易所買賣基金（ETF）均與 USDX 掛鈎，如 Powershares DB Bullish and Bearish 基金、 WisdomTree Bloomberg USDX 基金等。

另一例子為紐約期貨交易所（NYBOT，後來被洲際交易所（ICE）收購）於 2006 年 1 月推出的歐元指數（EURX 或 EXY）。指數計量歐元相對一籃子五種貨幣（美元、英鎊、日圓、瑞士法郎及瑞典克朗）的幣值，初期反映了歐洲央行計算歐元有效匯率所用的權重。但到 2011 年 5 月，ICE 美國期貨交易所終止該指數的期貨及期權交易，不久更停止 ICE 歐元指數的計算。

美元和歐元均屬主要的國際貨幣，有能力引發市場發展相應可作交易的指數的需求，但發展中國家的貨幣則少見有相關指數的研發。RXY是最新推出的可作交易的貨幣指數，是人民幣國際化進程中首次以人民幣為相關貨幣推出的可作交易的貨幣指數。該指數系列中的主要指數 —— TR/香港交易所RXY全球離岸人民幣指數 —— 尤具相關參考價值。

表 2：央行指數及其相應可作交易的指數的比較			
央行指數	美聯儲局的貿易加權 美元指數	歐洲央行的歐元有效 匯率指數	CFETS 人民幣匯率指數
推出時間	1973	1999	2015
發佈方	美聯儲局	歐洲央行	中國人民銀行
成分幣種數目	26	19/38	13
指數調整	每年	每年	每年
計算方法	幾何平均	幾何平均	幾何平均
權重	貿易加權	貿易加權	貿易加權
可交易性	無	無	無

12　The ICE, US Dollar Index Futures, Historical Monthly Volumes.

(續)

表 2：央行指數及其相應可作交易的指數的比較			
可作交易的指數	USDX / DXY	EURO / EXY	RXY 全球離岸人民幣指數
推出時間	1985	2006	2016
發佈方	ICE 美國期貨交易所	ICE 美國期貨交易所	湯森路透／香港交易所
成分幣種數目	6	5	14
指數調整	固定	固定	每年調整
可作交易的產品	期貨、期權、ETF	期貨、期權	尚無 *

* 可推出以RXY作基準的產品包括期貨、期權及ETF。香港交易所正考慮推出RXY期貨，待監管部門的批准。

資料來源：美國聯邦儲備銀行、歐洲央行、CFETS、ICE 美國期貨交易所、香港交易所。

2.3　其他指數

在人民幣市場中，部分市場參與者自行開發內部指數以計量人民幣匯價表現。銀行往往為了宏觀經濟分析而研發本身的內部基準，從中觀察匯市趨勢。就此而言，實際有效匯率（REER）指數（指經就國內通脹率調整的指數）能反映一國貨幣相對其他貨幣的強弱，所以是最佳指標。實際有效匯率以外，銀行亦會創設一些人民幣計價模式，側重若干具體範圍（例如出口或某些行業）又或人民幣在某些方面的發展（例以以全球化作主題）。有些銀行會使用其內部人民幣指數作為構建其外匯衍生產品策略的基準或參考匯率。

資產管理方面，內部開發的人民幣指數往往被市場機構的交易部用作人民幣走勢的指標，此外也常用於分析投資組合中人民幣計價資產的表現，或確定人民幣風險敞口的對沖策略。不過，基於投資經理之間的競爭，加上大家都追求更高回報，這些內部模式多不公開。

這些由銀行、證券行或資產管理公司等不同市場機構制定的內部指數，作為內部估值使用或屬良好基準，但卻作不了金融界的主要人民幣計價工具。部分原因是機構制定這些指數都是為了達到不盡相同的內部目標，其獨立性會是問題。其次，內部人民幣指數少有符合如國際證監會組織（IOSCO）所立原則的國際標準，部分原因是由於按國際原則調整、維持及認證內部基準所涉及的成本和人力都不少。因此，由於沒有獨立性、也不符合有關金融基準的國際規則，計算方法又往往不具透明度，內部工具始終無法成為業界的指標工具。

3　RXY 指數系列

RXY 指數由獨立機構香港交易所及湯森路透共同開發，旨在為市場參與者提供業內最高水平的人民幣表現指標。

RXY指數包括一隻主要指數（**TR/香港交易所RXY全球離岸人民幣指數**）以及三隻指數變體（**TR/香港交易所RXY全球在岸人民幣指數、TR/香港交易所RXY參考離岸人民幣指數及TR/香港交易所RXY參考在岸人民幣指數**）。該等指數的不同之處在於基準貨幣籃子（全球籃子包括14種貨幣，參考籃子包括13種貨幣，詳見下文第3.1節）以及計量的人民幣單位（在岸人民幣（CNY）及離岸人民幣（CNH））。RXY指數的基準日期為2014年12月31日，基數為100，與CFETS人民幣匯率指數相同，而RXY指數的歷史數據可追溯至2010年12月31日。圖3為系列四隻指數概要，圖4為其歷來表現。

圖3：四隻 RXY 指數

基準貨幣

		離岸人民幣	在岸人民幣
籃子貨幣	全球	TR／香港交易所 RXY全球離岸人民幣指數	TR／香港交易所 RXY全球在岸人民幣指數
	參考	TR／香港交易所 RXY 參考離岸人民幣指數	TR／香港交易所 RXY參考在岸人民幣指數

- ■ 主要指數
- □ 指數變體

基準日：　　　　2014 年 12 月 31 日
基準值：　　　　100
時間系列起始日：2010 年 12 月 31 日

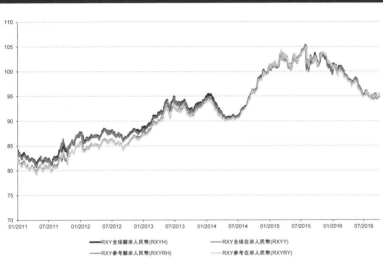

圖 4：四隻 TR／香港交易所 RXY 指數的歷來表現
（2010 年 12 月 31 日至 2016 年 9 月 30 日）

資料來源：香港交易所。

　　RXY指數按IOSCO對金融基準的原則[13]進行管理，確保指數管理客觀、有規可循，以及計算方法具透明度、所用匯率來自WM／路透。指數依循IOSCO原則編制，使其兼具雙重優勢 —— 不僅合資格用作估量人民幣價值的基準，還可作為期貨、期權及ETF等金融工具的相關參考。此外，自2008年全球金融危機後，針對投資產品的市場監管日趨嚴格，對產品以及其所依循的參考或相關資產的透明度和公正性均有更高要求。這樣的監管環境下，符合IOSCO原則的RXY指數可被市場用以發行投資產品，以切合不同類別機構及散戶投資者的需要。

　　湯森路透管理 RXY 指數。在由事務專家委員會與一位專責指數經理構成的框架下，湯森路透負責維持 RXY 指數的公正性及質素，以及進行以下例行工作及職能，包括：

- 闡釋指數編制方法，實施年度調整程序；
- 檢視來自指數利益相關者的意見；

13　資料來源：IOSCO 網站文件（www.iosco.org/library/pubdocs/pdf/IOSCOPD415.pdf）。

- 因應利益相關者意見或市場活動而研擬並實施對指數編制方法的修改方案；

- 就指數調整及指數編制方法修改事宜與**指數顧問小組**（**Index Advisory Group**，「IAG」）及**指數行動委員會**（**Index Action Committee**，「IAC」）溝通；及

- 向湯森路透基準監督委員會（**Thomson Reuters Benchmarks Oversight Committee**， 「TRBOC」） 報告。

在與 IAC 互動之後，如有需要，IAG 和指數經理負責確定對指數編制方法的任何更改。

IAC 是由湯森路透內部的指數專家和各資產類別的專家組成，為指數經理提供與指數編制方法的解釋或變更相關的建議。指數經理可以傳達從 IAG（小組成員包括香港交易所代表）和 / 或指數利益相關者處獲得的反饋以徵求 IAC 的意見。IAC 則向湯森路透基準監督委員會報告。

3.1　產品設計

按 RXY 指數的設計，基準貨幣（在岸人民幣或離岸人民幣）兌其基礎籃子貨幣升值時，指數會上升，當基準貨幣兌其基礎籃子貨幣貶值時，指數會下跌。指數每小時計算，並於交易日香港時間下午 4 時計算收市值。

每一 RXY 指數按以下公式計算其在任何時間點 (t) 的 I_t 值：

$$I_t = I_0 \cdot \prod_i \left(\frac{FX_{i,t}}{FX_{i,0}}\right)^{\omega_i}$$

其中，$FX_{i,t}$ 是籃子中貨幣 i 在時間 t 時其兌基準貨幣的即期匯率。I_0 及 $FX_{i,0}$ 分別是上一個重整時間的指數及即期匯率，而 w_i 是貨幣 i 的權重（使 $\Sigma w_i = 1$）。指數使用幾何平均算法及即期匯率轉換（以上公式所用的匯率全部源自兌美元的即期匯率）計算。

RXY 指數系列涉及兩個籃子：一是 14 種貨幣的**全球籃子**，一是 13 種貨幣的**參考籃子**。兩個籃子貨幣均按貿易加權，即個別貨幣在指數中的權重取決於其國家與中國的實際貿易量。

全球籃子中的 14 種貨幣為：

- AUD —— 澳元
- CAD —— 加元
- CHF —— 瑞士法郎
- EUR —— 歐元
- GBP —— 英鎊
- HKD —— 港元
- JPY —— 日圓
- KRW —— 韓元
- MYR —— 馬來西亞林吉特
- NZD —— 新西蘭元
- RUB —— 俄羅斯盧布
- SGD —— 新加坡元
- THB —— 泰銖
- USD —— 美元

參考籃子不包括韓元，即與現時 CFETS 人民幣匯率指數籃子中的 13 種貨幣相同，權重亦相若（見下圖 5）。RXY 指數中的貨幣權重每年重整以反映最新的貿易數據。計算權重的數據來自聯合國商品貿易統計數據庫（UN Comtrade）[14] 所提供的中國與其他國家之間的年度貿易量，但當中 UN Comtrade 所報中國內地與香港之間每年雙邊出口數據，會根據香港政府統計處 [15] 的貿易數據作出調整，因為中國內地對香港出口中有相當大部分都不是為香港所用，需要再作計算以得出實際由香港吸納的中國內地出口量數據。RXY 指數的貨幣權重參考國際貿易數據，其高透明度使得指數的成份貨幣的變動極具預測性。

---

14 聯合國商品貿易統計數據庫（www.comtrade.un.org）。

15 香港政府統計處（http://www.censtatd.gov.hk/home/）。

圖 5：RXY 指數及 CFETS 人民幣匯率指數的貨幣權重

RXY 全球人民幣指數中的貨幣權重 [1]

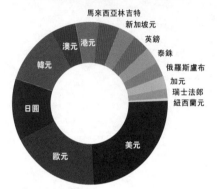

貨幣	貨幣權重		
	RXY 全球 人民幣指數[1]	RXY 參考 人民幣指數[1]	CFETS 人民幣匯率指數[2]
美元	24.69%	28.09%	26.40%
歐元	18.47%	21.03%	21.39%
日圓	12.27%	13.97%	14.68%
韓元	12.14%	0.00%	0.00%
澳元	5.02%	5.72%	6.27%
港元	4.89%	5.56%	6.55%
馬來西亞林吉特	4.29%	4.88%	4.67%
新加坡元	3.55%	4.04%	3.82%
英鎊	3.46%	3.93%	3.86%
泰銖	3.32%	3.78%	3.33%
俄羅斯盧布	2.99%	3.41%	4.36%
加元	2.45%	2.79%	2.53%
瑞士法郎	1.95%	2.22%	1.51%
紐西蘭元	0.51%	0.58%	0.65%

註：
（1）自 2016 年 10 月 3 日起生效的 RXY 指數中的權重，有效期至 2017 年 9 月 29 日。
（2）CFETS 人民幣匯率指數於 2015 年 12 月 8 日推出。指數內的貨幣權重來自該日 CFETS 在其網站上所發
　　的公告。
資料來源：香港交易所、CFETS。

　　調整 RXY 指數涉及每年更新各指數籃子中的成份貨幣權重，調整周期由每年
6 月在 UN Comtrade 及香港政府統計處發佈年度貿易統計數據後着手展開。IAG
研究初步權重計算後，在 6 月最後一個營業日公佈更新後的權重。新的權重於 10
月第一個交易日生效。

　　RXY 指數系列設計具有可作交易的指數的特點，包括符合監管規定、數據來
源可靠、指數編制公司聲譽良好、編制方法具透明度、發佈指數時間頻密（每小時
一次）等等。這些都是在公開市場上僅有的官方 CFETS 人民幣匯率指數所未能滿
足的。

3.2　使用及裨益

　　RXY指數是依據WM／路透所提供的匯率計算；WM／路透受英國金融市場行
為監管局（FCA）規管。WM／路透即期匯率受持續監控，確保符合行業最佳做法
的至高標準，故能將操縱或控制價格的機會減至最低。

　　從政策角度來看，RXY 指數為市場分析師及經濟學者提供了一個具透明度兼
每小時更新的人民幣匯率基準，有助他們分析人民幣匯率波動及前瞻人民幣走勢。

由於與 CFETS 人民幣匯率指數有高度相關性，RXY 指數預計會是作為 CFETS 人民幣匯率指數穩妥的代替品，因此亦會利便市場用來分析中國政府的外匯政策。市場參與者每日留意人行公佈的在岸人民幣兌美元的匯價變化，並與 RXY 指數對照，當不難窺見中國外匯政策的方向。

圖 6：RXY 指數及 CFETS 人民幣匯率指數的表現與在岸人民幣兌美元的匯價（2015 年 8 月 11 日至 2016 年 9 月 30 日）

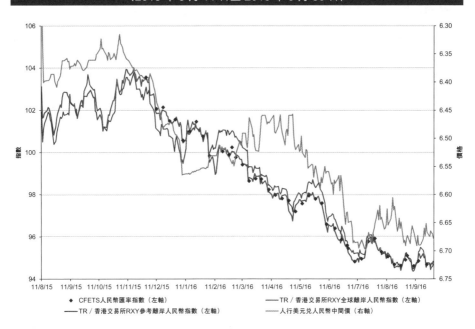

註：圖中 CFETS 人民幣匯率指數於 2015 年 11 月 30 日重整基準值為 TR／香港交易所 RXY 參考在岸人民幣指數的值，以作比較。

資料來源：香港交易所、CFETS、湯森路透。

　　從交易角度來看，RXY 指數可作為顯示源自市場推動的人民幣匯率走勢的尚佳指標，市場參與者可透過看漲／看跌人民幣匯率以作投資考慮。承受人民幣匯率風險的國際投資者亦可利用 RXY 指數更好地管理貨幣風險。

指數	回報						風險			與 CFETS 人民幣匯率指數的相關性 (30/11/2015-30/9/2016)	對 CFETS 人民幣指數的 Beta 系數 (30/11/2015-30/9/2016)
	2016 年 7 月		2016 年 8 月		2016 年 9 月		30 日波幅（截至）				
	按月	按年	按月	按年	按月	按年	29/7/2016	31/8/2016	30/9/2016		
TR／香港交易所RXY全球離岸人民幣指數（RXYH）	0.09%	-8.30%	-0.74%	-6.43%	-0.46%	-8.48%	3.44%	3.38%	3.95%	0.9963	0.6432
TR／香港交易所RXY全球在岸人民幣指數（RXYY）	-0.05%	-8.34%	-0.68%	-7.28%	-0.46%	-8.02%	3.40%	2.97%	3.43%	0.9817	0.6703
TR／香港交易所RXY參考離岸人民幣指數（RXYRH）	0.48%	-7.99%	-0.72%	-6.06%	-0.38%	-8.30%	3.31%	3.22%	3.90%	0.9933	0.6570
TR／香港交易所RXY參考在岸人民幣指數（RXYRY）	0.34%	-8.02%	-0.65%	-6.90%	-0.38%	-7.72%	3.55%	3.09%	3.57%	0.9860	0.6533

表 3：RXY 指數的風險及回報

資料來源：香港交易所、湯森路透、CFETS。

　　此外，RXY 指數亦可為中國政府所善用。從人行與匯市主要業者之間就匯率機制進行的公開對話中看到，中國有關當局可將由市場帶動的 RXY 指數作為價格發現工具來使用。

　　RXY 指數旨在為期貨、期權及 ETF 等金融工具提供參考基準，其設計旨在確保凡以其作為參考基準的衍生產品均可透過套戥交易獲得公平定價。面對人民幣國際化快速發展及市場隨之對人民幣金融產品的殷切需求，相信以 RXY 指數為相關資產的期貨及期權合約等人民幣對沖工具應可大派用場。RXY 指數有望成為外匯產品開發及估值領域中一個舉足輕重的工具。

第 7 章

香港交易所五年期中國財政部國債期貨

—— 全球首隻可供離岸投資者交易的人民幣債券衍生產品

2017 年 4 月 24 日

摘　要

　　經過數年迅速增長，中國債市已成為全球第三大債券市場，存量規模達人民幣 56.3 萬億元（約 8.1 萬億美元）[1]。中國在推進人民幣國際化及開放國內金融市場方面亦取得長足進展。目前儘管外資參與中國債市比例仍較低，境外資金已顯示出對中國主權債券的強烈興趣，並在人民幣獲納入國際貨幣基金組織的特別提款權貨幣籃子後快速增長。如果在不久將來內中國內地與香港推行「債券通」，外資投資中國債券比例將繼續上升，並將推動風險管理需求增長。

　　發展有效的風險管理工具和外匯交易服務對境外投資者增持人民幣資產至關重要。目前，境內市場現有的利率風險管理產品已為人民幣利率風險對沖提供了支撐手段，近期隨着境內外匯市場進一步開放，一些合格境外投資者也可直接使用境內的衍生品。香港交易所的國債期貨合約利用離岸市場的產品優勢為境外投資者提供差異化服務，並在產品設計中加入多項特性，以令該產品交易不太可能對在岸市場產生不利影響。

　　根據發達國家經驗，引進國債期貨在提高債券市場定價功能，促進現貨市場流動性，豐富債券投資者利率風險管理手段等方面將起到重要作用。大多數實證研究發現，引入國債期貨對現貨市場也不會有顯著影響或導致波動性下降。香港交易所推出國債期貨為境外投資者提供了對沖人民幣資產利率波動的有效工具，也是推動境外資本流入中國境內債券市場的重要步驟。銀行、資產管理公司、經紀公司和保險公司是本產品的主要目標客戶。

1　數據來源：Wind 資訊，截至 2016 年底。

1 中國境內債市逐步開放

　　中國債券市場規模急速擴張，過去五年債券存量以簡單年均增長率 21% 的速度增長，成為全球第三大債券市場，債券存量規模達人民幣 56.3 萬億元（約 8.1 萬億美元）（見圖 1）。

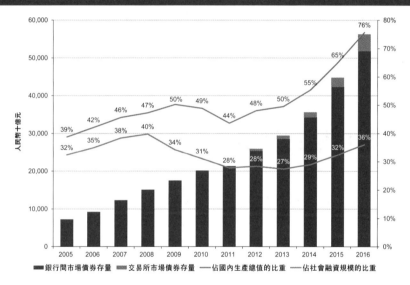

圖 1：中國債券存量規模佔國內生產總值和社會融資總量的比重（2005 至 2016 年）

資料來源：中央國債登記結算（中債登）的債券市場年度統計分析報告（2011 - 2016），因數據缺失不包括 2005-2011 年度交易所市場及上海清算所清算的債券數據。國內生產總值及社會融資總量的數據取自 Wind 資訊。

　　中國致力促進其債券市場在金融資源配置上擔當更重要角色，同時亦在推進人民幣國際化及開放境內金融市場方面取得長足進展。其中一個主要方向是鼓勵外資參與境內債市，推廣債券市場的多元化及多樣性，進一步擴展境內金融市場的規模及深度。其中包括 2010 年首度容許合格機構使用離岸人民幣投資於銀行間債券市場，翌年（2011 年）再推出人民幣合格境外機構投資者（RQFII）計劃，兩年後（2013 年）放寬合格境外機構投資者（QFII）的投資限制，進一步推動境內債市開放。

及至 2015 年，內地接連推出多項令人矚目的放寬措施，進一步便利境外投資者進入中國銀行間債券市場（CIBM），具體包括：2015 年 5 月底，中國人民銀行（人行）允許離岸人民幣清算行及參與銀行利用在岸債券持倉進行回購融資；2015 年 7 月中，人行進一步放寬合格債券交易範圍，容許合格主體參與銀行間債券市場的債券現券買賣、債券回購、債券借貸、債券遠期、利率互換及人行許可的其他交易；2016 年 2 月，人行發佈新規，放寬境外機構投資者進入銀行間債券市場的規則；以及 2016 年 5 月進一步頒佈詳細規則，釐清境外機構投資者在銀行間債市的投資流程。

這些政策舉措在一定程度上向市場表明，中國正逐步開放資本項目並鼓勵更多外資流入。

2 對中國主權債券需求持續增長

2016 年人民幣正式獲納入特別提款權貨幣籃子，各國央行及全球投資者開始考慮將資金重新配置到人民幣計價資產，市場對人民幣資產（特別是人民幣債券資產）的需求勢將穩步上揚。獲得特別提款權的地位提升了全球市場對人民幣作為全球投資及儲備貨幣的認受性，很大可能促進國際間政府及私人部門對人民幣計價資產的需求。根據我們的估算，若境外機構或個人持有的人民幣資產增至佔境內債市總存量 10%，預料未來數年將有逾人民幣 9.5 萬億元流入相關人民幣債券資產。

債券資產（特別是國債）一向是各國央行及全球資產管理基金經理首選的資產配置類別。目前儘管外資參與中國債市比例仍較低，境外資金已顯示出對中國國債的強烈興趣，近年外資所持主權債券亦大幅增長。2016 年，境外參與者所持的人民幣國債及政策性銀行債券增加了 2,330 億元人民幣，較 2015 年 350 億元飆升六倍。外資在中國主權債券市場上的佔比由 2015 年底的 2.62% 增至 3.93%（見圖 2）。由於目前主要發達國家的主權債券仍處於低（甚至負）的息率環境，若中國出台更多歡迎外資參與債市的舉措，有可能出現明顯的資本從其他金融領域轉投中國主權債券市場的轉換趨勢。

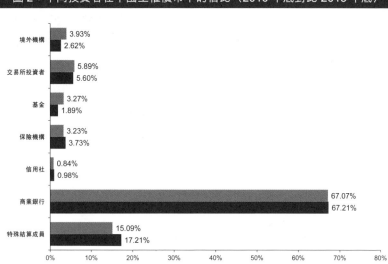

圖2：不同投資者在中國主權債市中的佔比（2016年底對比2015年底）

資料來源：Wind 資訊。

3 離岸市場在風險對沖及可參與性方面的優勢

發展有效的風險管理工具和外匯交易服務對境外投資者增持人民幣資產至關重要。目前中國境內定息衍生產品市場有一定深度和流通性，可供交易的外匯產品種類繁多（包括現貨、遠期、掉期及期權等），以及國債期貨產品。然而迄今為止，境內的國債期貨產品還未對境外機構完全放開用以風險對沖，並且境內保險公司及銀行這些國債現券的主要持有人，亦同樣未能參與國債期貨交易。境內債市分割可能影響流通性和市場深度。

　　如市場推出流通性好的債券期貨產品，可有助於中國境外投資者提高對沖利率風險的能力，增強他們增持中國債券資產的意願。2017 年 3 月 15 日，中華人民共和國國務院總理李克強公開表示準備在香港和內地試行「債券通」計劃[2]。如果在不久將來推出，外資投資中國債券比例將繼續上升，並將推動風險管理需求增長。

　　正是在這樣的背景下，香港交易所設計並推出五年期中國財政部國債期貨（國債期貨）。目前，境內市場現有的利率風險管理產品已為人民幣利率風險對沖提供了支撐手段，近期隨着境內外匯市場進一步開放，一些合格境外投資者也可直接使用境內的衍生品。香港交易所的國債期貨合約利用離岸市場的產品優勢為境外投資者提供差異化服務，它的推出將為境外投資者提供對沖人民幣資產利率波動的有效工具，也是推動境外資本流入中國境內債券市場的重要步驟。

4　在岸及離岸市場上用以對沖中國債券資產的產品

　　國債期貨合約是場內交易的利率衍生產品市場的重要組成部分之一，其設計原則使得期貨價格趨同於指定年期（如兩年、五年、十年或三十年）之最流通的主權債券。因此，國債期貨合約是對沖主權債券息率利率風險的重要工具。舉例而言，某公司按高於政府國債的固定息差借貸，或是投資於此公司債券的基金經理，都可使用國債期貨衍生產品進行對沖。

　　目前，中國境內市場的國債期貨為於中國金融期貨交易所（中金所）上市的五年期和十年期財政部國債期貨合約。五年期合約於 2013 年 9 月 6 日推出，十年期合約則於 2015 年 3 月 20 日推出。2017 年 3 月，該債券期貨的日均成交達 677.3 億元人民幣，未平倉合約為 845.7 億元人民幣（見圖 3）。然而，境內的國債期貨產品還未對境外機構完全放開用以風險對沖，並且由於缺乏境內的保險公司和銀

2　2017 年 3 月 15 日全國人大會議閉幕後，李克強總理在回答中外記者提出的問題時表示，準備今年在香港和內地試行「債券通」，也就是説允許境外資金在境外購買內地的債券，這是第一次。

行等主要參與者，境內的國債期貨的流動性有限。

離岸市場方面，在離岸國債期貨發行之前，市場也缺乏有效的中長期人民幣利率風險的對沖工具。過去對沖人民幣利率風險主要是不交收利率掉期與離岸人民幣利率掉期。前者的定價較易受到投機因素而非基本資金流動所影響，故一般來說不會視為有效的對沖人民幣利率工具，而後者則隨着離岸人民幣貨幣市場增長已不斷發展，令更多市場交易從不交收利率掉期轉移至離岸人民幣利率掉期。不過，離岸人民幣利率掉期的定價亦有其局限，由於市場流通量相對低加上欠缺借貸需求，離岸人民幣存款利率往往遠低於在岸利率，導致離岸人民幣利率掉期與境內利率產品存在價差（見圖 4）。香港交易所的財政部國債期貨可配合現有人民幣不交收利率掉期的收益率曲線，為離岸投資者提供了針對中國境內資產長期利率的基準工具。

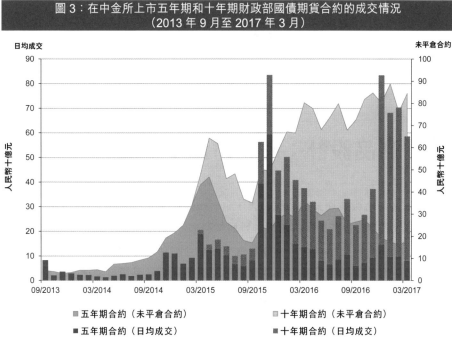

圖 3：在中金所上市五年期和十年期財政部國債期貨合約的成交情況
（2013 年 9 月至 2017 年 3 月）

資料來源：彭博。

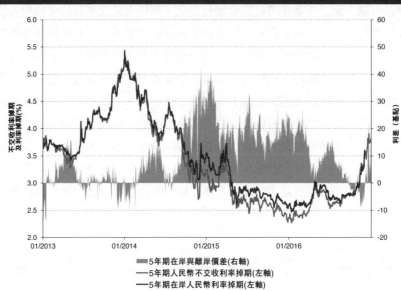

圖 4：不交收利率掉期與在岸人民幣利率掉期的表現（2013 年至 2016 年）

資料來源：彭博。

5 產品設計：釐定方法及應用

　　香港交易所國債期貨產品設計與中金所國債期貨合約的相似性在於，其相關資產皆為境內發行中國財政部國債、票面年利率皆為 3%，而不同之處在於境內的國債期貨採用現貨交割設計，容許短倉一方在合約到期時向長倉一方交付合資格債券中最便宜的債券。另外，香港交易所的國債期貨產品設計與澳洲證券交易所（澳洲證交所）及韓國證券交易所（韓國交易所）上市的政府債券期貨相也有一定相似之處[3]。

3　類似澳洲證交所及韓國交易所上市的政府債券期貨，香港交易所的國債期貨合約以債券籃子為相關資產並採取現金結算方式。澳洲證交所及韓國交易所會於合約開始交易日前公佈相關債券籃子。到最後交易日，合約均按根據債券籃子中的成份債券的平均收益率作現金結算。

5.1　建構債券籃子的原則

香港交易所國債期貨相關債券籃子在透明度、可預測性、流通量、易於追蹤複製及可靠性的原則下建構。

a) 透明度及可預測性：債券籃子及參考價均按既定規則釐定，包括定價、公式及算術模型在內的釐定方式皆屬公開資料。香港交易所仍保留權利，在有需要時因應中國國債發行政策的重大修訂而行使酌情權。

b) 成份債券流通量：債券籃子成份債券應為流通量普遍良好，以便對沖。因此，債券籃子成份債券必須為籃子建構當前按中債登[4]計算相對流通量最佳的三隻債券。

c) 易於追蹤複製：根據過去表現，債券籃子總成交量應佔合資格債券總成交量至少 50%。以債券籃子為基準的期貨，應能緊貼五年期財政部國債表現。因此，投資者可易於追蹤複製相關債券籃子以作對沖。

d) 價格可靠度：債券籃子參考價由中債登每日提供。中債登是中國財政部批准設立、以發展及運營國庫債券託管系統的國有獨資非銀行官方金融企業。

5.2　就每張期貨合約釐定的每日參考價

中債登按照香港交易所全資附屬公司香港期貨交易所提供的流程及演算法，確定債券籃子，以及計算每張期貨合約對應的債券籃子的每日參考價。

每張期貨合約（按季度）的債券籃子按下述安排釐定：

a) 釐定籃子日期：期貨合約開始交易日前 5 個工作日；

b) 債券必須是流通量最高的三隻債券（按中債登的相對流動性系數計算）；及

c) 流通量演算法：根據釐定籃子日期前 22 個工作日的交易數據計算。

類似於澳洲證交所及韓國交易所政府債券期貨設計，每張期貨合約的每日參考價均按以下公式計算：

a) 按成份債券當天中債登收益率，以 r_1、r_2、r_3 標示

4　中債登即中央國債登記結算有限責任公司，為中國境內債券市場的中央託管機構。

b) 計算債券籃子的算術平均到期收益率：

$$r = \frac{\sum_{i=1}^{3} r_i}{3}$$

c) 計算債券籃子的參考價格：按以下公式計算票面年利率 3% 的五年期債券：

$$\sum_{i=1}^{5} \frac{3\% \times 100}{(1+r)^i} + \frac{100}{(1+r)^5}$$

r 為（b）所計算的平均到期收益率。

5.3 假設性示例（僅作解說用途）[5]

例 1 —— 對沖利率變動

假設某基金經理擔心中國貨幣環境可能逐步收緊，希望對沖利率風險。2016 年 10 月 31 日，基金經理持有面值人民幣 1 億元的國債 160014.IB @101.813，久期 5.901。香港交易所國債期貨 2017 年 3 月合約成交價 102.282，年期 4.80。為消除貨幣久期效應，他賣出 245 張香港交易所國債期貨 2017 年 3 月合約，對沖持倉。2017 年 1 月 26 日，收益率上升，債券價值跌至 98.439 (-3.374)，錄得虧損人民幣 340 萬元。香港交易所國債期貨 2017 年 3 月合約價格跌至 99.480 (-2.802)，基金經理平倉，獲利人民幣 340 萬元。持現貨債券倉位虧損人民幣 340 萬元，由國債期貨倉位獲利人民幣 340 萬元所抵銷。

例 2 —— 久期管理

假設某組合經理持有市值人民幣 3 億元、久期 7.00 的多元化債券組合，根據基金聲明的投資目標，她可靈活調整久期至增減 10%。經理預期利率將下跌，因此擬將久期增至 7.70。香港交易所國債期貨合約當前成交價 102.282，年期 4.80。她可買入 86 張香港交易所國債期貨合約。

例 3 —— 合成債券

假設某境外機構投資者不可參與中國在岸債市，但基於中國市場的息差，希望有相類似債券投資安排作為替代，故擬構造合成現貨債券倉位。他可買入 100

5　這些例子並不構成投資建議，投資者應適時諮詢獨立投資意見。若涉及高風險的策略，投資者或會損失全盤投資。

張香港交易所國債期貨合約，造出面值人民幣5,000萬元的替代債券倉位。

　　例4 —— 信貸差價交易

　　假設某投資者預期企業債券收益率將偏離國債期貨收益率。如投資者預期信貸差價（企業債券收益率減去國債期貨收益率）將會收窄，則可買入企業債券同時賣出國債期貨。相反，如投資者預期信貸差價將會擴闊，則可賣出企業債券同時買入國債期貨。

5.4　歷史回溯分析

　　以2013年9月至2016年底為測算區間，香港交易所的模擬國債期貨參考價，與中金所以現貨交割為設計基礎的國債期貨的年化相關系數為92.1%（見圖5），表明香港交易所可滿足國際投資者有效應對日增的利率風險管理需求。銀行、資產管理公司、經紀公司及保險公司可為該產品的主要目標用戶。

圖5：香港交易所模擬國債期貨參考價與中金所期貨的相關性（2013年9月至2016年底）

資料來源：彭博、香港交易所。

　　再者，香港交易所國債期貨因與人民幣債券收益率指數高度相關，可視作人民幣債券收益率指數的替代。香港交易所的模擬國債期貨債券籃子的收益率表現緊貼中債登發佈的五年期國債收益率（見圖 6）。以 2011 年至 2016 年為測算區間，兩個系列的到期收益率的年化相關系數為 98.3%。因此，香港交易所國債期貨為測算中國債券資產價值，提供了相對便捷的工具。

圖 6：香港交易所模擬國債期貨參考價與中債登國債收益率的相關性
（2008 年 6 月至 2016 年底）

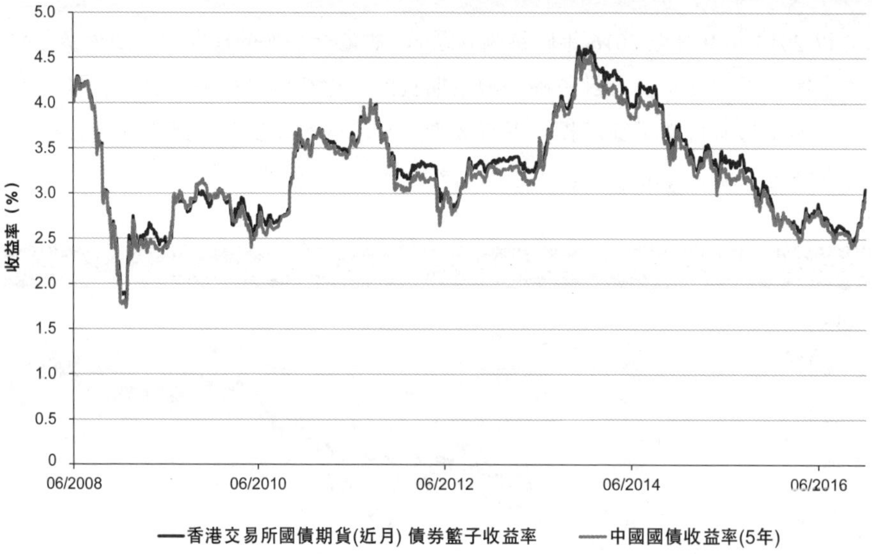

資料來源：Wind 資訊、香港交易所。

6 相互影響及有效性

　　根據發達國家經驗，引進國債期貨在提高債券市場定價功能，促進現貨市場流動性，豐富債券投資者利率風險管理手段等方面將起到重要作用。大多數市場文獻的實證研究發現[6]，引入國債期貨對現貨市場波動性不會有顯著影響，或導致波動性下降。

　　香港交易所的國債期貨經審慎設計，產品設計中加入多項特性，以令該產品交易不太可能對在岸市場產生不利影響。事實上，此產品對在岸定息產品市場的發展反而可起支持作用。

(a)　香港交易所的國債期貨合約是以離岸市場人民幣進行現金差額結算。合約到期後，以交易結算為目的的交易量在離岸市場進行，僅佔全部合同名義金額的一小部分。因此，與實物交割的期貨合約相比，該結算過程對流動性的影響要小得多。

(b)　香港交易所的國債期貨合約的結算價基於債券籃中三隻成份債券的平均收益率計算，該三個成份債券為債券範圍內流通量最高的三隻在岸國債。此最終結算價的設計減低了對任何單個債券的操縱風險（有關最終後結算價的詳情，請參閱第 5.1 及 5.2 節）。

　　此外，由於香港交易所的國債期貨合約價格在最後交易日必須和最終結算價一致，因此，它與在岸標的債券之間出現較大價格偏差將導致成本高昂。參考離岸人民幣利率掉期市場為例子，在岸與離岸息差，導致其流通量較薄弱。根據歷史資料進行回溯測試和類比分析，香港交易所國債期貨合約模擬國債籃子的平均收益率與境內五年期國債收益率（2011 年至 2016 年）的相關系數為 98.3％，每日參考結算價和境內五年期國債期貨價格（2013 年 9 月至 2016 年 12 月）的相關系數為 92.1％（詳見第 5.4 節），高度相關。因此，想透過持有一定規模的離岸國債期貨合約來影響在岸市場的穩定性，在實踐中是非常困難的。

6　見 "The Impact of Futures Trading on the Spot Market for Treasury Bonds"（Shantaram Hegde, 1994）及 "The Impact of Derivatives on Cash Markets: What Have We Learned?"（Stewart Mayhew, 2000）。

(c) 香港交易所的國債期貨合約在一個規範、集中和透明的交易平台上進行交易，從而提高市場透明度，並向市場參與者提供價格預期和未平倉合約的實用信息。

(d) 與其他香港交易所交易的期貨產品相同，香港交易所交易及結算規則及證券及期貨事務監察委員會（證監會）相關規則下的多項措施，可限制市場持有大額未平倉的國債期貨合約，減少市場上不必要的波動風險，例如：

- 要求持有相當未平倉合約水準的結算參與者提交額外集中抵押金，從而有效降低大額未平倉合約的槓桿；
- 要求交易所參與者（不論為其本身或代表任何客戶）向香港交易所匯報國債期貨合約的大額未平倉合約。香港交易所亦有權要求任何大額未平倉合約持有人提交額外資料，以説明其大額持倉需要；
- 實施持倉限額，為單一實益擁有人的持倉設定上限。持倉限額一概嚴格執行，違規可能構成違反相關香港交易所規則及《證券及期貨條例》，或可包括刑事責任。香港交易所及證監會均可對任何違規行為採取行動，包括要求參與者及時和有序地減持倉位。

中國是全球債券市場增長最快的國家，也是繼美國和日本後全球第三大債券市場。隨着銀行間債券市場進一步開放，人民幣的國際認可度提升，中國債市被納入新興市場政府債券指數，以及相對發達市場的債券息差等因素推動，中國債券市場的國際參與者亦會持續增加。**香港交易所推出中國財政部五年期國債期貨合約，為全球首隻對離岸投資者開放的在岸利率產品，是幫助境外投資者管理人民幣利率風險頭寸高效、透明及便捷的工具。**

第8章

進軍中國境內債券市場

—— 國際視角

2017 年 5 月 16 日

摘 要

　　人民幣要發展成為國際儲備貨幣,一個發展成熟並有外資高度參與的人民幣債券市場必不可少。基於中國經濟及人民幣債市規模龐大,外資持有人民幣債券的增長潛力亦會相當可觀。然而,受制於當前中國債市對境外投資者開放計劃的限制,目前外資參與中國債市的程度遠低於其他國際貨幣國家,甚至比不上部分新興市場。此情況顯示有必要推動創新措施,提升市場基礎設施、交易規則及金融產品,進一步推進人民幣國際化。

　　中國現時設有三項主要計劃允許境外投資者進入境內債券市場,分別為:合格境外機構投資者(QFII)計劃、人民幣合格境外機構投資者(RQFII)計劃及合資格機構進入內地銀行間債券市場(人行合資格機構計劃)[1]。雖然相關規例已經逐步放寬,有關額度管理、戶口管理或資金匯兌的規定仍然是限制境外參與者配置有效投資策略及資金的主要方面。此外,境內市場的一些制度特徵也是市場關注所在,妥善處理可促進境外投資者的活躍參與,包括市場分割、缺乏分散的市場結構、尚在發展中的信貸評級機制以及潛在信貸風險等等。

　　發展境內債券市場向來是內地發展資本市場及推進人民幣國際化的首要政策之一。為進一步推動境外參與中國境內債市,或可考慮以下改善措施:(1)進一步整合交易平台及外資參與機制;(2)加快跨境產品創新速度,將離岸匯市優勢與境內債市結合;及(3)按照中國人民銀行(人行)和香港金融管理局(香港金管局)所宣佈的債券通計劃,連接在岸與離岸債市,讓國際常規及準則逐步融入境內市場。跨境「債券通」平台可提供健全的金融基礎設施及與國際法規及監管標準接軌的市場規則,有助紓減監管壓力,為境外參與者及境內投資者提供更為便捷的交易環境,被視為國家加大開放資本市場及便利外資參與者交易人民幣計價資產的重要舉措之一,也將進一步鞏固香港作為連通內地市場和國際市場門戶的優勢地位。

1　計劃詳見第 4 節。

1 外資參與中國境內債市具發展潛力

　　過去十年，中國在利率市場化以至逐步放寬資本管制等方面不斷推出措施，債券市場發展取得重大進展，由此中國債市規模急速擴張，過去五年以簡單年均增長率 21% 的速度增長，成為全球第三大債券市場，債券存量規模達人民幣 56.3 萬億元[2]（約 8.1 萬億美元）（見本書第 7 章圖 1）。然而，相比其他國際貨幣國家，中國債市佔國內生產總值百分比仍然偏低。外資參與中國債市的程度依然微不足道，約佔整個市場 2.52% 及主權債市場 3.93%[3]，遠低於日本、美國甚至一些新興市場（見圖 1），顯示外資參與中國境內債市仍然有巨大增長空間。

　　人民幣獲納入國際貨幣基金組織的特別提款權貨幣籃子，為全球參與者提供了進軍中國債市的一個重要窗口。入選成為特別提款權籃子貨幣，意味着人民幣獲正式認可為國際金融體系的一部分，是中國融入全球金融體系的重要里程。特別提款權的重要性不止於其象徵意義；從投資角度而言，納入特別提款權雖不至於直接刺激大量投資需求，因為特別提款權貨幣籃子本身僅是一種補充性的國際儲備資產，約值 2,880 億美元，人民幣在其中的權重僅 10.92%[4]，但是，獲得特別提款權的地位可以提升人民幣作為全球投資及儲備貨幣的認受性，將很大可能促進國際政府及私人部門對人民幣計價資產的需求，從而導致全球資產配置逐漸由其他金融部門逐漸流入中國資產，特別是流入到人民幣計價的債券及相關金融產品。

2　資料來源：國際清算銀行、Wind 資訊，截至 2016 年底。
3　資料來源：中債登，截至 2016 年底。
4　資料來源：國際貨幣基金組織網站。

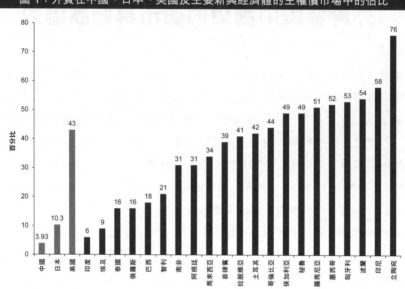

圖1：外資在中國、日本、美國及主要新興經濟體的主權債市場中的佔比

註：灰色為中國、日本及美國的外資佔比，藍色為主要新興市場經濟體的外資佔比。

資料來源：新興市場數據來自國際清算銀行及國際貨幣基金組織報告（2015）；中國數據來自 Wind 資訊，截至 2016 年底；日本數據來自亞洲債券在綫，截至 2015 年底；美國數據來自聯邦儲備局、美國財政部，截至 2016 年底。

　　政府部門方面，現時外國政府及半官方組織持有人民幣資產（包括債券、股票、貸款及存款）總值人民幣 6,667 億元[5]，相當於全球官方外匯儲備總值約 1%[6]，遠低於澳元或日圓（分別為 1.94% 及 4.48%（見圖2），截至 2016 年第三季末）。如國際政府部門持有人民幣的佔比可大致達到澳元水平，那意味將有 1,100 億美元的全球儲備轉移至人民幣資產；如進一步提升至日圓的佔比水平，流入人民幣資產的資金更高達 4,000 億美元。

　　私人部門方面，中國債券資產現時在國際基準指數中佔比不大。如中國資產獲納入若干國際指數，例如在國際定息產品市場中廣泛用作參考的摩根大通新興市場債券指數（EMBI Global Index），則根據國際貨幣基金組織的報告，中國在該指數的權重將約為三分之一（見圖3）。若還有相關政策助推機構及私人投資者加

5　見人行《人民幣國際化報告（2015）》。
6　根據國際貨幣基金組織數據，同樣地，2015 年人民幣佔官方外幣資產總值 1.1%。

大參與中國境內債市，相信外資持有的中國債券更可增持至與其他國際貨幣的相若水平，達債市總存量約 10%。

　　假設中國債市未來數年的增長率與社會融資總量過去五年的複合年增長率相同（即 14%），而且外資所持中國債券佔整個市場的 10%，那麼到 2020 年時外資所持中國債券可達人民幣 95,100 億元，佔國內生產總值的 9.93%（見表 1）。外資參與中國境內債市的增長潛力相當可觀。

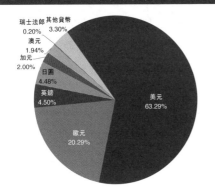

圖 2：官方外匯儲備貨幣組成
（2016 年第三季末）

瑞士法郎 0.20%
其他貨幣 3.30%
澳元 1.94%
加元 2.00%
日圓 4.48%
英鎊 4.50%
歐元 20.29%
美元 63.29%

註：人民幣歸入「其他貨幣」。
資料來源：國際金融統計（IFS）。

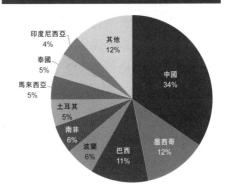

圖 3：EMBI Global Index
（若包括中國債券）

印度尼西亞 4%
其他 12%
泰國 5%
馬來西亞 5%
土耳其 5%
南非 6%
波蘭 6%
巴西 11%
墨西哥 12%
中國 34%

資料來源：國際貨幣基金組織全球金融穩定報告，
2016 年 4 月。

表 1：外資參與中國境內債券市場的預測（至 2020 年）	2016	2020
國內生產總值（人民幣十億元）	74,413	95,730
境內債市總值（人民幣十億元）	56,305	95,100
境內債市外資持有量（人民幣十億元）	853	9,510
佔國內生產總值百分比	1.15%	9.93%

註：計算時假設：（1）國內生產總值年增長率 6.5% 及中國債市年增長率 14%；（2）外資持有量佔未償還債務總額 10%。
資料來源：2016 年外資持有量數據來自人行；其他 2016 年數據取自 Wind 資訊；2020 年估算由作者計算。

2 增加外資參與度 對境內債市有所裨益

首先，市場匯集不同投資目標的投資者可激發更廣泛的投資策略，有助將資金推向更具生產力的行業。因此，鼓勵不同類別的境外投資者進入境內債市可助建立更多元化的投資者結構，激活市場交易，促進形成更具競爭力的市場，並進一步增加境內金融市場的規模和深度。

第二，促進外資持有中國債券，是增加人民幣在國際間使用量的一個關鍵。增加境外持有量是評估貨幣是否獲得廣泛使用的重要因素之一。然而，目前中國債券境外持有量遠低於國際貨幣的經濟體。以美國國債市場為例，美國國債市場投資者層面廣泛，包括金融機構、私人投資者及境外實體等。截至 2016 年底，政府實體（聯邦儲備局及地方政府）持有美國國債總值的 23%。此外，互惠基金及境外投資者亦是主要參與者，尤其是境外參與者持有量佔總存量超過 40%，其餘則由銀行（少於 5%）、保險公司以及信託和其他類別的投資者所持有（見圖 4）。由於債券通常是央行及全球基金經理的首選資產類別之一，對境外投資者而言，中國債券資產的交易及使用性，對於推進人民幣國際化及支持人民幣作為有意義的儲備貨幣至為重要。

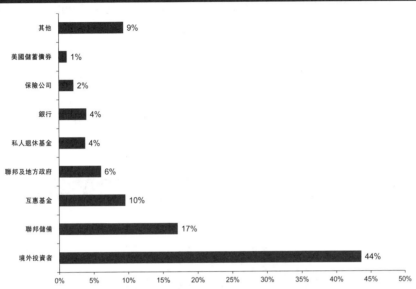

圖 4：多元化投資者基礎所持美國主權債的存量佔比（至 2016 年底）

註：債務包括短期國債、票據、債券，以及特別的國家及地方政府系列證券。
資料來源：聯邦儲備局、美國財政部。

第三，具市場深度及可提供多元類別工具、吸引長期投資者的債券市場可吸收境外資金流波動帶來的影響，提升全球投資者對持有人民幣計價資產的信心。2016 年人民幣匯率走弱，境外持有中國境內債券仍持續上升（見下文第 3 節）。即使外資涉足其他資產類別出現下跌，境外資金對債券資產始終「不離不棄」。因此，中期而言，外資參與中國債市預期將會持續增長，可抵銷資金流出的影響，對人民幣匯價起支持作用。

3　目前外資持有境內債券的結構現狀

中國正逐步擴大接受外資參與境內債市的程度。隨着離岸人民幣中心在全球分佈日漸擴闊，中國與多個國家簽定雙邊貨幣掉期，過去五年審批 RQFII 及 QFII

計劃的合格投資者及投資額度不斷提速。同時人行亦加快審批境外機構進入銀行間債市。因此，流入中國在岸債市的境外資金一直穩步上揚。於 2016 年底，外資持有中國境內債券已創下人民幣 8,526 億元新高，較前一年增長 13%[7]。

圖 5 顯示，2016 年底外資所持中國境內資產（包括債券、股票、貸款及存款）達人民幣 30,300 億元。其中外資持有的債券及股票持續上升，而存款及貸款則大幅下滑。值得注意的是，債券資產佔整體外資持有資產由 2015 年底的 20% 升至 28%，同期存款佔比則由 41% 下跌至 30%（見圖 5），反映境外資金有大幅轉移至債券資產的配置趨勢。

圖 5：外資持有的各類中國境內資產規模（2013 年 12 月至 2016 年 12 月）

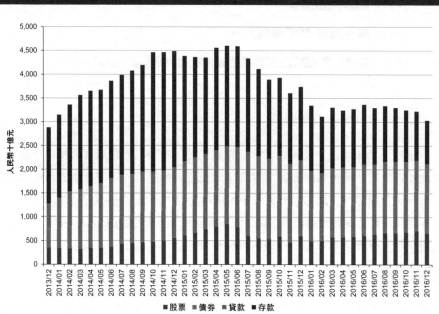

資料來源：人行。

7　資料來源：Wind 資訊。

　　在債券配置中，大部分境外資金流向利率而非信用債。2016 年境外參與者所持政府及政策性銀行債券增加了人民幣 2,330 億元，較 2015 年 350 億元人民幣飆升六倍[8]。外資在中國主權債券市場的佔比由 2015 年底的 2.62% 增至 3.93%（見圖 6）。

　　在 2016 年增持主權債券的投資者中，境外投資者佔總增量的 13%，僅次於全國性商業銀行（38%）及城市商業銀行（19%），是 2016 年中國主權債券第三大買家。相反，外資持有的信用債跌至人民幣 494 億元新低，只佔 2016 年底外資所持債券資產總值的 6%（見圖 7 及 8）。

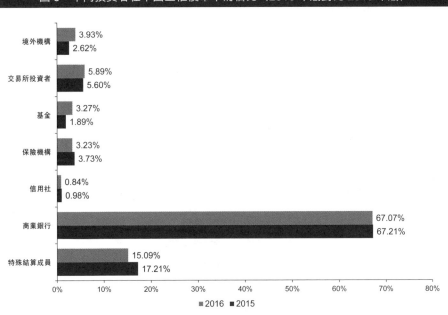

圖 6：不同投資者在中國主權債市中的佔比（2016 年底對比 2015 年底）

資料來源：Wind 資訊。

8　資料來源：Wind 資訊。

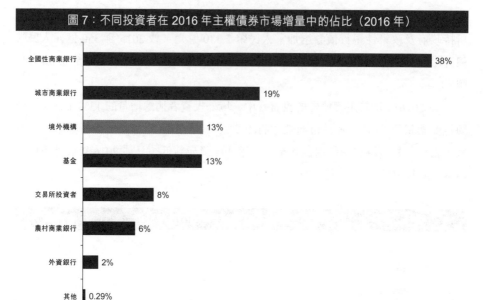

圖 7：不同投資者在 2016 年主權債券市場增量中的佔比（2016 年）

註：不包括以下於 2016 年減持主權債券的投資者類別：特殊成員、農村合作銀行、信用社、證券公司及保險機構。「其他」包括村鎮銀行、其他商業銀行、非銀行金融機構、非金融機構、個人及其他機構。

資料來源：中債登網站。

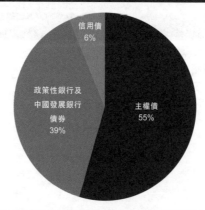

圖 8：外資持有債券按類別價值的分佈佔比（2016 年底）

註：境外持有信用債券包括中債登記錄的企業債券、中債登及上海清算所記錄的中期票據、以及上海清算所記錄的短期融資券及超短期融資券。

資料來源：Wind 資訊。

外資所持主權債券比例增加，或反映出在近期中國債市信貸違約上升的情況下，境外投資者對中國資產取態較為審慎。鑒於中國市場基礎設施薄弱，特別是欠缺可信的信用評級公司，外資機構傾向持有主權債及高評級債券作為外匯儲備。然而，由於主要發達市場目前處於低（甚至負）的息率環境，將資金配置至收益率較高的資產及信用債券的誘因將會增強。在此前提下，只要中國的市場基礎設施及債市信用情況大幅改善，信用債券可能會較政府債券增長更快。

總體而言，2016 年底外資持有中國債券佔債券總存量增至 2.52%（2015 年底：2.03%）[9]，在中國債市登記的境外機構共 411 家[10]。在現行市場開放計劃仍存限制的情況下，外資參與中國債市仍處於早期階段，顯示有必要推動創新措施，改良市場基礎設施、交易規則及金融產品，進一步拓展中國債市的闊度和深度。

4　有關現行計劃的最新政策發展

過去，中國對金融市場開放曾持審慎態度，避免資金進出影響境內金融體系穩定性。因此，中國債市對外資參與一度較為封閉，令境內債券市場上的外資佔比一直處於偏低水平。及至近年，中國推出多項措施開放債市，配合資本賬市場化及人民幣國際化的快速發展。

現時中國允許境外投資者進入境內債券市場有三條渠道，分別為 QFII、RQFII 及人行合資格機構計劃，闡釋如下。

4.1　合格境外機構投資者（QFII）計劃

QFII 計劃於 2002 年推出，最初只容許境外投資者進入交易所市場，包括買賣在交易所交易的債券，其後 QFII 計劃進行了多次大幅改革，降低境外機構的准入

9　依據中債登網站數據計算。
10　資料來源：中債登網站。

門檻並擴大投資範圍。2013 年 3 月，監管機構放寬了 QFII 投資限制，允許 QFII
進入銀行間債券市場。2016 年進一步放鬆管制，簡化了投資額度、資金匯入匯出
安排的管理，並縮短本金鎖定期（詳見表 2）。至 2016 年底，共向 276 家 QFII 發
出 873 億美元投資額度 [11]。

4.2　人民幣合格境外機構投資者（RQFII）計劃

作為 QFII 計劃的延伸，2011 年 12 月推出 RQFII 計劃。在此計劃安排下，境
外投資者可運用離岸人民幣資金投資於在岸資產。在首階段，中國基金管理及證
券公司的香港附屬公司可申請 RQFII 牌照及投資額度投資於中國資本市場。其後，
RQFII 計劃擴展至更多國家及地區，包括發展中及發達國家。截至 2016 年底，總
額度由初期的人民幣 2,700 億元增至人民幣 15,100 億元，已向 175 家 RQFII 發出
合共人民幣 5,280 億元額度 [12]。

與 QFII 計劃類似，RQFII 計劃的規定已逐步放寬。2013 年進行主要的政策
調整，取消了「20% 股市 / 80% 債市」的資產配置限制，投資範圍亦擴展至股指期
貨及銀行間債市定息產品等。

2016 年，RQFII 及 QFII 兩個計劃均獲大幅放寬。2 月，國家外匯管理局（外
管局）發佈《合格境外機構投資者境內證券投資外匯管理規定》。9 月，人行及外
管局共同發佈《關於人民幣合格境外機構投資者境內證券投資管理有關問題的通
知》。「新規」的修訂主要涉及 QFII 及 RQFII 的額度管理及資金匯兌等，這些向來
都是影響外資參與者執行有效投資策略及資金配置的主要方面。修訂大致包括：

(1) 針對投資額度的新管理辦法：根據新規，投資額度根據資產規模一定百分
比而非按原定投資額度限制所計算。此外，境外主權財富基金、中央銀行及貨幣
當局的投資額度則不受限，可按實際需要申請額度。

(2) 放寬本金額匯入及相關鎖定期：新規移除了須於投資額度獲批之日起計
六個月內匯入本金的限制，並將鎖定期縮短為累計匯入投資本金達人民幣 1 億元
（RQFII）或 2,000 萬美元（QFII）之日起計三個月。

11　資料來源：外管局網站。
12　資料來源：外管局網站。

(3) 放寬資金匯出：新規允許 RQFII 於相關鎖定期屆滿後辦理資金匯出手續。
QFII 方面，匯出本金毋須再經外管局事先批准，不過資金匯出規模仍設月度限制。

4.3　合資格機構進入內地銀行間債券市場（人行合資格機構計劃）

2010 年人行推出該項試點計劃，容許合格境外機構使用離岸人民幣投資於銀行間債券市場。推出時合格機構包括境外央行或貨幣當局、離岸人民幣清算行及參與銀行等三類。同時，主權財富基金及國際組織亦可據此安排進入銀行間債券市場。

自 2015 年，此計劃再推出多項重要放寬措施，進一步便利境外投資者進入中國銀行間債市：2015 年 5 月底，人行允許離岸人民幣清算行及參與銀行利用在岸債券持倉進行回購融資；2015 年 7 月中，人行進一步放寬合格債券交易範圍，容許合格主體參與在岸銀行間債券市場的債券現券買賣、債券回購、債券借貸、債券遠期、利率互換及人行許可的其他交易，而毋須經人行事先批准或受任何額度限制。

2016 年 2 月，人行發佈「第 3 號公告」[13]，進一步放寬境外機構投資者進入銀行間債券市場的規則。首先，合格境外機構參與者類別擴展至所有合格境外機構投資者，包括商業銀行、保險公司、證券公司、基金管理公司、其他類別金融機構及人行認可的中長期機構投資者。第二，公告進一步放寬對境外投資者施行的外匯管理。第三，第 3 號公告實施宏觀審慎管理，因此不對個別投資者施加額度限制。為配合「第 3 號公告」的實施，2016 年 5 月中國進一步頒佈詳細規則，釐清境外機構投資者在銀行間債市的投資流程。

2016 年的措施進一步開放境內債市，但仍有改進空間，例如：現時合格投資者的範圍仍限於金融機構；關於債券產品及額度等限制仍然存在；境內銀行間債券市場的准入程序可進一步簡化及釐清，以吸引更多外資參與。

13　見〈中國人民銀行公告 2016 年第 3 號〉。

表2：目前 QFII、RQFII 及人行合資格機構計劃的主要架構			
	QFII	RQFII	人行合資格機構計劃
監管批准	• 中國證監會：QFII/RQFII牌照 • 外管局：QFII額度 • 人行：進入銀行間債券市場事先備案		向人行事先備案
投資額度	• 如申請的額度於基礎額度內，只需向外管局事先備案；如要求的額度超出基礎額度，須經審批 • 基礎額度根據資產規模一定比例計算		• 對境外機構投資者實施宏觀審慎管理 • 无明确投資額度要求，需向人行備案擬投資規模
合資格 定息產品	• 交易所市場：政府債券、企業債券、公司債券、可換股債券，等等 • 銀行間市場：債券現券		• 外匯儲備機構：所有債券現券、債券回購、債券借貸、債券遠期、利率互換、遠期利率協議等等 • 其他金融機構：所有債券現券及人行許可的其他產品，離岸人民幣清算行/參與銀行亦可買賣回購
外匯管理	在岸與當地託管商進行兌換	須匯入離岸人民幣（取自離岸）	在岸/離岸
匯入本金 鎖定期	三個月	三個月，開放式基金則不設限	沒有
匯出頻次及 限制	每日（僅開放式基金）及月度匯出限制	每日（僅開放式基金）	累計匯出金額需符合一定比例規定

資料來源：截至2016年底資料。最新規則及政策見人行、中國證監會及外管局網站。

5　約束外資參與的制度特徵

除了准入限制外，境內債市若干制度性特徵也值得關注及需要妥善處理，以便吸引更多外資積極參與境內債券市場。

5.1　交易平台及產品相對分割

境內債市相對分割主要體現在交易工具、平台及其監管架構上。不同市場（主要為證券交易所及銀行間債券市場）提供不同的債務工具，負責監管的機構亦不相同。視乎交易的產品及市場，外資參與者須經不同監管機構審批。

表 3：中國境內兩大債市		
	銀行間債券市場	交易所市場
監管機構	人行	中國證監會
交易平台	中國外匯交易中心	上海／深圳證券交易所
中央證券登記	中央國債登記結算有限責任公司（中債登）／上海清算所	中國證券登記結算有限責任公司（中國結算）
可選工具	中央政府債、地方政府債、政策性銀行債、央行票據、企業債券、中期票據、短期融資券、商業銀行債券、金融機構債券、銀行間可轉讓定期存單、資產支持證券、回購、債券借貸、債券遠期、利率互換等等	中央政府債券、地方政府債券、企業債券、公司債券、可換股債券、資產支持證券、中小型企業發行的私募債
主要投資者	機構投資者（銀行、證券公司、保險公司、基金、財務公司、企業、離岸機構等等）	證券公司、保險公司、基金、金融公司、個人投資者、企業

資料來源：人行、中國證監會。

　　境內的機構投資者主要在銀行間債市進行交易，導致逾 90% 債券交易量都發生在銀行間市場，在滬深交易所進行的交易少於 10%（2016 年數據）[14]。兩個交易平台各有其本身規定，並非所有產品都可同時在兩個市場上交易。基本上只有數類債券（政府債、企業債及公司債）可在銀行間及交易所市場同時交易，其餘（例如政策性銀行債、商業銀行債券、央行票據、中期票據、短期融資券、回購、債券借貸等等）只可在銀行間市場交易。可換股債券及私募債則在交易所市場交易。

　　由於債市分割並涉及不同監管機構，導致流通量分散和市場深度有所限制。此外大部分對沖產品只在銀行間債券市場交易，給大部分外資參與者帶來風險，特別是基金及證券公司，因為他們大多數主要透過是 QFII 及 RQFII 安排進入交易所市場。

14　資料來源：Wind 資訊。

5.2　境內債市投資者基礎較為集中

另一個牽制中國債市流通量的主要因素是投資者結構高度集中。2016 年底，商業銀行持有債券存量佔整體比重為 58.5%。如計及特殊機構（主要為人行及政策性銀行），則銀行所持債券合計超過市場總存量的 60%。政府債券的持倉更為集中，銀行持有佔總存量約 80%。相比之下，其他非銀行金融機構（包括保險公司、基金及交易所參與者，他們買賣較為活躍）持有的債券存量佔整體的 32%[15]。

相對一面倒的投資者基礎不利於促進債市流通性。2016 年中國債券流通率為 2.79，遠遠低於美國，與當年日本、韓國進行貨幣國際化時，以及 90 年代商業銀行壟斷債市時的水平相比，亦有所不及[16]。中國債市流通比率低，可歸咎於投資者組合欠缺多元化以及商業銀行主導市場。投資者基礎多元化及因此而來的流通性，是中國債市健全發展以至人民幣國際化的兩大必要元素。

5.3　信用評級欠缺差異及透明度

現時，境內評級機構將近 90% 境內債券評為 AA 或以上級別[17]。中國債券（特別是信用債）的信用利差不足以抵償相關信用風險。與國際準則相比，境內評級機構在信用評級及評核指標方面，與國際評級機構都存有重大差異，令境外投資者難以識別中國公司債券的信用價差。有必要將境內與國際評級標準及慣例接軌，以及容許國際評級機構參與境內市場，境外投資者方可更容易追蹤中國信用質素，從而更準確判斷相關的信用風險。

6　可行改善方案

為進一步推動外資參與中國境內債市，或可考慮以下改善方案：

15　資料來源：Wind 資訊。

16　詳見 "People's Republic of China's financial market: are they deep and liquid enough for RMB internationalization?"，亞洲發展銀行工作論文，2014 年 4 月。

17　資料來源：Wind 資訊。

6.1　整合交易平台及現有外資參與計劃

市場規模及流通量是決定債市交易及定價效率的主要因素。如第 5 節所述，現時大部分中國境內債券的發行及交易仍分為銀行及交易所兩個市場，只有小部分可在兩個市場同時交易。此外，兩個市場的成交量亦相對不平衡，銀行間市場的成交佔比超過 90%，而交易所債券流通量則相對極低。

通過 QFII 及 RQFII 計劃的外資參與者 (例如證券公司、基金或中小型機構投資者) 大多只可進入交易所市場，因為銀行間市場的准入規定及交易成本相對較高。交易所市場流通量低且規模小，導致該市場的信用利差較高、對沖能力也較弱。整合交易平台可促使流通量達到足夠規模以改善定價能力。

此外，近期 QFII 及 RQFII 計劃的政策改革，使彼此在投資額度及資金匯兌方面的政策更為相近，日後兩項制度有可能更為劃一或整合，以減低交易成本及更有效形成更多元的投資者基礎。

6.2　加快跨境產品創新，將離岸外匯產品優勢與境內債市有效連接

外資增持中國債券，相關風險管理的需求也就上升。為便利分散人民幣債券投資風險，境內債市必須推出更多工具。此外，境外投資者投資人民幣債券時需要對沖人民幣匯率風險，這方面外匯工具亦很重要。雖然近期境內匯市進一步向境外投資者開放，但利用離岸市場對沖優勢及充裕的產品供應，也是另一種對沖境內債券資產的有效方式。香港交易所於 2017 年 4 月 10 日推出五年期中國財政部國債期貨 (國債期貨) 合約，為全球首隻對離岸投資者開放的在岸利率產品，為境外投資者提供了管理人民幣利率風險頭寸高效、透明及便捷的工具 [18]。

離岸人民幣市場的優點在於任何人均可自由進出，包括私人部門主體。離岸外匯市場的流通量亦大幅改善，離岸人民幣交易相對在岸交易的比重已顯著提高。香港離岸人民幣市場為持續發展人民幣衍生產品及對沖工具提供了穩健基礎，可便利外資參與者對沖持有中國債券資產及外匯波動風險，進行相應的風險管理。

18　見本書第 7 章〈香港交易所五年期中國財政部國債期貨 —— 全球首隻可供離岸投資者交易的人民幣債券衍生產品〉。

6.3　接通在岸與離岸債市，以實現境內市場與國際規範的深度融合

　　一如滬 / 深港通計劃，設立連接在岸與離岸債市的跨境平台，開展內地與香港債券市場互聯互通合作（債券通）是進一步便利人民幣債券交易及提高定價效率的可行方案，通過香港與內地債券市場基礎設施機構聯接，境內外投資者可以買賣兩個市場流通的債券。

　　從交易角度而言，債券通可促成債市跨境融合，提升流通量，實現交易所債券市場與銀行間債券市場對接。此外，市場整合後可帶來更多標準化的工具，為人民幣計價資產開發出更多價格基準，改善中國債券資產的定價效率。

　　雖然國際投資者目前可以直接參與境內人民幣市場（包括匯市及債市），離岸市場仍然是支撐人民幣作為全球貨幣的主要場所。基於香港的離岸金融環境及基礎設施發展成熟，跨境債券通計劃可為境外投資者紓減監管壓力，並提供更便捷的制度條件，例如提供符合國際標準的信用評級以及更佳的投資者保障。

　　對內地投資者而言，債券通亦可提供一系列的國際債券，配合他們的全球資產配置策略。通過與專業國際投資者共同參與國際交易平台，內地投資者亦可增加面對國際市場慣例及規例的經驗。從這個角度，債券通將助推提升內地境內債市的深度及廣度，培養更成熟及專業的投資者基礎。

第 9 章

香港交易所美元兌人民幣（香港）期權合約

—— 人民幣貨幣風險管理的工具

2017 年 8 月 17 日

摘 要

香港交易所推出人民幣貨幣期權,主要是有見於市場參與者對買賣和管理離岸人民幣匯率的多元化工具需求日股,希望滿足市場需要。

香港交易所的人民幣貨幣期權合約與人民幣貨幣期貨合約系列相輔相成。這些期權合約可用作針對非線性敏感度的風險管理工具,讓投資者可利用人民幣匯率進行波幅交易,正可切合以往人民幣貨幣期貨未能滿足的市場對沖需求[1]。隨着人民幣匯率朝向自由浮動的方向發展,以及相關的政策發展,人民幣匯率正由政策主導走向市場主導,預期會增加美元兌人民幣(香港)匯率的波動。2015 年 8 月人民幣匯率改革前一個月,美元兌人民幣(香港)的一個月引申波幅還是 1% 至 2% 左右[2],其後一年已攀升至 4% 至 10%。美元兌人民幣(香港)即期匯價波幅增加,可説造就了推出人民幣貨幣期權合約的機遇,令市場參與者可進行波幅交易,也利便作對沖匯率風險。

此外,全球場外人民幣期權市場 2016 年的平均每日成交金額已高達約 180 億美元[3],每宗交易平均金額為 1.5 億美元[4]。此外的場外人民幣(香港)衍生產品市場,現時的波動率頭寸大都以單純的普通期權形式存在(無特別結構設計的標準認購 / 認沽期權),與兩三年前以結構性遠期持倉為主導的市場結構大不相同。

鑒於場外市場相對欠缺透明度、新監管法規涉及的保證金要求以及交易對手方風險等考慮因素,人民幣貨幣期權交易轉往場內進行的需求日益增加。在香港期貨交易所(「期交所」)上市、並透過香港期貨結算有限公司(「期貨結算公司」)進行中央結算的美元兌人民幣(香港)期權合約,可為這個重要且持續增長的人民幣(香港)期權市場提供價格透明度及減低交易對手方風險。

---

1 相對表現與相關貨幣匯率呈線性關係的貨幣期貨而言,貨幣期權的表現與非線性的風險敏感度有關,譬如相關資產價格變化的變化率(即 gamma)、波動率(即 vega)及時間(即 theta)。

2 資料來源:彭博。

3 國際結算銀行 2016 年「三年一度外匯及場外結算衍生工具市場活動央行調查」。

4 滙豐銀行「2016 年新興市場貨幣指引」(Emerging Markets Currency Guide 2016)。

1 宏觀市場環境：市場需求及現行支援

1.1　雙向波動推動人民幣風險管理工具的需求

　　2015 年 8 月 11 日，中國人民銀行（人民銀行）推出以市場供求為基礎、有管理的浮動匯率制度，人民幣匯價參考上一個交易日銀行間外匯市場收盤匯率，以及外匯供求情況及一籃子貨幣匯率變化進行調節（改革）。2015 年 8 月改革前一個月，美元兌人民幣（香港）的一個月引申波幅還是 1% 至 2%，改革後一年已增至 4% 至 10%。

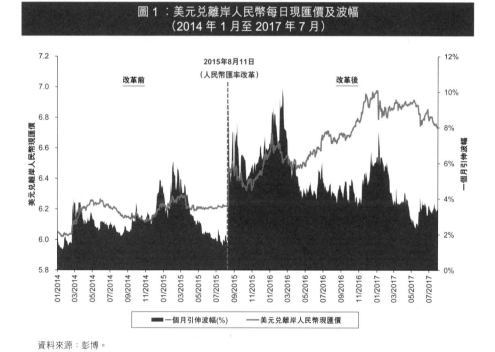

圖 1：美元兌離岸人民幣每日現匯價及波幅（2014 年 1 月至 2017 年 7 月）

資料來源：彭博。

人民幣國際化已邁進全新階段。作為連通內地與環球市場最為關鍵的離岸人民幣樞紐，香港擔當的角色越來越重要。香港應繼續加強自身優勢，在「共同市場」的新角色之上，容納更多創新的人民幣計價的產品，以鞏固香港作為金融中心的長遠發展。這個新角色可分為三方面：（1）優化的離岸人民幣市場；（2）人民幣風險管理中心；（3）建立成為一個「門戶市場」，詳述如下。

（1）就離岸人民幣市場而言，過往市場增長動力主要依賴市場對人民幣升值的預期，及在岸與離岸市場之間的套利交易。現在，市場上的相關金融產品越來越多，市場上有更多風險管理工具可用於更多不同的投資組合管理策略，市場規模不斷深化和擴大。這一切正好切合人民幣資產全球配置和相關的跨境資本流動的需求。在日本、英美等發達國家，信貸、股本及債券市場的規模都是其國內生產總值的 5 倍以上[5]，而這等比值在中國只是約 2.1 倍，顯示出中國在金融市場深化和金融產品多元化發展方面仍有很大的發展空間。再加上國際貨幣基金組織已將人民幣納入特別提款權貨幣籃子（SDR），預期人民幣匯價雙向波動將成為新常態。越來越多投資者日益關注這個市場變動，開始管理匯率方面的風險。

進一步發展和豐富多元化金融產品及相關金融服務類別，**有利香港繼續發展其離岸人民幣市場**，成為跨境投資（尤其是源自內地的海外投資）及相關風險管理的主要地點。現在正是香港進一步拓展其離岸人民幣市場深度及效能的良機。如果能夠發展更有效和合理的在岸及離岸人民幣市場定價基準，兩個市場的人民幣定價「各自為政」的情況可望有所改善，令在岸與離岸人民幣市場價差維持在合理水平。要達到這個目標，在岸與離岸人民幣債券市場、外匯市場和衍生產品市場需要進一步互聯互通、提高市場流通量，以及增加市場參與者數目和多樣化程度。

（2）香港已具備成為風險管理中心的優勢，有助加快內地現行經濟及金融體系轉型調整的進程。譬如：市場普遍預期，人民幣國際化邁進下一個新階段時，人民幣匯率的靈活性料將進一步提高，那時候，匯率風險管理的需求必定進一步增加。另一方面，內地公司正紛紛拓展全球網絡，也積極參與一帶一路沿線國家的不同項目[6]。對此香港亦恰可「把握機遇」，切合這些公司對海外投資風險管理及拓展國際業務網絡的需求。

---

5 資料來源：國際貨幣基金組織數據。
6 中國國家主席習近平 2013 年 9 月提出有關歐亞國家互聯互通和合作的發展方案。

(3) 隨着中國金融市場逐步開放，香港不再只是內地實體投資的活躍的產品匯集市場，而亦逐漸成為內地實體投資其他市場的**門戶市場**。內地與香港市場交易互聯互通機制試點計劃推出後，這個趨勢更為明顯。2014 年 11 月推出滬港股票市場交易互聯互通機制（滬港通），2016 年 12 月推出深港股票市場交易互聯互通機制（深港通），以及 2017 年 7 月推出的內地與香港債券市場互聯互通合作（債券通），香港、深圳及上海已連結成一個大規模的共同市場。在香港發揮橋樑作用下，這個共同市場一方面支持內地資金進行全球資產配置，另一方面為國際資金投資內地資本市場提供穩健的基礎設施和平台。可以預見的是，若互聯互通框架延伸至其他產品類別，香港作為連接市場的主要角色將會進一步加強。按現時跨境投資活動越趨頻繁的勢頭，市場的風險管理需求料將增加，增幅甚至可能數以倍計。

1.2　香港交易所人民幣產品及平台的支援

在上文分析的宏觀背景下，海外市場對中國定息及貨幣產品市場的興趣日增，自然對風險管理和投資的需求亦上升。對此，香港交易所一直多方面加大投入，冀能成為離岸人民幣產品交易及風險管理中心。現時，香港交易所旗下的交易平台提供不同類型的人民幣產品，包括債券、交易所買賣基金、房地產投資信託基金、股本證券、人民幣定息及貨幣衍生產品以及大宗商品衍生產品等，整個產品系列均以切合市場需求為目標。

香港交易所於 2012 年推出美元兌人民幣（香港）期貨合約後，產品的成交量自 2015 年開始穩步上升，現已成為全球買賣最活躍的人民幣期貨合約之一[7]。及後香港交易所進一步發展更加多元化的產品組合，於 2016 年 5 月 30 日推出人民幣兌其他貨幣的期貨合約（日圓、歐元及澳元）交易，利便交叉貨幣對沖。除人民幣貨幣風險管理工具外，香港交易所亦於 2017 年 4 月 10 日推出五年期中國財政部國債期貨（國債期貨），豐富旗下的人民幣利率風險管理工具。這隻國債期貨可有效對沖利率，尤其是 2017 年 7 月 3 日債券通之後。債券通是連接中國銀行間債券市場與全球市場的試點計劃，令國際投資者首次可經「北向通交易」直接在內地銀行間債券市場的交易平台 —— 中國外匯交易中心 —— 買賣債券。

7　見本書第 14 章〈香港交易所邁向成為離岸人民幣產品交易及風險管理中心〉。

另外，隨着人民幣已成儲備貨幣，其與國際貨幣之間的關係受人注目，市場對人民幣貨幣指數基準亦存在龐大的需求潛力。2016 年 6 月，香港交易所推出與湯森路透聯合開發的湯森路透／香港交易所人民幣貨幣系列指數（RXY 指數或 RXY 系列指數），便利市場參與者留意人民幣匯價走勢。香港交易所亦計劃未來推出該指數的期貨及期權產品，為市場提供更多有效的人民幣風險管理工具。

此外，香港交易所亦計劃完善旗下的人民幣產品系列，於 2017 年 7 月 10 日先在大宗商品市場推出其雙幣（美元及人民幣定價及結算）交易、實物交收的黃金期貨合約，為黃金生產商、用家及投資者提供風險管理及投資的有效解決方案，管理黃金現貨與期貨市場之間以及人民幣與美元之間的價差所產生的風險。

香港交易所亦優化旗下基礎設施平台，為人民幣衍生產品在香港市場的進一步發展奠下穩健基礎。香港交易所旗下附屬公司香港場外結算有限公司（場外結算公司）2013 年開業，提供了重要的資本市場基礎設施，滿足定息及貨幣產品市場參與者對結算服務的需求，尤其是人民幣計價衍生產品等區內交易產品。

1.3 香港交易所美元兌人民幣（香港）期貨：全球流通量最高的美元兌人民幣（香港）期貨合約之一

現在市場已越來越意識到對沖人民幣匯率風險的必要性以及其益處。首隻在香港交易所平台上買賣的人民幣衍生產品是 2012 年 9 月推出的美元兌人民幣（香港）期貨。不論是個人或機構投資者，都開始認識到人民幣匯率波動如何在人民幣資產、負債及現金流三方面影響他們的投資組合。人民幣雙向走勢已成為投資者風險管理框架的重要度量標準之一。

2016 年，香港交易所的美元兌人民幣（香港）期貨錄得全年成交合約 538,594 張，年度增幅 105%，年底未平倉合約達 45,635 張，年度增幅 98%，均創歷史紀錄[8]。此外，2016 年 12 月的日均成交量攀升至 4,325 張。現金結算的人民幣（香港）兌美元期貨於 2016 年下半年的合約成交量亦見增長，未平倉合約自推出以來穩步上揚。2016 年 12 月平均每日成交 95 張合約，年底未平倉合約達 1,494 張的高位。

踏入 2017 年，香港交易所美元兌人民幣（香港）期貨產品更創下多項新紀錄：

8　資料來源：香港交易所。

- 2017 年 1 月 5 日單日成交 20,338 張合約（名義價值 20 億美元）；第二及第三大成交則分別是 5 月 31 日及 6 月 1 日，同樣超過 8,600 張合約（名義價值超過 8.6 億美元）；
- 2017 年 1 月 4 日未平倉合約 46,711 張合約（名義價值 47 億美元）；
- 2017 年 1 月 4 日夜期成交 3,642 張合約（名義價值 3.6 億美元）；
- 市場參與度增加：曾參與買賣此產品的交易所參與者總數增至 112 名。

圖 2：香港交易所美元兌人民幣（香港）期貨交易表現（2012 年至 2017 年上半年）

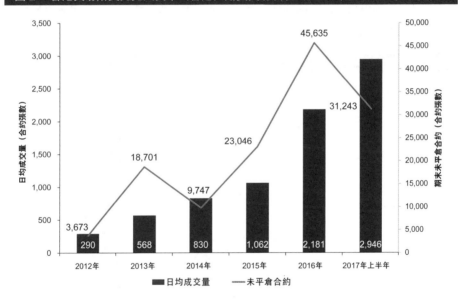

資料來源：香港交易所。

2 香港交易所美元兌人民幣（香港）期權：場內交易的風險管理工具

現時的場外人民幣期權市場已十分龐大，日均成交量達 180 億美元 [9]，每宗交易平均金額為 1.5 億美元 [10]（見圖 3）。有別於兩、三年前場外人民幣（香港）衍生工具市場上以結構性遠期倉盤為主導，現時幾乎所有新出現的波動風險均以標準化期權（即無特別結構設計的標準認購／認沽期權）來管理，足見市場對以標準期權對沖貨幣風險的需求日增。

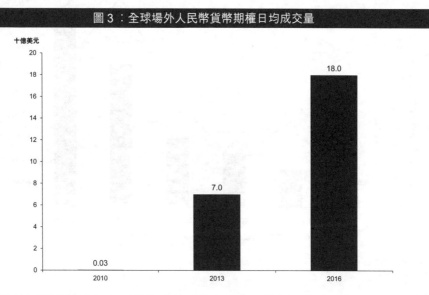

圖 3：全球場外人民幣貨幣期權日均成交量

十億美元

資料來源：國際結算銀行「三年一度外匯及場外結算衍生工具市場活動央行調查」（Triennial Central Bank Survey of foreign exchange and OTC derivatives markets）。

　　與場外的人民幣貨幣期權市場相比，香港交易所的貨幣期權市場有其若干特色，闡釋如下。

9　國際結算銀行 2016 年「三年一度外匯及場外結算衍生工具市場活動央行調查」（Triennial Central Bank Survey of foreign exchange and OTC derivatives markets）。

10　《新興市場貨幣指南 2016》（*Emerging Markets Currency Guide 2016*），滙豐銀行。

2.1　持續報價

按慣例，場外的人民幣貨幣期權市場以雙邊交易和報價請求（RFQ）模式運作。投資者必須逐一聯絡可以提供價格的市場參與者，與之協商並要求期權報價，再自行比較價格。這交易模式影響了價格發現的效率。

相比之下，香港交易所的美元兌人民幣（香港）期權不論在交易執行或價格發現方面都有顯著的優勢。一般情況而言，香港交易所人民幣貨幣期權的指定流通量提供者可提供大約 150 個期權系列的持續報價，期限較短的平均買賣價差為 12 至 40 點子，期限較長的則為 80 至 160 點子[11]。這樣的買賣盤報價串流讓投資者可按指定行使價及持有期限進行交易，有利流通量建立。除持續報價外，投資者亦可要求指定的流通量提供者就特定行使價及合約期提供報價。

與場外人民幣衍生工具相比，交易所場內交易的其中一個特點就是可匯集流通量，能提供持續不斷的流通量及買賣差價較窄等優勢。此外，買賣上市產品的監管及資本效率（下文第 2.2 節）亦已愈見明顯。簡而言之，場內市場能在公平基礎上提供一個井然有序兼高度透明的交易環境。

2.2　資本效益

歐美地區的新規則（如歐洲的 EMIR[12]、美國的 CFTC[13]）正影響着場外市場參與者。由 2017 年 3 月 1 日起，所有涉及到的對手方（主要為金融實體及具系統重要性的非金融實體）須就其所持有的未結算場外組合每日交換變動保證金，這對許多場外市場參與者而言是相對新的規定。另外有關交換初始保證金規定，正準備分階段強制實施，並於 2020 年 9 月或之前全面實施。

場內的人民幣衍生產品能為投資者提高資本效益，主要源自其在多方面均較場外市場有相對優勢。以下表 1 為場內交易人民幣衍生產品與場外人民幣衍生產品的對照比較。

11　資料來源：香港交易所（2017 年 7 月）；「點」指小數點後第四位數（0.0001），是貨幣對的最低價格波幅。

12　《歐洲市場基礎設施監管規則》（「EMIR」）是根據歐洲議會及歐盟理事會第 11 條歐洲市場基礎設施監管規則（歐盟）第 648/2012 號制定的監管技術標準。

13　美國商品期貨交易委員會（「CFTC」）的商品交易法。

表 1：人民幣場內場外衍生產品比較		
項目	場外人民幣衍生產品	場內人民幣衍生產品
價格透明度	相對不透明——需逐一聯絡對手方詢問以獲取價格	非常透明——香港交易所網站、資訊供應商及經紀的交易平台均可提供期權價格
中央結算	只有雙邊結算，並無中央結算	香港交易所充當交易雙方的中央結算對手
授信及抵押品	需與銀行磋商信貸額及抵押品安排	以保證金為基準，接受現金抵押品
結算風險	人民幣不是持續聯繫結算銀行（CLS[14]）的合資格貨幣，因此不能利用該系統進行持倉淨額計算	香港交易所交易的衍生產品 可以進行持倉淨額計算

資料來源：香港交易所分析。

2.3　獨有而靈活的風險管理工具，適用於不同的人民幣市場情況

香港交易所的美元兌人民幣（香港）期權合約的設計特意反映其美元兌人民幣（香港）期貨合約的特色，使其能提供跨產品對沖及跨產品保證金計算，以及為相同名義金額提供獨有的回報架構。

(1)　跨產品對沖

美元兌人民幣（香港）期權直接與香港交易所現有的美元兌人民幣（香港）期貨互補。兩者並用，投資者能就不同市況部署交易及對沖策略，對手方風險卻比場外衍生產品為低。在人民幣自由化進程及政策持續朝市場主導方向發展之際，人民幣匯價持續波動，兩種產品正好在這方面為投資者提供對沖工具。（期權及期貨的比較見表 2。）

(2)　跨市場計算保證金

香港交易所的美元兌人民幣（香港）期權合約按期貨結算公司採用的 SPAN 方

14　持續聯繫結算及交收系統（「CLS」）是處理跨境外匯交易的全球結算及交收系統。

法 [15] 以保證金基準交易，當中相關資產相同的期貨及期權計算保證金要求時，其淨對沖值是一重要因素。因此，同時持有美元兌人民幣（香港）期貨及期權持倉的投資者可享跨市場計算保證金之利，須支付的保證金會較獨立單邊持有為少。

　　從風險管理的角度來看，期權合約獨有的風險及回報模式使期權合約有許多用途。配合不同的期權／期貨策略，投資者可使用期權合約涉足多種市場參數（例如現貨匯率、波動率及時間等）。

　　期權合約適合多種人民幣市場狀況，提供了靈活的策略應對不同市況，可用於牛市、熊市、區間震盪或波動的市場。（見第 3.2 節有關產品的基本應用。）

表 2：期權期貨對比	
期權	期貨
• 買方有權利（但沒有責任）在預先釐定日或之前以預先釐定價（行使價）買入（或出售）相關資產；若買方行使權利，賣方有責任以行使價出售（或買入）資產 • 期權價格與相關資產有非線性關係，亦有獨有的風險回報架構 • 買方要即時支付一個價格作為期權金	• 買方有責任在日後指定時間以預先釐定價買入資產，賣方有責任在日後指定時間以預先釐定價出售資產 • 期貨價格和相關資產有線性關係 • 並無需即時繳付的費用（保證金及其他交易相關費用除外）
例子	
• 期權：投資者買入美元兌人民幣（香港）匯率的（歐式）認購期權，行使價為 7.0 元，於三個月後到期。三個月後期權到期，投資者有權利（但沒有責任）以每美元兌人民幣 7.0 元的匯價買入美元。 • 期貨：投資者以價格 7.0 買入美元兌人民幣（香港）期貨長倉，於三個月後到期。三個月後期貨到期當日，投資者有責任以每美元兌人民幣 7.0 元的匯價買入美元。 • 貨幣期權中，一貨幣的認購期權亦是另一貨幣的認沽期權，所以貨幣期權較其他資產類別的期權複雜。	

資料來源：香港交易所分析。

2.4　場內交易其他特色

- **成本效益較高：**買賣場內期權合約一般只支付期權金及保證金，要即時支付的僅佔合約名義價值一小部分，提供槓桿效應 [16] 及較高的成本效率。以

15　SPAN 指標準組合風險分析。請參閱下列網頁的保證金計算方法文件：www.hkex.com.hk/eng/market/rm/rm_dcrm/rm_dcrm_clearing/dmrm_clearing_settlement.htm

16　貨幣期權及槓桿效應風險頗高，不適合經驗不多的投資者或風險承擔能力不高的人士，詳情請參閱香港交易所網站。

香港交易所的美元兌人民幣（香港）期權來說，由於豁免首六個月（2017年 3 月 20 日至 2017 年 9 月 29 日）的交易費，且毋須支付證監會徵費，相關交易費更低。

- **交易透明度高**：場內買賣的期權合約均為標準合約，交易有序透明。投資者可透過從資訊供應商（市場資訊供應商名單見附錄三）及經紀的交易平台上取得實時的場內期權價格。
- **市場進入更簡便**：交易所一般均向不同類型投資者（包括但不限於散戶、公司用戶、資產管理公司及對沖基金）開放。以香港交易所為例，投資者可透過現時逾 120 家可買賣人民幣產品的交易所參與者買賣這期權產品。相對之下，場外的人民幣貨幣期權市場只向機構用戶開放。

基於具備上述各種特點，香港交易所的美元兌人民幣（香港）期權的累計成交量及未平倉合約穩步上揚。截至 2017 年 7 月 31 日為止，產品推出以來的總成交量為 4,914 張合約（以名義金額計算為 491 億美元），未平倉合約續創新高。於 2017 年 7 月 31 日，所有合約月份合計的未平倉合約達 1,727 張（以名義金額計算為 1.73 億美元）。

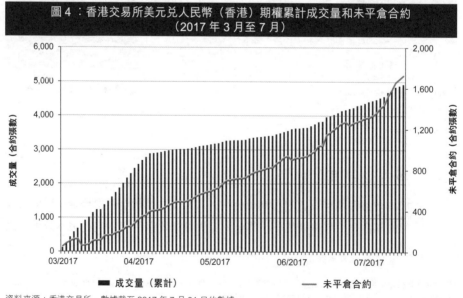

圖 4：香港交易所美元兌人民幣（香港）期權累計成交量和未平倉合約
（2017 年 3 月至 7 月）

資料來源：香港交易所，數據截至 2017 年 7 月 31 日的數據。

3 香港交易所美元兌人民幣（香港）期權：產品設計及應用

香港交易所的美元兌人民幣（香港）期權合約是歐式現貨期權，只可於到期當日而非之前行使，主要以當前場外市場慣例（大部分貨幣期權均為歐式）為設計基礎。到期日行使期權後，行使一方須按行使價交付與人民幣本金全額相等的美元，有關安排切合期權用戶對人民幣本金匯兌的需求。

3.1　定價行為及須關注的風險定價因素

期權金即期權價格是包含相關資產行使價及現貨價、兩種相關貨幣的利率、期限和波幅等若干因素的一個函數。經濟學者 Garman 和 Kohlhagen 將通常用於股票期權的 Black-Scholes 定價模型延伸至貨幣期權的定價，稱為加曼柯爾哈根模型[17]。

貨幣期權為可牽涉多層面的工具，二級市場的貨幣期權價格對不同市場參數均有變量回應。基於期權條款的多樣性（合約期、行使價等等），市場參與者要在市場上找完全相同的期權進行對沖並不可行。因此，買賣期權的人士要密切監測各種市場參數，做好風險管理。現時測量期權價值對不同市場參數的變動有特定方法，統稱「Greeks」。Greeks 的分析對期權估值和風險管理極為重要。

Greeks 將期權價格或期權組合所包含的風險分解成多個組成部分，讓買賣期權的人可決定保留及對沖哪些風險。Greeks 包含的不同風險計量指標包括：

- **Delta** 值：相關資產現貨價格變動下的期權價格變動
- **Gamma** 值：相關資產現貨價格變動下的 delta 值變動
- **Theta** 值：期權時間值的損耗，即時間推移下期權價格的變動
- **Vega** 值：相關資產波幅變動下的期權價格變動

---

[17] M.B. Garman 與 S.W. Kohlhagen 合著 "Foreign currency option values"，*Journal of International Money and Finance*，1983 年第 2 期。

- **Phi 值**：基準貨幣無風險利率變動 [18]
- **Rho 值**：定價貨幣無風險利率變動 [18]

3.2　產品應用

人民幣貨幣期權及期貨的主要用戶包括企業、資產管理公司和基金公司、自營交易公司、經紀行和專業投資者等，他們使用人民幣貨幣產品的目的不盡相同。

以下為人民幣貨幣期權產品應用的多個假設性例子（有關分析不包括交易成本，另過往表現不代表將來的表現）。

3.2.1　基本應用

(a)　人民幣貶值時的風險管理策略

背景假設

投資者擔心人民幣貶值。他需要在三個月後沽出人民幣資產換回美元。

可應用期權

例子是購買三個月認購期權（即買美元沽人民幣），行使價 6.8500。

情境假設

期權到期時，如美元兌人民幣（香港）匯價升至 6.7000，期權到期而不行使，投資者可沽出人民幣資產，按 6.7000 這個較佳匯價將人民幣兌換為美元。如美元兌人民幣（香港）匯價降至 7.0000，投資者行使期權，將沽出人民幣資產所得款項按原行使價（即 6.8500）兌換為美元。

潛在風險與回報

潛在回報：若人民幣貶值，投資者仍可按較佳匯價將人民幣兌換為美元。

潛在風險：投資者須付期權金。

(b)　人民幣升值時的風險管理策略

背景假設

投資者擔心人民幣升值。他需要在三個月後沽出美元資產換回人民幣。

---

[18] 在個別貨幣對中，用作報價參考的貨幣稱「定價貨幣」（下），被報價的貨幣稱「基準貨幣」（上）。例如歐元兌美元，歐元為基準貨幣，美元為定價貨幣。

市場策略

例子是購買三個月認沽期權（即沽美元買人民幣），行使價 6.8500。

情境分析

期權到期時，如美元兌人民幣（香港）匯價降至 7.0000，期權到期而不行使，投資者可沽出美元資產，按 7.0000 這個較佳匯價將美元兌換為人民幣。如美元兌人民幣（香港）匯價升至 6.7000，投資者行使期權，將沽出美元資產所得款項按原行使價（即 6.8500）兌換為人民幣。

潛在風險與回報

潛在回報：若人民幣升值，投資者仍可按較佳匯價將美元兌換為人民幣。

潛在風險：投資者須付期權金。

3.2.2　進階應用

(a)　提高收益率 —— 沽出備兌認購期權

背景假設

出口商三個月內有應收美元賬款，希望屆時以高於期貨價格的匯價將美元轉換成離岸人民幣。出口商無須在收到賬款時立即沽出美元，因此可待匯價較佳時才沽售，亦希望利用這筆預期收到的現金提高收益率。

市場策略

出口商沽出 2017 年 3 月到期、行使價 7.1000 的美元兌人民幣（香港）認購期權，收取人民幣（香港）775 點子的期權金。美元兌人民幣（香港）即期匯率：6.9300；2017 年 3 月期貨價：7.0450；波幅：7.40 買入。

情境分析

期權到期時，如美元兌人民幣（香港）匯價低於 7.1000，期權於價外到期而不行使。出口商保留人民幣（香港）期權金作為持倉的額外回報。如美元兌人民幣（香港）匯價高於 7.1000，期權被行使，出口商按 7.1000 的匯價將美元兌換為人民幣（香港），仍較若然三個月前進行對沖的期貨價為佳。由於出口商收了人民幣（香港）期權金，其兌出美元的匯價實際為 7.1775。

潛在風險與回報

潛在回報：出口商透過沽出期權利用閒置資金獲得額外回報。即使買方行使期權，出口商兌出美元的實際匯價也較佳。

潛在風險：如美元兌人民幣（香港）大幅升值，出口商就需承擔按當前匯價兌出美元的機會成本。

(b)　減低成本 ── 買入認購價差期權組合

背景假設

投資組合經理所管理的投資包括人民幣資產。他計劃購入美元兌人民幣（香港）認購期權對沖人民幣（香港）貶值，對沖時間為一年。然而，基於時間值、向上傾斜的波幅曲線及期貨曲線，遠期的美元兌人民幣（香港）認購期權非常昂貴。例如，2017 年 12 月到期、行使價 7.2500 的美元兌人民幣（香港）認購期權的價格為 2,515 點子。（美元兌人民幣（香港）即期匯率：6.9300；2017 年 12 月期貨價格：7.2650；波幅：8.85 賣出）

市場策略

投資組合經理可沽出 2017 年 12 月到期、行使價 7.5000 的美元兌人民幣（香港）認購期權，收取 1,585 點子期權金，對美元兌人民幣（香港）看跌，但不超出 7.5000。（行使價 7.5000 波幅 9.06 買入）。來自行使價 7.5000 認購期權的期權金可減少有關對沖策略（購入 7.2500 較低行使價的美元兌人民幣（香港）認購期權）的淨成本。投資組合現支付淨期權金 930 點子。

情境分析

期權到期時，如美元兌人民幣（香港）匯價低於 7.2500，兩隻期權均於價外到期而不需行使。投資組合經理要承擔期權金淨額作為對沖成本，但成本仍較不採取此策略為低。如美元兌人民幣（香港）匯價高於 7.2500 但低於 7.5000，經理行使購入的期權並讓沽出的期權到期。這是最佳情況，因經理保留對沖的同時又減低了對沖成本。如美元兌人民幣（香港）匯價高於 7.5000，兩隻期權均被行使。經理雖失去對沖，但卻獲得人民幣（香港）2500 點子的現金流淨額，有助補償其於現貨市場的對沖。

潛在風險與回報

潛在回報：此策略在持有特定人民幣匯價走勢看法下可減少對沖成本。

潛在風險：此策略在某些情況下可能僅能對沖部分風險。

(c)　風險逆轉組合

情境分析

交易員預期美元兌人民幣（香港）即期匯價未來三個月上升。他購入 2017 年 3 月到期、行使價 7.1500 的美元兌人民幣（香港）認購期權，並支付 715 點子的期權金（波幅 8.35 賣出）。不過，他不想承擔期權金全數，也不想過於進取地持倉，於是選擇沽出 2017 年 3 月到期、行使價 6.9500 的美元兌人民幣（香港）認沽期權，收取 525 點子的期權金（波幅 6.70 買入）。他的成本淨額為 190 點子。

結果

美元兌人民幣（香港）即期匯價及遠期／期貨曲線皆移向上。假設即期匯價及期貨價格皆平行上漲 600 點子，認購期權的價格為 955 點子，認沽期權為 355 點子。此策略的淨值為 600 點子。交易員因看對走勢獲利 200%。另一選擇是交易員可待到期日方行使認購期權並讓認沽期權到期。

潛在回報

透過風險逆轉組合獲取潛在回報有多種方法：

- 如美元兌人民幣（香港）即期匯價及遠期匯價／期貨價格向上，認購期權將較認沽期權的價值為高，交易員可選擇平倉獲利。
- 如市場對美元兌人民幣（香港）認購期權的需求較對認沽期權大，就引伸波幅而言，認購期權將較認沽期權的價值為高（稱為波幅偏差）。

潛在風險

如美元兌人民幣（香港）的匯價走勢不利，交易員不但失去用以購入認購期權的期權金，其認沽期權淡倉亦虧損。在這情況下，初期的成本雖然較低，但虧損卻擴大。

(d)　波動性交易 —— 馬鞍式組合（同一行使價的兩隻期權）

情境分析

交易員預期美元兌人民幣（香港）即期匯價短期內將繼續波動，波幅曲線或會向上。他買入 2017 年 12 月到期、行使價 7.2500 的美元兌人民幣（香港）認購期權，以及 2017 年 12 月到期、行使價 7.2500 的美元兌人民幣（香港）認沽期權。認購期權的定價為 2,490 點子（波幅 8.85 賣出），認沽期權為 2,350 點子。期權金合計 4,840 點子。

結果

馬鞍式組合的 vega 倉位為 550 點子（每個期權 275 點子）。假設 2017 年 12 月到期、行使價 7.2500 的引申波幅增至 10.00，認購期權的價值為 2,810 點子，認沽期權為 2,670 點子。此策略現時的定價為 5,480 點子，價值變幅 640 點子（vega 約 1.15）。

潛在回報

遠期馬鞍式組合提供最大風險因素為波幅風險，是交易雙方買賣及落實其波幅曲線走勢預測的最直接方法。（馬鞍式組合在交易成立時的 delta 值通常為中性）。短期馬鞍式組合可用以交易相關資產之波幅（gamma 交易）。

潛在風險

買賣馬鞍式組合須承受波幅風險。如相關資產波動但波幅變動不大，期權交易者須管理 delta 但未能從波幅中獲利。

其他可能應用

波幅較高時可應用勒束式組合（不同行使價的一個認購及一個認沽期權），亦稱「兩側交易」。如預期若干價格範圍會有波動但波幅不會太大，可進行蝶式買賣（馬鞍式組合長倉加勒束式組合短倉），即以沽出勒束式組合來資助馬鞍式組合。

3.3　期權行使時實物交收

3.3.1　認購期權

假設：

行使價（k）= 6.90；正式結算價（s）= 6.95

如結算價 > 行使價，期權被行使；如結算價 ≤ 行使價，期權到期時的價值是零。

實物交收流程（見圖 5）：

如認購期權被行使，實物交收時，買方向結算所繳付最後結算價值，即合約金額（100,000 美元）× k（6.90）= 人民幣（香港）690,000 元，並自結算所收取相等於合約金額（100,000 美元）的相關貨幣幣值。

另一方面，賣方向結算所交付相等於合約金額（100,000 美元）的相關貨幣幣值，並向結算所收取最後結算價值，即合約金額（100,000 美元）× k（6.90）＝ 人民幣（香港）690,000 元。

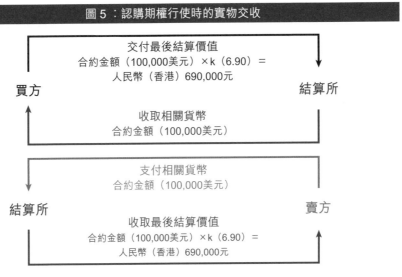

圖5：認購期權行使時的實物交收

資料來源：香港交易所。

3.3.2　認沽期權

假設：

行使價（k）＝ 6.90；正式結算價（s）＝ 6.85

如結算價＜行使價，期權被行使；如結算價≥ 行使價，期權到期時的價值是零。

實物交收流程（見圖6）：

如認沽期權被行使，實物交收時，買方向結算所收取最後結算價值，即合約金額（100,000 美元）× k（6.90）＝ 人民幣（香港）690,000 元，並向結算所交付相等於合約金額（100,000 美元）的相關貨幣幣值。

另一方面，賣方自結算所收取相等於合約金額（100,000 美元）的相關貨幣幣值，並向結算所繳付最後結算價值，即合約金額（100,000 美元）× k（6.90）＝ 人民幣（香港）690,000 元。

圖 6：認沽期權行使時的實物交收

支付相關貨幣
合約金額（100,000美元）

買方　　　　　　　　　　　　　　　　　　**結算所**

收取最後結算價值
合約金額（100,000美元）×k（6.90）＝人民幣（香港）690,000元

交付最後結算價值
合約金額（100,000美元）×k（6.90）＝
人民幣（香港）690,000元

結算所　　　　　　　　　　　　　　　　　　**賣方**

收取相關貨幣
合約金額（100,000美元）

資料來源：香港交易所。

附錄一

香港交易所美元兌人民幣（香港）期權合約細則

項目	香港交易所美元兌人民幣（香港）期權合約特點			備註
合約標的	美元兌人民幣（香港）貨幣對			
合約金額	100,000 美元			與期貨相同
期權金報價單位	以每美元兌人民幣以小數點後第四個位報價（0.0001）			沿用現行場外市場報價方法
行使價	行使價間距訂於 0.05			匯集流通量於指定行使價
正式結算價	由香港財資市場公會在到期日上午 11 時 30 分左右公佈的美元兌人民幣（香港）即期匯率			人民幣（香港）現貨市場的基準
行使時的結算	行使時實物交收			切合期權用戶的本金兌換需求
		持有人	沽出人	
	認購期權	以人民幣支付最後結算價值	交付美元	
	認沽期權	交付美元	以人民幣支付最後結算價值	
行使方式	歐式			場外結算市場最普遍形式
合約月份	現月、下三個月及之後的四個季月			與期貨相同（最遠第五個季月除外）
最後結算日	合約月份的第三個星期三			與期貨相同
到期日	最後結算日之前兩個香港營業日			與期貨相同
持倉限額	美元兌人民幣（香港）期貨合約、人民幣（香港）兌美元期貨合約和美元兌人民幣（香港）期權合約合計，以所有合約月份持倉合共對沖值 8,000（長倉或短倉）為限，並且在任何情況下： • 直至到期日（包括該日）的五個香港營業日內，現月美元兌人民幣（香港）期貨合約及現月美元兌人民幣（香港）期權合約持倉對沖值不可超過 2,000（長倉或短倉）；及 • 所有合約月份的人民幣（香港）兌美元期貨合約淨額之倉位不可超過 16,000 張（長倉或短倉）			人民幣貨幣期貨及期權的持倉對沖值合併計算
大額未平倉合約	於任何一個系列為 500 張未平倉合約			

附錄二

香港交易所美元兌人民幣（香港）期權合約的交易及結算安排

買賣盤手數上限

買賣盤手數上限為 1,000 張合約。交易所參與者應根據其業務需求和風險管理要求向交易所申請設定其買賣盤張數上限。

大手交易

交易所衍生產品交易系統支援大手交易設施。大手交易的成交量門檻為 50 張合約（名義價值為 500 萬美元）。0.4 或以上的價格，其價格幅度限制為 10%，0.4 以下的價格，其價格幅度限制為 0.0400。

莊家

部分流動性提供者將對常見行使價合約提供連續報價；部分流動性提供者將對報價請求提供報價。

結算安排

在結算方面，結算所參與者須加以安排，確保其能夠處理人民幣及美元交收。他們需在期貨結算公司 2 指定的交收銀行設立人民幣及美元帳戶，並與之設有相關授權。此外，結算所參與者須確保銀行帳戶運作正常及可進行現金交付。同時，非結算所參與者應委任合資格的全面結算參與者代其結算美元兌人民幣（香港）期權合約。

詳見香港交易所網站。

附錄三

檢索香港交易所美元兌人民幣（香港）期權的市場訊息

(1)　資訊供應商存取編號資料

供應商	存取編號
阿斯達克（AAStocks）	340900
Activ Financial	CUS/1701/9999P.HF
亞富資訊科技（AFE Solution）	873181-7
彭博（Bloomberg）	CSX Curncy OMON <GO>
CQG	C/P.CUS
財經智珠網（DBPower）	CUS
東方財富（Eastmoney）	CUS
易盛（Esunny）	CUS
經濟通（ETNet）	CUS
Fidessa	CUS_Osmy.HF
FIS Global	CUS+<STRIKE PRICE>+<MONTH CODE>+<LAST DIGIT OF THE YEAR>
浙江核新同花順網絡信息股份有限公司（Hexin Flush Financial Information Network Ltd）	CUS
匯港資訊（Infocast）	CUS（Menu > Derivatives > Options > Select "CUS"）
Interactive Data	O:CUS\MYYDD\[Strike Price]
Market Prizm	CUS <Strikes> my
報價王（QPI）	P11370-P11375

(續)

供應商	存取編號
SIX Financial	CUSmy
上海大智慧（Shanghai DZH）	CUS[mmyy][C/P][Strike]
上海澎博財經（Shanghai Pobo）	CUSyymm-C/P-SSSSS
電資訊（Telequote）	CUSOmy
Tele-Trend	Open->Options->CUS
路透社（Thomson Reuters）	0#HCUS*.HF
萬得（Wind）	Quant -> CUSO.HK

(2) 香港交易所網站的實時價格

https://www.hkex.com.hk/chi/ddp/Contract_RT_Details_c.asp?PId=388

(3) 提供香港交易所美元兌人民幣（香港）期權交易服務的交易所參與者名單

http://www.hkex.com.hk/chi/prod/drprod/rmb/EP-FXO_c.htm

(4) 可進行人民幣衍生產品交易的交易所參與者名單

http://www.hkex.com.hk/chi/prod/drprod/rmb/brokerlist_c.htm

第10章

債券通的制度創新及影響

—— 助推中國金融市場開放

2017 年 11 月 5 日

摘 要

「債券通」是指境內外投資者通過香港與內地債券市場基礎設施機構連接,買賣兩個市場交易流通債券的機制安排。債券通是深化內地和香港市場互聯互通的重要里程碑,是與現有債券市場開放渠道並行的、更富有效率的開放渠道,在許多環節實現了明顯的創新和探索,可吸引更多境外投資者參與中國銀行間債券市場,是更為適應國際投資者交易習慣的機制安排。

債券通的創新性具體體現在交易前的市場准入環節,交易中的價格發現與資訊溝通,以及交易後的託管結算環節,實現了以更低的制度成本、更高的市場效率,將國際慣例與中國債市的有效對接。7 月 3 日,債券通「北向通」通道正式開通,當天交易金額超過人民幣 70 億元,開通後三個月內,境外機構持有的境內人民幣債券餘額較開通前的 8,425 億元人民幣大幅增加至 10,610 億元人民幣 [1]。這或受益於債券通的渠道開放,在一定程度上反映出債券通對外資參與中國債券市場的追推作用。

債券通以可控的方式進一步提升中國債市的開放程度,從而為國際投資者參與中國債市、促進中國債市的改革和開放、人民幣國際化帶來新的動力。通過債券通,香港可成為境外投資者進入內地債市提供便利化的視窗,這將進一步鞏固香港作為離岸人民幣中心的地位,形成圍繞債券通的在岸和離岸人民幣產品生態圈,進一步強化香港作為國際金融中心的角色和資金進出內地的中介功能。

1　數據來源:中央國債登記結算有限責任公司、上海清算所網站。

1　中國金融市場對外開放的新突破

「債券通」是指境內外投資者通過香港與內地債券市場基礎設施機構連接，買賣兩個市場交易流通債券的機制安排。2017 年 5 月 16 日，中國人民銀行（人行）和香港金融管理局（香港金管局）發佈有關批准開展「債券通」的聯合公告。7 月 3 日，債券通的「北向通」通道正式開通[2]，推動中國金融市場進一步開放。近年來，兩地互聯互通試點計劃不斷推進，滬港通和深港通[3]（在此文中統稱為「滬深港通」）相繼落實，兩地股票市場已基本達到互聯互通，而債券市場是資本市場的另一重要組成部份，債券通建基於內地債券市場的龐大發展空間和國際資本對人民幣的需求增長，可視為中國金融市場對外開放的又一突破性創新。

1.1　中國內地債市發展空間巨大

隨着中國金融轉型的持續推進，債券市場的開放正在成為中國金融市場開放和人民幣國際化的重要推動力。截至 2017 年 3 月末，中國債券市場以人民幣 66 萬億元的存量規模成為全球第三大債券市場，僅次於美國和日本，公司信用類債券餘額位居全球第二、亞洲第一[4]。但是總體上看，中國債券市場上的外資參與率還處於相當低的水平，如果可採取適當的中國債市開放舉措，吸引更多的外資投資中國債市，不僅在短期內可促進國際收支的流入端改革，提高調節國際收支波動的能力，而且中長期也會促進中國債市流動性的提升。

2　根據現階段安排，債券通初期先開通「北向通」，即香港及其他國家與地區的境外投資者（以下簡稱境外投資者）經由香港與內地基礎設施機構之間在交易、託管、結算等方面互聯互通的機制安排，投資於內地銀行間債券市場。

3　滬港通及深港通是內地與香港股票市場互聯互通機制試點計劃，讓兩地市場的投資者可直接進入對方股票市場進行投資。2014 年 11 月滬港通正式開通，之後 2016 年 12 月深港通正式開通。

4　資料來源：中國人民銀行網站。

1.2　近年來中國推動外資參與債市開放步伐不斷加快

　　2010 年中國首次對境外合格機構開放銀行間債券市場，翌年（2011 年）再推出人民幣合格境外機構投資者（RQFII）計劃，兩年後（2013 年）允許合格境外機構投資者（QFII）進入銀行間債券市場；2015 年實施多項措施，對便利境外投資者進入銀行間債券市場起到實質性推動作用，具體包括：2015 年 6 月，人行允許已進入銀行間債券市場的境外人民幣業務清算行和參加行利用在岸債券持倉進行回購融資；2015 年 7 月，人行對於境外央行類機構（境外中央銀行或貨幣當局、主權財富基金、國際金融組織）參與銀行間債券市場推出了更為便利的政策，並明確其業務範圍可擴展至債券現券、債券回購、債券借貸、債券遠期以及利率互換、遠期利率協定等交易；2016 年 2 月，人行發佈新規，放寬境外機構投資者進入銀行間債券市場的規則，以及 2016 年 5 月進一步頒佈詳細規則，拓寬了可投資銀行間債券市場的境外機構投資者類型和交易工具範圍，取消了投資額度限制，簡化了投資管理程序。截至 2017 年 7 月債券通開通前，已有 473 家境外投資者進入銀行間債券市場，總投資餘額超過人民幣 8,000 億元[5]。

1.3　債券通對吸引國際資本具正面積極意義

　　這些中國債市領域的開放探索，為當前推出債券通奠定了市場基礎。但是截至 2016 年底，外資持有中國債市的比率依然低於 2%，明顯低於發達經濟體債市開放的平均水平（見圖 1）。前述中國債市的開放渠道，主要適應於對中國債市較為了解、能夠承擔較高的運作成本來參與中國債市的外國央行和大型機構，而對於為數眾多的中小型海外投資者來說，需要探索新的開放渠道吸引他們的參與，需要解決他們在參與中國債市時所面臨的一些挑戰。債券通正是在這樣的背景下推出的。

5　資料來源：人行網站。

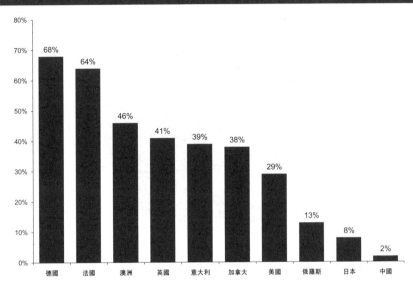

圖 1：海外投資者持有量在各發達國家債市存量中的佔比（按金額計）（2016 年底）

資料來源：彭博、國際清算銀行、人行。

2016 年 10 月人民幣被正式納入國際貨幣基金特別提款權的貨幣籃子，佔比為 10.92%，這會為人民幣計價的債券資產帶來新的參與主體和資本流量，也相應提升了全球市場對人民幣作為全球投資及儲備貨幣的認受性，無論是官方層面還是私人投資層面，都可以促進國際機構對人民幣計價資產的需求。但是，無論是從人民幣在官方外匯儲備中的佔比，還是在外匯市場交易中的佔比，迄今為止都遠遠低於 10.92% 的水平。這也意味着，下一步人民幣國際化的主要推動力，會來自於發展國際投資者可以投資的、多樣化的人民幣計價的離岸與在岸的金融資產，而中國債券市場的開放將是最為關鍵性的環節之一。債券市場更高程度的開放，還能促進監管機構與國際市場的聯繫更為緊密，使在岸的金融基礎設施的參與主體更為國際化，中國境內的金融機構也可通過債券通與境外機構投資者產生更為密切的業務聯繫，為中國金融機構更深入地參與海外市場奠定基礎。

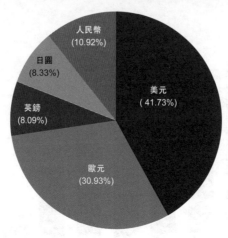

圖 2：人民幣在國際貨幣基金組織特別提款權的貨幣籃子中的佔比（2016 年底）

資料來源：國際貨幣基金組織。

2　債券通與中國債市有效對接

在債券通推出前，中國主要有三條途徑供境外投資者進入境內債券市場，分別為 QFII 計劃、RQFII 計劃以及境外機構直接進入內地銀行間債券市場（CIBM）計劃（見上文第 1.2 節）。與這些現有渠道相比，債券通在交易前、中、後三個環節實現了多方面的創新和突破，在技術操作環節回應了國際投資者參與中國債市的期待和訴求。

2.1　交易前：市場准入與並行通道

目前債券通的境外參與者與已有渠道的投資者範圍一致，參照 2016 年人行發佈的 3 號公告，以注重資產配置需求為主的央行類機構和中長期投資者為主要參與者，體現出中國穩步持續推進人民幣在資本和金融賬戶開放的戰略，同時在市場准入、備案程序、資格審核等方面，為長期資本流入中國債券市場提供了新的選擇，開闢了便捷的渠道。如果說現有的各種渠道能滿足外國央行和大型機構投資中國債市的需求的話，債券通的機制設計，則針對希望投資中國債市、又可能不願意承擔過高參與成本的機構投資者，或者說，在債券通的投資渠道下，境外投資者不必對中國債市的交易結算制度以及各項法律法規制度有很深入的了解，只需沿用目前熟悉的交易與結算方式，這就降低了外資參與中國債市的門檻與成本，使得債券通對於海外的機構投資者來說是更為「用戶友好」。具體表現在：

第一，在債券通開通之前，境外投資者參與內地銀行間債券市場主要通過代理結算的方式，即「丙類戶」方式進入銀行間市場，外資機構須委託境內的銀行間債券市場結算代理人來完成備案、開戶等手續，需要經過一定的入市備案程序。這些程序在一定程度上成為機構投資者參與中國債市的障礙。而在債券通的開放機制下，境外機構可利用境外的基礎設施，「一點接入」境內債券市場，境外投資者並不需要開立境內的結算、託管賬戶，也不需要在市場准入、交易資格等環節與境內主管部門直接接觸，而是完全可以利用其在香港已經開立的現有賬戶直接接入內地債券市場，保證了從交易流程一開始就沿用其已經熟悉的國際法則和交易慣例，利用境外的金融基礎設施來完成市場准入和備案流程，而不必重新熟悉與其長期交易結算習慣不同的中國內地市場運行慣例。

在具體操作中，境外由香港交易及結算所有限公司（香港交易所）和中國外匯交易中心合資成立的債券通公司，可以為債券通承擔專業的入市輔導、材料審核等輔助性入市備案，人行受理入市備案的流程所需的時間將會大大縮短。通過境外債券通實體進入中國在岸的債券市場，運作程序更為符合國際投資者、特別是希望參與中國債市但是又不熟悉中國債市規則的機構投資者的交易習慣，推動其入市速度和效率明顯提升。

第二，通過 QFII、RQFII 以及 CIBM 渠道投資內地債券市場，在市場准入時，根據現有的監管要求，對境外投資者有資金先期匯入、鎖定期等要求，並且需要預先說明預算投資金額，並在後續交易中滿足（見表 1），這在一些場合可能會與一部份境外機構靈活運用資金的投資策略不一致，也是現實交易中影響境外機構參與境內債市意願的因素之一。而債券通的市場准入中並沒有這些約束要求，使得境外機構在市場准入時面臨更少的入市阻礙，且境外機構可直接自行操作中國在岸的債券交易，在配置人民幣資產時獲得更大空間，無疑會明顯提升境外機構、特別是中小機構投資者參與中國債市投資的積極性。

表 1：目前 QFII、RQFII 及 CIBM 計劃的主要架構			
	QFII 計劃	RQFII 計劃	CIBM 計劃
監管批准	• 中國證券監督管理委員會（中國證監會）：QFII/RQFII 牌照 • 國家外匯管理局（外管局）：QFII/RQFII 投資額度 • 人行：進入銀行間債券市場事先備案		向人行事先備案
投資額度	• 如申請的額度於基礎額度內，只需向外管局事先備案；如要求的額度超出基礎額度，須經審批。 • 基礎額度根據資產規模按一定比例計算。		• 對境外機構投資者實施宏觀審慎管理 • 無明確投資額度要求，需向人行備案擬投資規模
合資格定息產品	• 交易所市場：政府債券、企業債券、公司債券、可換股債券等 • 銀行間市場：債券現券		• 外匯儲備機構：所有債券現券、債券回購、債券借貸、債券遠期、利率互換、遠期利率協議等 • 其他金融機構：所有債券現券及人行許可的其他產品；離岸人民幣清算行 / 參與銀行亦可參與回購
	QFII 計劃	RQFII 計劃	CIBM 計劃
外匯管理	在岸與當地託管商進行兌換	須匯入離岸人民幣	在岸 / 離岸
匯入本金鎖定期	三個月	三個月，開放式基金則不設限	沒有
匯出頻次及限制	每日（僅開放式基金）及月度匯出限制	每日（僅開放式基金）	累計匯出金額需符合一定比例規定

資料來源：截至 2016 年底資料。最新規則及政策見人行、中國證監會及外管局網站。

　　第三，債券通的入市渠道與現有的 QFII 計劃、RQFII 計劃及 CIBM 計劃並行不悖，境外投資者可以在這多重渠道之間進行靈活選擇。可以預計，債券通開通後，境外投資者可以更好地根據自身策略選擇不同的投資渠道，進行多元化的中國在岸金融市場的資產有效配置和產品開發。滬深港通開通之後，境外投資者的投資渠道選擇就出現了類似的微調，這說明現有的開放渠道是相互補充、並服務於不同的投資需求和不同類型的投資者，並不能說是簡單的相互替代關係。債券通開通後，使中國內地債券市場的對外開放得以繼續深化，對人民幣國際化和中國資本項目開放帶來重要的推動作用。

2.2　交易中：價格發現與資訊便捷

從交易方式來看，當前中國境內債券市場主要提供了詢價、點擊成交和請求報價（RFQ）交易等三種方式。由於中國債市的詢價模式以線下交易為主，對境外機構而言，債券交易可以說是相對不太容易深入了解的市場領域。而在債券通機制下，境外投資者可通過境外平台與境內做市商 以 RFQ 方式進行銀行間現券買賣，由境外投資者發起請求報價，做市商據以報出可成交價格，境外投資者選擇做市商報價確認成交，這個價格形成過程對於那些對中國債市還不是十分了解的境外機構投資者而言，交易資訊更為簡單易行，而且相對來說更透明對稱，更有利於價格發現。

另外，在代理行模式下，境外投資者不能直接與中國境內的對手方進行交易，只能委託中國境內的代理行代為交易。而在債券通機制下，境外投資者可以運用其熟悉的海外電子交易平台、操作介面和交易方式，自主選擇做市商報價，自主決定買賣時點進行交易。因此，這些境外投資者在通過債券通參與中國債市投資時，在具體操作時並沒有甚麼明顯的轉換成本，這對於那些對交易成本十分敏感的境外中小機構投資者來說十分重要。目前，Tradeweb 是債券通下第一家可供投資者使用的境外電子交易平台，彭博等其他電子平台亦在積極推進中，在系統準備就緒後可接入債券通的交易平台。境外機構在不改變交易習慣的情況下，可以直接與境內機構進行詢價、交易，使得整個交易過程更加透明高效。

從整個市場運行的不同環節看，債券通渠道在之前的「丙類賬戶」的代理交易模式繼續行之有效的同時，為境外投資者又提供了另一種直接交易的模式選擇，對於境外投資者、特別是對中國市場不太了解的機構投資者來說，在一定程度上降低了代理成本和溝通成本，交易效率明顯提高，有利於改善市場流動性。

2.3 交易後：託管結算

目前中國內地債券市場採用的是「一級託管制度」，這是經過長期實踐探索得出的、符合中國債券市場特點的重要市場制度。不過，目前境外市場長期形成的交易慣例是名義持有人制度和多級託管體系，這種制度差異為境外機構參與中國債券市場帶來了一定困難。國際市場經過多年的融合發展形成而來的多級託管體系和名義持有人結構，使得境外機構投資者已存有較強的路徑依賴。如果操作模式出現顯著變化，境外機構投資者所在的市場監管部門、機構內部的法律合規與後台運作都將面臨很大的調整困難，從而有可能制約一部分中小型海外機構參與內地債券市場的進程。

債券通以國際債券市場通行的名義持有人模式，並且疊加上中國的託管制度下所要求的穿透性模式，實現了「一級託管」制度與「多級託管體系」的有效連接。香港金管局的債務工具中央結算系統（HKMA-CMU）作為香港市場的中央債券存管機構，與作為內地中央債券存管機構的中債登和上清所進行連接，為境外投資者辦理債券登記、託管和結算。這樣，境外機構就可以在不改變長期沿襲的業務習慣、同時有效遵從中國內地市場制度的前提下，實現操作層面與國際慣例接軌，有效降低了不同市場體系對接的交易成本，也有利於在債券通開通後進一步發展與之相關的金融產品和商業模式（債券通的北向通系統設置見圖3）。

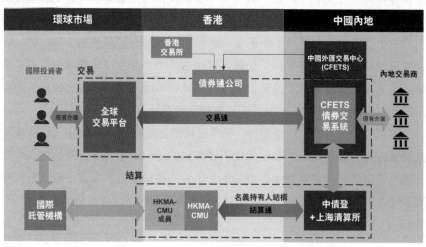

圖3：債券通的北向通系統設置

資料來源：香港交易所。

從法律框架相容性角度來看，債券通的北向通明確了相關交易結算活動將遵守交易結算發生地的監管規定及業務規則，在名義持有人制度下，境外投資者應通過名義持有人，即 HKMA-CMU 行使對債券發行人的權利。如果發生債券違約，HKMA-CMU 作為境外投資者的名義持有人，登記為債券的持有人，可以行使相關債權人權利、提起訴訟。同時，境外投資者作為債券實際權益擁有人，在提供相關證據後，也可以以自己的名義在內地法院提起法律訴訟。

3　中國金融市場開放和 香港離岸人民幣中心發展

3.1　以可控的方式進一步提升中國債市的開放程度

如同已經成功運行的滬深港通框架一樣，債券通的總體框架設計，實現了相對封閉的系統設計，使得由債券通推動的市場開放進程是總體可控的，可以說是以創新的方式提高了中國債市的開放程度。滬深港通突破性地實現了內地與香港股票市場之間資本的雙向流動。相比合格機構投資者計劃（QFII、RQFII），滬深港通擁有投資者主體更加寬鬆、額度管理更加靈活、交易成本更低、轉換成本低等優勢，取消了總額度，閉環式資金流動降低了資金大幅進出中國金融市場所引致的風險。而債券通於 2017 年 7 月啟動，先啟動「北向通」，沒有總額度限制，且是閉環式管理推動現貨債券市場的互聯互通，有助於支援國際資金流入中國債券市場，順應了中國資本市場國際化的發展趨勢。

3.2　進一步強化和鞏固香港離岸人民幣中心地位

香港已經是全球離岸人民幣業務樞紐[6]，對於人民幣的使用已不僅僅局限在最初跨境貿易結算，境外人民幣的投資功能、融資功能、風險對沖功能及外匯儲備

6　參見 *Hong Kong :The Global Offshore Renminbi Business Hub*，香港金管局網站，2016。

管理都得到了較大發展，離岸人民幣外匯交易量持續增長[7]。滬港通和深港通先後啟
動，以及其交易總額度的取消等為內地資本市場開放提供了新的渠道。對於香港
市場來說，債券通更大的價值在於彌補了香港作為國際金融中心在債市發展上的
短板。未來將適時研究擴展的南向通，預料更多內地資金將會進入香港市場，對
香港債市、以及香港整個金融體系的帶動作用將更為明顯。

3.3　圍繞債券通構建在岸和離岸人民幣產品生態圈

由於債券市場的交易特點，可以預計，債券通公司將發揮一種市場培育者和
組織者的公共職能。債券通的開通，以及債券通公司的平穩運行，對於中國在岸
和離岸債券市場的影響會是深遠的，其突出作用預計表現在：同時帶動離岸和在
岸的金融市場形成一個與債券配置相關的生態圈。債券市場作為主要由機構投資
者參與的市場，其交易與金融衍生品的交易和風險管理需求密切相關。債券通的
啟動，可促進更多國際債券投資者進入香港市場，投資香港債券市場上涉及中國
資產標的離岸債券產品，這不僅可以激活香港債券市場交易，更大程度上會對債
券通相關的人民幣金融衍生品市場、風險管理等專業服務起到帶動作用，這將使
整個市場得益更多。

目前，中國境內的衍生產品市場已經具備一定深度和流動性，可供交易的衍
生品種類繁多（包括遠期、掉期及期權、國債期貨等），為對沖人民幣相關風險提
供了支撐手段，近期隨着境內外匯市場進一步開放，一些合格境外投資者也可直
接使用境內的衍生品。與此同時，香港的場外市場已推出了包括人民幣外匯現貨、
遠期、掉期及期權等一系列產品，香港交易所亦有提供包括人民幣期貨、期權、
國債期貨[8]在內的場內交易產品，可便利境外參與者對沖持有的中國債券資產及外
匯波動風險。可以預計，隨着債券通的啟動，香港圍繞債券通的風險管理、人民
幣計價的金融創新等專業服務也會隨之獲得巨大的發展動力。

7　參見 "Triennial Central Bank Survey of Foreign Exchange and OTC Derivatives Markets"，國際清算銀
　　行，2016。

8　香港交易所的人民幣國債期貨試點計劃將於 2017 年 12 月合約到期以後暫停。香港交易所現正全面準備與債券通
　　相配套的風險管理工具，將適時推出新的人民幣利率產品。

3.4　平穩運行助推外資參與境內債券一級市場發行

2017 年 7 月 3 日，債券通的「北向通」正式開通，首日交易活躍，共有 19 家報價機構、70 家境外機構達成 142 筆、人民幣 70.48 億元交易。截至 9 月末，共 184 家境外機構接入債券通，境外機構對內債市持有量亦明顯增加。境外機構持有的境內人民幣債券餘額較開通前 6 月底的人民幣 8,425 億元增加至 9 月底的人民幣 10,610 億元[9]，這或受益於債券通的渠道開放與機制創新。

債券通開通後，也首次實現了境外投資者直接參與境內的債券發行認購。債券通開通首日，非金融類企業發行了 5 隻債務融資工具，總額為人民幣 70 億元。開通首月有 4 隻金融債、14 隻短期融資券通過債券通面向境內、外投資者完成發行，兩類債券的發行金額分別為人民幣 606.8 億元及人民幣 155 億元[10]。7 月 26 日，匈牙利政府發行了首隻 3 年期、人民幣 10 億元熊貓債[11]，也可通過債券通銷售。截止 2017 年 8 月底，在銀行間市場利用債券通的發債主體範圍已包括央企、地方國企及境外主權政府機構，行業分佈涵蓋電力、電信業務、交通運輸、金屬、農林牧漁等，顯示出債券通正逐步成為境外機構參與境內債券一級市場發行的重要通道，促進了境內一級市場的投資者多元化。

4　總結

整體而言，債券通第一次允許境外資金直接通過香港的離岸平台買賣內地債券，借鑒滬深港通的成功經驗，沿用海外投資者長期形成的交易和結算習慣，增加了海外投資者投資中國債市的渠道，提高了中國債券市場的開放程度。具體而言，債券通的創新性體現在交易前的市場准入，交易中的價格發現與資訊溝通，以及交易後的託管結算環節，以更低的制度成本、更高的市場效率，將國際慣例

9　數據來源：中債登網站。
10　數據來源：Wind 資訊。
11　熊貓債是指境外機構在境內債市發行的人民幣計價債券。

與中國債市的有效對接。

　　作為區內主要的國際金融中心，香港為債券通提供了一個符合國際慣例的交易、結算平台，實現了內地與全球資本市場的連接。這將吸引更多主體參與到香港金融市場活動，為香港引入更多資金，進一步豐富香港的人民幣金融產品，有助於香港進一步發展成為人民幣資產配置中心，強化和鞏固香港作為離岸人民幣中心的地位。

　　從長遠看，債券通將使跨境投資資本的流動效率大幅提高，同時提升內地市場的國際化水平。可以預計，債券通的實施將促進內地人民幣債券產品和服務進一步完善，有利於內地培育多元化的投資者基礎和構建更加開放的債券市場。

英文縮略詞

CFETS　　　　中國外匯交易中心（China Foreign Exchange Trade System）

CIBM　　　　中國銀行間債券市場（China Interbank Bond Market）

HKMA-CMU　香港金融管理局債務工具中央結算系統
（Central Moneymarkets Unit of the Hong Kong Monetary Authority）

QFII　　　　合格境外機構投資者（Qualified Foreign Institutional Investor）

RFQ　　　　請求報價（Request for Quote）

RQFII　　　人民幣合格境外機構投資者
（RMB Qualified Foreign Institutional Investor）

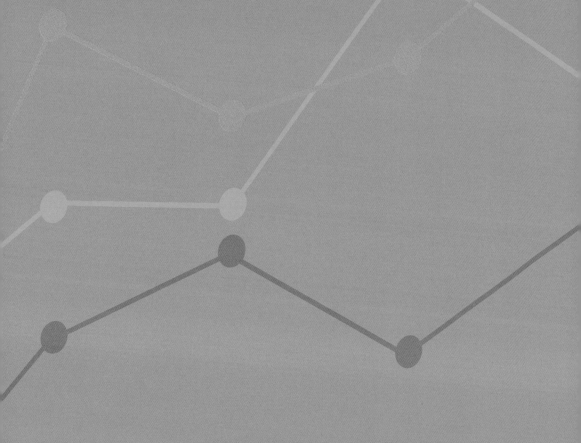

第三部分

大宗商品

第11章

香港邁向亞洲黃金定價中心

2017 年 7 月 10 日

摘 要

香港交易及結算所有限公司（香港交易所）於 2017 年 7 月 10 日透過其附屬公司香港期貨交易所（期交所）推出實物交收雙幣（以美元及人民幣定價及結算）黃金期貨合約（黃金合約）。

香港交易所於 2012 年收購倫敦金屬交易所（LME）後，一如其集團《戰略規劃2016-2018》的願景所勾劃，香港交易所冀可將集團轉化成為一家提供全方位產品及服務、縱向全面整合的全球交易所；其四管齊下的多資產戰略中，大宗商品乃是核心支柱之一。香港交易所推出黃金合約，清楚證明其銳意在亞洲提供具吸引力的大宗商品產品。

香港金市雖有超過 100 年歷史，但在建立基準、流動性和產品服務的整全性方面，仍然落後紐約和倫敦等其他國際黃金交易中心。然而，位處中國這個全球第二大經濟體和全球最大黃金消費國之門戶，香港作為國際金融中心已具備成為亞洲黃金定價中心的成熟條件。

挾着自由市場和轉口貿易中心的優勢，香港擁有活躍的現貨金交易市場，是世界主要金市之一。另外，香港是全球最大離岸人民幣中心，在促進人民幣國際化方面有獨特角色。香港推出實物交收黃金期貨將是實現這一目標的踏腳石。

中國對買賣黃金有基本的市場需求，世界其他地區亦有黃金交易需求，加上相關風險管理需求，種種因素均有利香港發展成為亞洲黃金定價中心。香港要培植新的黃金定價基準，其必須建設一個運作完善的市場連接現貨與期貨交易、提供高效渠道在香港為這些交易提供服務，並加入其他黃金相關金融產品和服務（如黃金租賃和相關衍生產品），完善整個黃金市場生態系統。待這個生態系統內透過上述渠道匯聚的黃金流動性不斷增加，新的亞洲基準便會自然而然在香港形成。

1　黃金的性質與用途

黃金是密度高及明亮的橙黃色貴金屬，柔軟、有韌性及可伸展。由於其相對稀少和不易銹蝕，黃金無論是用作首飾及其他裝飾、作為投資，又或歷史上作為金錢的一種形式，其價值均非常高昂，時至今日仍是央行儲備的主要組成部份（見表 1）。

排名	經濟體 / 多邊組織	公噸	佔中央銀行儲備百分比
1	美國	8,133.5	75%
2	德國	3,377.9	69%
3	國際貨幣基金組織	2,814.0	—
4	意大利	2,451.8	68%
5	法國	2,435.9	64%
6	中國內地	1,842.6	2%
7	俄羅斯	1,680.1	17%
8	瑞士	1,040.0	6%
9	日本	765.2	2%
10	荷蘭	612.5	64%
11	印度	557.8	6%
12	歐洲中央銀行	504.8	27%
13	土耳其	427.8	16%
14	中國台灣地區	423.6	4%
15	葡萄牙	382.5	55%
16	沙特阿拉伯	322.9	2%

表 1：二十大據報官方持金量（2017 年 3 月）

(續)

表 1：二十大據報官方持金量（2017 年 3 月）			
排名	經濟體 / 多邊組織	公噸	佔中央銀行儲備百分比
17	英國	310.3	9%
18	黎巴嫩	286.8	21%
19	西班牙	281.6	17%
20	奧地利	280.0	46%

資料來源：國際貨幣基金組織國際金融統計資料庫、世界黃金協會。

　　人類認識黃金的歷史可追溯到超過五千多年前的古埃及時代。從那時候開始，黃金就與人類發展有着不可分割的緊密關係。

　　化學上，黃金的符號是 Au（來自拉丁文 aurum），原子序數是 79 —— 是天然元素中最高的元素之一。黃金最常以自由元素形式呈現，例如岩石和沖積礦床中的礦塊或粒狀物；海中也有大量黃金。1880 年代以來，南非一直是世界黃金供應的主要來源，迄今產量可能已達累計產量的 50% 左右[1]。不過，南非最近已被其他生產國尤其是中國超越。

　　世上第一枚金幣在公元前 600 年左右在小亞細亞麗迪亞（Lydia）鑄造，自此人類歷史大部份時候均以黃金為貨幣體系基礎，世界各地到 1971 年才捨棄金本位，瑞士繼續沿用黃金支持其 40% 的貨幣價值，直到 1999 年。今天，許多央行儲備仍然包含黃金。黃金也仍然是重要投資工具 —— 形式為金塊、紙黃金、衍生產品和交易所買賣基金；在動盪時期，黃金更被視為避險投資。金價傾向在戰爭時期上升，最近一次大升是 2008 年全球金融危機。然而，今天投資領域已擴大，黃金的相對重要性隨之下降。

　　黃金能耐大部份酸性及大部份鹼性物質，傳電能力良好，使得其在電腦化設施和電機設備中持續用作耐腐蝕導體，是其主要工業應用。一個典型的流動電話可能就含 50 毫克黃金，今天市價計算約 2.00 美元。黃金也可用於紅外線屏蔽、彩色玻璃生產和金箔。攝入黃金對人體無害，故有時更用以裝飾食品；醫學上金鹽仍用作抗炎藥。

1　資料來源：世界黃金協會。

如上文所述，在整個人類歷史中，黃金一直作金錢使用。全球的黃金貿易和交易始於 200 年前。下圖 1 顯示現代歷史上黃金貿易及交易發展的若干關鍵事件——大部份發生於 1971 年布雷頓森林體系崩潰金本位被放棄之後。

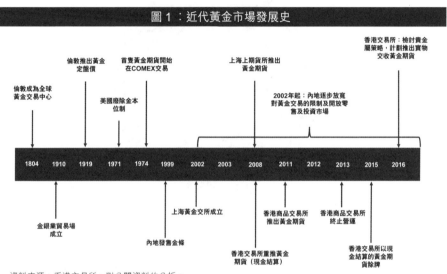

圖 1：近代黃金市場發展史

資料來源：香港交易所，對公開資料的分析。

2　黃金基本面

2.1　黃金的供求

2016 年全球約 53% 的黃金消費用於首飾，37% 用於投資，10% 作工業用途（見圖 2）。 由於黃金不會腐朽或輕易與其他物質產生反應，人類幾千年來開採出來的黃金今天大部份仍然存在，儘管很多相信已散失、埋於墓中，或是（如黃金作工業用）嵌藏於垃圾堆填區中的某些垃圾。不過，來自既有私人首飾和金條塊的黃金廢料、來自回收工業產品的黃金廢料，都與新開採的黃金一樣，同是每年黃金供應的重要來源。

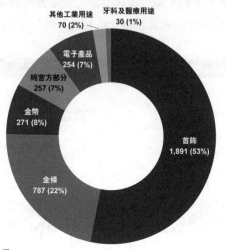

圖2：全球對黃金的實物需求（2016年）

其他工業用途 70 (2%)
牙科及醫療用途 30 (1%)
電子產品 254 (7%)
純官方部分 257 (7%)
金幣 271 (8%)
金條 787 (22%)
首飾 1,891 (53%)

單位：公噸
（總計：3,560 公噸）

資料來源：黃金礦業服務公司。

　　一如其他大宗商品，黃金的價格也是由供（來自金礦場及黃金廢料的實體供應）與求（首飾及其他用途對黃金的需求）的平衡帶動。然而，由於黃金的地上可用存貨不少，實物供應有剩還是不足對定價的影響不若其他大宗商品重要（儘管可能影響價值鏈上的交貨時間、溢價和利潤）。由於黃金仍是投資工具，貨幣情況和公眾對經濟的信心是金價重要決定因素。回顧歷史，黃金一直是環境不明朗及市場動盪時期的避險大堂。如圖3所示，1980年惡性通貨膨脹和能源危機期間，金價上漲至每盎司870美元，2008年全球金融危機之後，金價更曾於2011年高見每盎司1,895美元。

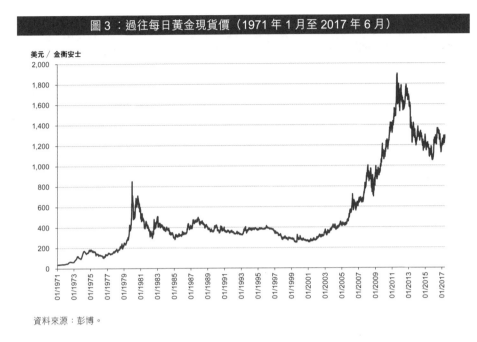

圖 3：過往每日黃金現貨價（1971 年 1 月至 2017 年 6 月）

美元／金衡安士

資料來源：彭博。

如下表 2 所示，金礦產量過去 10 年不斷上升，現時年產超過 3,000 噸。2016年，廢金供應按年增加 8% 至 1,268 噸，符合金價近期升勢。如下表 2 所示，這逆轉了 2014 年廢金供應跌至 1,158 噸最低點的下降軌。

表 2：全球黃金供求（公噸）（2007 年至 2016 年）										
	2007	2008	2009	2010	2011	2012	2013	2014	2015	2016
供應										
礦產	2,538	2,467	2,651	2,775	2,868	2,883	3,077	3,172	3,209	3,222
廢金	1,029	1,388	1,765	1,743	1,704	1,700	1,303	1,158	1,172	1,268
淨對沖供應	-432	-357	-234	-106	18	-40	-39	108	21	21
總供應	3,135	3,498	4,182	4,412	4,590	4,543	4,341	4,438	4,402	4,511
需求										
首飾	2,474	2,355	1,866	2,083	2,091	2,061	2,610	2,469	2,395	1,891

(續)

表 2：全球黃金供求（公噸）（2007 年至 2016 年）									
	2007	2008	2009	2010	2011	2012	2013	2014	2015	2016
工業製造	492	479	427	480	471	429	421	403	366	354
電子產品	345	334	295	346	343	307	300	290	258	254
牙科及醫療用途	58	56	53	48	43	39	36	34	32	30
其他工業用途	89	89	79	86	85	83	85	79	76	70
純官方部份	-484	-235	-34	77	457	544	409	466	436	257
零售投資	449	937	866	1263	1616	1407	1873	1164	1162	1058
金條	238	667	562	946	1,247	1,056	1,444	886	876	787
金幣	211	270	304	317	369	351	429	278	286	271
實物需求總量	2,931	3,536	3,125	3,903	4,635	4,441	5,313	4,502	4,359	3,560
實物盈餘 / 赤字	204	-38	1,057	509	-45	102	-972	-64	43	951
ETF 囤積存貨	253	321	623	382	185	279	-880	-155	-125	524
交易所囤積存貨	-10	34	39	54	-6	-10	-98	1	-48	86
淨結餘	-39	-393	395	73	-224	-167	6	90	216	341

資料來源：黃金礦業服務公司。

　　如上表 2 顯示，黃金實體需求總額持續下降至 2016 年三年低位 3,560 噸，跌幅 18%，所有需求領域均顯示需求下降。首飾始終是需求的最大來源，繼而是散戶投資。不過，首飾需求跌 21%，主要是印度和中國消費下跌[2]。工業製造繼續下降 3% 至 354 噸，是十年來最低水平，因所有主要行業需求均疲弱，特別是電子業（持續減少使用黃金，代之以其他物質）以及牙科和 裝飾用途，但與此同時，隨着

2　資料來源：世界黃金協會。

投資者配置更多資金作交易所買賣基金及期貨交易，黃金的非實物投資總額增至 610 噸（主因是年內交易所買賣基金的購買量很大，有別於對上一年該等基金錄得贖回淨額）。

　　根據黃金礦業服務公司數據，2016 年的地上黃金存貨總額（礦產累計歷史總額）按年增加 1%，達 187,200 噸，相等於 2017 年 6 月 21 日 7.6 萬億美元左右。如圖 4 所示，首飾的存貨量最大，佔 48% 左右，繼而是私人投資和官方的持有量，約佔 38%。

圖 4：全球地上黃金庫存總額（2016 年）

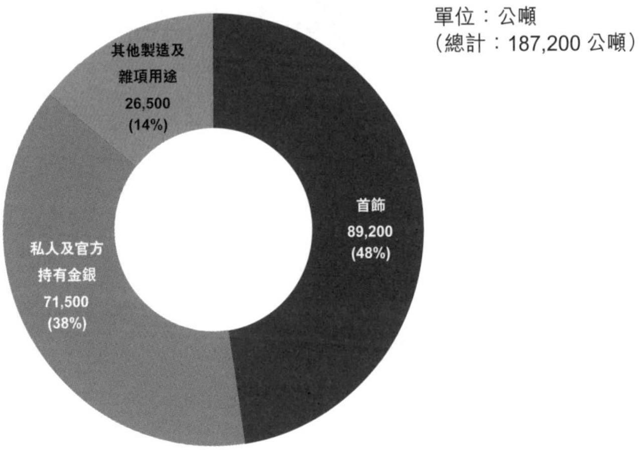

單位：公噸
（總計：187,200 公噸）

其他製造及
雜項用途
26,500
(14%)

私人及官方
持有金銀
71,500
(38%)

首飾
89,200
(48%)

資料來源：黃金礦業服務公司。

2.2　黃金價值鏈中的主要參與者

　　如下圖 5 所示，黃金價值鏈是一個由不同參與者組成的生態系統。採礦公司提取礦石進行加工，取得黃金原料後提供予加工商作進一步提煉和分銷予消費者，消費者再將黃金製成可分銷予終端用戶（零售消費者、投資者、工業用戶和央行）的產品。價值鏈由許多服務供應商支持，例如檢測機構負責證明金條的質量和重量、託管商以金庫保管黃金、資訊供應商發佈黃金價格，交易所提供市場會員買賣黃金標的合約。

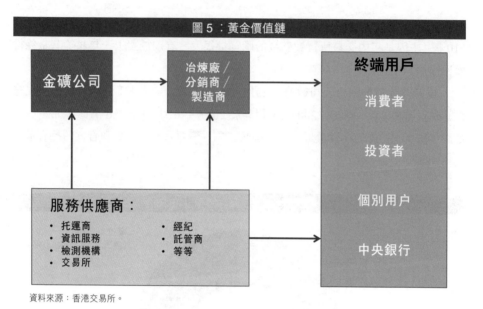

圖 5：黃金價值鏈

資料來源：香港交易所。

3　全球黃金市場

　　全球黃金市場分佈廣泛，但由以倫敦為中心的場外交易主導，不過芝加哥商業交易所（CME，簡稱芝商所）集團的 COMEX（紐約商品交易所）及上海期貨交易所（上期所）等數家期貨交易所亦錄得顯著成交。左右黃金價格的信息來自不同渠道，包括礦場、首飾需求、央行交易以及宏觀經濟發展。西方國家的黃金市場生態圈已非常成熟，由現貨交易以至遠期交易、黃金租賃、以及融資及其他衍生產品等等琳琅滿目。目前，全球黃金價格以倫敦現貨定價及紐約期貨交易價格為首。下文闡述該兩個西方市場及中國內地和香港這兩個東方市場。

3.1　西方市場

3.1.1　以倫敦為基礎的場外市場

倫敦黃金市場發展歷史可追溯至 200 多年前，從 1804 年倫敦超越阿姆斯特丹成為全球的黃金交易中心之時說起。場外金銀批發市場 —— 倫敦金銀市場協會（「LBMA」）於 1919 年開始投入運作，每日市場價格由五大金銀批發商及銀行釐定。倫敦價格亦影響着紐約及香港的黃金價格。現時，倫敦仍是世界最大黃金現貨交易市場，全球金銀交易仍參考倫敦金價作為基準價格。

於 2016 年，LBMA 結算會員轉移及結算了約 157,828[3] 噸黃金，價值 6.3 萬億美元。隨着 LBMA 的活動（大部份為商業銀行之間）而產生的，可以是黃金實物轉移或純粹合約轉移。LBMA 結算會員於 2016 年轉移的該 157,828 噸黃金佔倫敦市場約八分之一的成交。隨着中國、泰國及新加坡等交易中心的崛起，2016 年倫敦成交佔全球總成交由 90% 下跌至 65%。2016 年全球總成交估計達 1,867,000 噸，價值 75 萬億美元，相當於礦產量的 580 倍。由於流通量龐大，黃金（一如大部份貨幣）以持倉成本全覆蓋（full carry）的方式交易。

3.1.2　蘇黎世黃金市場

相對倫敦而言，蘇黎世黃金市場並無任何正式規模，主要是由三大瑞士銀行為場外市場提供流通量及結算服務。瑞士是全球最大黃金中轉樞紐，匯集了全球若干最著名的黃金冶煉廠，如 PAMP 及 Metalor。此外，基於其特有法制為黃金擁有者提供了額外保障，瑞士也是世界最大的私人黃金儲存中心。

3.1.3　芝商所集團的 COMEX 期貨市場

COMEX（現隸屬芝商所集團）前稱 Commodity Exchange Inc.，於 1974 年成立，那是美國放棄金本位制、黃金對美元改為靈活定價後，再加上美國大部份法人機構對於套期保值或投資增值的需要增加等因素所推動而成。

受美元大幅波動及其他因素所推動，美元黃金期貨市場於 1978 年至 1980 年

3　LBMA 結算會員的轉移及結算數字來自 LBMA 網站，本段其餘數字由香港交易所根據多個渠道估算，包括黃金新聞報道及諮詢黃金行業的主要業界人士。

間急速擴展。時至今日，COMEX 按成交量計仍是全球最大黃金期貨交易中心，對現貨金價有龐大影響。COMEX 黃金期貨為每月合約，於交割月份採用每日交割機制。所有交割點均位於紐約市及附近的特拉華州。2016 年，COMEX 黃金期貨總名義成交量為 179,000 噸[4]。

3.2　東方市場

3.2.1　中國內地黃金市場

1950 年，中國將黃金行業納入國家管制。自 1978 年經濟改革首階段起，黃金市場開始審慎開放，主要體現形式為用於深圳經濟特區生產首飾。1983 年頒佈的《金銀管理條例》，確認了中國人民銀行（人行）在規管、監督及管制金銀採購及分佈的中央角色。人行亦負責管理中國的黃金儲備。2001 年人行取消其黃金統購統配及定價管制後，私營市場對首飾的需求及近年對黃金投資的需求急劇增長。

人行放寬直接管制後，由人行作為主要組建人兼主要權益持有人的上海黃金交易所（上金所）肩負起定價責任。上金所於 2002 年 10 月 20 日開始交易。所有冶煉黃金均於上金所出售，使該所成了業界及金融機構購買黃金的唯一市場。所有進口金銀經由上金所出售。2003 年進一步放寬措施，廢除了經營金銀產品業務的牌照制度；2004 年，中國更容許私人擁有及買賣金銀。

中國黃金市場仍受國家間接管制。雖然私人黃金買賣已大幅度放寬，中國與國際市場之間的互動仍然受限，也是資本賬管制的一項主要措施。上金所 2014 年在上海自由貿易區（自貿區）推出黃金國際板，開放予國際交易參與者參加，是進一步審慎開放的措施，不過至今交易量仍然甚少。

下表 3 顯示中國黃金政策及市場發展的主要里程碑。

4　資料來源：美國期貨業協會。

表 3：中國黃金政策的里程碑及主要市場發展	
年份	描述
1950	• 金業受國家管控 • 嚴禁私人持有金銀
1983	• 人行發佈黃金管理條例
1995	• 黃金首飾的消費稅由 10% 減半至 5%
1996	• 新首飾定價架構 ── 將材料費與手工費分開
1998	• 中國人民銀行深圳分行開始從瑞銀、滙豐及景順進口黃金
2001	• 中國黃金協會成立 • 國家物價局廢除零售價限制
2002	• 上海黃金交易所開始正式買賣，交易免收增值稅
2004	• 解除持有金銀的禁令
2007	• 中國成為全球最大黃金生產國
2008	• 上海黃金交易所接受外國銀行成為會員：滙豐、Scotia Mocatta、ANA、瑞銀、渣打 • 上海期貨交易所推出黃金期貨
2010	• 中國工商銀行推出金積存計劃 • 再有 4 家銀行獲發黃金進口牌照
2011	• 上海期貨交易所首次允許外國銀行買賣黃金：澳新銀行、滙豐
2012	• 銀行間市場參與者獲許進行場外黃金交易，透過上海黃金交易所進行結算
2013	• 中國成為全球最大黃金消耗國 • 首隻中國黃金 ETF 在 7 月推出 • 外國銀行獲發黃金進口牌照：澳新銀行、滙豐
2014	• 在上海自貿區推出金交所國際板
2016	• 推出上海黃金交易所的上海金定價

資料來源：世界黃金協會。

　　1978 年，中國黃金年產量少於 10 噸，至 2016 年已增至 453 噸。為提高產量，所採取的措施包括建立中國黃金集團公司的前身公司，以及成立人民解放軍特別黃金礦業組負責勘探黃金及開發金礦。1981-1985 年及 1986-1990 年的兩個五年計劃先後推出了多項便利政策及投資，將中國黃金年產量提升至 1990 年初的 100 噸，此後增長率持續強勁。2007 年中國成為世界最大黃金生產國，2016 年佔全球總產量 14%。

　　雖然近年源自國內生產及循環再用的黃金供應一直上漲，但國內需求也直線上升並超出供應量，以致中國從錄得黃金盈餘走向重大黃金赤字。即使中國並無刊發黃金進口數字，但中國黃金進口量自 2010 年起一直顯著增長，大部份經香港進口[5]。

中國的黃金交易主要集中於上金所及上期所。上金所是中國唯一可合法進行實物黃金交易的場所，接通黃金產量與消耗需求。上期所於 2008 年建立，是中國內地唯一的黃金期貨市場。

上金所主要買賣現貨，較近期則加人現貨延期合約。散戶投資者可到本身是上金所會員的銀行開立戶口買賣黃金。這計劃最初由工商銀行試點經營，但 1 千克合約額對散戶投資者來說實在過大。2007 年 7 月，散戶投資者獲准開始經銀行買賣 Au9999 及 Au100g 合約。根據上金所資料，於 2010 年，約 180 萬散戶投資者佔該所交易的 19%[6]。那年上金所成交 5,715 噸，至 2016 年成交已超過 23,000 噸[7]。散戶投資者現可買賣上金所的所有黃金期貨合約。合約以實物交割結算，但大部份均為投機性質，於結算前早已平倉。

2014 年 9 月，上金所在上海自貿區開設人民幣計價的期貨合約國際板。由推出至 2014 年底期間名義黃金期貨合約成交為 78 噸，交投淡靜，至 2015 年首兩個月稍升至約 50 噸。2016 年 4 月，上金所推出有史以來首個人民幣黃金定價。

根據美國期貨業協會（Futures Industry Association，簡稱 FIA），上期所的黃金期貨交易量自 2013 年起按合約張數計排名全球第二，僅次於 COMEX，下圖 6 顯示近年兩家交易所的成交量對照。

5　資料來源：Metals Focus。
6　資料來源：上海黃金交易所。
7　資料來源：上海黃金交易所。

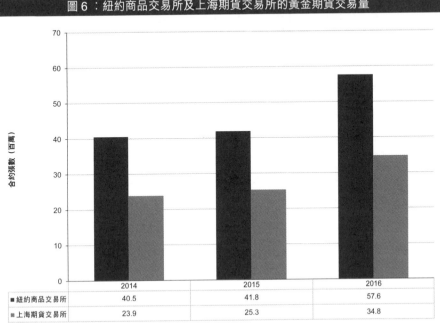

圖 6：紐約商品交易所及上海期貨交易所的黃金期貨交易量

	2014	2015	2016
■ 紐約商品交易所	40.5	41.8	57.6
■ 上海期貨交易所	23.9	25.3	34.8

資料來源：美國期貨業協會及上期所。

3.2.2　香港黃金市場

香港毗鄰深圳這個國家黃金加工中心，是中國進口黃金的主要來源，2016 年錄得 867 噸[8]，佔中國黃金總進口量約 86%。

在過去一個多世紀，香港一直是重要的黃金交易中心。隨着中國內地市場開放，香港在作為中國與世界之間一個實物黃金交易中心方面也擔當着要角。

香港的黃金市場自 1910 年金銀業貿易場成立起漸見雛形。1974 年香港政府撤銷了黃金進出口管制後，香港黃金市場蓬勃發展，至今已成為亞洲的重要黃金投資和交易樞紐，連接歐美其他時區。有見香港在全球黃金交易的重要，倫敦五大黃金批發商及瑞士三大銀行均在香港設立交易櫃台及分行。基於有很多海外主要金銀業人士參與，香港市場的黃金定價自此參考倫敦基準價。

8　資料來源：Metal Focus。

4 建立香港基準價的可行性

4.1 香港推出實物交收黃金期貨的條件是否成熟？

香港金市雖有超過 100 年歷史，但在建立基準、流動性和產品服務的整全性方面，仍然落後紐約和倫敦等其他國際黃金交易中心。然而，位處中國這個全球第二大經濟體和全球最大黃金消費國之門戶，香港作為國際金融中心已具備成為亞洲黃金定價中心的成熟條件。香港擁有成熟穩健的金融市場，只要香港交易所與各金融及監管機構合作，共同透過滿足市場需求、順應國際金市趨勢、逐步增加香港金市在歐美交易時段以外的定價權，理應可以達成上述這個獨特使命。從全球主要大宗商品市場定價的發展歷程可見，新的定價中心的形成必須符合以下兩大基本條件。

(1) 市場對交易及風險管理的基本需求很大

市場需要進行大量實體交易，是形成價格基準的重要先天因素之一。如圖 7 所示，中國進口黃金佔全球供應約四分一，當中約 70% 均經香港進入中國。

圖 7：全球現貨黃金流動情況（2016 年）

註：以公噸 (t) 為單位。

資料來源：香港交易所，根據 Metals Focus 數據分析。

大規模的交易量代表着市場的參與者對於該市場價格的認可程度，是一個定價中心必不可少的元素。無論是亞洲時段的黃金和其衍生品交易量，還是受中國需求推動的黃金產品交易量的支持，都應當可以支撐一個新的定價圈子的形成。紐約和倫敦的金市過去亦是這樣發展起來。

(2)　優良的市場 —— 有效使用現貨及衍生產品市場

香港要培植新的黃金定價基準，其必須建設一個運作良好的市場連接現貨與期貨交易，並提供高效渠道在香港為這些交易提供服務，不用說也還要提供其他黃金相關的金融產品和服務（如黃金租賃和相關衍生產品等），就如倫敦和紐約現時所提供的一樣。

4.2　香港的黃金機會

過去百年，香港的現貨金市場一直非常活躍，從冶煉、加工、檢驗，到批發、零售，再到交易和對沖，一應俱全，一直推動香港的黃金進口及轉口貿易。然而，香港金市向來被動及透明度不高，而由於倫敦金作為基準有全球主導地位，香港的黃金實物交易傳統上一直以倫敦金為定價基礎。

儘管上金所及上期所已主導中國內地黃金市場（在岸市場），但基於在岸與離岸市場法律法規的差異，加上資金及黃金進出口管制，香港的黃金買賣未能參照內地黃金基準價。因此，香港交易所有獨特的時間和地理優勢開發合適的產品迎合市場需要。

所以，現在正是香港採取行動、無論如何也要爭取定價權的時候。由香港交易所推出雙幣（美元及人民幣定價及實物結算）實物交收黃金期貨產品可以是其中一個有效的方法。香港交易所推出黃金期貨市場後，香港將同時齊備黃金現貨及期貨交易。主要的金銀交易商看到市場上的買賣契機，自會經不同渠道進行交易，屆時整個黃金生態系統就可進一步伸延，將中國內地與西方發達市場的黃金現貨及衍生工具交易連接起來。待流動性增加及全球認受性逐漸建立，結果將會是香港得以確立新的亞洲黃金基準價。

此外，香港交易所集團的倫敦全資子公司倫敦金屬交易所（「LME」）於 2017 年 7 月 10 日亦在倫敦推出了黃金（及白銀）期貨合約，等同為香港交易所集團的客戶提供 24 小時不間斷的兩地黃金期貨交易，配合客戶的商業需要。這樣將現貨金的交易活動「金融化」及「期貨化」並納入 LME 資金池，可有助提升倫敦金的買賣。

隨着黃金期貨（以美元及人民幣定價）市場不斷增長，支持黃金交易生態系統的其他相關領域之間的互動也可愈趨完善。這些領域包括利率、外匯價格及黃金租賃市場。黃金市場生態系統逐漸成熟，加上黃金期貨市場活躍，將為離岸人民幣利率市場以至人民幣國際化的最終實現提供重要支持。最後，新的亞洲黃金基準價將水到渠成。

5 香港交易所黃金期貨：產品設計及主要技術特點

5.1 新市場形勢

香港交易所 2012 年收購 LME 後，一如其集團《戰略規劃 2016-2018》的願景所勾劃，香港交易所冀可將集團轉化成為一家提供全方位產品及服務、縱向全面整合的全球交易所；其四管齊下的多資產戰略中，大宗商品乃是核心支柱之一。香港交易所推出黃金合約，清楚證明其銳意在亞洲提供具吸引力的大宗商品產品。

正如上文所述，挾着自由市場和轉口貿易中心的優勢，香港擁有活躍的現貨金交易市場，是世界主要金市之一。

另外，香港是全球最大離岸人民幣中心，在促進人民幣國際化方面有獨特角色。由於實物黃金可對個別法定貨幣起「支持」作用（以人民幣而言，透過人民幣利率與黃金租賃利率機制，類似現行美元與黃金的關係），香港推出實物交收黃金期貨將是實現這一目標的踏腳石。

5.2 需留意的產品設計要點

為符合客戶所需，合約設計考慮了下列主要因素：

(a) 合約標的及單位 —— 亞洲客戶（尤其在大中華地區）常買賣金含量不小於

99.99% 的千克黃金；實物交易機制確保市價貼近真正的實物現貨金價，建立新的香港基準價，為終端用戶提供有力的風險管理工具。

(b) 交易及結算貨幣 —— 以美元及離岸人民幣作為交易及結算貨幣，會同時吸引美元及離岸人民幣投資者。雙幣的黃金合約由於涉及同一標的，其將會產生一個美元兌離岸人民幣的引申匯率。此引申匯率與匯市其他匯率之間差異提供套戥機會，故此可提高美元兌離岸人民幣的整體流動性，同時提升及拉平有關市場的遠期曲線。

(c) 合約月份 —— 即月及後續 11 個曆月將涵蓋國內外期貨市場流動性最強的交易月份，並為實物市場提供更多對沖工具。

5.3　香港交易所黃金期貨的產品應用及用戶

向投資者推出黃金合約的增值效益包括 (但不限於)：

(a) 利便內地及國際投資者透過香港交易所在一個亞洲時區內的穩健交易樞紐參與黃金市場；

(b) 為投資者及終端用戶提供對沖及風險管理選擇；

(c) 為增長中的離岸人民幣存款提供更多投資出路；及

(d) 吸引想將黃金加入投資組合的投資者。

黃金合約的潛在用戶及客戶是：

(a) 實物業者，例如黃金冶煉廠、製造商及珠寶商等需要對沖金價風險的人士；

(b) 金融業者，例如銀行及基金利用期貨市場與其黃金相關投資產品連繫掛鈎，以及套戥者透過在岸離岸市場之間 (指紐約、倫敦、上海及香港市場之間) 價格差異進行買賣，並就匯價及利率差異部署其他交易戰略；及

(c) 其他有意涉足黃金交易的投資者及交易商。

附錄一

香港交易所黃金期貨合約細則[9]

項目	美元黃金期貨	人民幣（香港）黃金期貨
合約標的	金含量不小於99.99%，帶有認可的精煉廠標簽及其序列號的1公斤黃金	
合約單位	1千克	
交易貨幣	美元	人民幣（即人民幣（香港））
合約月份	現貨月及後續11個月	
最低波幅 / 最小變動價位	每克0.01美元	每克人民幣0.05元
交易時間 （香港時間）	上午8時30分至下午4時30分（日間交易時段）及 下午5時15分至翌日凌晨1時（收市後期貨交易時段）	
最後交易日	合約月份的第三個星期一，如該日為香港公眾假期，則延至下一營業日	
最後結算日	最後交易日後的第二個香港營業日	
結算方式	實物交收	
交易所費用[10]	交易費：每邊每張合約1.00美元 結算費：每邊每張合約2.00美元	交易費：每邊每張合約人民幣6.00元 結算費：每邊每張合約人民幣12.00元

9　專有名詞的定義見下列有關黃金產品的交易及結算規則修訂本的連結：
　　http://www.hkex.com.hk/eng/rulesreg/traderules/traderuleupdate-hkfe/Documents/49-17-HKFE-Star_e.pdf
　　http://www.hkex.com.hk/eng/rulesreg/clearrules/clrruleupdate_hkcc/Documents/50-17-HKCC-Star_e.pdf
10　費用金額不時可予更改。

附錄二

交易及結算安排及規定[11]

1. 交易及結算安排

交易及結算安排如下圖所示。

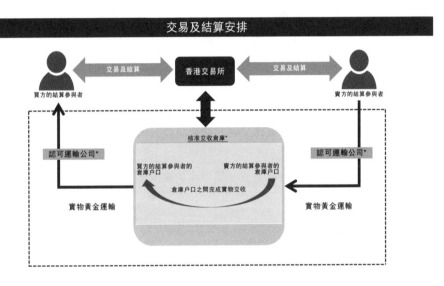

* 香港交易所網站將不時更新相關名單：核准交收倉庫及認可運輸公司。

資料來源：香港交易所

2. 交易及結算規定

香港期貨結算有限公司（期貨結算公司）參與者要進行實物交收，需在交收銀行開立美元及 / 或人民幣（香港）交收戶口，及在每個核准交收倉庫開立戶口，或與另一可進行交割的結算參與者簽署交收協議。

---

11 專有名詞的定義見下列有關黃金產品的交易及結算規則修訂本的連結：
http://www.hkex.com.hk/eng/rulesreg/traderules/traderuleupdate-hkfe/Documents/49-17-HKFE-Star_e.pdf
http://www.hkex.com.hk/eng/rulesreg/clearrules/clrruleupdate_hkcc/Documents/50-17-HKCC-Star_e.pdf

附錄三

誠信產業鏈[12]

　　為建立及維持穩健機制，確保在香港交易所市場交收的金條質素，香港交易所一如全球其他金銀市場，設置了有關的誠信產業鏈保障交易所及其參與者，詳情如下。

　　香港交易所規定所有交付金屬都須得到認可冶煉廠認證，並須附以期貨結算公司參與者或其認可運輸公司簽發的文件，證明該等送至核准交收倉庫的交付金屬，乃從另一核准交收倉庫、認可檢測機構、認可冶煉廠或認可交收倉庫，經由認可運輸公司運送至該處。

　　布林克香港有限公司（布林克）是首家獲委任支援實物交收黃金期貨合約的核准交收倉庫。所有擬進行實物交收的期貨結算公司參與者須於布林克開設倉儲戶口。

　　認可運輸公司、認可檢測機構、認可冶煉廠及認可交收倉庫的最新名單將登載於香港交易所網站。

12　專有名詞的定義見下列有關黃金產品的交易及結算規則修訂本的連結：
　　http://www.hkex.com.hk/eng/rulesreg/traderules/traderuleupdate-hkfe/Documents/49-17-HKFE-Star_e.pdf
　　http://www.hkex.com.hk/eng/rulesreg/clearrules/clrruleupdate_hkcc/Documents/50-17-HKCC-Star_e.pdf

第12章

香港發展鐵礦石
衍生產品市場的機遇

2017 年 11 月 13 日

摘　要

　　香港交易及結算所有限公司（香港交易所）於 2017 年 11 月 13 日在香港期貨交易所有限公司（期交所，為香港交易所附屬公司）平台上推出其首隻黑色金屬產品 —— 美元計價、現金結算的 TSI CFR 中國鐵礦石 62% 鐵粉期貨合約，希望借由期交所的綜合電子交易及結算平台，提高境外鐵礦石衍生產品市場的價格透明度、優化價格發現效率。

　　鐵礦石是煉鋼的主要原材料，按交易金額計亦是排在原油之後的全球第二大大宗商品 [1]。中國是全世界最大的鐵礦石進口國及消費國。由於中國的鐵礦石需求極其依賴進口，以及其經濟增速較快，中國主要經濟政策及國家發展戰略（例如供給側改革及「一帶一路」倡議）等「中國因素」對鐵礦石的潛在需求及定價都有着顯著影響。

　　全球鐵礦石衍生產品市場（包括內地市場）近年來發展迅速，成交屢破紀錄，但仍有進一步增長的空間。內地市場交投活躍，但外資暫時無法參與。境外市場方面，透過傳統的口頭商議撮合成交的場外掉期市場仍為最主要的交易方式，而交易所的場內市場雖提供高效的電子交易方式，卻欠流通量及市場深度。香港交易所看準當中契機推出鐵礦石期貨電子盤，填補現有市場空缺、優化鐵礦石的價格發現方式：

- 香港交易所鐵礦石期貨合約是經電子交易平台集中撮合的交易所買賣產品。相對於場外市場，電子盤的流動性可使交易更便利、更透明，及價格發現過程更優化。

- 鐵礦石衍生產品市場發展歷史較短，增長潛力仍待全面發掘。香港交易所在其電子交易衍生產品市場推出鐵礦石期貨，可在便利和透明的基礎設施上擴闊投資者群體及優化市場准入條件，有助市場進一步擴容。

- 現貨鐵礦石的定價模式過去十年已經起了很大變化，目前指數掛鈎定價模式日後會否沿用還是再次演變仍有待關注。香港交易所鐵礦石期貨將有助建立流通透明的期貨市場，或有助市場定價模式另闢新徑。

　　儘管「中國因素」對鐵礦石市場舉足輕重，但內地與海外鐵礦石衍生產品市場之間的連繫仍待加強。香港作為位處中國門戶的全球金融中心，具有重要的戰略地位，因此在香港建立透明度高且流動性好的的離岸鐵礦石期貨市場不單可滿足現貨商和貿易商對鐵礦石價格風險管理的需要，更為機構投資者乃至個人投資者提供具吸引力的中國相關投資標的。

1　資料來源：《經濟學人》2012 年 10 月 13 日 "The lore of ore"（http://www.economist.com）。

1 中國在全球鐵礦石市場的影響力

1.1　全球最大的鐵礦石終端市場

鐵礦石是煉鋼的主要原材料，按貿易額計亦是排在原油之後的全球第二大大宗商品[2]。

鋼鐵在房地產、運輸、汽車製造、能源供應網絡、機械製造、造船及家用電器等眾多下游產業用量極大。過去 20 年中國經濟迅速發展，鋼鐵需求不斷增加，中國的粗鋼產量大幅增長至 2016 年的 8.08 億噸，為之前的 8 倍，佔全球粗鋼總產量半數之多[3]。鐵礦石是煉鋼的主要原料，中國的用量在過去 20 年增長至 2016 年的 13 億噸[4]，為之前的 20 倍以上。

中國是全世界最大的鐵礦石進口國，2016 年進口量達 10.24 億噸，佔全球海運貿易 70%[5]。基於中國國產鐵礦石儲備品位低[6]、雜質多，要滿足國內對中高品位鐵礦石的龐大需求，唯有大量從澳洲、巴西、南非及印度等地進口（進口依存度一度高達 84%[7]）。

由於中國對鐵礦石有龐大的實際消費需求，故其於鐵礦石現貨及衍生產品中的交投一直活躍。許多中國國有及民營鋼廠和鋼鐵貿易公司為了業務發展的需要在海外設辦事處，當中許多在香港、新加坡或其他離岸稅務港口設有貿易及融資業務，有些甚至於澳洲、西非、南美及北美等有豐富鐵礦石儲量的地區進行礦產投資。

2　資料來源：《經濟學人》2012 年 10 月 13 日《The lore of ore》（http://www.economist.com）。

3　資料來源：Wind，2016 年數據。

4　資料來源：中國鋼鐵工業協會、彭博，2016 年數據。

5　資料來源：中華人民共和國海關總署，2016 年數據。

6　中國鐵礦石的鐵粉含量約為 30%。

7　中國極其依賴進口鐵礦石 —— 每消耗 100 噸鐵鋼石，有 84 噸來自進口。　資料來源：中國鋼鐵工業協會，2015 年數據。

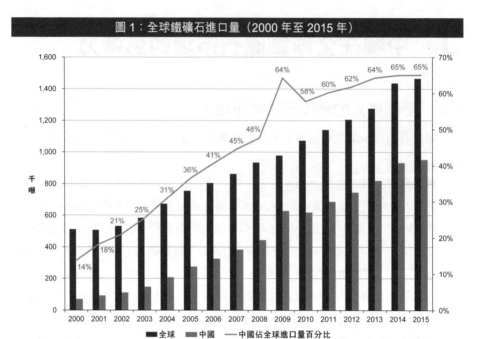

圖 1：全球鐵礦石進口量（2000 年至 2015 年）

資料來源：Wind 資訊。

圖 2：世界主要海運鐵礦石貿易參與者（2016 年）

註：mt = 百萬噸；bnt = 十億噸。
資料來源：Wind 資訊、麥格里研究（2016）。

1.2　鋼鐵行業深受中國相關戰略政策的影響

中國 2017 年經濟增長調整至 6.5% 左右的水平（2016 年增速為 6.7%），目前正密鑼緊鼓進行結構性改革，以期解決產能過剩及提高生產效率，當中對中國鋼鐵業影響尤大的，是以下兩項戰略及政策：

(1)　「一帶一路」倡議

「一帶一路」倡議包括「絲綢之路經濟帶」及「21 世紀海上絲綢之路」，當中包括輸出過剩資金及產能，促進貿易以及建立基礎設施網絡，沿傳統貿易路線加強中國與亞洲、歐洲及非洲的連繫。這政策可帶來超過 60 項總值 1,000 億美元的雙邊合作協議，覆蓋達約 65 個國家 [8]。根據中國《2017 年政府工作報告》（見表 1），該倡議亦會為鋼鐵行業帶來歷史性契機，可支持中國出口貿易及緩解產能過剩的問題。

表 1：中國《2017 年政府工作報告》—— 鋼鐵業相關成果及目標	
2016 年成果	**2017 年目標**
• 去產能、去庫存、去槓桿、降成本及補短板 • 成功削減鋼鐵業過剩產能 6,600 萬噸、煤炭（製造鋼鐵的原材料）產量 2.9 億噸，超越年度目標 • 供給側結構性改革初見成效 • 「一帶一路」倡議見重大進展，順利推出多個主要的國際工業合作項目、創造協同效益及加強與其他參與國家的聯繫	• 繼續推進供給側結構性改革、減少過剩供給及擴大有效供給，以更高效滿足產業需求 • 繼續削減過剩產能及庫存、去槓桿、降成本及補短板：進一步削減鋼鐵產量 5,000 萬噸及減少煤炭產量 1.5 億噸 • 嚴格執行環境、能耗、質量及安全法規；推動企業兼併重組；減少低效及過剩產能 • 擴大內需並提高有效性，使供給側改革和需求側改革相輔相成、相得益彰，充分釋放國內的發展潛力

(2)　供給側改革

「十三五」規劃訂明鋼鐵業將着眼於整合鋼廠、去除過剩產能及提高產能利用率。中國《國務院關於鋼鐵行業化解過剩產能的意見》嚴禁備案新增產能，及促進環保質量標準嚴格遵守相關規則及法規 [9]。鋼鐵行業 2016 年成功削減 8,500 萬噸產能，目標於 2020 年前將產能減少 1 億至 1.5 億公噸 [10]。

8　資料來源：中國國家發展和改革委員會（發改委），2016 年數據。
9　資料來源：中國工業和信息化部。
10　資料來源：中國國務院《2017 年政府工作報告》。

　　「一帶一路」倡議及供給側改革旨在為鋼鐵行業解決產能過剩、刺激內需及提高行業利潤率，但亦對鋼鐵及生產鋼鐵所需的原材料（鐵礦石、焦煤及焦炭）價格有深遠影響，故業界及其他市場參與者越來越需要就相關商品的價格波動進行風險管理。

2 鐵礦石現貨市場的變遷及其衍生產品市場的發展

2.1　鐵礦石現貨市場的歷史發展

　　鐵礦石現貨交易於 20 世紀 60 年代起採用年度長協機制，每年由全球主要礦山（鐵礦供給側代表）與主要鋼廠代表（需求側代表）一年一度釐定全年的鐵礦供應價，定出來的價格即成為業界指標在其他貿易談判中被參考採用。這個傳統的年度長協模式欠靈活，忽略了年內現貨市況的變化，一旦市價偏離基準價格，違約就會屢有發生。

　　2010 年中國拒絕接受巴西淡水河谷、澳洲必和必拓與日本鋼廠釐定的年度基準價格成為事件的轉捩點，沿用數十年的年度長協機制宣告結束，轉為採用季度以至最終月度以指數掛鈎的定價模式。鐵礦石並非唯一改變定價模式的商品。事實上，其他大宗商品都曾出現過類似的轉變，如動力煤（21 世紀初）、鋁（20 世紀80 年代初）及原油（20 世紀 70 年代末），均廢除了年度長協定價機制，改為較短期較靈活的定價模式。

　　現貨鐵礦石以指數為基準的定價，是基於 Platts（普式）、TSI（鋼鐵指數公司）、Metals Bulletin（金屬導報）等西方指數供應商或一些中國指數供應商所發佈的一個或多個市場認可現貨價格指數月均價的定價方式。這個方法較靈活，可確保價格符合現貨市況及反映當前市場供求。此後，一直受相對不靈活的年度長協定價機制所限制的鐵礦石現貨市場，開始逐步發展壯大。

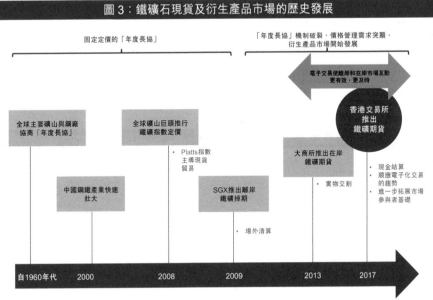

圖 3：鐵礦石現貨及衍生產品市場的歷史發展

註：大商所 —— 大連商品交易所；SGX —— 新加坡交易所。
資料來源：湯森路透、中國鋼鐵工業協會。

2.2 離岸鐵礦石衍生產品市場的發展

隨着指數計價的興起及鐵礦石現貨市場的發展，現貨價格日趨波動，產業鏈上所有參與者 —— 生產商、消費者、海運公司、貿易商及融資機構（銀行）—— 都更加關注價格風險管理的需求，鐵礦石衍生產品市場由此誕生。全球首隻鐵礦石掉期合約由新加坡交易所（新交所）於 2009 年推出。

過去九年，全球鐵礦石衍生產品市場高速增長，年增長率高達 89%（不包括中國內地）[11]。2016 年，中國境外集中清算的鐵礦石衍生產品全年成交量及年末未平倉合約分別約 14.2 億噸及 7,200 萬噸[12]。迄今，新加坡交易所、芝加哥商品交易所

---

11 資料來源：新加坡交易所及芝加哥商品交易所網站。
12 資料來源：期貨業協會，2016 年數據。

（芝商所）、洲際交易所、LCH Clearnet[13] 及 Nasdaq Clearing[14] 等多個海外交易所和
清算所都有在中國內地以外的離岸市場提供鐵礦石衍生產品交易及 / 或結算服務。
在眾多離岸衍生產品標的之中，TSI CFR 中國鐵礦石 62% 鐵粉價格指數（TSI 62
指數）是最常見的參考基準指數，代表美元計價的中國北方港口現貨鐵礦石的到岸
價格[15]。

圖 4：鐵礦石現貨歷史價格 —— TSI CFR 中國鐵礦石 62% 鐵粉價格指數（美元 / 噸）

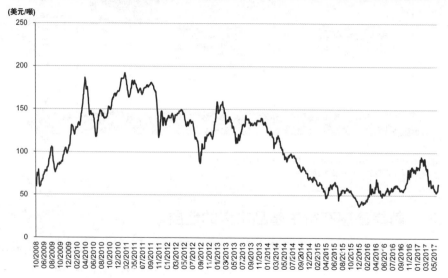

資料來源：TSI 鋼鐵指數公司。

　　經過十年發展，中國境外的全球鐵礦石衍生產品市場成功吸引了鋼廠、貿易
商、礦山及銀行的參與。然而，據從市場參與者了解所得，市場仍然由場外經紀
口頭商議的場外撮合模式所主導，真正在場內電子化撮合交易的流通量及市場深

13　LCH Clearnet 是領先的多資產類別結算所 LCH Group 旗下成員公司，服務對象包括多個主要交易所和交易平台
　　以及場外市場。
14　Nasdaq Clearing 是歐洲市場基礎設施監管規則 (EMIR) 認可的主要結算所，為多個市場及資產類別提供中央交易
　　對手結算服務。
15　TSI CFR 中國鐵礦石 62% 鐵粉價格指數參照中國北方港口（不包括青島港）的交割價格。

度都相對薄弱（場內成交量僅為場外成交量約 10%[16]）。場外市場較適合商議大宗交易或自訂條款交易，但對天然不乏大量買家賣家的普通衍生產品，場外交易的模式便有欠效率和成本效益。若沒有交投活躍的電子交易平台，就限制了多元化的投資羣體進入市場，這在某種程度上也限制了中國內地以外鐵礦石衍生產品市場的進一步發展。

2.3　內地鐵礦石衍生產品市場迅速發展

大連商品交易所（大商所）2013 年 10 月推出中國內地首隻鐵礦石衍生產品——人民幣計價實物交割的鐵礦石期貨合約（連鐵）。基於中國對該品種風險管理及投機需求龐大，大商所迅速發展為全球最大的鐵礦石衍生產品市場，2016 年全年成交量及年末未平倉合約分別高達 340 億噸（是同期整個離岸市場總成交量的24 倍以上）及 5,500 萬噸[17]，超過全球各地的交易平台的交易規模。

連鐵吸納了龐大的個人投資者、金融機構及現貨用戶入市參與，交投活躍。但暫時只是封閉的國內市場，尚未開放予海外投資者直接入市，因此其價格的國際化進程仍有待展開。此外還有若干主要挑戰仍有待解決，例如建立連續月份的流動性，及提高產業用戶的參與度。

不過，在岸衍生產品市場的成立及快速發展始終有助增進離岸市場的流通量，也對優化整個鐵礦石市場的價格發現機製作出了重要貢獻。連鐵推出後，新加坡交易所等海外市場鐵礦石衍生產品的成交量也出現了倍增，並且出現了於大商所交易時段內的交投是最為活躍之現象[18]。根據市場觀察所得，跨市場價格聯動亦變得更及時，證明內地在岸期貨市場健康發展對整個市場的價格發現效率和離岸市場發展具有重要意義。

16　是次研究中結算經紀的非正式估計。
17　資料來源：期貨業協會。
18　資料來源：新加坡交易所網站。

3 香港發展鐵礦石衍生產品市場的機遇

3.1 離岸市場期待更加透明的定價機制

境外鐵礦石衍生產品市場於 2009 年建立，內地則於 2013 年末才開設有關品種。然而，現時內地在岸市場與離岸市場的規模約為 24:1 [19]。在岸市場出現複式增長的原因可能包括境內投資資金充裕及投機性風格強等多種因素。但在岸市場採用電子交易模式，使價格更為透明、進而成功吸引各類參與者入場，無疑也是推動市場高速增長的重要因素。相比之下，據從市場人士了解所得，離岸市場的交易（以鐵礦石掉期為主）仍主要是透過經紀下盤的場外撮合交易模式為主，以電子盤集中撮合交易的比例仍很低。那麼，若然能將離岸交易搬上屏幕、用電子化交易模式提升市場的流通量及透明度，是否對市場發展更加有利？答案絕對是肯定的。

表 2：場內場外衍生產品交易的比較		
特點	場內衍生產品	場外衍生產品
價格透明度	透明度高：買賣差價透明	透明度低：買賣差價不透明
交易效率	效率高：集中在電子平台集中配對成交；能適時、公平地配對大量買賣盤	效率低：雙邊透過經紀口頭協定撮合成交
對手方風險	中央結算，交易一刻立即作出對手方變更，對手方風險減至最低	若中央結算，對手方風險可減至最低，但口頭確認交易與實際作出對手方變更並不同步，兩者之間常有時間差
信貸及抵押品	接納按金及現金抵押品	需與銀行商討信用額度及抵押品安排
文件	只需開戶文件	須有國際掉期業務及衍生投資工具協議等雙邊文件

---

19 參考期貨業協會統計數字中有關申報交易所的 2016 年成交量（噸）數據。

　　其實場外掉期市場亦有其優點，包括 (1) 大額交易可在場外雙邊進行（避免對市場價格的衝擊）；及 (2) 可靈活商議、定制交易的結構。然而，場外市場的價格發現及成交一直是通過場外經紀操作執行，論速度、準確度及效率，都遠遠比不上現代化的電子交易平台的中央自動配對及結算。此外，由於不是所有的市場參與者都有機會獲得市場價格，價格有欠透明度，交易成本會較高，並有市場信息不對稱的弊處。正如上文表 2 所示，場內市場的電子盤交易有明顯好處。基於這些相對優勢，近年場內衍生產品的數量及未平倉合約均呈升勢。

　　有鑒於電子交易的諸多優勢，市場對其接受度和認可度也逐步提高，香港交易所推出場內鐵礦石期貨合約，為市場提供集價格發現、交易及結算等功能於一身的綜合平台，可望更加有利於提高鐵礦石市場的透明度，使價格發現過程無障礙，並大幅度減低交易成本。

3.2　鐵礦石衍生產品市場仍具有極大增長潛力

　　過去數年，鐵礦石衍生產品市場從無到有迅速增長至 2016 年全球（包括中國內地）共 360 億公噸的規模 [20]。然而，相比黃金及銅等較成熟商品的衍生產品交易量對現貨交易量的比例約為 80 至 100 倍 [21]，2016 年全球鐵礦石的衍生產品交易量對鐵礦現貨交易量的比例約 25 倍；若只計算海外的美元計價的鐵礦石衍生產品規模，該比例更只有 1.25 倍 [22]。究其原因，應該與鐵礦石衍生產品市場的發展時間尚短有關。相比之下，基礎金屬、能源及貴金屬等商品的衍生產品交易已有數十年甚至超過百年的歷史，可見鐵礦石衍生產品仍處於發展的初級階段，增長潛力尚有極大發揮空間。（見圖 5）

　　出於對沖需要，通常最早也是最積極參與大宗商品衍生產品市場的都是從事現貨貿易的羣體。隨着市場進一步發展，就會有愈來愈多不同類型的參與者（包括各類金融機構、投資基金及個人投資者）加入、進行對沖或投機交易。隨着投資者來源逐步趨於多樣化，市場也會逐步趨於成熟，推動市場的容量及流通量齊升。

20　按期貨業協會統計數字中有關申報交易所的名義成交量數據。

21　特定商品的比例乃全球（包括中國內地）衍生品交易量的噸數除以該商品的全球（包括中國內地）現貨交易量的噸數計算得出。（資料來源：期貨業協會、世界黃金協會、彭博。）

22　內地鐵礦石衍生品市場只提供期貨合約，離岸衍生品市場則提供掉期、期貨及期權合約。

這正是鐵礦石衍生產品市場正在經歷的變化。電子盤交易方便透明，大大提高價格發現的效率，有利於市場的進一步發展。

回想 2009 年鐵礦石衍生產市場剛剛出現時，其增長勢頭無人能料。展望將來，可以肯定的是，鐵礦石衍生產品的市場結構及動態、市場參與者的類型以至產品種類都會逐步演變及改進。

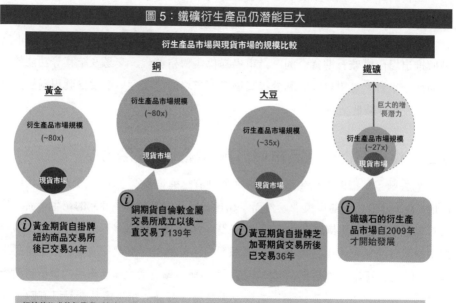

圖 5：鐵礦衍生產品仍潛能巨大

衍生產品市場與現貨市場的規模比較

黃金

衍生產品市場規模
（~80x）

現貨市場

銅

衍生產品市場規模
（~80x）

現貨市場

i 銅期貨自倫敦金屬交易所成立以後一直交易了 139 年

i 黃金期貨自掛牌紐約商品交易所後已交易 34 年

大豆

衍生產品市場規模
（~35x）

現貨市場

i 黃豆期貨自掛牌芝加哥期貨交易所後已交易 36 年

鐵礦

巨大的增長潛力

衍生產品市場規模
（~27x）

現貨市場

i 鐵礦石的衍生產品市場自 2009 年才開始發展

相較其他成熟的商品（如銅、黃豆和黃金）衍生產品市場規模，鐵礦石衍生產品仍具有巨大的市場增長潛力

市場規模的資料來源：FIA，數據截至 2016 年。

3.3　鐵礦石定價的未來發展走向？

自 2010 年「年度長協機制」被指數定價模式取代後，鐵礦石現貨市場出現了巨變。指數定價模式再加上指數相關衍生工具興起，從根本上改變了現貨定價及市場管理價格風險的方法。然而，當市場不斷演化，日後是否會出現更符合市場需要的新型定價模式？

參考部份歷史悠久的大宗商品衍生產品市場（如大豆、銅及原油）的發展，「點價」機制是實物現貨貿易中廣泛採用的一種定價方式，「點價」是買賣雙方以某月份的期貨價格為計價基礎，以期貨價格加上或減去雙方協定同意的升貼水從而確定雙方買賣現貨商品的價格。因此當商議現貨貿易價格時，所協定的條款並非一個固定售價，而是一個協定的基差（實物現貨價與期貨價之間的價差）加一個期貨市場價格。買方有權根據特定商品期貨市場買賣的期貨合約在協定點價期內進行點價。點價模式被視為一個有效釐定價格且反映市場狀況的定價機制。由於買方有一定程度的靈活性決定何時進行「點價」（因此一定程度上決定買賣價格），這模式可將違約風險減至最低。其次，參考期貨市場價格來釐定實物價格，會消除定價時實物及期貨市場之間的基差風險，因而有助無縫對沖。

價格發現是期貨市場核心功能之一。因為是使用期貨價格作為實物現貨貿易的基準，這種建基於期貨價格的定價模式可說是期貨市場價格發現功能的終極體現。

圖 6：以期貨價格為基礎的商品定價模式

期貨市場價格　＋　協定基差　＋　點價期　➡　實物商品的「點價」

就鐵礦石而言，值得一提 2016 年 11 月出現了試行點價模式的個案。當時北京鐵礦石交易中心股份有限公司（北鐵中心）有一宗 10,000 公噸的現貨鐵礦石交易，價格就是按大商所鐵礦石期貨價格加一個基差來釐定。

一般情況下，點價機制要能有效運作，其必須事先滿足若干主要條件：

(1) 關鍵是擁有功能齊全、**流動性高及透明度高**的期貨市場。期貨價格在任何時候均可讓所有市場參與者及時知悉與獲取，使定價於日內任何特定時段內均可進行。

(2) 買賣雙方必須**信任及認可**期貨市場價格是能有效反映及代表相關現貨市場。

市場的發展和進步需要時間。市場會決定甚麼定價方式最符合行業需求、需否在現有基礎上變化出新的定價模式又或索性另行自創出一套更適應自身發展的模式，這些都值得我們密切關注。

3.4　香港市場和香港交易所的戰略定位

中國是全世界第二大經濟體及最大的鐵礦石進口國和主要消費國，香港作為位處中國門戶的全球金融中心，一直是中國內地與全球其他地區之間的「超級聯繫人」，有很好的條件建立鐵礦石衍生產品市場，為中資企業、區域大宗商品貿易公司以至它們的全球商業夥伴在風險管理需求方面提供更佳的服務。

香港交易所 2012 年收購倫敦金屬交易所 (LME) 後，亦一直準備為更好的服務實體經濟而積極發展大宗商品業務。香港交易所 2017 年 11 月 13 日推出鐵礦石期貨合約，正是其中一項具建設性的舉措。

(1)　香港的戰略定位：通往中國的主要門戶及中國與世界之間的「超級聯繫人」

以中國的門戶作為戰略定位，香港是中國與世界之間的「超級聯繫人」。香港在金融市場方面聯通兩方已完成多項突破和創新，當中包括推出滬深港通[23] 及多項促使香港成為世界最大離岸人民幣中心的計劃。

香港是亞洲重要航運中心和國際貿易中心，約半數 (2016 年為 4,540 億美元) 貿易額為轉口貿易[24]。內地的國際貿易約兩成經香港進行。航運繁忙再加上港口及物流高效，都是香港繁榮昌盛的原因。香港亦是世界級國際金融中心，提供全面的金融產品和服務，也是商業中心，有大量國際及內地企業在此間設立辦事處。奉行自由市場經濟和法治的香港是內地對外投資最大單一目的地市場，提供廣泛的金融服務及一站式投資方案。香港同時擁有全世界最大的離岸人民幣資金池。另一方面，香港也是國際投資者投資內地市場的不二之選的目的地市場和門戶市場。香港交易所證券市場是 2015 年及 2016 年全球首次公開招股融資額最高的市場。香港交易所上市公司中，逾半數為內地企業[25]，於 2016 年 12 月 31 日，自然資源相關行業的上市公司超過 150 家[26]。

香港又能提供必須的金融基建，為企業提供切合業務需要的出入口業務、融資、貿易融資、資產管理和金融風險管理等多方面的服務。這當中又以金融風險

23　滬深港通是內地與香港市場互聯互通機制試點計劃。中國內地與香港市場的投資者首次可直接進入對方的股票市場。

24　資料來源：香港貿易發展局研究網站；同段其後引用的香港經濟數據亦來自這網站。

25　於 2016 年 12 月 31 日，香港交易所主板及創業板合共 1,973 家上市公司中，1,002 家是內地企業。（資料來源：香港交易所）

26　資料來源：香港交易所。

管理 (特別是資產價格風險管理) 對從事大宗商品交易的企業尤為重要，因此大宗商品的價格風險管理被視為香港金融市場一個重要發展領域。這些增值服務一方面是大宗商品業羣體在香港蓬勃發展所必須，另一方面亦有利中國爭取大宗商品的國際定價權。

(2)　「一帶一路」等戰略計劃提供歷史契機，讓香港可協助中資企業在海外擴大其市場及業務

如上文第 1.2 節所述，中國的「一帶一路」政策倡議為中國鋼鐵業帶來了扶持出口及輸出過剩產能的歷史性契機。中資鐵礦石企業向外擴張市場和服務將會刺激鐵礦石業參與者的風險管理和投資需求。因此，香港獲得了歷史性的機遇，可利用其在金融、貿易及物流方面的優勢、借助其在方方面面的龐大專業人才隊伍，為滿足上述需求貢獻本身的力量。香港的專業優勢及作為中國門戶的獨特地位，將使其在融資、管理風險、項目主導及輸出專業服務等方面繼續發揮重要作用。因此，香港有極佳的條件發展離岸鐵礦石衍生產品市場，協助中資企業在離岸市場管理資產價格風險。

(3)　香港交易所作為香港金融市場的營運機構，透過建立穩定及流動性高的大宗商品衍生產品市場，為滿足內地、本地及國際企業的資產價格風險管理需求而提供卓越服務，以期提升香港的國際金融中心地位

基於香港金融市場在聯通中國內地與世界方面的獨特地位，加上中國戰略發展計劃所提供的歷史性機遇，香港交易所作為香港金融市場的基礎設施營運機構，在服務中國與全球大宗商品衍生產品市場的需要方面，有以下相對優勢可使其扮演更強的角色：

- 所營運的證券及衍生產品交易、結算和交收系統是全球最穩健者之一；
- 提供多方面的產品和服務，涵蓋股本證券、股本證券衍生產品、定息及貨幣產品和大宗商品衍生產品；
- 營運監管制度完善，市場規則及規例符合國際最高標準，重視投資者保障。

香港交易所近年推出了一系列革新及戰略部署支持內地金融市場逐步對外開放。在股本證券方面，2014 年 10 月與上海證券交易所聯合推出滬港通，2016 年

12 月與深圳證券交易所推出深港通，基本上形成了一個跨境共同市場[27]；在定息及貨幣產品方面，2012 年 9 月推出全球首個場內交易人民幣可交收貨幣期貨產品 —— 美元／離岸人民幣（美元兌人民幣（香港））貨幣期貨，2017 年 7 月推出債券通北向交易[28]；在大宗商品方面，2012 年收購了全球最大基礎金屬市場 LME，2017 年 7 月推出其首隻實物交割雙幣計價的黃金期貨。香港交易所還進軍鐵類金屬產品系列，鐵礦石期貨是首隻推出的產品；此新產品旨在服務區內現貨交易業及金融機構對鐵礦石價格風險管理的需要。（有關產品的主要特徵，見附錄一；有關產品的合約細則，見附錄二。）

27　見本書第 2 章〈滬港通與深港通下的互聯互通 —— 內地及全球投資者的「共同市場」〉。
28　債券通是透過在內地與香港的機構金融基礎設施之間建立連接，容許海外投資者買賣內地中國銀行間債券市場的債券（北向交易）及內地投資者買賣香港市場債券（南向交易）的一個互聯互通計劃。開通初期僅限於北向交易。

附錄一

香港交易所鐵礦石期貨合約的主要特色

(1)　場內交易的期貨合約

鐵礦石期貨合約的價格透明度高，價格發現過程高效。

(2)　季度合約

此產品是全球首度提供場內交易季度合約的產品，為市場參與者提供一個較場外掉期合約更為透明及方便的平台，使市場參與者可執行買賣盤並對沖其季度交易持倉，也便於釐定遠期價格曲線中的價格。

(3)　日間交易時段及收市後交易時段

交易時段為上午 9 時至翌日凌晨 1 時（日間交易時段：上午 9 時至下午 4 時 30 分；收市後交易時段：下午 5 時 15 分至翌日凌晨 1 時）。交易時段安排覆蓋中國內地及主要海外市場的營業時間，方便全球各地的市場參與者。

(4)　大宗交易

鐵礦石期貨合約設有大宗交易機制，方便申報場外成交進入到交易所結算，降低對手方風險。

(5)　所追蹤指數是認受性最高的衍生工具基準

鐵礦石期貨合約按 TSI CFR 中國鐵礦石 62% 鐵粉指數結算，該指數為鐵礦石現貨交易中最獲廣泛參考的指數，大部份以美元計價的鐵礦石衍生工具合約也是以之作為結算價。

附錄二

香港交易所鐵礦石期貨合約細則

TSI CFR 中國鐵礦石 62% 鐵粉期貨合約		
項目	月度合約	季度合約
交易代碼	FEM	FEQ
合約單位	100噸	
最低波幅	每噸0.01美元	
相關指數	TSI CFR 中國鐵礦石62%鐵粉指數（TSI 62 指數）[29]	
結算方式	以現金結算	
合約月份	現貨月及後續23個曆月	現貨季及後續7個曆季（即1月至3月、4月至6月、7月至9月及10月至12月）
交易時間（香港時間）	T時段：上午9時至下午4時30分； T+時段：下午5時15分至翌日凌晨1時 （到期合約於最後交易日下午6時30分停止交易）[30]	
最後交易日	每個曆月非新加坡公眾假期的最後一個香港營業日	每個曆季中最後一個月度合約的最後交易日
最後結算價	該合約月份公佈的所有指數的算術平均值	該合約季度相應的三個月度合約的最後結算價的算術平均值
最後結算日	通常為最後交易日後第二個香港營業日[31]	
交易所費用[32, 33]	交易費：每邊每張合約1.00美元； 最終結算費：每邊每張合約1.00美元	
徵費[34, 35]	每邊每張合約0.07美元	
大宗交易的最低合約交易量	最少50張合約	
假期	與香港交易所假期表相同	

29 　根據 Platts 於 2017 年 7 月 6 日發佈的公告，TSI 62 指數將於 2018 年 1 月 2 日起與 Platts 的 IODEX 指數合併。
　　有關詳情，請參閱 Platts 的訂戶備註及指數計算方法及細則。
30 　如最後交易日為新年或農曆新年前最後一個香港營業日，以及為新年或農曆新年前 TSI 62% 指數最後一次公佈的
　　日子，該日交易時段將不會超過下午 12 時 30 分。
31 　除非：（1）最後交易日為新年或農曆新年前最後一個香港營業日，（2）現貨月合約及現貨季合約的交易時間於下
　　午 12 時 30 分結束，及（3）其他合約月份的日間交易時段於 下午 4 時 30 分結束，則最後結算日為最後交易日後
　　首個香港營業日。有關進一步詳情，請參閱香港交易所網站相關的規例及合約細則。
32 　費用可不時更改。
33 　2017 年 11 月 13 日至 2018 年 5 月 11 日期間豁免（不包括 2018 年 5 月 11 日收市後期貨交易時段）。
34 　目前收費率訂為每張合約 0.54 港元，其等值美元由交易所不時釐定。
35 　2017 年 11 月 13 日至 2018 年 5 月 11 日期間豁免（不包括 2018 年 5 月 11 日收市後期貨交易時段）。

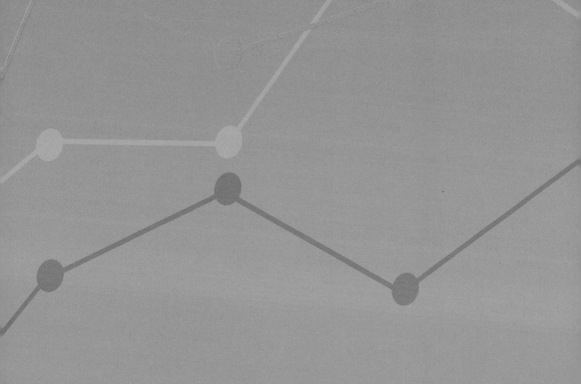

第四部分

人民幣離岸產品中心

第13章

離岸人民幣流動性供應機制的現狀、影響及改善方向

2017 年 1 月 4 日

摘 要

離岸市場人民幣流動性大致分為長期及短期兩個層面。長期流動性主要通過實體經濟（跨境貿易結算渠道）從在岸市場獲得資金；短期流動性主要包括監管機構的貨幣互換及市場融資。

離岸人民幣流動性供應機制的現狀和結構特點：長期流動性主要依賴跨境貿易結算渠道，易受人民幣匯率波動影響；短期流動性供應機制在效率、規模、及運作時間上，與市場發展存有一定差距；另外，離岸人民幣資金投資結構有待改善。

離岸人民幣短端利率的大幅波動對離岸債券市場的穩定擴張帶來壓力，同時加大了境外機構持有人民幣資產的風險對沖成本，亦可能誘發投機性的短期跨境資本流動。

隨着人民幣已正式加入國際貨幣基金組織特別提款權（SDR），全球投資者對人民幣資產配置需求不斷上升，充足的離岸人民幣流動性對提升市場深度，滿足跨境貿易、離岸投融資、外匯交易等經濟活動至關重要。不斷擴寬雙向跨境資本渠道，對現有市場機制加以改善調整，將為人民幣作為國際可兌換貨幣以及在國際投資領域中廣泛使用鋪平道路。

1　兩層次的離岸人民幣流動性供應機制

基於離岸市場的特殊發展歷史，離岸人民幣供應機制可分為長期及短期兩個層面。

長期流動性方面，離岸人民幣市場主要通過實體經濟（跨境貿易結算渠道）從在岸市場獲得人民幣資金。自 2009 年 7 月推出跨境貿易結算以來，內地對外人民幣結算收付比一直呈現實付大於實收狀態，儘管該比率逐步下降，從 2011 年第一季度人民幣收付比為 1:5，下降至 2015 年底 1:0.96，但人民幣總體處於淨流出狀態，因此，跨境貿易結算成為了境內市場向離岸市場輸出人民幣資金的主要渠道。2014 年末香港人民幣資金池規模達到 1.15 萬億元 (離岸人民幣存款及存款證)，台灣人民幣存款餘額為 3,022 億元，加上新加坡人民幣存款兩千億餘元，全球離岸人民幣存款規模達 1.6 萬億元的歷史高點，基本上通過跨境貿易結算渠道從境內市場獲得。

在結算過程中，境內外進出口商根據匯率波動，相應利用人民幣及美元進行不同地點的結算，從而推動離岸市場人民幣池不斷擴大，機制具體表現為：當人民幣匯率處於升值預期時，香港離岸市場美元兌人民幣匯率 (CNH) 較境內市場美元兌人民幣匯率 (CNY) 升值更為明顯。CNH 升水意味着離岸市場人民幣價格更貴，企業使用人民幣進行進口貿易結算能夠獲取額外收益，因此有動力通過對進口支付人民幣取代美元，導致境內人民幣流動性外溢至境外 [1]。

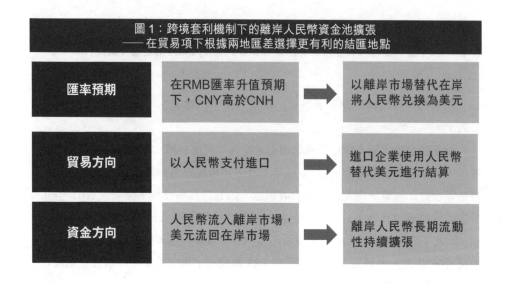

圖 1：跨境套利機制下的離岸人民幣資金池擴張
—— 在貿易項下根據兩地匯差選擇更有利的結匯地點

匯率預期	在RMB匯率升值預期下，CNY高於CNH	以離岸市場替代在岸將人民幣兌換為美元
貿易方向	以人民幣支付進口	進口企業使用人民幣替代美元進行結算
資金方向	人民幣流入離岸市場，美元流回在岸市場	離岸人民幣長期流動性持續擴張

　　短期流動性方面，離岸市場的人民幣資金供應渠道分為官方及市場融資兩個層面。官方提供的短期流動性，包括香港金融管理局（金管局）向市場提供的一日及一周期限流動資金安排（兩者均為翌日交收），以及隔夜流動資金安排（即日交收）；2014 年為回應滬港通的開通，香港金融管理局又推出每日上限 100 億元的拆借資金，以滿足滬港通開放後離岸市場對人民幣的即時需求。另外，還同時推出香港離岸人民幣市場的一級流動性提供行（CNH Primary Liquidity Providers，簡稱 PLPs）服務，由金管局為 7 間 PLPs 提供各 20 億元人民幣回購服務（repo facility），以支持擴展離岸人民幣市場的莊家活動及其他業務[2]。

2　2016 年 10 月 27 日，香港金管局宣佈將一級流動性提供行由 7 間增加至 9 間，該計劃總額度也由原來的人民幣 140 億元增加至人民幣 180 億元。

表 1：官方提供的短期流動性供給機制（截至 2016 年 10 月）		
資金期限	定價方式	規模及來源
1 星期期限的資金（翌日交收）	參考當前市場利率	貨幣互換項下資金
1 日期限的資金（翌日交收）	參考當前市場利率	貨幣互換項下資金
隔夜資金（即日交收）	最近 3 次財資市場公會隔夜人民幣香港銀行同業拆息定價（包括回購協議交易當日的定價）的平均數加 50 基點，最少 0.50%	推出時預計規模不超過 100 億元人民幣
日間資金（即日交收）	最近 3 次財資市場公會隔夜人民幣香港銀行同業拆息定價（包括回購協議交易當日的定價）的平均數，最少 0%，按當日使用有關流動資金的實際時間每分鐘計算利息	規模不超過 100 億元人民幣
一級流動性提供行（PLPs）	市場運作原則	共 180 億元人民幣

資料來源：香港金融管理局。

　　另外，離岸市場短期流動性缺口還可通過外匯掉期市場加以解決。貨幣掉期是指在外匯市場上買進即期外匯的同時又賣出同種貨幣的遠期外匯，或者賣出即期外匯的同時又買進同種貨幣的遠期外匯。目前，離岸人民幣外匯掉期期限一般為當天、隔夜到一年。

　　此外，離岸市場上的參與者還可以通過離岸銀行間拆借、清算行回購（REPO）等方式獲得短期流動性資金，這些與官方渠道形成較完整的供給機制向離岸市場提供短期流動性資金。

2 運作現狀和結構特點

2.1 長期流動性供應主要依賴跨境貿易結算渠道，自811匯改以來出現收縮

如前所述，跨境人民幣結算是離岸市場獲得長期流動性的主要渠道，也導致離岸資金的流向和規模易於受到人民幣匯率波動的影響。自2015年811匯改以來[3]，CNH表現出較CNY更大的貶值趨勢，市場套利機制反向運作，人民幣資金由一貫地流向離岸市場轉為回流在岸市場。具體表現為：當CNH較CNY貶值超過一定基點時，意味着人民幣在境內市場價值更高，因此貿易商有動力在離岸市場以更便宜的價格買入人民幣，同時在在岸市場上以更貴的價格賣出人民幣，並通過跨境貿易結算途徑將離岸人民幣資金輸送回在岸市場，即可賺取匯差。與此同時，境外投資者對持有人民幣匯率信心有所減弱，部份人民幣存款轉回美元、港幣資產。兩者共同作用之下，香港人民幣存款已從高峰時期的10,035億元跌至6,529億元人民幣[4]，較2015年底下跌約23%。

2.2 現有的短期流動性供應機制在使用效率、規模及運作時間上，與市場發展存有一定距離

第一，離岸市場以即日交收的日間資金規模有限，相比之下目前香港離岸市場日均交易金額已增加至平均每天7,700億元人民幣，部份時間超過了港幣清算量（見下圖2及3）。根據國際清算銀行（BIS）2016年統計數據，離岸人民幣即期、遠期和外匯互換的場外日均交易量達2,020億美元，因此，目前來看，市場對即日交收的短期流動性需求殷切。

---

3 2015年8月11日，內地央行啟動人民幣對美元匯率中間價報價機制改革，此次改革被普遍視為人民幣匯率市場化改革的重要一步。

4 數據來源：金管局，截至2016年8月底。

圖 2：「811」匯改後香港人民幣資金池出現收縮

資料來源：彭博。

圖 3：2015 年下半年以來人民幣日均結算金額部份時間超過港幣

資料來源：香港銀行同業結算有限公司。

　　第二，官方渠道向市場提供的資金，有相當部份來自與內地央行貨幣互換協議，使用時需要參考內地銀行間市場和清算系統的運作時間。內地長假期間，內地資金結算暫停，令離岸市場面臨人民幣流動性來源壓力。

　　第三，離岸人民幣掉期市場作為離岸獲得人民幣短期流動性的主要渠道，在進入美元加息週期後波動增加。

　　第四，兩地貨幣市場之間缺乏有效的短期資金跨境渠道。如前文所述，現有人民幣跨境資金的流動主要集中在經常項目和資本項下的中長期資金層面，包括人民幣跨境貿易結算、人民幣直接投資（FDI）渠道、三類機構投資於境內銀行間債券市場、人民幣合格境外機構投資者（RQFII）等。由於境外非居民參與內地貨幣市場方面仍屬於不可兌換項目[5]，除了貨幣互換渠道以外，短期資金、特別是隔夜至一周流動性較強的短期資金，境內、外市場之間尚未建立起有效的跨境渠道。

5　詳見中國人民銀行調查統計司課題組報告（2012）《我國加快資本賬戶開放的條件基本成熟》。

2.3 目前離岸人民幣主要配置於境內長期性資產，投資結構有所失衡

目前離岸人民幣主要配置於：RQFII 渠道下 2,700 億元人民幣額度；點心債餘額約 5,000 億元人民幣；人民幣貸款 2,816 億元人民幣[6]。儘管人民幣貸款帶來的乘數效應可以進一步放大人民幣資金池，但上述人民幣資金配置已基本盡數使用離岸資金。離岸人民幣配置的這些資產交投量並不活躍，也不方便進行回購質押，一旦市場短時間內出現較大人民幣資金流轉需求，若部份金融機構一時難以調整資產期限組合，有可能引起短期流動性緊張。

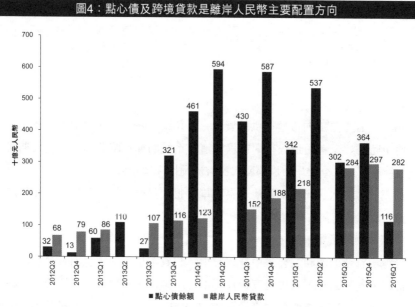

圖4：點心債及跨境貸款是離岸人民幣主要配置方向

■點心債餘額　■離岸人民幣貸款

資料來源：彭博。

在目前離岸人民幣市場整體規模收縮的情況下，資產配置失衡帶來的擴散效應可能有所放大，尤其是臨近季末假期（中秋、國慶等時點），內地銀行間市場因

6　數據來源：金管局，截至 2016 年 3 月底。

節日休市，季節性因素導致資金回流在岸，這也是 2016 年 9 月底至 10 月初國慶前，離岸人民幣市場拆息出現大幅波動的主要背景。

3 離岸人民幣短端利率波動對市場帶來的影響

3.1　對離岸債券市場穩定擴張帶來壓力

香港一直是全球最大的離岸人民幣債券市場，也是海外機構投資人民幣債券的主要場所，但是自 2015 年 811 匯改後，離岸市場資金池收縮，融資成本逐步抬升，銀行人民幣一年期存款利率平均上升至 4% 以上，三年期點心債融資成本顯著上漲近 200 基點。相比之下，內地貨幣政策則穩中有鬆，流動性充裕，在岸、離岸債券利差逐漸拉闊，導致大部份點心債發行主體回到在岸，離岸債券市場發行規模大幅收縮。以點心債市場上一向活躍的房地產企業為例，2015 年陸續有近 6 成境外房地產企業回流至在岸市場發債融資，導致離岸房地產板塊債券發行量由 2014 年的 248 億美元下降至 96 億美元 [7]。

3.2　加大了境外機構持有人民幣資產的風險對沖難度

人民幣正式加入 SDR 以後，對人民幣資產的需求持續上升，各國央行和全球投資者將考慮增配人民幣資產。SDR 規模僅佔國際儲備 2.4%，人民幣入籃本身僅具備小幅增量資金效應，但加入 SDR 將提升人民幣作為國際儲備和投資貨幣的國際認可度。我們預計如果國際機構或個人持有人民幣金融資產的佔比達到日圓佔全球外匯儲備資產佔比水平，將有超過 2 萬億人民幣配置相關金融資產。

持續增加的投資需求對風險管理需求日益迫切，如果離岸人民幣短期波動性相對較大，會導致離岸機構在開發人民幣浮息貸款、人民幣資產定價、風險管理

7　參見〈近期離岸人民幣債券市場的發展態勢、原因及趨勢〉，《中國貨幣市場》2016 年第 1 期。

產品方面上缺乏合適的利率基準，加大國際投資者對沖利率風險的難度，因此需要市場開發更多利率避險工具，引導市場主體調整外匯交易策略，進而促進境外資金對離岸市場的參與。

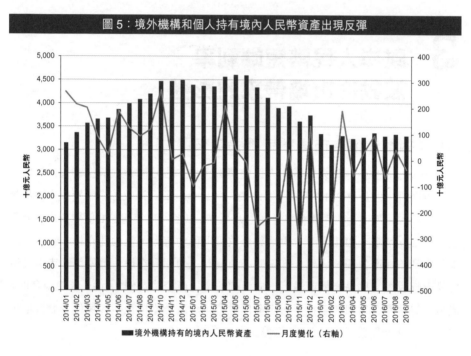

圖5：境外機構和個人持有境內人民幣資產出現反彈

資料來源：Wind 資訊。

3.3　兩地利差拉大可能誘發投機性的短期跨境資本流動

伴隨內地多次降息降準、內地人民幣資產收益持續下跌，使得境內資本需要尋找相對有價值的資金投向。目前利差套利型資本流動並不明顯，如果離岸與在岸人民幣利差持續時間較長，有可能引起資本的非正規渠道流出，對在岸市場的貨幣流動性帶來壓力。

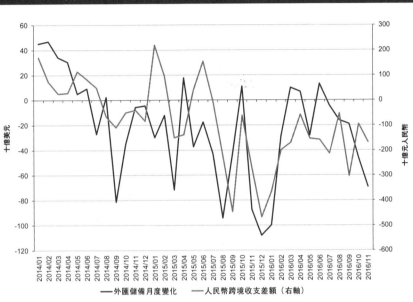

圖 6：人民幣持續對外淨支付，可能加劇資本外流壓力

資料來源：Wind 資訊。

4 可能的改善方向

隨着離岸市場日均交易金額快速增長，RQFII、滬港通、深港通等投資活動不斷活躍，無論是應對金融產品交易還是長期性融資需求，離岸市場都需要獲得充足的人民幣流動性作為支持。充足的人民幣離岸市場流動性對提升市場的深度，滿足跨境貿易、投資、外匯交易等經濟活動需求至關重要。

值得注意的是，儘管目前國際投資者可更為直接地參與在岸市場交易，市場交易重心有轉向境內的跡象，但以美元等國際貨幣的發展經驗來看，貨幣國際化需要同步發展離岸市場，將對促進貨幣的境外循環和廣泛國際使用具有重要作用。

為進一步改善離岸人民幣流動性，可以考慮以下的改善方向。

4.1　穩步推動人民幣國際化，
　　　逐步放開雙向跨境資本流動渠道

如前述，人民幣匯率是影響離岸人民幣整體資金池和長期流動性擴張的主要因素。為支持人民幣匯率改革，前期在岸以穩定匯率為主。然而，在市場逐漸適應新的人民幣匯率機制和政策效果逐步顯現後，適度促進內地人民幣資金向離岸市場流動，進一步放開跨境雙向渠道，增加境外人民幣資金池規模，將有利於離岸市場的發展。

從境外循環渠道來看，目前全球經濟不振，中國進出口乏力，利用經常項目和貿易結算推動人民幣全球使用已經遭遇瓶頸。如果更多利用資本項目直接投資等渠道向外輸送人民幣，特別是通過人民幣對外直接投資 ODI，內地企業走出去，「一帶一路」等區域合作戰略，可望提升人民幣的國際接受度，解決離岸人民幣市場規模停滯不前的問題[8]。

4.2　充分利用現有政策，打通兩地債券回購市場

2015 年內地央行推出了債券回購交易的新政[9]，該政策允許境外機構在境內銀行間市場進行回購交易，且資金可以用於境外。此舉在一定程度上聯通了境內外資金市場，緩解離岸市場流動性不足的問題。

如若進一步提升交易便利及效率，可考慮建設連通境內外債券市場的「債券通」跨境平台。上文提到的債券回購新政，允許境外機構進行債券回購獲得流動性，是以境內持有債券為限，境外人民幣債並不能進入境內回購市場進行質押融資；通過 RQFII 或者三類機構購買的內地人民幣債券，也不能在境外進行回購；目前境外人民幣債平均餘額約 5,000 億元人民幣[10]，規模接近外資持有的境內債券規模。建設跨境債券通平台，可使境外機構利用持有的離岸債券，到境內進行回購交易並獲得融資。這不僅可提升境外的人民幣流動性，也可改善離岸人民幣資產的交易便利性和保持離岸人民幣市場穩定性。

8　〈香港離岸市場在調整中前行〉，《中國外匯》2016 年第 15 期。
9　詳見《中國人民銀行關於境外人民幣業務清算行、境外參加銀行開展銀行間債券市場債券回購交易的通知》(2015)。
10　數據來源：彭博。

4.3　為發展利率互換、掉期等衍生產品提供市場基準，進一步強化離岸人民幣市場的定價效率及風險管理能力

　　為推進利率市場化改革，內地央行更加強調利率市場化形成機制，逐步減低利率曲線各期限之間差距過大的情況，繼而加強兩地市場關聯性，強化離岸人民幣拆息（CNH Hibor fixing）的定價效率。CNH Hibor fixing 穩定性和效率的提高，也會有利於離岸市場開發出更多的債券回購、利率互換產品，從而進一步深化離岸人民幣市場的風險對沖功能，為人民幣境外交易創造更有深度的市場環境。

圖 7：2015 年下半年開始離岸人民幣拆息 CNH Hibor 與在岸銀行間拆借利率（Shibor）走勢分化

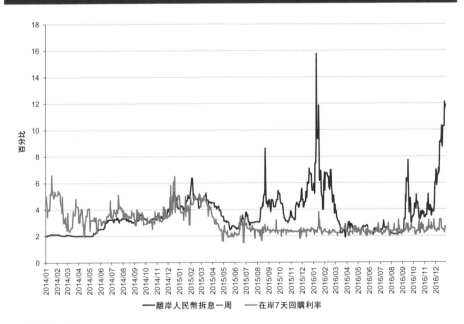

資料來源：彭博。

4.4　拓寬離岸市場的產品規模和類別，
　　　進一步擴大離岸人民幣資金池

　　隨着滬港通、深港通總額度取消，合格境外機構投資者（QFII）、RQFII 投資規模繼續擴大，將不斷拓寬離岸人民幣投資渠道，推動更多人民幣資金在海外市場流轉。除此之外，隨着人民幣國際化程度和資本項目開放逐步加深，互聯互通模式將進一步拓寬到更大範圍市場，包括債券、商品等，吸納更多海外客戶羣通過香港平台進入內地市場，為離岸人民幣資金提供更多類別的投資工具，吸引更多資金在離岸沉澱。

第14章

香港交易所邁向成為離岸人民幣
產品交易及風險管理中心

2017 年 4 月 19 日

摘 要

　　香港於 2003 年獲內地政府批准，成為全球首個開展離岸人民幣業務的市場。在隨後內地政策的繼續開放及中央政策支持的推動下，香港場內和場外市場的人民幣金融產品蓬勃發展。到今天，香港交易所的人民幣產品包括債券、交易所買賣基金（ETF）、房地產投資信託基金（REIT）、股票、人民幣定息及貨幣衍生產品以及大宗商品衍生產品。香港交易所證券市場方面，以人民幣 ETF 的成交最為活躍；而衍生產品市場方面，則以人民幣貨幣期貨最受熱捧，其 2016 年的成交更刷新紀錄。

　　環顧全球交易所，不論在證券或衍生產品市場，香港交易所的人民幣產品在上市及交易方面均獨佔鰲頭。市場上提供人民幣證券產品的其他交易所寥寥可數，產品種類不多、成交量低。另一方面，人民幣貨幣期貨及期權則是全球多家交易所頗為常見有提供的產品，然而成交量集中在亞洲區的交易所，其中以香港交易所的交投最為活躍。事實上，成交數據證明，香港交易所的人民幣期貨合約於人民幣匯率極度波動之時盡展其作為人民幣貨幣風險管理工具的功能。

　　香港交易所在離岸人民幣的產品交易及風險管理享有相對優勢，原因眾多，包括香港位處「一帶一路」戰略計劃的中心、龐大的人民幣資金池支持活躍的香港人民幣業務、香港交易所旗下市場匯聚國際投資者及高效率的市場基礎設施等等。

　　人民幣產品和風險管理工具供應充裕的離岸市場，是支援人民幣國際化並維持穩定匯率水平的基礎。為此，香港交易所旗下證券及衍生產品市場的人民幣產品配套將會持續豐富，以迎合在人民幣穩步推進國際化之下日益增長的投資者需求。除了近期推出的美元兌人民幣（香港）期權及內地國債期貨外，日後亦可能推出其他人民幣風險管理工具。可見香港交易所具備優越條件，足以成為全球投資者的離岸人民幣產品交易及風險管理中心。

1　背景

2003 年 11 月，中國央行中國人民銀行（人行）與香港金融管理局（金管局）簽署備忘錄，准許香港銀行開辦個人人民幣業務，香港於 2004 年正式開展人民幣業務。初期服務範圍只限於匯款、兌換及人民幣信用卡。到 2007 年 1 月中國國務院批准香港擴充人民幣業務，允許內地金融機構在港發行人民幣金融債券，香港始出現人民幣投資產品，首先是人民幣債券（俗稱「點心債」）的發行。有關實施此國策的辦法[1]於 2007 年 6 月頒佈，同月稍後一家內地國營政策性銀行即在港發售首隻人民幣債券[2]。

及後政策放寬，香港人民幣債券市場提速發展。2010 年 2 月，根據政策釐清文件[3]，香港人民幣債券的合資格發債體範圍、發行安排及投資者主體可按照香港的法規和市場因素來決定。同月，人行批准金融機構就債務融資在港開設人民幣戶口，這使香港得以推出人民幣債券基金。2011 年 10 月再有新規則容許以合法渠道（例如境外發行人民幣債券及股票）取得的境外人民幣在內地作直接投資[4]。

隨着香港金融業界的合格人民幣業務範圍日漸擴充，加上內地政府中央政策的支持[5]，香港場內外的人民幣金融產品蓬勃發展，不再只限於人民幣債券。交易所的證券市場方面，人民幣債券是數量最多的上市產品，人民幣交易所買賣基金（ETF）則是交投最活躍的產品。交易所的衍生產品市場方面，人民幣期貨合約於匯率極度波動之時盡展其作為人民幣匯率風險管理工具的功能。（見下文第 2 節）

本報告闡述香港交易所相對環球其他交易所在人民幣產品發展方面的情況，指出香港交易所在全球市場的場內人民幣產品中有着領先地位，推動其邁向成為離岸人民幣的產品交易及風險管理中心。

1　人行與國家發展和改革委員會（發改委）於 2007 年 6 月 8 日聯合頒佈的《境內金融機構赴香港特別行政區發行人民幣債券暫行辦法》。

2　由國家開發銀行發售的兩年期人民幣 50 億元人民幣債券，票面息率 3%，至少 20% 售予散戶投資者。

3　金管局《香港人民幣業務的監管原則及操作安排的詮釋》，2010 年 2 月 11 日。

4　人行頒佈的《外商直接投資人民幣結算業務管理辦法》；商務部頒佈的《關於跨境人民幣直接投資有關問題的通知》。

5　2011 年 8 月，時任國務院副總理李克強於訪港期間公佈一系列有關香港發展的中央政策。具體而言，將提供政策支持香港發展成為離岸人民幣業務中心，包括鼓勵香港發展創新的離岸人民幣金融產品、增加赴港發行人民幣債券的合格機構主體數目，並擴大發行規模。2012 年 6 月，內地政府正式宣佈一套政策措施，加強內地與香港之間的合作，包括支持香港發展為離岸人民幣業務中心的政策。

2 香港交易所的人民幣產品發展

2.1 證券產品

中國境外首隻人民幣債券發行的三年後，2010 年 10 月 22 日見證首隻人民幣債券 —— 由國際金融機構 (亞洲開發銀行) 發行的 10 年期債券 —— 在香港交易所證券市場上市。首隻由內地機構 (中國農業銀行) 境外發行上市的人民幣債券則於 2012 年 1 月上市。首兩隻人民幣內地政府債券於 2012 年 7 月上市，與 2009 年 9 月首隻境外人民幣內地政府債券在香港發行相距約三年。

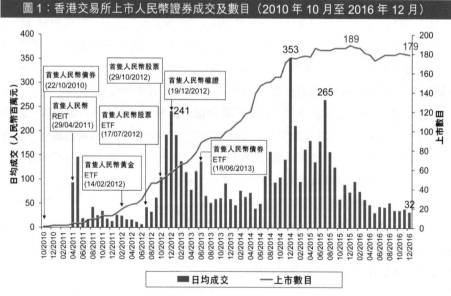

圖 1：香港交易所上市人民幣證券成交及數目 (2010 年 10 月至 2016 年 12 月)

資料來源：香港交易所。

上市人民幣證券的種類不久即進一步擴展至人民幣債券以外的券種：首隻人民幣房地產投資信託基金 (REIT) 於 2011 年 4 月上市、首隻人民幣 ETF (以黃金為標的物) 於 2012 年 2 月上市、首隻人民幣股票於 2012 年 10 月上市，以及首隻人民幣權證於 2012 年 12 月上市。至 2016 年底，上市人民幣證券總數已增至 179

隻。圖 1 顯示香港交易所人民幣證券的歷史日均成交及數目增長。

　　2016 年底，人民幣證券數目於所有主板上市證券的佔比增至 2%。2016 年底的人民幣證券主要為人民幣債券（75%）及 ETF（23%）。人民幣 ETF 中，股票指數 ETF 佔比最多（佔所有人民幣證券的 20%）。人民幣證券數目雖然在主板所有證券類別的佔比偏低，但於 ETF 中則佔顯著比重（31%），亦佔債務證券相當比重（15%）。（見圖 2 至 4）

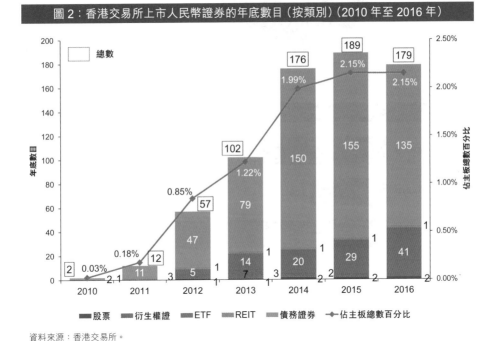

圖 2：香港交易所上市人民幣證券的年底數目（按類別）（2010 年至 2016 年）

資料來源：香港交易所。

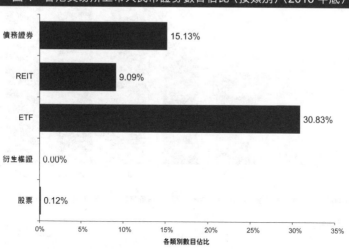

圖 3：香港交易所上市人民幣證券數目（按類別）（2016 年底）

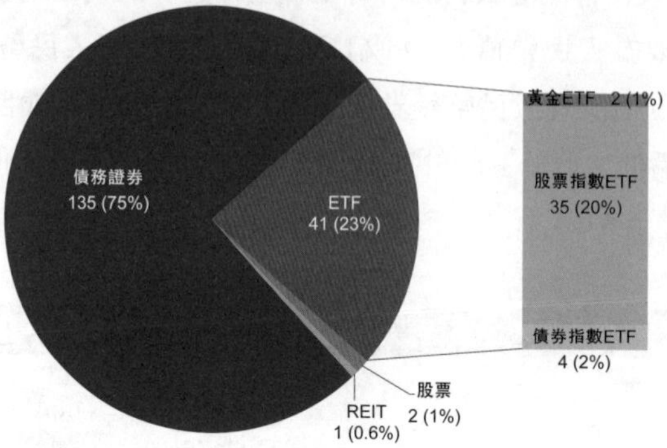

債務證券
135 (75%)
ETF
41 (23%)
黃金ETF 2 (1%)
股票指數ETF
35 (20%)
債券指數ETF
4 (2%)
股票
2 (1%)
REIT
1 (0.6%)

資料來源：香港交易所。

圖 4：香港交易所上市人民幣證券數目佔比（按類別）（2016 年底）

債務證券
15.13%
REIT
9.09%
ETF
30.83%
衍生權證
0.00%
股票
0.12%
0%
5%
10%
15%
20%
25%
30%
35%
各類別數目佔比

資料來源：香港交易所。

人民幣證券成交自 2011 年起連續四年增長，於 2016 年才告回落。由於上市數目少，以人民幣買賣的證券於主板市場總交易額的佔比仍然微不足道。其中，人民幣 ETF 自 2012 年推出以來每年均佔最高比重——2016 年佔 77%，主要為股票指數 ETF 的交易。唯一的一隻 REIT 排行第二（2016 年佔 20%）。人民幣債券的上市數目雖然最多，但其成交佔比卻尚低（3%）。（見圖 5 及 6）

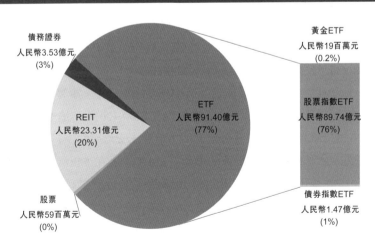

圖 5：香港交易所上市人民幣證券每年人民幣成交（按類別）（2010 至 2016 年）

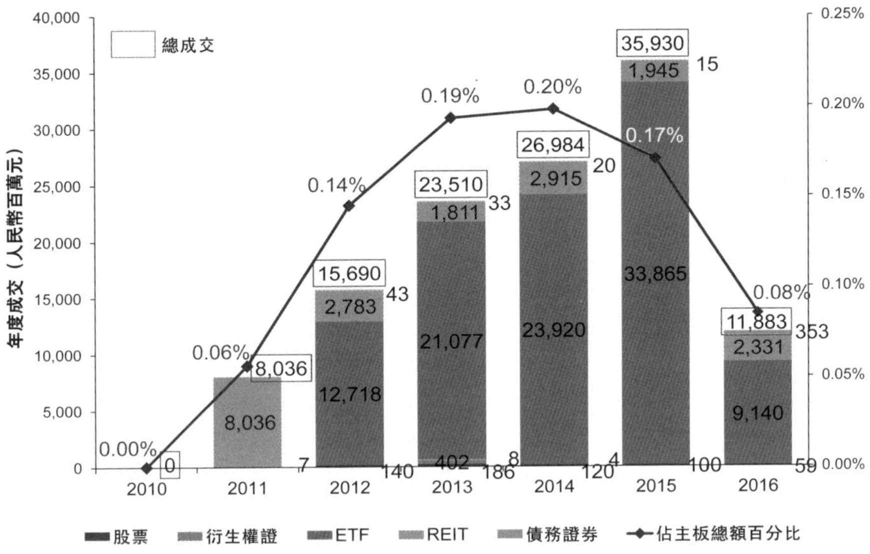

註：人民幣成交指人民幣證券的人民幣交易櫃枱的成交金額。

資料來源：香港交易所。

圖 6：香港交易所上市人民幣證券的人民幣成交佔比（按類別）（2016 年）

債務證券
人民幣3.53億元
(3%)

REIT
人民幣23.31億元
(20%)

ETF
人民幣91.40億元
(77%)

股票
人民幣59百萬元
(0%)

黃金ETF
人民幣19百萬元
(0.2%)

股票指數ETF
人民幣89.74億元
(76%)

債券指數ETF
人民幣1.47億元
(1%)

註：人民幣成交指人民幣證券的人民幣交易櫃枱的成交金額。

資料來源：香港交易所。

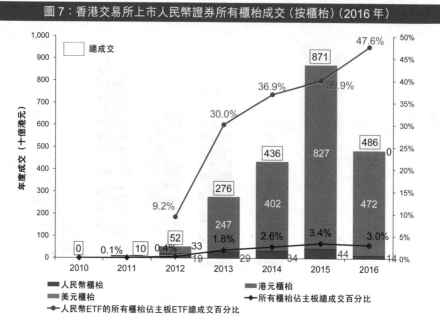

圖 7：香港交易所上市人民幣證券所有櫃枱成交（按櫃枱）（2016 年）

註：首隻雙櫃枱人民幣證券於 2012 年 10 月 12 日上市。

資料來源：香港交易所。

為人民幣證券提供同時以其他貨幣交易的雙櫃枱交易於 2012 年 10 月推出，其他交易貨幣主要為港元，其次為美元。2016 年底，香港交易所人民幣證券共有 41 隻港元雙櫃枱及 9 隻美元雙櫃枱 —— 兩隻人民幣股票中有一隻設有港元櫃枱；除一隻黃金 ETF 外，所有人民幣 ETF 均設港元櫃枱，其中 9 隻另設美元櫃枱。與人民幣櫃枱相比，人民幣證券的港元櫃枱成交遠為活躍。事實上，大部份設有不同櫃枱的人民幣證券，其成交均集中於港元櫃枱（2016 年逾 97%）。2015 年及 2016 年人民幣證券所有櫃枱合計成交佔主板總成交增至約 3%。特別是人民幣 ETF 的所有櫃枱合計成交（幾乎全來自股票指數 ETF），在主板所有 ETF 總成交的佔比自其 2012 年推出以來一直上升，於 2016 年達 48% 歷史高位（相對於按數目計的佔比 31%）。（見圖 7）

總括而言，香港交易所的人民幣證券正穩步發展。推出港元雙櫃枱交易為投資者提供交易便利，吸引了相當的交易量。人民幣 ETF（主要為股票指數 ETF）對投資者特別吸引，縱使大部份成交集中於港元櫃枱。

2.2　衍生產品

香港交易所首隻人民幣衍生產品為 2012 年 9 月推出的美元兌離岸人民幣期貨（即美元兌人民幣（香港）期貨）。該產品旨在於人民幣漸趨國際化下為市場提供貨幣風險管理工具及投資工具。該產品初期成交不太活躍，及後人行於 2015 年 8 月 11 日推出政策措施，改革銀行間外匯市場上人民幣兌美元匯率中間價的報價機制，促進貨幣匯率更趨市場化，刺激該產品活躍交投。2016 年人民幣匯率波幅加劇，促使該產品交投進一步上揚。有見全球以人民幣計價的經濟活動日增，預期市場對人民幣貨幣衍生產品的需求日漸增長，香港交易所於 2016 年 5 月 30 日推出三隻全新離岸人民幣分別兌歐元、日圓及澳元的人民幣計價貨幣期貨 —— 歐元兌人民幣（香港）期貨、日圓兌人民幣（香港）期貨及澳元兌人民幣（香港）期貨，以及推出美元計價的人民幣（香港）兌美元期貨。

香港交易所另於 2014 年 12 月推出人民幣計價的大宗商品期貨合約，作為支持人民幣國際化用途及為實體經濟作人民幣定價的另一產品計劃。首批推出的產品為鋁、銅及鋅的倫敦金屬期貨小型合約。鋁、銅、鋅這三種金屬是中國佔全球耗用量重要比重的金屬[6]，也是香港交易所附屬公司倫敦金屬交易所（LME）交投最活躍的期貨合約[7]。及至一年後，香港交易所再推出另外三隻金屬期貨小型合約（鉛、鎳及錫）。該六隻人民幣計價金屬合約為對應 LME 現貨結算合約的現金結算小型合約，是中國境外首批針對相關資產人民幣風險的金屬合約產品，對人民幣作為亞洲時區內相關金屬的定價標準起支持作用。

圖 8 顯示香港交易所人民幣衍生產品自推出以來的歷史日均成交及未平倉合約。

---

[6] 中國佔全球金屬耗用量：2015 年鋁為 36%（69,374 千噸中佔 24,960 千噸，資料來源：World Aluminium，http://www.world-aluminium.org）；2015 年銅為 46%（21.8 百萬噸佔 9,942 千噸，資料來源：The Statistics Portal，https://www.statista.com）；2014 年鋅為 45%（13.75 百萬噸中佔約 6.25 百萬噸，資料來源：Metal Bulletin、The Statistics Portal）。

[7] 2016 年，LME 鋁、銅及鋅的期貨合約成交量佔 LME 期貨總成交量的 35.5%、24.7% 及 18.0%（資料來源：LME）。

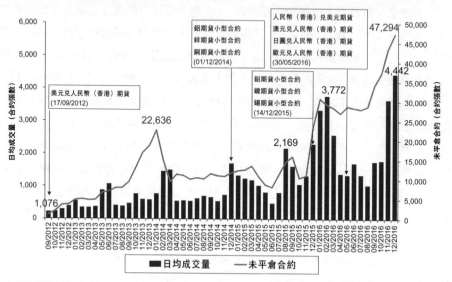

圖 8：香港交易所人民幣衍生產品成交及未平倉合約（2012 年 9 月至 2016 年 12 月）

資料來源：香港交易所。

　　香港交易所的人民幣衍生產品中，以兩隻離岸人民幣與美元貨幣對的匯率期
貨合約的成交最為活躍。其中**美元兌人民幣（香港）期貨**於 2016 年表現卓越：全
年成交合約 538,594 張，年度增幅 105%；年底未平倉合約達 45,635 張，按年增幅
98%，雙雙創卜歷史紀錄。此外，2016 年 12 月的日均成交量攀升至 4,325 張。新
推出的**人民幣（香港）兌美元期貨**於 2016 年下半年的合約成交量亦見增長，未平
倉合約自推出以來穩步上揚。2016 年 12 月平均每日成交 95 張合約，年底未平倉
合約達 1,494 張的高位。（見圖 9）

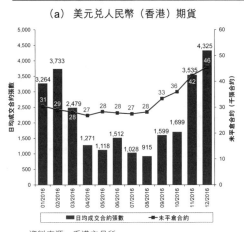

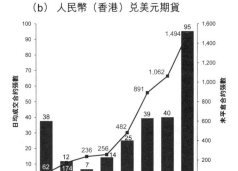

圖9：香港交易所美元兌人民幣（香港）及人民幣（香港）兌美元合約成交及未平倉合約（2016年）

（a）美元兌人民幣（香港）期貨

（b）人民幣（香港）兌美元期貨

資料來源：香港交易所。

　　香港交易所這兩隻期貨合約的成交及未平倉合約同見增長，正正反映產品發揮着作為人民幣貨幣對沖工具的功能。離岸人民幣兌美元匯率極度波動之時均見兩隻合約成交高企（見下圖10及11）。當離岸人民幣匯率於2017年1月5日大幅波動之際，隔夜同業拆息（Hibor）飆升至33.335%，美元兌人民幣（香港）期貨全日成交及未平倉合約分別達20,338張及46,711張新高。在一日前（2017年1月4日）的收市後期貨交易時段，該產品合約已錄得3,642張的成交新高。2016年該兩隻產品的單日成交合約張數及未平倉合約張數與離岸人民幣的隔夜同業拆息之間的相關度呈現中度但統計學上為顯著的相關性（相關系數約0.4至0.5），亦印證了該等人民幣貨幣產品的風險管理功能。換言之，離岸人民幣的流動性問題愈大，該等期貨產品的成交愈趨活躍，未平倉合約數字亦愈趨上升。

圖 10：香港交易所美元兌人民幣（香港）期貨單日成交及未平倉合約與美元兌離岸
人民幣匯率（2016 年 1 月 4 日至 2017 年 1 月 6 日）

資料來源：期貨數據源自香港交易所；美元兌人民幣（香港）匯率源自湯森路透。

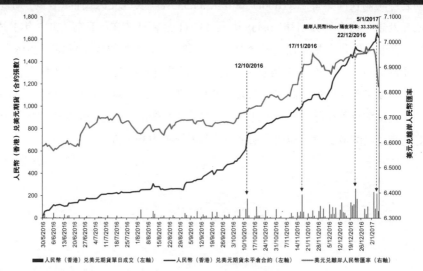

圖 11：香港交易所人民幣（香港）兌美元期貨單日成交及未平倉合約與美元兌離岸
人民幣匯率（2016 年 5 月 30 日至 2017 年 1 月 6 日）

資料來源：期貨數據源自香港交易所；美元兌人民幣（香港）匯率源自湯森路透。

2017 年 3 月 20 日，香港交易所推出其首隻人民幣貨幣期權合約 —— **美元兌人民幣（香港）期權**，進一步豐富旗下人民幣貨幣風險管理工具。產品的首日成交量為 109 張合約，至 2017 年 3 月底日均成交量為 122 張合約。產品種類的增多，可供投資者因應本身的人民幣風險敞口採納不同投資策略。

隨着人民幣推進國際化及人民幣市場逐步開放，全球投資者對人民幣風險的對沖需求與日俱增。為更迎合投資者的需要，香港交易所繼續推進新產品計劃，2016 年 6 月推出與湯森路透聯手開發的**湯森路透／香港交易所人民幣貨幣指數系列**（RXY 指數或 RXY 指數系列）[8]。RXY 指數系列是中國內地境外首個可供買賣的人民幣指數系列。以 RXY 為**相關資產**的**指數期貨**可於時機成熟時推出。

此外，以中國財政部發行的國債為相關資產的期貨合約（**國債期貨**）剛於 2017 年 4 月 10 日推出。這些人民幣衍生產品都是有效對沖利率的工具，特別是在 2017 年 3 月 15 日全國人民代表大會閉幕後，國務院李克強總理在新聞發佈會上表示準備在內地與香港的債券市場試行「**債券通**」計劃，預料這類產品未來將日益重要。

總括而言，全球投資者在人民幣國際化的進程中對人民幣貨幣產品需求殷切，香港交易所的人民幣貨幣衍生產品切合投資者需要，備受歡迎。香港交易所的人民幣衍生產品系列將會持續豐富，會有更多定息及貨幣產品，以迎合投資者日益增長的需求。

3　環球交易所提供的人民幣產品

研究顯示全球主要交易所中不太多交易所有人民幣交易的證券或衍生產品在旗下市場上市[9]。以下兩節闡釋相關結果。

8　見本書第 6 章〈湯森路透／香港交易所人民幣貨幣指數〉。
9　有關資料是在全球交易所官方網站上盡力而為搜索所得，不保證全面及準確。

3.1 證券產品

按主要交易所的官方網站顯示，有人民幣交易證券上市的交易所包括德意志交易所、日本交易所集團（日本交易所）、倫敦證券交易所（倫敦證交所）、新加坡交易所（新交所）和台灣證券交易所（台證所）[10]。

德意志交易所與上海證券交易所（上交所）及中國金融期貨交易所（中金所）組成的合資公司**中歐國際交易所（中歐所）**於 2015 年 11 月 18 日開業，定位為歐洲的離岸人民幣資產交易及定價中心，初期產品發展集中於人民幣交易及結算的現貨證券產品，條件成熟將擴展至衍生產品。2016 年底，中歐所上市的人民幣交易證券包括兩隻 ETF 及三隻債務證券，另外亦有約 14 隻以中國資產為相關資產、但以歐元交易的 ETF。中歐所於 2017 年 2 月 20 日推出其首隻 ETF 衍生產品——以內地指數 CSI300 為相關資產的 ETF 期貨，但以歐元交易。相應的 ETF 期權會於稍後推出，但亦以歐元交易。

日本交易所旗下的專業投資者債券市場 2016 年底有兩隻人民幣交易債券，第一隻於 2015 年 7 月上市；該交易所同時亦有數隻中國相關的 ETF，但全以日圓交易。

倫敦證交所 2016 年底有兩隻 ETF 及超過 100 隻債券以人民幣交易。首隻人民幣 ETF 於 2015 年 3 月上市，第二隻於 2016 年 9 月上市；另外亦有其他中國相關的 ETF，惟全部以英鎊或美元交易。

新交所 2016 年底有一隻同時提供新加坡元及人民幣交易的雙櫃枱股票——中資公司揚子江船業（控股）有限公司，以及 96 隻人民幣交易債券。該交易所並無以人民幣交易的 ETF，但有 6 隻中國相關的 ETF，當中 5 隻以美元交易，一隻以新加坡元交易。

台證所 2016 年 8 月 8 日推出雙幣 ETF 交易機制，是台證所首次開放櫃枱供證券產品以外幣交易。2016 年底，台證所雙幣 ETF 交易機制有兩隻以人民幣交易的櫃枱。

研究發現部份其他交易所亦有與中國相關的 ETF，但都以本國貨幣交易，當中包括澳洲證券交易所（澳交所）、韓國交易所（韓交所）、紐約證券交易所（紐交

10　有關香港交易所及海外交易所已知的人民幣交易證券名單，見附錄一。

所）和納斯達克交易所。

　　至於人民幣債券，研究發現有逾 400 隻離岸產品在其他交易所交易，其中包括法蘭克福證券交易所、MarketAxess、盧森堡證券交易所、台北證券櫃枱買賣中心等 [11]。

3.2　衍生產品

　　研究發現，於香港交易所以外的其他交易所買賣的人民幣衍生產品僅限於人民幣貨幣期貨及期權。這些交易所包括美洲的芝加哥商業交易所集團（芝商所集團）及巴西證券期貨交易所（巴西期交所）；亞洲的新交所、ICE 新加坡期貨交易所（ICE 新加坡期交所）、台灣期貨交易所（台灣期交所）和莫斯科交易所；非洲的約翰內斯堡證券交易所（約翰內斯堡證交所）；以及中東的杜拜黃金及商品交易所（杜拜商交所）[12]。香港交易所以外未有發現有交易所買賣以人民幣交易的商品合約。

　　在所列的交易所中，**芝商所集團**提供的人民幣貨幣產品數目最多——2016 年底有 8 隻期貨及兩隻期權合約以在岸或離岸人民幣為相關資產，都是在芝商所旗下兩間交易所交易：

- 芝加哥商業交易所（**CME**）——4 隻期貨及 2 隻期權。各有兩隻在岸人民幣期貨合約（一隻標準合約，一隻小型合約）於 2016 年 5 月除牌。
- **CME** 歐洲交易所（**CMED**）——4 隻期貨合約。

　　於 2016 年底，**新交所**有 5 隻在岸或離岸人民幣兌美元、新加坡元及歐元的貨幣期貨合約，以及一隻人民幣貨幣期貨的期權合約；**ICE 新加坡期交所**有兩隻在岸或離岸人民幣兌美元的貨幣小型期貨合約；而台灣期交所則有兩隻期貨及兩隻期權合約以離岸人民幣兌美元為基礎——標準合約及小型合約各有一隻期貨及一隻期權。

　　其他交易所——**巴西期交所、杜拜交易所、約翰內斯堡證交所及莫斯科交易所**——2016 年底時各有一隻人民幣貨幣期貨合約。

11　資料來源：2017 年 1 月 6 日湯森路透。由於同一人民幣債券可能在多個交易所交易，故有關數字包括重複點算。須注意名單未能與交易所的官方來源核證。

12　有關香港交易所及海外交易所已知的人民幣衍生產品名單，見附錄二。

4 香港交易所與全球交易所的人民幣產品比較

4.1 證券產品

與全球其他交易所比較，香港交易所提供的人民幣交易證券數目最多 [13]。在中**國境外市場，最受歡迎的場內人民幣交易產品種類是 ETF**。雖然也有不少的上市人民幣計價債券，但其場內交易即使有亦微不足道 [14]。下表 1 比較香港交易所與全球其他所知有提供人民幣交易證券之交易所的人民幣交易證券產品數目，表 2 則比較各交易所的人民幣 ETF 的成交。2016 年在香港交易所交易的人民幣 ETF 的平均每日成交金額（日均成交）為人民幣 37 百萬元，即使按每隻證券計亦高於其他交易所。

表 1：香港交易所及個別交易所以人民幣交易的上市證券（2016 年 12 月）					
交易所	股票	ETF	REIT	債務證券	合計
香港交易所	2	41	1	135	179
中歐所	0	2	0	3	5
日本交易所	0	0	0	2	2
倫敦證交所	0	2	0	101	103
新交所	1	0	0	96	97
台證所	0	2	0	0	2

註：數據乃盡力而為編制。
資料來源：香港交易所及相關交易所的網站。

---

[13] 根據現有所知數據及資料。

[14] 債券交易通常在場外而非交易所內進行。發行商安排債券於交易所上市，或是為配合一些按其授權規定必須投資於認可證券交易所上市證券的投資者及基金經理，使他們也可買賣其債券。

表 2：2016 年人民幣交易所買賣基金總成交及日均成交		
交易所	總成交（人民幣百萬元）	日均成交 （人民幣百萬元）
香港交易所	9,140	37.0
中歐所*	74	0.3
倫敦證交所	0.5	0.0
台證所	141	1.4

* 人民幣產品於德意志交易所平台上交易。

4.2　衍生產品

　　人民幣貨幣期貨已成為全球最普及的人民幣交易衍生產品，香港交易所以外至少有 8 家其他交易所有提供這類產品。投資者對美元 / 離岸人民幣合約的興趣最大，從該類產品的成交相對較高可見一斑。人民幣對另一國際貨幣歐元及其他本國貨幣如新加坡元於過去兩年的合約交易則無足輕重，甚或全無交易（據研究中各交易所的官方數據）。下表 3 列出香港交易所及據知有提供人民幣衍生產品交易的全球其他交易所的人民幣衍生產品數目。

表 3：香港交易所及個別交易所的人民幣衍生產品（2016 年 12 月）							
交易所	貨幣		商品		合計		總數
	期貨	期權	期貨	期權	期貨	期權	
香港交易所	5	0	6	0	11	0	11
巴西期交所	1	0	0	0	1	0	1
芝商所集團 (1)	8	2	0	0	8	2	10
杜拜商交所	1	0	0	0	1	0	1
ICE新加坡期交所	2	0	0	0	2	0	2
約翰內斯堡證交所	1	0	0	0	1	0	1
莫斯科交易所	1	0	0	0	1	0	1
新交所	5	1	0	0	5	1	6
台灣期交所	2	2	0	0	2	2	4
合計	26	5	6	0	32	5	37

(1) 由芝加哥商業交易所（CME）提供 4 隻期貨及 2 隻期權產品，以及 CME 歐洲交易所（CMED）提供 4 隻期貨產品。

註：數據乃盡力而為編制。

資料來源：香港交易所及相關交易所的網站。

　　2016年香港交易所為全球交易所中人民幣貨幣衍生產品平均每日名義成交金額（2.19億美元）及年底未平倉合約數目（47,294張）最高的交易所。新交所與台灣期交所的人民幣貨幣衍生產品成交量頗大，這或與新加坡及台灣與中國的經濟活動程度有關[15]。表面上，台灣期交所2016年的人民幣衍生產品合約成交張數較香港交易所為高，但成交集中於小型合約，而且其於2015年達致的相對較高的名義成交金額並未能持續至2016年。至於新交所，合約成交及名義成交金額於2016年均相對較高，而2016年底未平倉合約則不足香港交易所的40%。（見圖12）

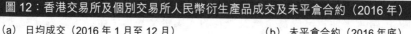

圖12：香港交易所及個別交易所人民幣衍生產品成交及未平倉合約（2016年）

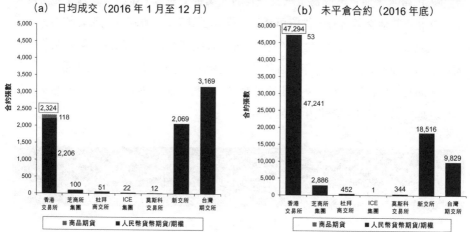

（a）日均成交（2016年1月至12月）　　　　（b）未平倉合約（2016年底）

15　根據中國人民銀行《2015年人民幣國際化報告》（2015年6月），新加坡及台灣是繼香港以後人民幣貿易結算金額最高的地區。

（續）

圖12：香港交易所及個別交易所人民幣衍生產品成交及未平倉合約（2016年）

（c）人民幣貨幣衍生產品日均名義成交金額（2016年對比2015年）

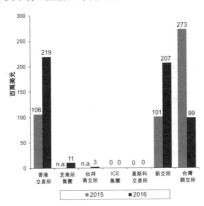

n.a.：沒有資料。

資料來源：香港交易所及相關交易所的網站。

人民幣貨幣期貨中，有四家其他交易所提供相類似的產品，合約金額與香港交易所標準期貨合約美元兌人民幣（香港）期貨相同。香港交易所在此等標準合約的交易上亦處於領先地位。（見圖13）

圖13：香港交易所及個別交易所標準美元兌人民幣合約的成交及未平倉合約（2016年 對比 2015年）

（a）日均成交（2016年對比2015年）　　（b）年底未平倉合約（2016年對比2015年）

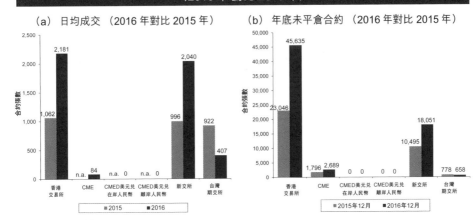

n.a.：沒有資料。

註：標準合約金額為100,000美元。除另有註明外，貨幣對為美元兌離岸人民幣。

資料來源：香港交易所及相關交易所的網站。

（有關不同交易所每隻人民幣貨幣產品 2016 年的日均成交，見附錄三。）

5 香港交易所作為離岸人民幣的 產品交易及風險管理中心

香港交易所在上市及買賣人民幣產品方面領先中國內地以外的全球交易所，其中人民幣 ETF 更是迄今內地境外最受歡迎的人民幣證券產品。雖然離岸人民幣 ETF 的交易量只是一般，離岸人民幣貨幣衍生產品的交投卻呈現動力 —— 全球名義交易金額約為 1,340 億美元（約人民幣 9,350 億元），與 2015 年約 830 億美元（約人民幣 5,390 億元）相比，增幅超過 60%[16]。

隨着 2016 年 8 月人民幣匯率機制改革及中國匯率體制愈益趨向市場化，市場愈來愈能接受人民幣匯率有更大的波動性。根據全球銀行間金融電信協會（SWIFT）統計，人民幣在國際支付中以金額計為第五或第六活躍貨幣 —— 2017 年 1 月佔比 1.68%（第六位），2015 年 12 月為 2.31%（第五位）。離岸匯市方面，有人說人民幣有潛力晉身五大交易最活躍的貨幣[17]，人民幣外匯風險管理工具如掉期及期權亦已愈來愈普及。此外，人民幣貨幣的客戶羣亦日趨多元化，除最先那批與中國經商的企業外，現在的人民幣客戶還包括所有類別的銀行、機構投資者（如合格境外機構投資者）、對沖基金等投機份子以及散戶投資者。人民幣貨幣期貨及期權的供應，使得這些投資者進行人民幣外匯交易時，有更多對沖工具可選擇。

現在世界各地交易所爭相推出各種人民幣風險管理工具，冀在一定程度上支持市場上與中國日益增多的人民幣經濟活動。香港交易所在人民幣產品尤其是人民幣衍生產品方面成為領先交易所，原因有多項：

---

16 資料來源：香港交易所及各相關交易所的網站，見第 4 節的分析。

17 引述芝商所集團外匯產品一名執行董事所說，見 2015 年 4 月 24 日 *Global Capital Euroweek* 周刊〈中國離岸人民幣最終局勢第三回：從影子銀行到網絡空間〉（"China's offshore RMB endgame, Part III: From shadow banking to cyberspace"）一文。

(1)　地理因素

根據國際結算銀行三年期調查，香港在內地境外人民幣外匯工具場外交易總額中的佔比最大 (39%)[18]，其次是新加坡 (22%) 和美國 (12%)。中國領導人於 2013 年提出「一帶一路」倡議，預期中國現時及短期內與世界其他地區的經濟發展將以亞洲為重。一帶一路由「絲綢之路經濟帶」及「21 世紀海上絲綢之路」組成。絲綢之路經濟帶由中亞、西亞、中東延至歐洲，另一支線貫通南亞和東南亞。海上絲綢之路則經過東南亞、大洋洲和北非。

為支持一帶一路的基建項目及經濟活動，人民幣必將進一步國際化及普及，因此預料人民幣期貨作為風險管理工具在亞洲區的交易需求將會最大。香港是位處一帶一路中央的國際金融中心，這裡的場內及場外人民幣產品及服務勢將繼續蓬勃發展。

(2)　香港人民幣資金池及人民幣業務

自 2004 年首辦離岸人民幣業務至今，香港已成為人民幣貿易結算、融資和資產管理的全球樞紐。 如上文第 4.2 節所述，香港於 2015 年的人民幣貿易結算額為全球最高。香港的點心債市亦是中國內地境外最大市場[19]。

現時香港離岸人民幣服務包括人民幣個人銀行及企業銀行、人民幣資本市場、人民幣貨幣及外匯市場以及人民幣保險等。這些服務背後有賴龐大的人民幣資金池支持，兩者聯動有助不同種類人民幣產品發展。人民幣產品和服務切合顧客或投資者的需要而更多元化及活躍，反過來亦將提振或促進人民幣資金池持續擴大。

雖然最近一浪人民幣貶值潮導致全球離岸人民幣資金池萎縮，香港的離岸人民幣存款仍穩列世界首位：人民幣 5,225 億元 (2017 年 1 月底)，相對於台灣的人民幣 3,107 億元 (2017 年 1 月底) 及新加坡的人民幣 1,260 億元 (2016 年 12 月底)[20]。龐大的人民幣資金池既支援所有人民幣業務，亦是人民幣證券及衍生產品市場交投活躍的支柱。

18　按美元每日平均名義成交額計。資料來源：國際結算銀行網站的三年期調查報告有關成交量的統計數據。

19　引述自 2016 年 1 月香港金融管理局《香港：全球離岸人民幣業務樞紐》。

20　資料來源：香港金管局、中國銀行 (香港) 離岸人民幣快報 2017 第三期、新加坡金融管理局網站。

(3) 國際投資者羣

香港是知名國際金融中心，香港交易所的市場參與者來自世界各地。根據香港交易所的調查研究，國際投資者的交易金額佔證券市場成交額的 39%（與本地投資者的比重相同）以及衍生產品市場成交的 28%（本地投資者為 21%）[21]。國際投資者的積極參與有助人民幣產品定價國際化。

(4) 基建效率

香港市場基礎設施效率高而穩健，是支持香港各類人民幣業務和服務有高活躍度與持續蓬勃發展的基礎。有關基建包括香港金管局以 SWIFT 為基礎的**人民幣即時支付結算系統（人民幣 RTGS）**，該系統可協助世界各地市場參與者處理人民幣交易，不論是與中國內地進行的，還是離岸市場之間的，都可通過人民幣 RTGS 系統處理。香港的人民幣 RTGS 受香港法律規管[22]，且直接連接中國內地的「中國現代化支付系統（CNAPS）」，能處理所有與中國內地的人民幣交易，亦與港元、美元及歐元 RTGS 系統有連繫。人民幣 RTGS 不但以即時支付結算方式處理銀行同業人民幣支付項目，亦處理人民幣批量結算及交收支付項目，功能類似港元 RTGS 系統。按 SWIFT 的統計，香港銀行處理的人民幣收付交易量佔全球離岸人民幣交付總量（包括與中國內地的交付及純離岸的交付）的 70% 左右[23]。

除了人民幣 RTGS 系統外，香港金管局並提供**人民幣流動資金安排**，為銀行提供短期資金（即日、隔夜、一日、一星期），加強離岸人民幣市場短期資金流動性，減少短期資金因季節性因素或資本市場活動而受到影響。此外，2016 年 10 月 27 日起，香港金管局委任了九家銀行為離岸人民幣市場的一級流動性提供行。

除了獲得香港整個金融體系的強力支持外，**交易所的基礎設施**亦為離岸人民幣市場提供高效支援。香港交易所的衍生產品結算所 —— 香港期貨結算有限公司（期貨結算公司）—— 為交易所買賣的衍生產品提供中央結算。期貨結算公司符合支付及結算系統委員會與國際證監會組織[24] 的規定，在國際認可監管機制及香港法律保護下運作。期貨結算公司是巴塞爾協議III下「合格中央結算對手」，國

---

21 資料來源：香港交易所《現貨市場交易研究調查 2014/15》及《衍生產品市場交易研究調查 2014/15》。
22 款項交收的終局性受香港《結算及交收系統條例》保障。
23 引述自香港金管局 2016 年 1 月《香港：全球離岸人民幣業務樞紐》。
24 支付及結算系統委員會以及國際證券事務監察委員會組織（國際證監會組織）技術委員會。

際參與者須支付的資本費用較低。2016年9月底，期貨結算公司放寬現金抵押品政策，容許結算參與者以可接受現金及/或非現金抵押品履行不超過人民幣10億元的人民幣按金要求[25]。政策放寬有助減輕投資者買賣人民幣計價衍生產品的資金成本。此外，香港交易所的**場外結算公司**於2013年投入服務，香港從此在場外衍生產品領域亦擁有對等的基建設施。場外結算公司服務現涵蓋若干利率掉期、不交割利率掉期、跨貨幣利率掉期及不交割貨幣遠期合約。

上述各平台為香港人民幣衍生產品進一步發展奠下堅實基礎。

如中國政府於 2017 年 3 月 5 日提交的工作報告所述，人民幣將成為國際貨幣體系中的重要貨幣，人民幣匯率將處於基本穩定水平。人民幣產品和風險管理工具供應充裕的離岸市場，是支援人民幣國際化而同時維持穩定滙率水平的基礎。為此，香港交易所旗下證券及衍生產品市場的人民幣產品配套將會持續豐富，以迎合人民幣穩步推進國際之化下日益增長的投資者需求。除了近期推出的美元兌人民幣（香港）期權及內地國債期貨外，日後亦可能推出其他人民幣風險管理工具。可見香港交易所具備優越條件，足以成為全球投資者的離岸人民幣產品交易及風險管理中心。

25　若超出此數，期貨結算公司的交易對手須以人民幣現金履行人民幣按金要求。

附錄一

在香港交易所及海外交易所以人民幣交易的股票、ETF 及 REIT 名單

（2016 年底）

香港交易所		
類別	證券代號	產品
股票	80737	合和公路基建有限公司
股票	84602	中國工商銀行人民幣 6.00% 非累積、非參與、永續境外優先股
ETF	82808	易方達花旗中國國債 5-10 年期指數 ETF
ETF	82811	海通滬深 300 指數 ETF
ETF	82822	南方富時中國 A50 ETF
ETF	82828	恒生 H 股指數上市基金
ETF	82832	博時富時中國 A50 指數 ETF
ETF	82833	恒生指數上市基金
ETF	82834	iShares 安碩納斯達克 100 指數 ETF
ETF	82836	iShares 安碩核心標普 BSE SENSEX 印度指數 ETF
ETF	82843	東方匯理富時中國 A50 指數 ETF
ETF	82847	iShares 安碩富時 100 指數 ETF
ETF	83008	添富共享滬深 300 指數 ETF
ETF	83010	iShares 安碩核心 MSCI 亞洲（日本除外）指數 ETF
ETF	83012	東方匯理恒生香港 35 指數 ETF
ETF	83074	iShares 安碩核心 MSCI 台灣指數 ETF
ETF	83081	價值黃金 ETF
ETF	83095	價值中國 A 股 ETF
ETF	83100	易方達中證 100 A 股指數 ETF
ETF	83107	添富共享中證主要消費指數 ETF
ETF	83115	iShares 安碩核心恒生指數 ETF
ETF	83118	嘉實 MSCI 中國 A 股指數 ETF
ETF	83120	易方達中華交易服務中國 120 指數 ETF
ETF	83122	南方東英中國超短期債券 ETF
ETF	83127	未來資產滬深 300 ETF

(續)

香港交易所		
類別	證券代號	產品
ETF	83128	恒生 A 股行業龍頭指數 ETF
ETF	83129	南方東英滬深 300 精明 ETF
ETF	83132	添富共享中證醫藥衛生指數 ETF
ETF	83136	嘉實 MSCI 中國 A 50 指數 ETF
ETF	83137	南方東英中華 A80 ETF
ETF	83139	iShares 安碩人民幣債券指數 ETF
ETF	83146	iShares 安碩德國 DAX 指數 ETF
ETF	83147	南方東英中國創業板指數 ETF
ETF	83149	南方東英 MSCI 中國 A 國際 ETF
ETF	83150	嘉實中證小盤 500 指數 ETF
ETF	83155	iShares 安碩歐元區 STOXX 50 指數 ETF
ETF	83156	廣發國際 MSCI 中國 A 股國際指數 ETF
ETF	83162	iShares 安碩 MSCI 中國 A 股國際指數 ETF
ETF	83168	恒生人民幣黃金 ETF
ETF	83170	iShares 安碩核心韓國綜合股價 200 指數 ETF
ETF	83180	華夏中華交易服務中國 A80 指數 ETF
ETF	83188	華夏滬深 300 指數 ETF
ETF	83199	南方東英中國五年期國債 ETF
REIT	87001	匯賢產業信託
海外交易所	類別	產品
中歐國際交易所（中歐所）[產品於德意志交易所平台上買賣]	ETF	BOCI Commerzbank SSE 50 A Share Index UCITS ETF
	ETF	Commerzbank CCBI RQFII Money Market UCITS ETF
倫敦證券交易所（倫敦證交所）	ETF	Commerzbank CCBI RQFII Money Market UCITS ETF
	ETF	ICBC Credit Suisse UCITS ETF SICAV
新加坡交易所（新交所）	股票	揚子江船業（控股）有限公司
台灣證券交易所（台證所）	ETF	富邦上證 180 證券投資信託基金
	ETF	群益深證中小板證券投資信託基金

資料來源：香港交易所產品源自香港交易所；其他交易所的人民幣產品源自相關交易所網站。

附錄二

在香港交易所及海外交易所買賣的人民幣貨幣期貨／期權名單

（2016 年底）

交易所	產品	合約金額	交易貨幣	結算方式
香港交易所	人民幣貨幣期貨——美元兌人民幣（香港）期貨	10 萬美元	離岸人民幣	可交收
	人民幣貨幣期貨——歐元兌人民幣（香港）期貨	5 萬歐元	離岸人民幣	現金結算
	人民幣貨幣期貨——日圓兌人民幣（香港）期貨	600 萬日圓	離岸人民幣	現金結算
	人民幣貨幣期貨——澳元兌人民幣（香港）期貨	8 萬澳元	離岸人民幣	現金結算
	人民幣貨幣期貨——人民幣（香港）兌美元期貨	人民幣 30 萬元	美元	現金結算
	倫敦鋁期貨小型合約	5 噸	離岸人民幣	現金結算
	倫敦鋅期貨小型合約	5 噸	離岸人民幣	現金結算
	倫敦銅期貨小型合約	5 噸	離岸人民幣	現金結算
	倫敦鉛期貨小型合約	5 噸	離岸人民幣	現金結算
	倫敦鎳期貨小型合約	1 噸	離岸人民幣	現金結算
	倫敦錫期貨小型合約	1 噸	離岸人民幣	現金結算
巴西期交所	人民幣期貨	人民幣 35 萬元	巴西雷亞爾	現金結算
芝加哥商業交易所（CME）	標準規模美元／離岸人民幣（CNH）期貨	10 萬美元	離岸人民幣	可交收
	E-微型美元／離岸人民幣（CNH）期貨	1 萬美元	離岸人民幣	可交收
	人民幣／美元期貨	人民幣 100 萬元	美元	現金結算
	人民幣／歐元期貨	人民幣 100 萬元	歐元	現金結算
	人民幣／美元期貨期權	人民幣 100 萬元	美元	可交收
	人民幣／歐元期貨期權	人民幣 100 萬元	歐元	可交收

(續)

交易所	產品	合約金額	交易貨幣	結算方式
CME 歐洲交易所 (CMED)	歐元／離岸人民幣（EUR/CNH）實物交收期貨	10 萬歐元	離岸人民幣	可交收
	美元／人民幣（USD/CNY）現金結算期貨	10 萬美元	在岸人民幣	現金結算
	美元／離岸人民幣（USD/CNH）實物交收期貨	10 萬美元	離岸人民幣	可交收
	歐元／人民幣（EUR/CNY）現金結算期貨	10 萬歐元	在岸人民幣	現金結算
杜拜商交所	美元人民幣期貨	5 萬美元	離岸人民幣	現金結算
ICE 新加坡期交所	小型離岸人民幣期貨	1 萬美元	離岸人民幣	可交收
	小型在岸人民幣期貨	人民幣 10 萬元	美元	現金結算
約翰內斯堡證交所	人民幣／蘭特貨幣期貨	人民幣 1 萬元	南非蘭特	現金結算
莫斯科交易所	人民幣／盧布匯率期貨	人民幣 1 萬元	俄羅斯盧布	現金結算
新交所	人民幣／新加坡元外匯期貨	人民幣 50 萬元	新加坡元	現金結算
	人民幣／美元外匯期貨	人民幣 50 萬元	美元	現金結算
	歐元／離岸人民幣外匯期貨	10 萬歐元	離岸人民幣	現金結算
	新加坡元／離岸人民幣外匯期貨	10 萬新加坡元	離岸人民幣	現金結算
	美元／離岸人民幣外匯期貨	10 萬美元	離岸人民幣	現金結算
	美元／離岸人民幣外匯期貨的期權	10 萬美元	離岸人民幣	現金結算
台灣期交所	美元兌人民幣期貨	10 萬美元	離岸人民幣	現金結算
	小型美元兌人民幣期貨	2 萬美元	離岸人民幣	現金結算
	美元兌人民幣期權	10 萬美元	離岸人民幣	現金結算
	小型美元兌人民幣期權	2 萬美元	離岸人民幣	現金結算

資料來源：香港交易所產品源自香港交易所；其他交易所的人民幣產品源自相關交易所網站。

附錄三

在香港交易所及主要海外交易所
買賣的人民幣貨幣產品的日均
成交量及年底未平倉合約

（2016 年對比 2015 年）

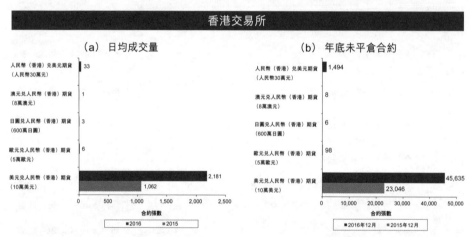

註：歐元兌人民幣（香港）、日圓兌人民幣（香港）、澳元兌人民幣（香港）及人民幣（香港）兌美元期貨合約
　　於 2016 年 5 月 30 日推出。

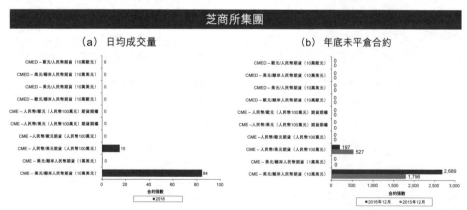

註：2015 年無數據。

(續)

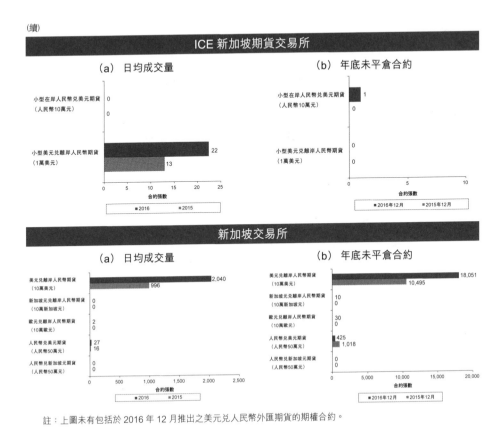

註：上圖未有包括於 2016 年 12 月推出之美元兌人民幣外匯期貨的期權合約。

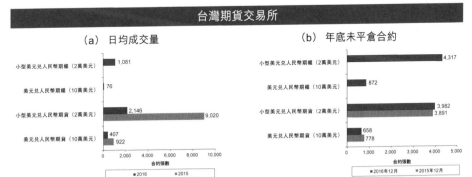

註：美元兌人民幣期權及小型美元兌人民幣期權於 2016 年 6 月 27 日推出。

資料來源：香港交易所產品源自香港交易所；其他交易所的人民幣產品源自相關交易所網站。

第15章

配合中國內地跨境衍生產品交易
與日俱增的場外結算方案

2017 年 11 月 14 日

摘　要

在中國內地市場經濟改革的過程中，內地金融機構為風險管理目的而進行愈來愈多的場外衍生產品交易。在國內，利率市場化改革令利率波動增加，促使更多投資者以利率掉期（即利率互換交易）作對沖用途，令銀行間市場的利率掉期交易金額大升。此外，人民幣匯率的市場化改革令外匯風險增加，市場紛紛透過買賣外匯衍生產品作為對沖，帶動境內銀行間市場的外匯衍生品交易急劇增長。

與此同時，在中國市場開放及經濟發展進程中，內地金融機構亦愈來愈多與海外參與方進行國際業務。由於國際貿易、一帶一路倡議基建項目以至內地企業發展國際業務莫不需要資金，內地金融機構的資產負債表中，以外幣計價的佔比不斷增長，當中又會以美元這個在國際貿易金融中最常用的環球貨幣佔最大部份。利率風險及匯價風險同樣會影響到外幣資產的利息收益。因此，內地金融機構除愈來愈多利用境內場外市場上的風險管理工具外，亦需要與境外機構進行場外利率及外匯衍生產品交易，為其日益增加的外幣資產作對沖，使得相關需求亦有增無減。

2008 年全球金融危機過後，世界各地都在收緊對場外衍生產品風險管理的監管，金融機構或是被強制要求為標準場外衍生產品進行中央結算，或是必須就其作雙邊結算的場外衍生產品遵守更高的資本和保證金要求。如屬後者，有關金融機構或會自願選擇中央結算以求減低交易成本。歐美的國際參與者一般選用其國內監管機構認可的主要結算所，如歐洲的 LCH 或歐洲期貨交易所清算公司（Eurex Clearing），美國的 CME 結算公司（CME Clearing）。內地銀行與境外機構進行場外衍生產品交易，必須透過認可的中央結算對手（CCP）進行中央結算。由於中華人民共和國（稱「中國」或「內地」）作為淨額結算的身份並不明確，內地銀行往往無法成為外國 CCP 的直接結算會員，因此通常是經由身為境外 CCP 結算會員的結算經紀作為內地銀行的代理人進行結算。

在亞洲時區，香港交易及結算所有限公司（香港交易所）的附屬公司「香港場外結算有限公司」（香港交易所場外結算公司）是香港一家獲認可的場外衍生產品結算所，也是可讓歐美機構在處理 CCP 風險敞口中獲優惠資本待遇的 CCP。它亦是唯一可接受有香港分行的內地銀行成為直接結算會員的境外 CCP，歐美的結算所並無此例。由於許多內地銀行在香港已有分行或附屬公司，讓它們成為香港交易所場外結算公司直接會員，將較委託結算經紀在境外 CCP 為它們進行衍生產品交易結算更具成本效益。按照香港交易所場外結算公司的方案，內地銀行可透過其香港附屬公司或分行直接進行結算。因此，相比其他結算所，香港交易所場外結算公司為內地銀行的場外衍生產品交易進行中央結算，將更為方便，也更具成本效益。再者，除為美元及其他主要貨幣的場外交易提供服務外，香港交易所場外結算公司亦較海外同業有更大能力支援內地以至全球金融機構在場外的離岸人民幣交易，相信此等交易的增長潛力會相當高。

1 內地金融機構場外 衍生產品交易與日俱增

場外衍生產品在中國的歷史只稍過十年。事實上，除了在中國人民銀行（人行）及國家外匯管理局（外管局）監督下的**中國外匯交易中心（CFETS）**所營運的場外衍生產品市場外，中國內地可以說並無其他場外衍生產品市場可言。CFETS首隻衍生產品為 2005 年 6 月推出的債券遠期合約，時至今日，產品系列已擴闊至外匯、利率及信貸衍生品。

1.1 CFETS 及其產品

CFETS 亦稱全國銀行間同業拆借中心，是人行 1994 年 4 月 18 日成立的直屬事業單位，主要職能是提供銀行間外匯交易、人民幣借貸、債券（包括短期融資券）交易及衍生產品交易服務，以及相關的結算、資訊、風險管理和監察服務。

貨幣市場方面，CFETS 營運外匯市場和人民幣市場。在外匯市場，CFETS負責計算及發佈人民幣兌主要貨幣（包括美元、歐元、日圓、英鎊和港元）的中間價。人民幣市場包括同業拆借市場、存款證、貸款轉讓、債券市場（包括資產支持證券）及人民幣衍生產品市場。

銀行間市場首隻正式推出的衍生產品是 2005 年 6 月的**債券遠期合約**，其時與 1996 年銀行間債券市場成立並同年推出國債回購合約相隔已有九年之久。由於利率市場化改革加快，內地金融機構的利率風險增加，但欲以既有工具（如債券回購及遠期合約）對沖風險卻愈不容易。為提供更多利率風險管理工具，2006 年 2月銀行間市場開展**人民幣利率掉期（或利率互換）（IRS）**交易試點，容許若干合格機構在某些規限下進行人民幣 IRS 交易，最後試點計劃終成就人民幣 IRS 產品於2008 年 2 月全面推出。其他新增的對沖及風險管理工具還有 2007 年 11 月推出的**人民幣遠期利率協議（FRA）**、2014 年 11 月的**標準利率衍生產品** [1]，以及 2010 年的**信用風險緩釋憑證（CRMW）**和 2016 年 9 月的**信貸違約掉期（CDS）**等信用風險緩

1　標準利率衍生產品包括具標準化到期日及標準化利率期的 IRS 及 FRA 產品。

釋（CRM）工具。

　　銀行間外匯市場方面，2005 年 8 月推出**貨幣遠期合約**（人民幣對外幣）。2006 年 4 月貨幣掉期（人民幣對外幣）試點展開[2]，相關規定發佈後貨幣掉期合約於 2007 年 8 月正式推出。**貨幣期權**其後於 2011 年 4 月推出，接着是 2015 年 2 月推出**標準化貨幣掉期**及 2016 年 5 月推出**標準化貨幣遠期合約**。這些產品為銀行提供對沖工具，使銀行可更靈活管理其外幣持倉。

　　CFETS 參與者包括銀行及非銀行金融機構，如證券公司、保險公司、信託投資公司、基金及基金管理公司、資產管理公司和社會保障基金等。銀行間債券市場是首個（於 2010 年）向境外參與者開放的銀行間市場。初期的合格參與者包括央行類機構、人民幣結算銀行和參與銀行、合格境外機構投資者（QFII）和人民幣合格境外機構投資者（RQFII）。認可境外參與者其後擴展至包括所有合法註冊金融機構及其投資產品、養老金和慈善基金。結合其後的政策放寬，若干境外機構投資者（包括央行類機構、國際金融機構和主權基金）現可利用銀行間市場廣泛的產品，包括現貨債券和外匯市場、債券衍生產品和利率衍生產品。

　　CFETS 場外衍生產品交易的清算及交收由三家機構提供，CFETS 本身外，還有中央國債登記結算有限責任公司（中債登）和上海清算所。CFETS 提供交易確認及直通式處理，透過人行的支付及交收系統支援其外匯及人民幣市場交易的清算及交收，同時亦提供 IRS 及外匯掉期的交易沖銷／壓縮服務。中債登為 CFETS 的債券及債券衍生產品交易提供清算及交收服務，上海清算所則為不同類型衍生產品的交易提供中央結算服務。2014 年 2 月人行特別指定上海清算所為 CFETS 人民幣 IRS 合約交易強制進行中央結算的 CCP[3]。

1.2　境內場外衍生產品交易

　　如圖 1 所示，IRS 自 2010 年以來一直主導債券／利率衍生產品交易。債券遠期合約交易 2009 年後開始式微，後來更幾乎完全被 IRS 交易取代。IRS 交易額於 2016 年達到人民幣 9.92 萬億元（約 1.4 萬億美元），2006 年至 2016 年間複合年增

2　銀行與客戶之間的人民幣對外幣掉期合約早於 2005 年 8 月已推出。

3　人行 2014 年 2 月 21 日發佈的《中國人民銀行關於建立場外金融衍生產品集中清算機制及開展人民幣利率互換集中清算業務有關事宜的通知》。

長率為 76%。然而，相比主要國際市場，交易水平仍然偏低——2016 年 4 月中國場外單一貨幣利率衍生產品的平均每日成交金額約為 40 億美元，僅為美國的每日平均成交金額 12,410 億美元及英國的 11,800 億美元的 0.3%[4]。貨幣種類而言，2016 年 4 月人民幣場外貨幣利率衍生產品的平均每日成交金額為 100 億美元，大約為美元相關產品的 0.7%、歐元的 1.6% 及英鎊的 4%[5]。

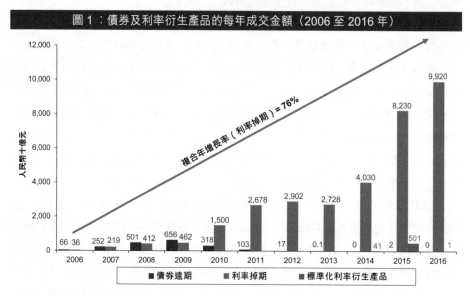

圖 1：債券及利率衍生產品的每年成交金額（2006 至 2016 年）

資料來源：人行 2006 至 2016 年年報。

　　外匯衍生產品方面，人民幣外匯掉期（包括交叉貨幣掉期）[6] 按名義本金交易額計，是最主要的產品類別，其於 2016 年的交易總額達到 10 萬億美元，2006 年至 2016 年間複合年增長率為 121%（見圖 2）。相比主要國際市場，2016 年 4 月人民幣場外外匯衍生產品（遠期、掉期及期權）的平均每日成交金額為 1,340 億美元，

4　資料來源：國際結算銀行（BIS）有關場外衍生產品的三年期調查統計（2016 年 4 月），載於 BIS 網站；每日平均數按「淨 - 毛」（net-gross）基準計算。

5　2016 年 4 月美元、歐元和英鎊的場外單一貨幣利率衍生產品按「淨 - 淨」（net-net）基準計算的平均每日成交金額分別為 13,570 億美元、6,410 億美元及 2,370 億美元。資料來源：BIS 有關場外衍生產品的三年期調查統計（2016 年 4 月），載於 BIS 網站。

6　人民幣外匯掉期涉及兩種貨幣（人民幣對外幣）在特定日期按合約協定匯價進行的實際本外幣交換，及相同貨幣在未來一指定日期按另一匯價進行的相反本外幣交換。人民幣外匯交叉貨幣掉期涉及兩種貨幣（人民幣及外幣）在協定期間的利息交換，亦可涉及兩種貨幣的本金在未來指定時間按預定匯率進行交換。

約為美元相關產品的 4%、歐元的 12%、日圓的 19% 和英鎊的 30%[7]。

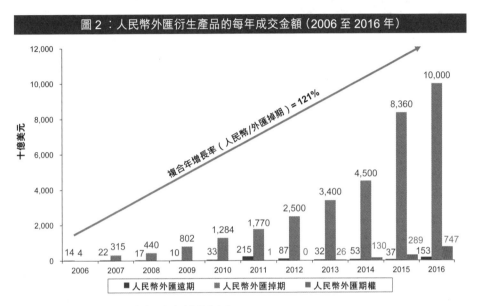

圖 2：人民幣外匯衍生產品的每年成交金額（2006 至 2016 年）

註：限於銀行間交易，不包括銀行與客戶之間的交易。

資料來源：人行 2006 至 2016 年年報。

　　隨着外匯衍生產品特別是人民幣外匯掉期工具的交易急速增長，外匯衍生產品的交易額已超過現匯市場交易額，2016 年外匯衍生產品交易總額為現匯市場交易的 1.3 倍（見表 1）。

　　除了在 CFETS 受規管的場外市場上交易的特定衍生產品外，銀行之間亦有按其業務需要進行其他場外衍生產品的交易。正如全球其他市場的情況一樣，這些場外衍生產品是買賣雙方為滿足其特定需要而創設的，條款度身定制，交易及交收雙邊進行。例如，**內地銀行的公司客戶與外資夥伴有業務往來，內地銀行就可能對美元等外幣的貨幣或利率對沖工具有所需求**。不過，這些交易現時沒有正式統計數字。

7　2016 年 4 月美元、歐元、日圓及英鎊場外外匯衍生產品的平均每日成交金額分別為 30,530 億美元、10,720 億美元、7,010 億美元及 4,380 億美元。資料來源：BIS 的場外衍生產品三年期調查統計（2016 年 4 月），載於 BIS 網站；每日平均數按「淨 - 淨」（net-net）基準計算。

表 1：人民幣外匯市場交易額（2016 年）			
（萬億美元）	銀行 - 客戶	銀行間市場	合計
人民幣外匯市場	3.4	16.8	20.3
現匯	2.9	5.9	8.8
外匯衍生產品	0.5	10.9	11.5
人民幣外匯遠期	0.2	0.2	0.4
人民幣外匯掉期及交叉貨幣掉期	0.1	10.0	10.1
人民幣外匯期權	0.2	0.7	1.0

註：由於四捨五入的關係，數字相加未必等於總數。
資料來源：人行 2016 年年報。

1.3　內地場外衍生產品交易增長潛力巨大

　　IRS 交易急速增長，原因可追溯至過去十年內地持續進行利率市場化改革。隨着市場經濟改革開展，早於 2000 年代初已見若干利率自由化措施出台。拆息的浮動區間於 2004 年 1 月放寬，2004 年 10 月進一步擴闊，同時金融機構獲准將人民幣存款利率降至基準利率以下。國務院常委 2013 年 6 月確定利率市場化改革為支持經濟結構重整的主要金融政策。同年 9 月 24 日，市場利率定價自律機制成立，是內地金融機構的自律監管及協調機制，負責按國家的相關利率規定對金融市場利率進行自律管理。

　　2014 年 3 月，中國（上海）自由貿易試驗區成為內地全面實行外幣存款利率市場化的先驅。2015 年 5 月 1 日，存款保險條例生效，為內地全面落實利率自由化奠定基礎。採用市場利率的大額可轉讓存款證於 2016 年 6 月推出[8]，象徵利率進一步邁向全面開放。

　　利率市場化是指拆借利率將因應市場及經濟情況而變動。**利率波動日增，金融機構對場外市場利率對沖工具（如 IRS）的需求將不斷增加**。在銀行間市場交易的人民幣 IRS 主要以 7 天回購固定利率和上海同業拆借利率（SHIBOR）為浮動端

8　大額可轉讓存款證是銀行向非金融機構投資者發出的人民幣記賬式存款憑證。

的參考利率。圖 3 為一周期 SHIBOR 的每日變動，顯示 2016 年 1 月至 2017 年 7 月期間利率的上升趨勢。

圖 3：一周期 SHIBOR 的每日變動（2006 年 1 月 4 日至 2017 年 7 月 31 日）

資料來源：SHIBOR 網站（http://www.shibor.org）。

　　除了需要進行人民幣利率風險管理的境內人民幣拆借業務外，內地金融機構以外幣進行的海外金融事務亦愈來愈多。首先，利率自由化的同時，內地經濟亦日益對外開放，與世界上其他地區進行的商業及金融活動愈來愈多。結果是，**內地金融機構愈來愈多海外國際業務**，外幣資產及負債不斷增加。據外管局統計數字顯示，短短一年內，內地銀行（不包括央行即人行）對外債券資產由 2015 年底的 484 億美元差不多倍增至 2016 年底的 952 億美元，佔海外金融資產總額的比率由 7% 升至 11%（見圖 4）。人行數據顯示內地金融機構的境外貸款由 2015 年 1 月的人民幣 23,280 億元（約 3,710 億美元）增至 2017 年 4 月的人民幣 35,000 億元（約 5,070 億美元）（見圖 5）。

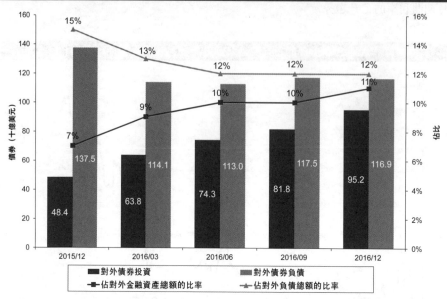

圖4：內地銀行（不包括央行）對外債券資產及負債（2015年12月至2016年12月）

資料來源：來自 Wind 資訊的外管局數據。

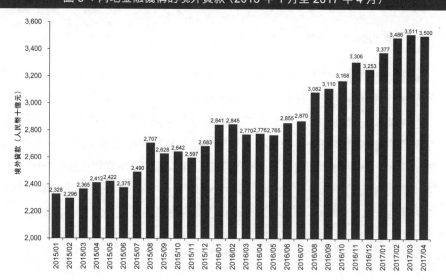

圖5：內地金融機構的境外貸款（2015年1月至2017年4月）

資料來源：人行網站。

　　第二，中國領導人 2013 年推出「一帶一路」倡議[9]，一帶一路沿線國家將開展大量基建項目，預計中國在資助這些項目方面將發揮重要作用。普華永道估計，2016 年一帶一路沿線國家新公佈的項目總額上漲至 4,000 億美元左右，同比增幅2.1%，最終價值漲幅可能高達 10%[10]。亞洲開發銀行一份報告[11] 表示，直至 2030年，發展亞洲地區將需要投資合共 26 萬億美元或每年 1.7 萬億美元建設基礎設施，方可維持其增長勢頭、消減貧窮及應對氣候變化（若不計氣候變化紓減及適應的支出，則為 1.5 萬億美元）。基建投資的落差（投資需求與現行投資水平之間的差額）在 2016 年至 2020 年的五年期內估計佔預計國內生產總值的 2.4%。亞洲開發銀行預計這一落差通過財政改革可填補 40%，透過私營部門可填補 60%。

　　內地金融機構（包括政策性銀行和商業銀行）以及中國推動的專業投資基金[12]和多邊金融機構[13] 將在資助一帶一路項目方面擔當重要角色。2016 年，中國與一帶一路沿線國家的貿易額約為 9,535.9 億美元，工程合約總額約為 1,260.3 億美元[14]。一帶一路項目的融資渠道包括優惠貸款、銀團貸款、出口信用保險、行業基金、債券投資、委託資產管理、持有股權等等。

　　第三，內地政府一直鼓勵境內企業特別是國企「走出去」作為企業改革的部份手段。因此，**企業發展國際業務**包括進行併購是主要發展方向之一。**內地銀行會透過發行債券為這些企業提供外幣資金。**

　　如上所述，由於國際貿易、一帶一路基建項目及企業發展國際業務在在需要資金，內地金融機構的資產負債表中，外幣的佔比將不斷增加，當中又會以美元這個在國際貿易金融中最常用的環球貨幣佔最大部份。**外幣資產的利息收益會受到利率風險及匯價風險的影響。**2015 年 8 月人民幣匯率機制的改革，讓銀行間外匯市場人民幣兌美元匯率的中間價形成機制更趨市場主導後，匯價風險已變得更為顯著。圖 6 顯示該次改革後離岸人民幣（CNH）兌美元匯率的波幅擴大。

9　「一帶一路」包括「絲綢之路經濟帶」和「21 世紀海上絲綢之路」兩項計劃。「絲綢之路經濟帶」連接中亞、西亞、中東至歐洲，延伸至南亞和東南亞。「21 世紀海上絲綢之路」則橫跨東南亞、大洋洲及北非。中國國際貿易研究中心於 2015 年 8 月發表的《「一帶一路」沿線國家產業合作報告》列出了 65 個會參與一帶一路計劃的沿線國家。

10　*China and Belt & Road Infrastructure, 2016 review and outlook*，普華永道 2017 年 2 月。

11　*Meeting Asia's infrastructure needs*，亞洲開發銀行 2017 年 2 月。

12　包括絲路基金、中國—東盟投資合作基金和中非發展基金。

13　包括亞洲基礎設施投資銀行和金磚國家開發銀行。

14　資料來源：《一帶一路資金支持再盤點》，招商證券 2017 年 5 月 14 日。

圖6：美元兌離岸人民幣每日現匯價及波幅（2014 年 1 月至 2017 年 7 月）

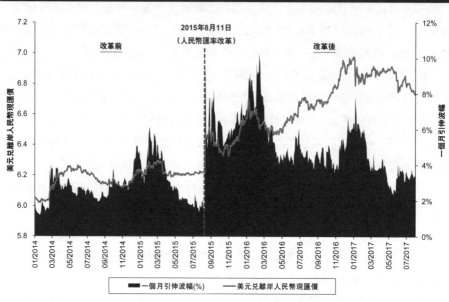

資料來源：彭博。

　　因此，內地金融機構對使用風險管理工具對沖外幣資產利率及匯率風險的需求將不斷增加。標準化場內工具如債券期貨未必適合這些機構的獨特需求，這些機構通常採用 IRS 及外匯掉期等具度身定制條款的場外產品來配對其外幣資產的付款期 15。內地金融機構這些以外幣計價的場外交易大部份以美元進行 —— 2016年 4 月中國場外單一貨幣利率衍生產品成交中，21% 以美元交易，79% 以人民幣成易，澳元及歐元的交易佔比微不足道 16。

15　境內現有的場內工具為在中國金融期貨交易所（CFFEX）交易的人民幣債券期貨 —— 5 年期及 10 年期國債期貨。
　　CFFEX 是中國內地唯一可買賣金融衍生產品的交易所。境內的內地銀行取得內地當局批准後，亦可在海外交易所
　　買賣外幣債券衍生產品。
16　資料來源：BIS 有關場外衍生產品的三年期調查統計（2016 年 4 月），載於 BIS 網站；每日平均數按「淨 - 毛」
　　（net-gross）基準計算。

2　場外衍生工具的風險管理規定

2.1　強制清算及按金要求

2008 年全球金融危機揭示了銀行及其他市場參與者於金融及經濟動盪時期的韌力嚴重不足。場外衍生工具市場之缺乏透明度、衍生工具之愈益複雜、其與金融市場內各個範疇之相互交錯，尤其被視為一大系統風險因素。場外衍生工具交易缺乏監管，加上本身屬於雙邊交易，導致市場缺乏透明度。為應對有關風險，20 國集團在 2009 年提出改革計劃，務求降低場外衍生工具的系統性風險。改革計劃最初包括下列各項 [17]：

- 所有標準化場外衍生工具應在交易所或電子交易平台 (視情況適合) 買賣；
- 所有標準化場外衍生工具應**透過 CCP 結算**；
- 場外衍生工具合約應向交易資料儲存庫匯報；
- **非中央結算的衍生工具合約應符合較高的資本要求。**

2011 年，20 國集團同意在改革計劃中加入對非中央結算衍生工具的保證金要求。預期 20 國集團的保證金要求將會於全球劃一執行，否則，保證金要求較低的區域的金融機構會有較大競爭優勢，造成不公，有損保證金要求的成效 (指招致監管套戥現象) [18]。

保證金要求的其中一項重要原則是，所有涉及非中央結算衍生工具交易的金融機構及具有系統重要性的非金融實體，必須因應交易對手的風險而交換適當的初始及變動保證金。交換初始保證金不可以淨額計算。這樣的雙邊交易保證金，將會較透過 CCP 為該等交易進行結算所收取的初始保證金為高。所規定的標準化初始保證金將取決於資產類別，介乎交易名義金額的 1% 至 15% 之間。根據美國、香港及歐盟監管機構的規定，變動保證金的規定於 2017 年 3 月普遍適用。屆時，上述地區的若干受監管衍生工具實體 (包括美國的掉期交易商及香港的授權機構) 必須收取及提供受涵蓋交易對手的變動保證金。

17　資料來源：*Margin requirements for non-centrally cleared derivatives*，由巴塞爾銀行監管委員會、國際證監會組織以及 BIS 於 2015 年 3 月發佈。

18　同上。

2.2 《巴塞爾協定三》的資本要求

《巴塞爾協定三》實施後，全球金融機構現要符合更嚴格的資本規定 [19]。《巴塞爾協定三》的框架旨在增強對資本及流動資金的監管，並提高個別銀行以至整個銀行業的穩定及復原韌力。銀行的資本對風險加權資產比率必須維持在最低要求或以上。風險加權資產的計算方法是將資產價值乘以權重因子（風險權重），給「較安全」的資產打折，減少所需資本。場外衍生工具一般被視為風險資產，其風險權重會因應交易對手的類型而有所不同。透過合資格中央結算對手（QCCP）[20] 結算的場外交易，其風險權重可低至 2% 至 4%；否則若按雙邊基準計算，風險權重可高達 20% 或以上 [21]。

2.3 全球性規定在美國、歐盟及香港的實施

在美國，商品期貨交易委員會按交易實體的分類分階段對場外衍生工具交易實施強制性中央結算，最初階段是從 2012 年 12 月起，對若干類別的 IRS 及 CDS 實施，至 2013 年 9 月 9 日完成 [22]。「最終用戶」結算的例外情況僅適用於並非金融實體的對手方。2013 年 7 月 2 日，聯邦儲備系統管治委員會（Board of Governors of the Federal Reserve System）通過一項最終規則，構建一個適用於所有美國銀行機構的新的全面監管資本框架，以實施《巴塞爾協定三》當中的資本規定。

在歐盟，歐洲委員會於 2015 年 8 月為《歐洲市場基礎設施監管規則》（EMIR）引入新規則，強制若干場外利率衍生工具交易須透過 CCP 進行結算，其後於 2016 年 3 月再強制若干場外信貸衍生工具交易亦須透過 CCP 進行結算。EMIR 的風險緩釋規定適用於所有非中央結算的場外衍生工具交易，當中要求交換抵押品及雙邊保證金。EMIR 第 4 條所規定的結算責任已於 2016 年 6 月 21 日生效，根據公司類別及衍生工具交易量分階段實施。結算責任適用於任何金融對手方之間或涉及

19　根據巴塞爾銀行監管委員會於 2010 年 12 月刊發的報告 *Basel III: A Global Regulatory Framework for More Resilient Banks and Banking Systems*，全球市場會於 2013 年至 2019 年間逐步採納《巴塞爾協定三》。

20　按《巴塞爾協定三》的界定，QCCP 為獲相關監管機構發牌可作為 CCP 經營的實體，受其所在司法權區的審慎監督，而該司法權區的相關監管機構就金融市場基礎建設實施的當地規則與國際證監會組織轄下支付及市場基建委員會為金融市場基建制定的原則一致。

21　參見美國聯邦儲備系統管治委員會就實施《巴塞爾協定三》而通過的「最終規則」（"Final Rule"）。

22　"OTC derivatives central clearing in the US"，Risk Advisors Inc.，2013 年 2 月。

交易量（按名義毛金額計）超過結算限額的非金融對手方之間又或兩者之間。場外交易中所有類別的公司須遵守該等結算責任的最後期限為 2018 年 12 月 21 日 [23]。

　　在香港，香港金融管理局（香港金管局）及香港證券及期貨事務監察委員會（香港證監會）聯同香港政府及其他持份者一直致力制定適用於香港場外衍生工具市場的監管架構，順應全球大勢。香港先後進行多次市場諮詢，並已就諮詢結果採取相關措施，部份已在實施中。在建立場外衍生工具監管機制的過程中，金管局設立**場外衍生工具交易資料儲存庫，並於 2013 年 7 月啟動其匯報功能**。香港交易及結算所有限公司（香港交易所）則成立了**香港場外結算有限公司（香港交易所場外結算公司）**，於 2013 年 11 月開展業務。首階段對場外衍生工具交易強制中央結算已於 2016 年 9 月開展 [24]，當中涉及主要交易商之間訂立的標準化 IRS 工具。

　　根據巴塞爾銀行監管委員會於 2017 年 4 月發佈的巴塞爾監管框架採納進度報告，美國針對非中央結算衍生工具的保證金要求已於 2016 年 9 月 1 日起逐步生效，並將於 2020 年 9 月 1 日全面生效。歐盟方面的初始保證金規定由 2017 年 2 月 4 日起按交易對手類型逐步生效，而變動保證金要求則於 2017 年 3 月 1 日起實施。香港的保證金規定於 2017 年 3 月 1 日起生效（設六個月過渡期）。

2.4　對內地金融機構的影響

　　因應上述加強監管規定的措施，內地參與場外衍生工具市場的金融機構（尤其在與境外對手方進行交易時）都要遵守海外各項強制結算及匯報規定，並須考慮為場外交易作非中央結算對成本的影響。下表 2 概述該等影響。

23　資料來源：英國金融行為監管局網站（https://www.fca.org.uk）。
24　在生效後，適用於 2017 年 7 月 1 日或之後訂立的附合範圍內的交易。

表 2：適用於場外衍生工具交易的全球性規定的影響		
監管規定	全球基準	影響
資本對風險加權資產比率	《巴塞爾協定三》	• 場外衍生工具雙邊交易的風險權重遠高於經 CCP 進行結算的交易，意味着對內地交易對手方而言，非中央結算的場外合約的定價會較高[25]。
強制性中央結算	20 國集團	• 若金融機構的交易對手所在的司法權區規定須為場外衍生工具交易作強制性中央結算，金融機構亦須遵從有關規定，否則無法與該等交易對手進行交易。 • 使用海外交易對手的當地監管機構認可的結算所進行結算時需要建立聯繫（如通過結算經紀），耗費可能不菲，因為除 CCP 收取的結算費之外，結算經紀亦會收取佣金。
對非中央結算的衍生工具收取保證金	20 國集團	• 與遵守 20 國集團規定的司法權區的海外交易對手進行雙邊場外衍生工具交易時，初始保證金及變動保證金須按非淨額結算基準計算，意味着資金及營運成本會高於選擇中央結算。 • 就雙邊結算而言，由於目前尚未確定中國為淨額結算司法權區[26]，內地銀行與海外交易對手在執行 ISDA 的抵押品協議[27] 時將會有難度，在執行場外衍生工具交易時未必獲海外機構接納為交易對手，又或即使獲接納，也會被收取較高的保證金[28]。

　　總括而言，按全球現時有關場外衍生工具的最新監管規定，**內地金融機構與境外交易對手買賣場外衍生工具時將會因為未有就其交易使用中央結算服務而須支付高昂費用**。再者，中央結算的多邊淨額結算程序可大大降低保證金要求，這將遠低於分別與多方交易對手進行雙邊結算所需的保證金總額。**因此，內地金融機構在其與境外交易對手的場外衍生工具交易日益增加之際，選擇中央結算當更理想可取，當中最重要的考量，只在如何選擇成本較低的中央結算所。**

25　交易對手向內地銀行報價時會將資本費用計算在內，內地銀行因而要承受較高價格，尤其是當交易並無訂立任何應對信貸風險的 ISDA（國際掉期與衍生工具協會）抵押品協議時。

26　若屬「淨額結算司法權區」，遇上交易對手破產時，可根據 ISDA 主協議就場外衍生工具交易強制執行終止交易的抵銷和淨額交割（close-out netting）。（參見 Derivatives Week 雜誌 2014 年 2 月 10 日第 5 期第 23 卷 "China — The New Netting Jurisdiction"，以及金杜律師事務所（King&Wood Mallesons）2017 年 4 月 3 日發佈的 "ISDA publishes updated memoranda on China close-out netting"）。

27　指 ISDA 主協議的「信用支持附件」（Credit Support Annex），附件中界定了掉期對手方之間提供或轉讓抵押品以減輕「價內」衍生倉盤信貸風險的條款或規則。

28　為釐清有關向內地金融機構強制執行主衍生工具協議的平倉淨額結算條款，中國銀監會發佈日期為 2017 年 7 月 4 日的答覆文件，以回應全國人民代表大會財政經濟委員會的相關問題。答覆文件中說明，中國《企業破產法》原則上與 ISDA（國際掉期與衍生工具協會）相關規定（主協議）的平倉淨額計算條文並不衝突，但中國司法機關有權決定終止交易 [平倉] 淨額結算條文的有效性。

3 可供內地銀行選用的場外結算服務

內地銀行在執行與境外對手方進行的場外衍生工具交易時或須跟從境外對手方的做法，透過一國際結算所為交易作中央結算。已發展市場中較有規模的結算所有歐洲的 LCH 及德意志交易所集團（Deutsche Börse AG）旗下的歐洲期貨交易所清算公司（Eurex Clearing），和美國的芝商所清算分部（CME Clearing）。與西方市場這些結算所相比，作為香港交易所在香港的場外結算分部 —— 香港交易所場外結算公司 —— 或是個更好的選擇。

3.1　LCH[29]

LCH 營運多個在 LCH.Clearnet Ltd 旗下的場外衍生產品結算分部，當中包括 **SwapClear** 及 **ForexClear**，分別結算 IRS 及外匯衍生產品，均為內地銀行交易活躍的產品。

SwapClear 服務範圍涵蓋不同掉期類別（一般 IRS／零息／基準／通脹／FRA）、指數掉期（隔夜指數掉期）及固定期限掉期（可變名義掉期）。2017 年始至 7 月 28 日為止，經 SwapClear 結算的場外交易名義金額達 526.44 萬億美元，當中 54% 為美元合約，122.69 萬億美元屬客戶結算，當中 62% 為美元合約。同期，IRS 佔名義總結算量的 29%（154.28 萬億美元，當中 58.69 萬億美元即 38% 為美元合約）；IRS 佔名義客戶結算量的 30%（37.04 萬億美元，當中 46% 為美元合約）。根據 SwapClear 所稱，所有場外 IRS 的交易逾半經其結算，而全球採用結算服務的場外 IRS 逾 95% 經其結算。

ForexClear 的服務對象為不交收遠期外匯合約（NDF）市場，於 2017 年 7 月底涵蓋美元兌 12 種貨幣，包括人民幣。ForexClear 於 2013 年 11 月推出客戶結算，使客戶可經期貨佣金商（FCM）[30] 進行結算。2017 年始至 7 月 22 日為止，

29　相關數據及資料來自 LCH 網站（http://www.lch.com）。

30　期貨佣金商接受客戶下單，為其買賣期貨合約、期貨期權、零售場外外匯合約或掉期，並接納客戶的款項或其他資產以進行交易。

ForexClear 的結算總量達 5.45 萬億美元，當中 13%（7,120 億美元）為美元兌人民幣的合約。

場外交易要交予 SwapClear 或 ForexClear 結算，必須經 LCH 規則手冊定義的認可買賣源頭系統（Approved Trade Source System）（例如彭博、 MarkitWire）提交 LCH。 SwapClear 的會員分為 SwapClear 結算會員（SwapClear Clearing Members）及期貨佣金商結算會員（FCM Clearing Members）兩類。 SwapClear 結算會員是直接結算會員，可結算自營業務及非美國居民客戶業務；期貨佣金商結算會員也是直接結算會員，可結算自營業務、美國居民客戶業務及非美國居民客戶業務。 ForexClear 的參與者包括外匯結算會員（FX Clearing Members）、期貨佣金商結算會員、交易商或客戶。交易商透過與外匯結算會員簽訂結算協議而在 ForexClear 登記其交易。客戶可經由外匯結算會員或期貨佣金商結算會員進行結算。

要申請成為 LCH.Clearnet 會員需達到一系列要求，包括最低淨資本要求及營運要求（例如恰當的系統配置），若要能在 SwapClear 或 ForexClear 結算場外交易則還需符合額外準則。若申請者是銀行，更必須獲所屬國家的銀行監管者的適當許可，同時符合英國銀行監管當局所訂的任何通知或許可／認可／授權要求。

LCH 在歐美及本港等監管體系上均符合巴塞爾協議 III 對 QCCP 的準則。換言之，銀行為其使用 LCH 中央結算服務的場外衍生產品交易活動所需的資本費用可較低。

由於中國未必獲接納為淨額結算司法權區，內地銀行若想成為 LCH 結算會員恐怕暫不可行（見表 2 及相關註釋）。內地銀行為其跨境場外交易使用 LCH 結算服務，通常要選用 SwapClear 或 ForexClear（視情況而定）的結算會員作其結算代理，還要與該結算會員簽訂客戶結算協議及擔保契據 [31] 等符合規管有關結算所的相關法規的法律文件。SwapClear 對客戶結算收取下單費及維持費（以每百萬元名義金額為計算單位）。交易壓縮服務會收取混合費及多邊交易壓縮費，交易壓縮服務讓市場參與者透過淨額結算其交易來減低所持組合中的整體名義金額及項目。ForexClear 也以每百萬元名義金額為單位向客戶收取結算費用，使用交易壓縮服務亦須支付客戶壓縮費。

---

31 若相關司法權區並無豁免客戶結算規則，結算所會要求提供擔保契據作為保護機制，好使萬一結算會員失責，結算所也有權處理客戶資產，保障客戶權益。

3.2　芝商所集團場外結算（CME Group OTC Clearing）[32]

在 IRS 市場方面，芝商所集團場外結算（CME OTC Clear）的產品類型與 LCH 的 SwapClear 類近（有定息／浮息掉期、隔夜指數掉期、基準掉期、FRA 等），涉及貨幣共 21 種。在外匯衍生產品市場方面，服務包括人民幣在內 12 種貨幣的 NDF 以及現金結算遠期合約。2017 年上半年，IRS 及外匯衍生產品的結算總量分別為 16.38 萬億美元及 2,100 萬美元。

為客戶結算場外衍生工具的場外結算會員必須向美國商品期貨交易委員會（CFTC）註冊成為期貨佣金商。在非美國司法權區註冊成立的場外結算會員必須受結算所接納的法律及清盤制度所規限。

芝商所集團的結算分部 CME Clearing 及 CME Clearing Europe 分別在美國和歐洲有結算所符合作為 QCCP 的準則，服務涵蓋場外衍生工具。此外，CME Clearing 的營運公司獲歐洲證券及市場管理局（ESMA）認可為 QCCP。因此，歐洲客戶可視 CME Clearing 為 QCCP。

一如在 LCH 進行結算，內地銀行若透過 CME OTC Clear 進行結算須在 CME OTC Clear 選用一家結算公司，並填交期貨戶口協議（連同用於採用結算服務的場外衍生工具的附錄）。結算費及維持費以每百萬元名義金額為收費單位。多邊交易壓縮服務也須收費。

3.3　歐洲期貨交易所結算公司（Eurex Clearing）[33]

Eurex Clearing 的場外結算服務（EurexOTC Clear）涵蓋 IRS、基準掉期、隔夜指數掉期、零息通脹及 FRA。2016 年 12 月，Eurex Clearing 宣佈 EurexOTC Clear 打算推出歐元兌美元及英鎊兌美元的場外外匯掉期、場外現匯及場外外匯遠期合約結算服務。2017 年上半年，EurexOTC Clear 結算的 IRS 總名義金額為 7,396.25 億歐元（約 8,450 億美元）。

Eurex Clearing 的場外結算會員分三種：全面結算會員（General Clearing Member）可結算自有業務及所有客戶的業務；直接結算會員（Direct Clearing Member）可結

32　相關數據及資料來自芝商所集團網站（http://www.cmegroup.com/clearing.html）。
33　相關數據及資料來自 Eurex Clearing 網站（http://www.eurexclearing.com/clearing-en/）。

算自有業務；基本結算會員（Basic Clearing Member）則結合了直接結算會員與客戶結算傳統服務關係的元素。基本結算會員與 CCP 是主事人的關係，但進行客戶結算則須經結算代理。 全面結算會員可擔任結算代理。場外交易結算的客戶須向 Eurex Clearing 披露並記錄為註冊客戶，並需就此而與 Eurex Clearing 及他們的結算會員訂立三方協議。

Eurex Clearing 獲美國 CFTC 按商品交易法有條件註冊為衍生產品結算機構，待 Eurex Clearing 符合 CFTC 的「直通式交易處理運作」規定後即可獲正式註冊。在此情況下，Eurex Clearing 可為美國結算會員的 IRS 自營倉盤提供結算服務，但尚未可為期貨佣金商客戶的倉盤進行結算 [34]。

Eurex Clearing 適用的收費包括下單費、維持費、其他行政費用以及買賣淨額結算或多邊交易壓縮等額外服務費。

3.4　香港交易所場外結算公司 [35]

香港交易所場外結算公司於 2013 年為呼應 G20 於 2009 年提出的改革計劃而成立，以 CCP 的角色為場外衍生工具提供結算服務。公司獲香港證監會認可為「認可結算所」，後於 2013 年 11 月 25 日開業，並於 2016 年 8 月獲香港證監會指定為替香港場外衍生工具監管制度所訂明的場外衍生工具交易進行強制性結算的 CCP，其後於 2017 年 3 月推出客戶結算服務。

香港交易所場外結算公司為 IRS、基準掉期、交叉貨幣掉期、不交收 IRS 及 NDF（所涵蓋產品列表見附錄一）同時提供自營結算及客戶結算服務。自推出起至 2017 年，名義結算量的複合年增長率達 343%，若從 2014 年首個全年運作年度起計更高達 484%（見圖 7）。

34　資料來源：CFTC 新聞稿第 7316-16 號（2016 年 2 月 1 日）。
35　資料來源：香港交易所。

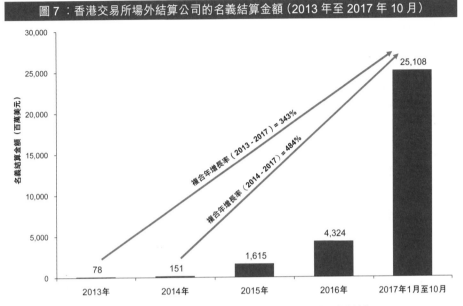

圖 7：香港交易所場外結算公司的名義結算金額（2013 年至 2017 年 10 月）

註：複合年增長率按 2017 年 1 月至 10 月數據按比例推算 2017 年全年十二個月備考數字計算。

資料來源：香港交易所。

　　香港交易所場外結算公司接納獲香港金管局簽發牌照的「認可機構」或獲香港證監會簽發牌照的公司（持牌法團）為結算會員。場外交易結算及交收系統接納來自兩個認可交易登記系統（MarkitWire 及 DSMatch）的場外衍生工具交易。

　　香港交易所場外結算公司於 2015 年 4 月獲歐洲的 ESMA 認可為「第三方國家CCP」，可進行場外衍生工具結算；2015 年 9 月獲澳洲證券與投資委員會（Australia Securities and Investments Commission）認可為澳洲強制性結算制度規定的 CCP 設施；2015 年 12 月獲得美國 CFTC 豁免，允許其不用註冊為美國「衍生產品結算機構」（Derivatives Clearing Organisation），可為美國人士的自營交易提供結算服務。香港交易所場外結算公司具備此等國際認可的場外 CCP 地位，足以為內地銀行與外國對手方進行的美元、歐元及人民幣場外交易合約提供中央結算服務。只要獲得監管機構批准，往後的產品涵蓋範圍更可擴闊至可交收外匯遠期貨幣合約及掉期以及外匯期權。交易壓縮服務亦預定於 2017 年末推出，迎合客戶所需。

除就結算會員使用的服務收取的費用外，就客戶結算服務的收費包括 IRS 的登記費及維持費以及 NDF 的登記費。

3.5 內地銀行的選擇

香港交易所場外結算公司是香港唯一可提供場外衍生產品結算服務的認可結算所，亦是歐美客戶的 QCCP。其營運地點為亞洲金融中心 —— 香港，位處亞洲時區。主要中資銀行都在這裏設有分行及營運多年。此外，香港是全球場外外匯及利率工具的主要交易市場 [36]，更是極其重要的離岸人民幣中心 [37]。

在香港，香港金管局以環球銀行金融電信協會（SWIFT）為基準的**人民幣即時支付結算系統（人民幣 RTGS）**利便全球市場參與者處理與中國內地及離岸市場的人民幣交易。人民幣 RTGS 系統由中國銀行（香港）有限公司擔任清算行。清算行於人行設有交收賬戶，並為中國國家現代化支付系統（CNAPS）的成員。香港的人民幣 RTGS 系統與 CNAPS 直接連繫，從技術層面而言可視為中國內地 CNAPS 的延伸，但受香港法例監管。香港的人民幣、港元、美元及歐元 RTGS 系統互為相連，讓銀行可進行同步交收，提高交收效率，並消除因交易時差及不同時區所引起的交收風險。SWIFT 的統計數字顯示，相對於中國內地及環球離岸市場，香港銀行經手的人民幣結算金額佔離岸人民幣付款總額約七成 [38]。

有鑒於人民幣國際化不斷向前推進，因應人民幣風險管理的需要，相信離岸人民幣衍生工具交易將會有高增長潛力 [39]。在此方面，內地機構的參與預期會增加，所帶來的人民幣流動性正正配合全球機構提高人民幣資產的需求。透過可在香港使用人民幣 RTGS 付款系統，香港交易所場外結算公司除為美元及其他主要貨幣的場外交易提供服務外，亦能夠支援內地與全球金融機構的離岸人民幣場外交

---

36 在場外外匯工具及場外單一貨幣利率衍生工具方面，香港的交投量分別位列全球第四及第三。資料來源：BIS 有關場外衍生產品的三年期調查統計（2016 年 4 月），載於 BIS 網站；每日平均數按「淨 - 毛」（net-gross）基準計算。

37 截至 2017 年 7 月底，香港的人民幣總存款為人民幣 5,347.3 億元（資料來源：香港金管局網站），新加坡截至 2017 年 6 月底為人民幣 1,380 億元（初步數據，資料來源：新加坡金融管理局網站）、英國截至 2017 年 3 月底為 90 億英鎊（約人民幣 770 億元）（資料來源：英國中央銀行網站）、台灣截至 2017 年 4 月底為人民幣 3,078 億元（資料來源：中國銀行（香港）離岸人民幣快報 2017 年第六期）。

38 引述自香港金管局 2016 年 1 月的報告 *Hong Kong — The Global Offshore Renminbi Business Hub*。

39 亦可見本書第 14 章〈香港交易所邁向成為離岸人民幣產品交易及風險管理中心〉。

易 [40]。相較之下，海外結算所使用 CLS 等海外平台提供的外匯結算服務，則未能提供人民幣結算服務 [41]。

作為亞洲區內服務內地市場參與者的場外結算所，香港交易所場外結算公司在戰略位置上佔有優勢，股東包括五家內地金融機構，分別為中國農業銀行有限公司、中國銀行（香港）有限公司、交通銀行股份有限公司香港分行、建銀國際證券有限公司及中國工商銀行（亞洲）有限公司。

內地註冊成立的銀行的香港分行如屬前述的認可機構或持牌法團，可在符合會員資格要求後成為香港交易所場外結算公司的會員。由於內地銀行在港有廣泛的業務，成為香港交易所場外結算公司的會員，會較直接申請成為歐美結算所會員來得具成本效益。內地銀行若與外國對手方進行場外衍生工具交易，因其並非歐洲 LCH 或美國芝商所集團的主要結算所的會員，便需經結算經紀進行客戶結算。使用此方式需支付較高昂的交易費及佣金予結算經紀，彌補其淨額結算及資金的成本，花費較高，亦要承受結算經紀違約的風險。此外，內地銀行亦可能擔心經結算代理進行客戶結算會有運作效率問題及一定障礙，包括每當有需要時再委聘結算代理會需時較長、須依靠結算代理的系統基建以及須倚賴代理能獲 CCP 批准接納替其有關交易進行中央結算等。

再者，香港交易所場外結算公司已在香港監管框架下制訂特別方案，可接納內地註冊成立的持牌銀行為其結算會員。在中國是否屬淨額結算司法權區尚未能確定之時，這樣的處理方案及方案下的結算服務，在歐美結算所尚未能提供。方案的詳情見第 4 節。

40　於 2017 年 9 月 25 日，作為付款系統營運者的香港銀行同業結算有限公司合共有 141 名本地人民幣結算會員及 68 名海外人民幣結算會員（資料來源：香港金管局網站）。

41　現時 CLS 透過在各貨幣央行開設賬戶，為 18 種主權貨幣提供外匯結算服務，幣種與央行不包括人民幣及人行。（資料來源：CLS 網站。）

4 香港交易所場外結算公司為內地銀行提供的解決方案

4.1 中國結算會員

在中國註冊成立並設有香港分行的銀行，只要是受香港金管局規管的認可機構，便可成為香港交易所場外結算公司的直接結算會員（「中國結算會員」）。

香港交易所場外結算公司已就其在香港及中國法律下執行結算規則（包括有關平倉及抵銷的條文）的效力取得法律意見，可按中國結算會員所結算的整個投資組合的淨額風險收取按金。此外，香港交易所場外結算公司對於中國結算會員結算合約所支付的款項，以及監管機構在香港法律（即規管結算規則及場外結算公司所持有中國結算會員所支付的抵押品的香港法律）下根據結算規則的違責條文對中國結算會員所採取的行動均享有終局性的保障。[42]

現時有四家中國註冊成立的銀行（中國農業銀行股份有限公司、交通銀行股份有限公司、中國民生銀行股份有限公司及上海浦東發展銀行股份有限公司）已透過各自的香港分行成為香港交易所場外結算公司的直接結算會員。直接結算會員亦包括另外三家中國註冊成立銀行的香港附屬公司。（香港交易所場外結算公司的結算會員見附錄二。）

4.2 與國際對手方作跨境結算的模式

成為直接結算會員後，內地銀行毋須委任結算經紀，便可直接在香港交易所場外結算公司進行結算，避免了結算經紀違責的風險，也省卻佣金成本和減輕交易費用。直接結算可由內地銀行的香港附屬公司（如其香港附屬公司已成為香港交易所場外結算公司的結算會員）或香港分行（如其本身已透過其香港分行成為香港交易所場外結算公司的結算會員）進行，具體運作分別見下圖 8 及 9。

---

42 資料來源：香港交易所。

圖 8：透過內地銀行香港子公司清算的清算模式

(a) 交易登記過程

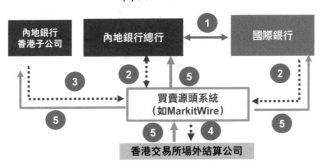

1. 內地銀行總行與國際銀行進行雙邊交易。
2. 內地銀行總行（作為客戶）和國際銀行通過買賣源頭系統提交清算要求。
3. 內地銀行香港子公司（作為代理行）在接受要求之前檢查總行信用額度。
4. 當交易被雙邊接受，買賣源頭系統將交易記錄發送到香港交易所場外結算公司進行產品、保證金和信用檢查。
5. 當交易成功註冊後，香港交易所場外結算公司將通過買賣源頭系統向內地銀行香港子公司、內地銀行總行和國際銀行通報交易清算狀況，先前的雙邊交易將更替為如下交易：
   - 香港交易所場外結算公司對國際銀行
   - 香港交易所場外結算公司對內地銀行香港子公司
   - 內地銀行香港子公司對內地銀行總行

(b) 保證金結算和抵押品管理過程

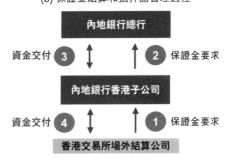

1. 香港交易所場外結算公司向內地銀行香港子公司發出初始保證金和價格變動保證金要求。
2. 內地銀行香港子公司通知內地銀行總行有關香港交易所場外結算公司要求的相關初始和價格變動保證金。
3. 內地銀行總行將所要求的初始保證金和價格變動保證金交付香港子公司。
4. 內地銀行香港子公司將內地銀行總行的初始保證金和價格變動保證金交付給香港交易所場外結算公司。

資料來源：香港交易所場外結算公司。

圖 9：透過香港分行清算的清算模式

(a) 交易登記過程

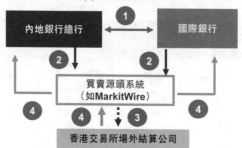

1. 內地銀行總行與國際銀行進行雙邊交易。
2. 內地銀行總行和國際銀行通過買賣源頭系統提交清算要求。
3. 當交易被雙邊接受，買賣源頭系統將交易記錄發送到香港交易所場外結算公司進行產品、保證金和信用檢查。
4. 當交易成功註冊後，香港交易所場外結算公司將通過買賣源頭系統向內地銀行香港分行、內地銀行總行和國際銀行通報交易結算狀況，先前的雙邊交易將更替為如下交易：
   • 香港交易所場外結算公司對國際銀行
   • 香港交易所場外結算公司對內地銀行總行（透過內地銀行會籍）

(b) 保證金結算和抵押品管理過程

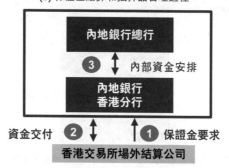

1. 香港交易所場外結算公司向內地銀行香港分行發出初始保證金和價格變動保證金要求，而總行和香港分行的所有交易將會計算在內。
2. 內地銀行香港分行將初始保證金和價格變動保證金交付給香港交易所場外結算公司。
3. 內地銀行總行與香港分行之間進行內部資金安排及拆賬。

資料來源：香港交易所場外結算公司。

　　以上結算模式相比透過結算經紀代理進行結算的模式（見圖 10），提供了更便利的解決方案，成本更低。

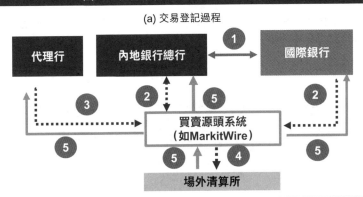

圖10：通過清算代理行清算的清算模式

(a) 交易登記過程

1. 內地銀行總行與國際銀行進行雙邊交易。
2. 內地銀行總行（作為客戶）和國際銀行通過買賣源頭系統提交清算要求。
3. 代理行接受要求之前檢查總行信用額度。
4. 當交易被雙邊接受，買賣源頭系統將交易記錄發送到場外清算所，進行產品、保證金和信用檢查。
5. 當交易成功註冊後，場外清算所將通過買賣源頭系統向代理行、內地銀行總行和國際銀行通報交易清算狀況，先前的雙邊交易將更替為如下交易：
   * 場外清算所對國際銀行
   * 場外清算所對代理行
   * 代理行對內地銀行總行

(b) 保證金結算和抵押品管理過程

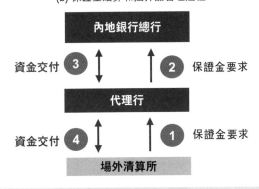

1. 場外清算所向代理行發出初始保證金和價格變動保證金要求。
2. 代理行向內地銀行總行發出場外清算所要求的相關初始和價格變動保證金的對賬單。
3. 內地銀行總行將所要求的初始保證金和價格變動保證金交付代理行。
4. 代理行將內地銀行總行的初始保證金和價格變動保證金交付給場外清算所。

資料來源：香港交易所場外結算公司。

4.3 相對其他 CCP 的優勢

內地銀行選擇不在歐美的結算所經結算經紀進行間接結算，而是通過香港交易所場外結算公司進行直接結算，將可享有同一時區之便、風險較低和成本也較低等等相對優勢（見表3）。

表 3：香港交易所場外結算公司與其他主要場外結算所的對照				
項目	香港交易所場外結算公司	CME OTC Clear	LCH （SwapClear / ForexClear）	EurexOTC Clear
服務時間	香港時間 08:30－19:00 07:30－23:00（接入網站）	每日 23 小時 45 分鐘	SwapClear 07:30－24:00（GMT） 14:30－07:00（香港時間） ForexClear 星期一至五24小時 星期日20:00（GMT）至 星期六01:00（GMT）	8:00－22:00 CET
會員分佈	歐洲、美國、中國、香港及其他亞洲銀行	主要為國際銀行		
產品範圍	多種IRS及NDF	多種IRS、NDF及現金結算遠期合約	多種IRS及NDF	多種IRS
貨幣範圍	歐元、美元、人民幣及其他亞洲貨幣	主要國際貨幣		
內地銀行的會員資格	• 香港附屬公司可成為直接會員 • 內地註冊成立的銀行可透過香港分行成為直接會員	銀行（而非分行）若符合相關司法管轄區的額外監管規定，便可申請成為直接會員		
提供給內地銀行的結算模式	透過成為結算會員的香港附屬公司進行直接結算，或本身透過其香港分行成為會員者可透過其香港分行進行結算	透過結算經紀（另一國際銀行）進行間接結算		
風險	透過同集團內的實體機構進行直接結算，風險較低	會面對結算經紀違責風險		
成本	交易成本較低，零佣金	委任結算經紀的交易成本及佣金較高，以彌補淨額結算及資金的成本		

註：GMT —— 格林威治標準時間；CET —— 歐洲中部時間。
資料來源：香港交易所場外結算公司及相關結算所網站。

香港既是匯聚內地和國際用戶的金融樞紐、又是離岸人民幣中心，以此為營運據點的香港交易所場外結算公司遂可為內地銀行和其國際結算對手提供平台，替雙方以國際貨幣和人民幣進行的場外衍生產品交易進行結算（見圖11）。

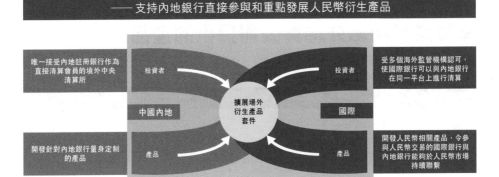

圖 11：場外結算公司的獨特定位
—— 支持內地銀行直接參與和重點發展人民幣衍生產品

5 總結

　　內地金融機構所持美元及其他外幣的境外資產越來越多，相關外幣的利率風險和外匯風險日增。為對沖風險，內地金融機構與境外對手方進行涉及以外幣計值的風險管理工具（如 IRS 及外匯衍生產品）的跨境場外交易的需求亦水漲船高。

　　2008 年全球金融危機過後，世界各地都在收緊對場外衍生產品風險管理的監管，金融機構或是被強制要求為標準場外衍生產品進行中央結算，或是必須就其作雙邊結算的場外衍生產品遵守更高的資本和保證金要求。如屬後者，有關金融機構或會自願選擇中央結算以求減低交易成本。

　　有別於歐美結算所，香港交易所場外結算公司接受中國註冊成立的內地銀行透過其香港分行成為直接結算會員，令內地銀行可經香港分行進行直接結算。這無疑為內地銀行提供了更便利、成本效益更高的場外衍生產品中央結算方案。

　　再者，在人民幣日趨國際化的進程中，為管理人民幣風險而進行的離岸人民幣衍生產品交易會具高增長潛力。與海外同業相比，香港交易所場外結算公司更有能力在結算美元及其他主要貨幣的場外交易之餘，同時為內地以至全球金融機構以離岸人民幣進行的場外交易提供結算服務。

附錄一

香港交易所場外結算公司提供的產品

產品	貨幣	最長剩餘年期
單一貨幣利率掉期	人民幣（離岸）	10年
	美元	
	歐元	
	港元	
單一貨幣基準掉期	美元	10年
	歐元	
	港元	
不交收利率掉期	人民幣	5年
	印度盧比	10年
	馬來西亞令吉	
	韓圜	
	泰銖	
	台幣	
交叉貨幣掉期	美元兌人民幣（離岸）	10年
不交收遠期外滙合約	美元 / 人民幣	2年
	美元 / 印度盧比	
	美元 / 韓圜	
	美元 / 新台幣	

附錄二

香港交易所場外結算公司結算會員

（2017 年 9 月）

香港

1. 東亞銀行有限公司
2. 恒生銀行有限公司
3. 香港上海滙豐銀行有限公司

中國內地

4. 中國農業銀行股份有限公司
5. 中國銀行（香港）有限公司
6. 交通銀行股份有限公司
7. 建銀國際證券有限公司
8. 中國民生銀行股份有限公司
9. 中國工商銀行（亞洲）有限公司
10. 上海浦東發展銀行股份有限公司

歐洲

11. 法國巴黎銀行
12. 德意志銀行
13. 渣打銀行

美國

14. 花旗銀行
15. 摩根大通銀行

亞太區

16. 星展銀行有限公司
17. 澳新銀行集團有限公司

英文縮略詞

BIS   國際結算銀行（Bank for International Settlements）

CCP   中央結算對手（Central counterparty）

CDS   信貸違約掉期（Credit default swap）

CFETS   中國外匯交易中心（China Foreign Exchange Trade System）

CNAPS   中國國家現代化支付系統
（China National Advanced Payment System）

CRM   信用風險緩釋（Credit risk mitigation）

CRMW   信用風險緩釋憑證（Credit risk mitigation warrant）

EMIR   歐洲市場基礎設施監管規則
（European Market Infrastructure Regulation）

ESMA   歐洲證券及市場管理局（European Securities and Markets Authority）

FRA   遠期利率協議（Forward rate agreement）

IRS   利率互換或利率掉期（Interest rate swap）

ISDA   國際掉期與衍生工具協會
（International Swaps and Derivatives Association）

NDF   不交收遠期外匯合約（Non-deliverable currency forwards）

QCCP   合資格中央結算對手（Qualified central counterparty）

QFII   合格境外機構投資者（Qualified Foreign Institutional Investor）

RQFII   人民幣合格境外機構投資者
（Renminbi Qualified Foreign Institutional Investor）

RTGS   即時支付結算（Real time gross settlement）

SHIBOR   上海同業拆借利率（Shanghai Interbank Offered Rate）

風險與免責聲明

買賣證券的風險

證券買賣涉及風險。證券價格有時可能會非常波動。證券價格可升可跌，甚至變成毫無價值。買賣證券未必一定能夠賺取利潤，反而可能會招致損失。

買賣期貨及期權的風險

期貨及期權涉及高風險，買賣期貨及期權所招致的損失有可能超過開倉時繳付的按金，令投資者或須在短時間內繳付額外按金。若未能繳付，投資者的持倉或須平倉，任何虧損概要自行承擔。因此，投資者務須清楚明白買賣期貨及期權的風險，並衡量是否適合自己。投資者進行交易前，宜根據本身財務狀況及投資目標，向經紀或財務顧問查詢是否適合買賣期貨及期權合約。

免責聲明

本書所載資料及分析只屬資訊性質，概不構成要約、招攬、邀請或推薦買賣任何證券、期貨合約或其他產品，亦不構成提供任何形式的建議或服務。書中表達的意見不一定代表香港交易及結算所有限公司（香港交易所）的立場。書中內容概不構成亦不得被視為投資或專業建議。儘管本書所載資料均取自認為是可靠的來源或按當中內容編備而成，香港交易所及其附屬公司、董事及僱員概不就有關資料（就任何特定目的而言）的準確性、適時性或完整性作任何保證。香港交易所及其附屬公司、董事及僱員對使用或依賴本書所載的任何資料而引致任何損失或損害概不負責。

後 記

發揮交易所在人民幣
離岸市場發展中的獨特作用

　　在迅速變化的全球金融體系中，交易所作為具有獨特功能的金融基礎設施，既是重要的交易平台，也是市場的一線組織者和創新的推動者。同時，國際主要的交易所也越來越突破傳統的交易所業務模式，逐步成為覆蓋更多產品領域、橫跨多個市場的多元化創新型的交易所。香港交易所作為全球金融體系中活躍的國際型交易所，近年來就明顯體現出這種趨勢，特別是近年來香港交易所基於新的戰略定位，推出了一系列多領域的產品創新，使得香港交易所的產品線更為豐富和多元。

　　從具體操作層面看，香港交易所每一種新產品的推出，從產品設計、內部和外部的系統調整、測試和溝通、產品推出後的市場交流等不同環節，都需要不同的機構和專業人士付出大量的努力。在這些新產品的推出過程中，香港交易所首席中國經濟學家辦公室以不同的方式、在不同的環節參與了一些工作，為了讓境內、外市場更好地了解香港的人民幣離岸產品創新，我們專門組織多個業務部門，共同對香港交易所近年來在互聯互通、人民幣產品方面的新進展進行了一個全面梳理，由此形成了《人民幣國際化的新進展 —— 香港交易所的離岸金融產品創新》一書。

　　根據我們的理解，要讓市場各方較為系統全面地了解一種新的產品，既要從宏觀和行業層面了解這一產品推出的大背景、國際市場相關行業和產品發展的趨勢、以及香港交易所推出這些產品的設計目標，同時也要從微觀和技術層面了解這種產品的風險收益特性、以及在實際投資決策中的可能應用。這就需要專業研究部門與不同業務部門、法律和監管合規部門等的共同合作。本書中的各篇文章，可以說是香港交易所不同業務部門及一些子公司共同合作的產物，首席中國經濟學家辦公室作為研究的組織者，與包括證券產品發展團隊、定息及貨幣產品發展團隊、大宗商品團隊、香港場外結算公司以及與中華證券交易服務有限公司等在內的多個部門和團隊保持了良好的合作，我們共同成為這些報告的起草者，在報告的修訂和發佈過程中，又針對了許多部門的意見。

在與市場各方的交流中我們了解到，隨着香港交易所的產品線日趨豐富，市場越來越不滿足於對單一產品的分散式的了解，而是更希望有更全面的多產品、一攬子的系統介紹。我們也希望，通過本書對於多個產品的系統跟蹤研究，即使金融機構、以及香港交易所的相關業務團隊在介紹某一種產品時，也可以通過參閱本書，由點及面地了解到香港交易所近年來的主要產品創新。

在目前的工作中，香港交易所首席中國經濟學家辦公室的職能定位主要包括幾個方面的內容，即：內部的研究智庫（think tank）、外部的專業意見引領者，和一系列香港交易所的戰略項目和產品創新的專業支持者。這幾個職能定位都程度不同地與產品創新密切相連，也使我們能夠更深切地了解到不同產品創新的來龍去脈。

在此，我要特別感謝香港交易所集團行政總裁李小加先生對本書一直以來的鼓勵和支持，以及香港交易所監管合規團隊、法務團隊、企業傳訊團隊對本書的大力協助，正是他們的通力合作和建議，本書才得以成功出版。

香港交易所成長與發展的歷程，從特定意義上也可以說集中體現了香港金融業發展的歷程，也從特定的角度反映了國際和中國內地的經濟金融風雲變幻。在不同的發展階段，香港交易所往往有當時非常活躍的金融產品，或者說，這些代表性的產品創新，也同時成為了香港交易所和香港金融市場發展特定階段的標誌。從這個意義上來說，本書嘗試梳理的香港交易所的產品創新，也正是體現了當前香港金融市場上正在進行的金融創新進程，同時也在一定程度上反映出全球金融體系的變化新趨勢。

由於內地發展一日千里、國際金融市場環境不斷變化，書中倘有缺點錯漏亦在所難免，敬請廣大讀者批評指正，以便我們在修訂時改進。

巴曙松 教授

香港交易及結算所有限公司 首席中國經濟學家

中國銀行業協會 首席經濟學家

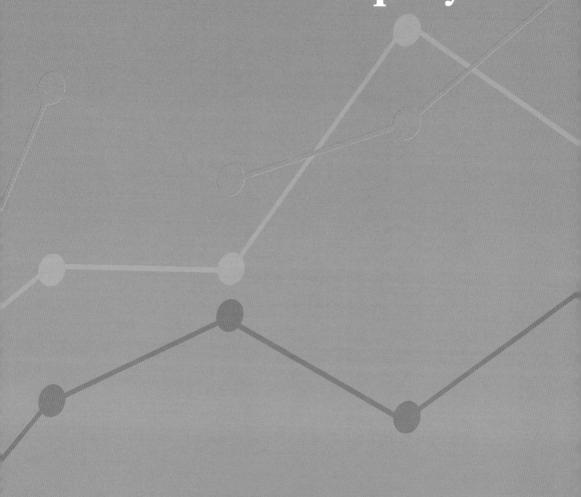

Part 1
Equity

Chapter 1

CES China 120 Futures

A useful offshore hedging tool
for cross-border investment

14 February 2017

Summary

Cross-border stock investment in the Mainland and Hong Kong markets achieved a new page upon the launch of the Shanghai-Hong Kong Stock Connect (Shanghai Connect) in November 2014. A breakthrough was made when the aggregate quota for the pilot programme was abolished immediately on the announcement of the Shenzhen-Hong Kong Stock Connect (Shenzhen Connect) on 16 August 2016 and further with the subsequent launch of Shenzhen Connect on 5 December 2016. The Shanghai Connect and Shenzhen Connect are collectively referred to as the "Stock Connect" scheme. By then, a Mutual Market platform across Shanghai, Shenzhen and Hong Kong is basically formed. This undoubtedly will facilitate and promote cross-border stock investment, along with which there will be increasing demand for risk management of cross-border stock portfolios.

However, relevant risk management tools such as index futures and options that make reference to Mainland A shares or cross-border stocks in the Mutual Market are scarce in global markets. Among China-related indices with derivatives (futures/options) traded on overseas exchanges outside Mainland China, the CES China 120 Index (CES 120) is the only index based on both Mainland A shares and Chinese stocks listed in Hong Kong. Other traded indices are either based on Mainland A shares alone (the FTSE China A50 Index) or overseas-listed (mostly Hong Kong) Chinese stocks (including FTSE China 50 Index and MSCI China Free Index). In comparison, the CES 120 has a distribution of constituents across the three exchanges (Hong Kong, Shanghai and Shenzhen) in the Mutual Market and a good coverage by exchange, stock type and industry sector of eligible stocks under Stock Connect. In addition, the index is highly correlated with the A-share indices. It had outperformed other traded indices in terms of return and volatility in 2016 and had a dividend yield comparable to the Mainland blue-chip index, SSE 50 Index.

Given the characteristics of the CES 120 Index, CES 120 Futures traded on the HKEX derivatives market can be an effective risk management tool for investors to hedge their stock investment or to gain investment exposure in Mainland A shares and the Mutual Market. Furthermore, compared to FTSE China A50 Futures traded on the Singapore Exchange, CES 120 Futures has a lower exchange fee per contract notional value, a smaller tick size relative to index level and a higher position limit. In consideration of all these, CES 120 Futures can be considered to be a convenient and cost efficient offshore market tool for A shares and Mutual Market investment. Referencing the success story of EURO STOXX 50 Index and its derivatives and structured products in serving the mutual European market, CES 120 Index and its derivatives and structured products can serve the same need for the Mainland-Hong Kong Mutual Market.

1 The need for cross-border investment risk management

1.1 Increasing cross-border stock investment facilitated by Stock Connect

A brand new official channel for overseas investors to invest in the Mainland stock market and for Mainland investors to invest in the Hong Kong stock market was opened in November 2014 when the **Shanghai-Hong Kong Stock Connect ("Shanghai Connect")** was launched as the first initiative under the Mutual Market Access pilot programme (the **"Pilot Programme"**) between Mainland China and Hong Kong. Prior to this, foreign participation channels in the Mainland stock market had been limited mainly to the Qualified Foreign Institutional Investor (QFII) scheme and the Renminbi Qualified Foreign Institutional Investor (RQFII) scheme and foreign retail investors could only participate through investment funds offered by QFIIs and RQFIIs[1]. In the opposite direction, the Qualified Domestic Institutional Investor (QDII) scheme and the Renminbi Qualified Domestic Institutional Investor (RQDII) scheme had been the only national official channels of Mainland participation in overseas stock markets[2].

The extended initiative of the Pilot Programme — the **Shenzhen-Hong Kong Stock Connect ("Shenzhen Connect")**, with an expanded scope of eligible securities was launched on 5 December 2016. The Shanghai Connect and Shenzhen Connect are collectively referred to as the "Stock Connect" scheme. The key breakthrough in the Stock Connect scheme is the abolition of the aggregate quota immediately upon the joint announcement made on 16 August 2016 by the China Securities Regulatory Commission (CSRC) and the Hong Kong Securities and Futures Commission (SFC) on the establishment of Shenzhen Connect. Effectively with the launch of Shenzhen Connect, the **"Mutual Market"** model across Shanghai, Shenzhen and Hong Kong is basically formed,

---

1   The B-share market (traded in foreign currency and separated from the A-share market) on the Shanghai and Shenzhen Stock Exchanges, which was launched in 1992 as the Mainland's first attempt of stock market opening to foreign investors, has become inactive in the new wave of market opening.

2   There are special pilot schemes launched by local governments, e.g. the Qualified Domestic Limited Partnership (QDLP) programme in Shanghai and the Qualified Domestic Investment Enterprise (QDIE) in Shenzhen Qianhai. However, these are limited to mainly privately offered funds or investment vehicles not widely accessible by general investors as these target mainly the institutional investors and high-net-worth individuals, compared to QDII products which target general investors.

under which the suites of financial products offered in the three markets, where permitted by regulation, could be traded by both Mainland and global investors across the border. This is effected through Mainland investors' "Southbound trading" of eligible products listed in Hong Kong by placing orders via the trading platforms of the Shanghai Stock Exchange (SSE) and the Shenzhen Stock Exchange (SZSE); and through global investors' "Northbound trading" of eligible products listed on the SSE and SZSE by placing orders via the trading platform of the Stock Exchange of Hong Kong (SEHK). **Without an aggregate quota, this Mutual Market will enable investors to conduct asset allocation across the Mainland and Hong Kong markets with a long-term investment perspective.**

Under Shanghai Connect and Shenzhen Connect, Northbound eligible stocks comprise constituents of the SSE 180 Index and SSE 380 Index; constituent stocks of the SZSE Component Index and of the SZSE Small/Mid Cap Innovation Index which have a market capitalisation of RMB 6 billion or above; and all SSE- or SZSE-listed A shares with H shares listed on SEHK[3]. In the opposite direction, Southbound eligible securities comprise the constituents of Hang Seng Composite LargeCap Index (HSLI) and Hang Seng Composite MidCap Index (HSMI); constituents of Hang Seng Composite SmallCap Index (HSSI) with a market capitalisation of HK$5 billion or above; and all H shares with A shares listed in the Mainland market, whether on SSE or SZSE[4]. HSLI and HSMI already cover up to 95% of the total market capitalisation of the Hang Seng Composite Index (HSCI), which in turn covers the top 95% of the total market capitalisation of the Hong Kong market[5]. As of the launch date of Shenzhen Connect (5 December 2016), Southbound eligible securities constituted 87% of the total market capitalisation of listed stocks on the SEHK Main Board while Shanghai Connect Northbound eligible securities constituted 81% of the total market capitalisation of listed A shares on the SSE and Shenzhen Connect Northbound eligible securities constituted 71% of the total market capitalisation of listed A shares on the SZSE markets (Main Board, SME Board and ChiNext)[6].

Albeit the scope of eligible securities is currently confined to a specified set under Shanghai Connect and Shenzhen Connect, the Pilot Programme potentially opens up a **Mainland-Hong Kong mutual stock market** of a combined equity market value of US$10,986 billion (as of end-November 2016) and an average daily equity turnover of

---

3   Except those which are not traded in RMB and those under risk alert treatment by the SSE or SZSE (including shares of "ST companies" and "*ST companies" and shares subject to the delisting process).

4   Except those which are not traded in HKD and H shares which have the corresponding A shares put under risk alert.

5   Source: Hang Seng Indexes Co Ltd website. The Hong Kong market universe of the HSCI refers to all stocks and real estate investment trusts ("REITs") that have their primary listings on the SEHK, excluding securities that are secondary listings, foreign companies, preference shares, debt securities, mutual funds and other derivatives.

6   Source: Based on data obtained from Thomson Reuters and the respective exchanges' websites and eligible stock lists obtained from the exchanges' websites.

about US$85 billion (2016 up to November), **ranking 2nd by market value (following New York Stock Exchange) and 2nd by equity market turnover among world exchanges**[7].

1.2   Risk management for cross-border stock investment

With the Mutual Market platform established, increasing cross-border trading activities are expected and along with that risk management for cross-border stock portfolios will become increasingly important. Risk management instruments commonly used to hedge portfolio investment risks include futures and options on listed stocks and/or related market indices. However, these hedging tools for Mainland investors in trading Hong Kong stocks are not readily available — there are currently no Hong Kong index/stock futures and options offered on any Mainland exchanges. Similarly in Hong Kong, there is also a lack of A-share hedging tools like A-share index futures and options. While derivatives may be included in the Mutual Market model in the future, investors may consider using proxy instruments in their home markets in the meantime.

For global investors trading Mainland A shares from Hong Kong, proxy index futures may be used for hedging purposes. Under Stock Connect where asset allocation can be made across both Hong Kong and Mainland (Shanghai and Shenzhen) markets, an index with constituent stocks listed on the three markets may serve as a good proxy. The CES China 120 Index developed by China Exchanges Services Company Ltd (CESC), a joint venture of HKEX, SSE and SZSE, is such a cross-border index with futures contracts traded on HKEX. In fact, the CES China 120 Index Futures (CES 120 Futures) is the only futures contract among global exchanges which is based on an underlying index that tracks both Mainland A shares and Chinese stocks listed in Hong Kong. The other China-related index futures available on global exchanges are either based on Mainland A shares alone or overseas-listed (mostly Hong Kong) Chinese stocks.

Section 2 below gives an overview of these indices and their derivative products, and Section 3 examines the CES China 120 Index and its futures in comparison with other Mainland-related stock indices.

---

7   World Federation of Exchanges (WFE) statistics, from WFE website, 22 December 2016. Average daily turnover was calculated from the combined shares turnover value for 2016 up to November from WFE statistics using the total number of trading days (225 days) for the Hong Kong market. Ranking was based on November 2016 year-to-month combined trading value.

2 China stock indices and their derivatives

FTSE Russell and MSCI are the two main index providers in the global market. Each of the two institutions compile some 20 China indices, some on domestic A shares alone, some on non-domestic overseas-listed Chinese stocks and some on both domestic and non-domestic listed Chinese stocks. However, only a few indices have futures (and options) products offered by exchanges outside Mainland China.

In the Mainland, the China Securities Index Co., Ltd (CSI), jointly established by the SSE and the SZSE, is the major index provider, producing single-market (SSE or SZSE) indices and cross-market (SSE and SZSE) indices. Of these, only three indices have futures products traded on the China Financial Futures Exchange (CFFEX). The major China stock indices with exchange-traded derivatives identified (referred to as "traded indices" hereinafter) are summarised in Table 1 below.

Table 1. Major China stock indices with exchange-traded derivatives				
Index	Short name	Constituents	Derivative products	Listed exchange^
On Mainland-listed stocks				
FTSE China A50 Index	FTSE A50	50 largest A-share companies listed on SSE and SZSE	FTSE China A50 Index Futures	SGX
CSI 300 Index	CSI 300	300 largest and most liquid A shares listed on SSE and SZSE	CSI 300 Index Futures	CFFEX
CSI 500 Index	CSI 500	500 small to medium sized A shares listed on SSE and SZSE	CSI 500 Index Futures	CFFEX
SSE 50 Index	SSE 50	50 largest and most liquid A shares listed on SSE	SSE 50 Index Futures	CFFEX
On Hong Kong-listed Chinese stocks				
Hang Seng China Enterprises Index	HSCEI	H shares listed on SEHK	H-Shares Index Futures and Options	HKEX
FTSE China 50 Index	FTSE China 50	50 of the largest, most liquid Chinese stocks (H shares, red chips, P chips*) listed and traded on SEHK	E-Mini FTSE China 50 Index Futures	CME
			FTSE China 50 Index Futures	JPX

(continued)

Table 1. Major China stock indices with exchange-traded derivatives				
Index	Short name	Constituents	Derivative products	Listed exchange^
On Hong Kong and overseas-listed Chinese stocks				
MSCI China Free Index	MSCI China Free	Large and mid-cap Chinese companies listed outside Mainland China, including H shares, red chips, P chips* listed on SEHK and foreign listed shares	MSCI China Index Futures and Options	SGX
			MSCI China Free Index Futures	Eurex
On Mainland and Hong Kong-listed Chinese stocks				
CES China 120 Index	CES 120	80 most liquid and largest A shares listed on SSE and SZSE and 40 most liquid and largest Mainland companies (H shares, red chips and P chips*) listed on SEHK	CES China 120 Index Futures	HKEX

Notes:

* H shares are issued by companies incorporated in Mainland China and are listed on HKEX; red chips are shares issued by companies listed in Hong Kong that are incorporated outside Mainland China and that are controlled by Mainland government entities through direct or indirect shareholding and/or representation on the company's board; P chips are shares issued by companies listed in Hong Kong that are incorporated outside Mainland China and that have operations in Mainland China run by private sector individuals in Mainland China.

^ Abbreviations of exchanges: CME - CME Group; JPX - Japan Exchange; SGX - Singapore Exchange

Source: Websites of FTSE Russell, MSCI, CESC, CSI and the respective exchanges.

Five major indices based on Chinese stocks are found to have futures products traded on major global exchanges outside Mainland China (referred to as "overseas traded indices"). Among them, one is based solely on Mainland-listed A shares — FTSE China A50 Index (FTSE A50), with futures (FTSE A50 Futures) traded on Singapore Exchange (SGX); three are based on Hong Kong-listed Chinese stocks — FTSE China 50 Index and Hang Seng China Enterprises Index (HSCEI), or including other foreign listed stocks — MSCI China Free Index; and **only one — the CES China 120 Index (CES 120) — is based on both Mainland-listed A shares and Chinese stocks listed in Hong Kong.** Figure 1 below gives a diagrammatic presentation of the China stock indices with exchange-traded derivatives by constituent type.

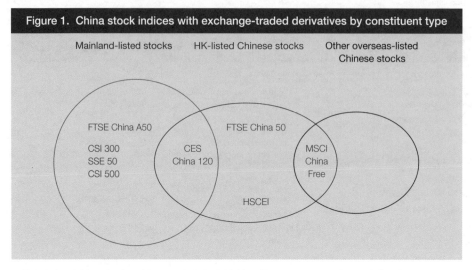

Figure 1. China stock indices with exchange-traded derivatives by constituent type

Note: Indices in blue colour have futures products traded in Hong Kong or overseas exchanges; indices in grey colour have futures products traded on CFFEX in the Mainland.

3 CES China 120 Index and index futures

This section firstly examines the composition of listed stocks across the border by the traded indices (indices with derivatives traded on the Mainland or overseas derivatives exchanges) in terms of market capitalisation (MC) and industry sectors. A comparison is made of the composition across the indices and against the eligible stocks under Stock Connect and, where applicable, with the key market indices in Shanghai — SSE A-Shares Index (SSE A), Shenzhen — SZSE A-Shares Index (SZSE A) and Hong Kong — Hang Seng Index (HSI). Correlations between the indices are also examined and a brief comparison of their performance is made.

This is followed by an overview of the CES China 120 Futures as a risk management tool for cross-border investment.

3.1   CES China 120 Index
— An A-shares and Mutual Market index proxy

In terms of weighting of constituents from the three markets — SEHK, SSE and SZSE — on the Mutual Market platform, CES 120 had almost equal weightings on SEHK and SSE stocks, with relatively less weighting on SZSE stocks as of 30 November 2016 (44%, 43% and 13% respectively). This is relatively more representative of the Mutual Market than the currently tradable indices on Chinese stocks — compared to A-shares indices, CES 120 had an equal weighting on SZSE stocks as FTSE A50 and a similar sharing of SSE and SZSE A shares (77:23) as the Shanghai and Shenzhen cross-market big-cap CSI 300. Other overseas-traded indices on Chinese stocks cover only Chinese stocks listed on SEHK (e.g. FTSE China 50 and HSCEI); or including those listed in the US (e.g. MSCI China Free). The **weighting composition of CES 120 by listing exchange is also the closest to the corresponding composition of the eligible stocks under Stock Connect.** (See Figure 2.)

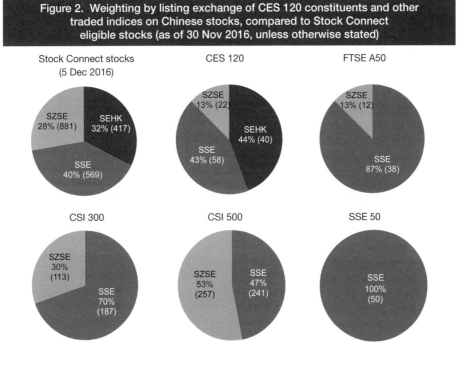

Figure 2. Weighting by listing exchange of CES 120 constituents and other traded indices on Chinese stocks, compared to Stock Connect eligible stocks (as of 30 Nov 2016, unless otherwise stated)

(continued)

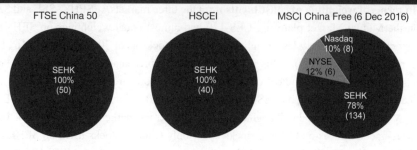

Figure 2. Weighting by listing exchange of CES 120 constituents and other traded indices on Chinese stocks, compared to Stock Connect eligible stocks (as of 30 Nov 2016, unless otherwise stated)

FTSE China 50 — SEHK 100% (50)

HSCEI — SEHK 100% (40)

MSCI China Free (6 Dec 2016) — Nasdaq 10% (8); NYSE 12% (6); SEHK 78% (134)

Note: Excluding constituents where the market capitalisation or weighting is not available. Number of stocks in brackets.

Source: HKEX, SSE and SZSE websites for Stock Connect eligible stock lists, with market capitalisation data from HKEX and Thomson Reuters; CESC website for CES 120 weightings; Bloomberg for MSCI China Free weightings; Thomson Reuters for others.

In addition, the **SEHK-listed constituents of the CES 120 comprise H shares, red chips and non-H share Mainland private enterprises (P chips)**[8], with weightings by stock type comparable to that for Stock Connect eligible stocks (see Figure 3).

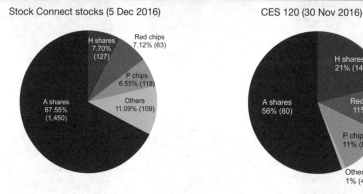

Figure 3. Weighting by stock type of CES 120 constituents compared to Stock Connect eligible stocks

Stock Connect stocks (5 Dec 2016) — H shares 7.70% (127); Red chips 7.12% (63); P chips 6.55% (118); Others 11.09% (109); A shares 67.55% (1,450)

CES 120 (30 Nov 2016) — H shares 21% (14); Red chips 11% (14); P chips 11% (8); Others 1% (4); A shares 56% (80)

Note: Number of stocks in brackets.

Source: HKEX, SSE and SZSE websites for Stock Connect eligible stock lists, with market capitalisation data from HKEX and Thomson Reuters; CESC website for CES 120 weightings.

---

8   See note in Table 1 for definitions of each of these stock types.

Looking more closely at each of these indices in respect of their **coverage of eligible stocks under Stock Connect**, the CES 120 has the largest coverage in terms of market capitalisation as of the launch date of Shenzhen Connect (5 December 2016). By market capitalisation, the CES 120 covers 41.3% of Stock Connect eligible stocks, compared to 22% for FTSE A50, the highest among other overseas-traded indices and 40.6% for CSI 300, the highest among Mainland-traded indices. In terms of number of stocks, CES 120 represents 6.4% of Stock Connect eligible stocks which, albeit not high per se, is the highest among overseas-traded indices, except MSCI China Free which have coverage of SEHK-listed stocks only. For coverage of Stock Connect stocks by listing exchange, other overseas-traded indices cover either Mainland A shares only, or SEHK-listed stocks only and not both. On the contrary, **CES 120 has stocks spread over the three exchanges of SEHK, SSE and SZSE**. (See Figure 4.)

In terms of **index composition by industry sector in comparison with Stock Connect stocks**, CES 120 has a heavier weighting on Financials (47% as of 30 November 2016 vs 25% for Stock Connect stocks) but a similarly balanced distribution of weighting on the other sectors like Stock Connect stocks. The sector distribution of the CES 120 also resembles CSI 300, the Shanghai-Shenzhen cross-market blue-chip index. On the contrary, some other overseas-traded indices have a much higher concentration on a single sector (68% and 71% on Financials for FTSE A50 and HSCEI respectively, 72% on Real Estate for FTSE China 50). In particular, **CES 120 is more balanced on stocks from Information Technology (IT) sector (12%) and Telecommunication Services sector (6%)** compared to FTSE A50 — which has no stocks from Telecommunication Services and low weighting on IT (1%), and compared to HSCEI — which has no stocks from IT and low weighting on Telecommunication stocks (2%). (See Figure 5.)

While SSE A shares and SZSE A shares have rather different industry compositions, **CES 120 has an industry composition closer to the Mainland cross-market index CSI 300** owing to the inclusion of both SSE and SZSE stocks. For the same reason, **it has a broader industry coverage than HSI of the Hong Kong market**. (See Figure 5.)

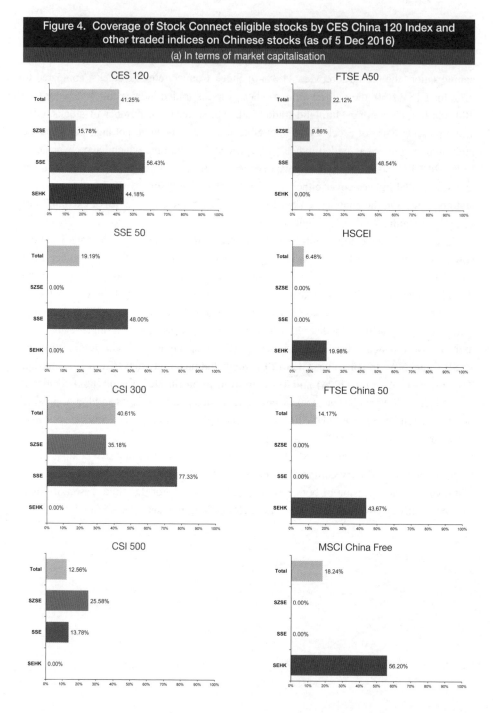

Figure 4. Coverage of Stock Connect eligible stocks by CES China 120 Index and other traded indices on Chinese stocks (as of 5 Dec 2016)

(a) In terms of market capitalisation

(continued)

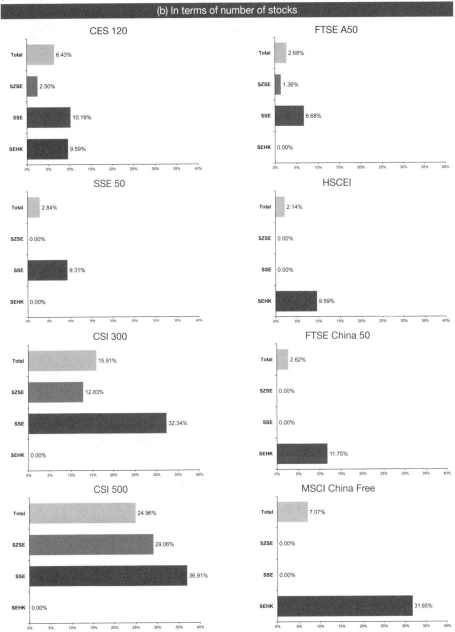

(b) In terms of number of stocks

CES 120
- Total: 6.43%
- SZSE: 2.50%
- SSE: 10.19%
- SEHK: 9.59%

FTSE A50
- Total: 2.68%
- SZSE: 1.36%
- SSE: 6.68%
- SEHK: 0.00%

SSE 50
- Total: 2.84%
- SZSE: 0.00%
- SSE: 9.31%
- SEHK: 0.00%

HSCEI
- Total: 2.14%
- SZSE: 0.00%
- SSE: 0.00%
- SEHK: 9.59%

CSI 300
- Total: 15.91%
- SZSE: 12.83%
- SSE: 32.34%
- SEHK: 0.00%

FTSE China 50
- Total: 2.62%
- SZSE: 0.00%
- SSE: 0.00%
- SEHK: 11.75%

CSI 500
- Total: 24.96%
- SZSE: 29.06%
- SSE: 36.91%
- SEHK: 0.00%

MSCI China Free
- Total: 7.07%
- SZSE: 0.00%
- SSE: 0.00%
- SEHK: 31.65%

Note: Titles are in blue for overseas-traded indices and in grey for Mainland-traded indices.

Source: HKEX, SSE and SZSE websites for Stock Connect eligible stock lists; index constituent stock lists are from CESC website, Thomson Reuters and Bloomberg; market capitalisation data are from HKEX and Thomson Reuters.

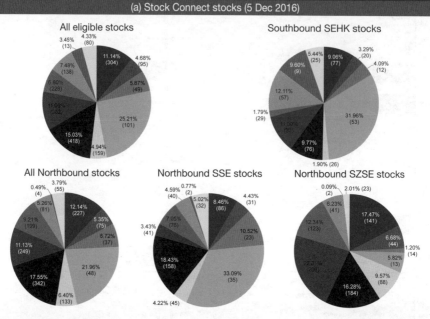

Figure 5. Weighting by industry sector of CES 120 constituents and other traded indices on Chinese stocks, compared to Stock Connect eligible stocks and Mainland and Hong Kong key market indices

(a) Stock Connect stocks (5 Dec 2016)

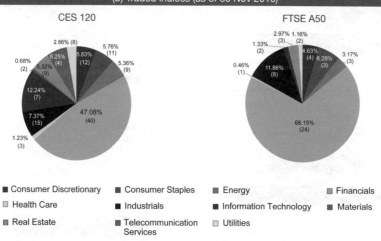

(b) Traded indices (as of 30 Nov 2016)

Legend:
- ■ Consumer Discretionary
- ■ Consumer Staples
- ■ Energy
- ■ Financials
- □ Health Care
- ■ Industrials
- ■ Information Technology
- ■ Materials
- ■ Real Estate
- ■ Telecommunication Services
- □ Utilities

(continued)

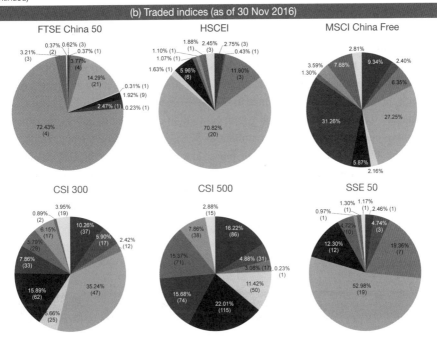

(b) Traded indices (as of 30 Nov 2016)

FTSE China 50   HSCEI   MSCI China Free

CSI 300   CSI 500   SSE 50

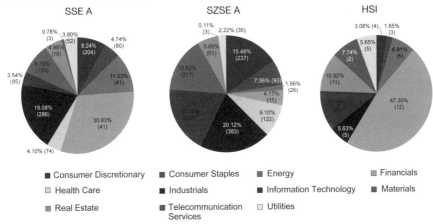

(c) Mainland and Hong Kong key market indices (as of 30 Nov 2016)

SSE A   SZSE A   HSI

- ■ Consumer Discretionary ■ Consumer Staples ■ Energy ▨ Financials
- ☐ Health Care ■ Industrials ■ Information Technology ■ Materials
- ■ Real Estate ■ Telecommunication Services ☐ Utilities

Notes: Excluding constituents where the market capitalisation or weighting is not available. Number of stocks in brackets. Titles are in blue for overseas-traded indices and in grey for Mainland-traded indices. Percentages may not add up to 100% due to rounding.

Source: HKEX, SSE and SZSE websites for Stock Connect eligible stock lists; CESC website for CES 120 constituent list; MSCI website for MSCI China Free industry composition; HKEX and Thomson Reuters for others.

Having a good representation of Mainland A shares in the index, the CES 120 would be a good tracker of the A shares market. As shown in Figure 6, the daily movement of CES 120 followed closely the trend of SSE 50 and CSI 300. Further examination found that **CES 120 is highly correlated with the A-share indices**. For the period from January 2011 to November 2016, the daily returns of CES 120 had a correlation coefficient of about 0.9 with those of FTSE A50 (coefficient: 0.904), SSE 50 (coefficient: 0.905) and CSI 300 (coefficient: 0.887), and a comparably high correlation coefficient (0.869) with that of SSE A, albeit a much lower correlation with that of SZSE A (coefficient: 0.671). Daily returns of the other overseas-traded indices on Chinese stocks (FTSE China 50, HSCEI and MSCI China Free), for which all or most of the constituents are Hong Kong listed stocks, were more highly correlated with each other and with HSI of the Hong Kong market (correlation coefficients of 0.944 or above during the period) than with the A-share indices (coefficients: about 0.6 or less). The patterns are similar across each year during the period.

(See Figure 7 for correlation of CES 120 with the indices and Appendix 1 for that of each index pair.)

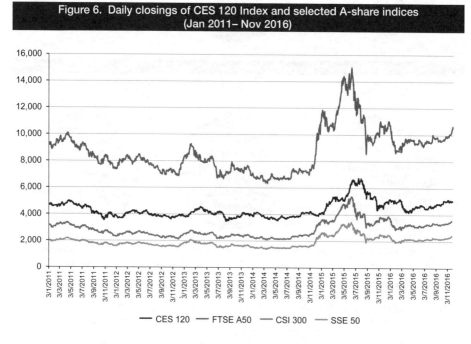

Figure 6. Daily closings of CES 120 Index and selected A-share indices (Jan 2011– Nov 2016)

Source: Thomson Reuters.

Figure 7. Correlation coefficients of daily returns of CES 120 Index with selected indices (Jan 2011–Nov 2016)

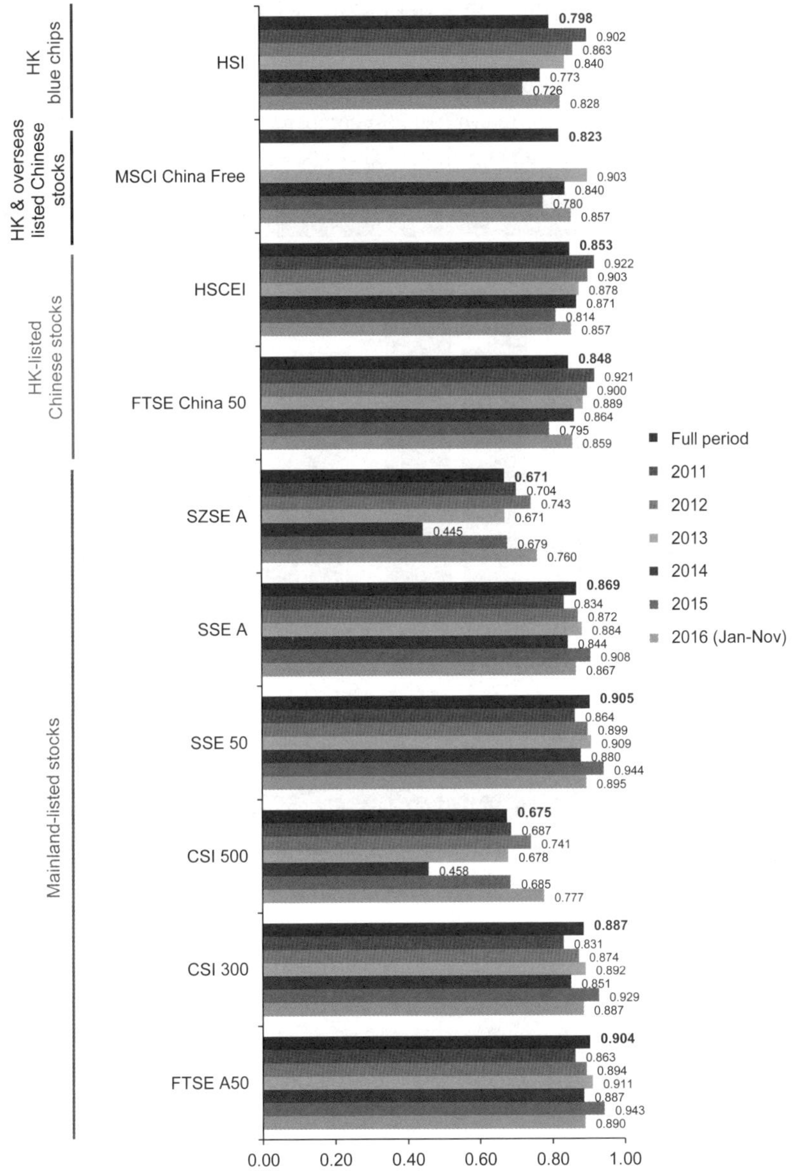

Source: Analysis based on daily index closings from Thomson Reuters.

While being a good tracker of the A shares market, CES 120 had a dividend yield comparable to SSE 50 and higher than those of CSI 300, CSI 500 and FTSE A50 in 2015 and 2016 (see Figure 8). In addition, it performed in the mid-stream in terms of return and volatility in comparison with the traded indices in 2015 — a low negative return (-2.66%) and a volatility in between those of the other indices; and outperformed all these indices in 2016 — a positive, albeit low, return (0.17%) compared to negative returns of all other indices and the lowest volatility among all the indices (see Figure 9).

Figure 8. Dividend yields of CES 120 and other traded indices (2015 & 2016)

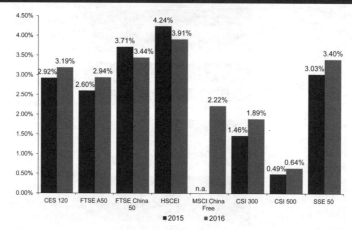

Note: Dividend yield of CES 120 for 2016 is as at 30 Sep 2016 (instead of as at year-end) from CESC website. Dividend yield of MSCI China Free Index for 2015 is not available.

Source: Thomson Reuters, CESC website, MSCI website.

Figure 9. Annual return and annualised volatility of daily returns of CES 120 Index and other traded indices (2015 & 2016)

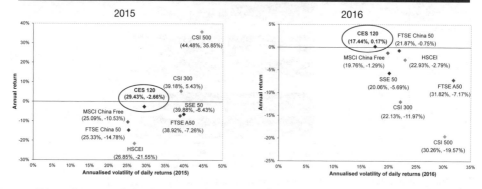

Note: Returns are calculated as natural logarithm returns.

Source: Calculation based on daily index closings from Thomson Reuters.

In summary, the analysis findings above reveal that CES 120 is an index which:

- Has a weighting composition by listing exchange closest to the corresponding composition of the eligible stocks under Stock Connect than other overseas traded indices;
- Comprises A shares and SEHK-listed stocks by stock type (H shares, red chips, P chips) comparable to that of Stock Connect eligible stocks;
- Has an inclusion, and a broad coverage, of constituents over the three exchanges of SEHK, SSE and SZSE among all Mainland and overseas traded indices;
- Has an industry composition closer to the Mainland cross-market big-cap index CSI 300 and a broader industry coverage than HSI of the Hong Kong market;
- Is highly correlated with the A-share indices; and
- Had outperformed other traded indices in terms of return and volatility in 2016 and a dividend yield comparable to the Mainland blue-chip index, SSE 50.

As a blue-chip Mutual Market index, the CES 120 may serve as the underlying of investment products offering exposure or risk management to stock investment in the Mainland-Hong Kong Mutual Market. Like the EURO STOXX 50 Index, Europe's leading cross-market blue-chip index for the Eurozone (see below), the CES 120 can become the leading cross-market blue-chip index for the Mainland-Hong Kong Mutual Market.

The EURO STOXX 50 Index was introduced on 26 February 1998 by the index provider STOXX owned by Deutsche Börse (DB) to meet the birth of the Eurozone with the official launch of the euro on 1 January 1999. The goal of the index is "to provide a blue-chip representation of Supersector leaders in the Eurozone". It is made up of the 50 largest companies among the 19 supersectors in terms of free-float market capitalisation in 11 Eurozone countries — Austria, Belgium, Finland, France, Germany, Ireland, Italy, Luxembourg, the Netherlands, Portugal and Spain, which are the initial eleven member states of the European Union admitted into the Eurozone in 1999. The index captures about 60% of the free-float market capitalisation of the EURO STOXX Total Market Index (TMI), which in turn covers about 95% of the free-float market capitalisation of the represented countries[9].

EURO STOXX 50 Index is licensed to financial institutions to serve as underlying for a range of investment products such as Exchange Traded Funds (ETFs), futures and options, and structured products worldwide. Eurex, an electronic marketplace operated by DB,

---

9   Source: Eurex website and Wikipedia.

introduced EURO STOXX 50 Index futures in June 1998. The futures contract is said to be the most liquid derivatives instrument in Europe, registering a yearly total turnover of 374 million contracts or an average daily volume of 1.46 million contracts on Eurex in 2016[10].

Referencing the EURO STOXX 50 Index and its derivatives and structured products in serving the mutual European market, CES 120 Index and its derivatives and structured products can serve the same need for the Mainland-Hong Kong Mutual Market.

3.2 CES China 120 Index Futures
— A useful tool for investment in A-shares and Mutual Market

The CES China 120 Index Futures (CES 120 Futures) contract was launched on 12 August 2013 on the Hong Kong Futures Exchange (HKFE), the HKEX's derivatives market. **It is the only futures contract in global markets based on an underlying index that tracks both A shares listed in the Mainland and Chinese stocks listed in Hong Kong.** Other China-related index futures traded outside Mainland China are based either on A shares only or overseas-listed Chinese stocks (mostly stocks in Hong Kong) (see Table 1 in section 2). Among the other index futures, only SGX's FTSE A50 Futures contract offers exposure to Mainland A shares. Table 2 below gives a brief comparison of CES 120 Futures with SGX's FTSE A50 Futures.

Table 2. Comparison of HKEX's CES 120 Futures and SGX's FTSE A50 Futures		
Feature	CES 120 Futures	FTSE A 50 Futures
Underlying	Mainland A shares and Chinese stocks listed on SEHK (including H shares, red chips and P chips)	Mainland A shares
Contract size	HK$50 per index point (~HK$263,844 or US$34,014 as of 30 Nov 2016)	US$1 per index point (~US$10,500 as of 30 Nov 2016)
Contract months	Spot, the next calendar month and the next two calendar quarter months	2 nearest serial months and Mar, Jun, Sep and Dec months on 1-year cycle
Block trade	Minimum 100 contracts	Minimum 50 lots
Position limit	300,000 contracts	15,000 contracts
Minimum tick (price fluctuation)	0.5 index point (HK$25 or ~US$3)	2.5 index point (US$2.50)
Minimum tick to index	0.0095%	0.0237%

---

10 Source: EURO STOXX 50 Index Quanto Futures, March 2016, Eurex website; Eurex Monthly Short Statistics, Dec 2016, Eurex website.

(continued)

Table 2. Comparison of HKEX's CES 120 Futures and SGX's FTSE A50 Futures		
Feature	CES 120 Futures	FTSE A 50 Futures
Margins	(Effective from 1 Dec 2016) Initial: HK$16,450 (~US$2,121) Maintenance: HK$13,170 (~US$1,698)	(As of 4 Nov 2016) Initial: US$495 Maintenance: US$450
Margin to notional value (%)	Initial: 6.23% Maintenance: 4.99% Total: 11.22%	Initial: 4.7% Maintenance: 4.3% Total: 9%
Exchange fee	HK$10 (discounted to HK$5 until 31 Dec 2017) (~US$1.3, discounted to US$0.6)	US$0.80 (Clearing fee)
Exchange fee to notional value (%)	0.0038% (0.0019% at discounted fee)	0.0076%

Note: Calculations are based on index closings on 30 November 2016 — 5276.87 for CES 120 and 10537.38 for FTSE A50. Exchange rate used is US$1 = HK$7.757 as at end of November 2016 from Hong Kong Monetary Authority (HKMA) website.

Source: HKEX and SGX websites.

As analysed in section 3.1 above, the CES 120 Index has a composition that represents Mainland A shares and Hong Kong-listed Chinese stocks on the cross-border Mutual Market platform and has a performance highly correlated with the A shares market. With these characteristics, the CES 120 Futures can be **an effective risk management tool for investors to hedge their positions in the Mainland A shares market, including exchange traded funds (ETFs) based on the A-share indices of FTSE A50 and CSI 300, and in the Mutual Market through Stock Connect**. It would also serve as a trading tool to offer investors simultaneous exposure to the Mainland and Hong Kong stock markets.

As an A-share hedging or exposure tool, CES 120 Futures would incur **a lower exchange fee** in trading than FTSE A50 Futures per contract notional value (see Table 2 above), albeit at somewhat higher margin costs. Given the higher position limit and the smaller tick size relative to the underlying index level in comparison with FTSE A50 Futures, CES 120 Futures can offer alternative cost-effective trading opportunities to gain investment exposure.

Moreover, CES 120 Futures would be a convenient and cost efficient offshore market tool for risk management of portfolio investment in the Mutual Market across Mainland China and Hong Kong. Investors with simultaneous exposure to Mainland A shares and Chinese stocks in Hong Kong may consider using CES 120 Futures to hedge their Mutual Market positions based on the relevance of the CES 120 to their portfolio holdings.

Appendix 1

Correlation of daily returns of CES 120 Index and selected indices

(Jan 2016 — Nov 2016)

Note: Statistics are computed for indices with data from the first available date in the given period.
Daily return = LN (current-day closing index / previous-day closing index)

Index	First date of daily return
CES 120	03/01/2011
FTSE A50	04/01/2011
CSI 300	04/01/2011
CSI 500	04/01/2011
SSE 50	04/01/2011
SSE A	04/01/2011
SZSE A	04/01/2011
FTSE China 50	03/01/2011
HSCEI	03/01/2011
MSCI China Free	11/01/2013
HSI	03/01/2011

Pearson correlation coefficients
(all coefficients are statistically significant at 0.1% level)

(Number of cases)

Period	Index	Mainland-listed stocks						HK-listed Chinese stocks		HK & overseas-listed Chinese stocks	HK key index
		FTSE A50	CSI 300	CSI 500	SSE 50	SSE A	SZSE A	FTSE China 50	HSCEI	MSCI China Free	HSI
Full period	CES 120	0.904 (1,438)	0.887 (1,436)	0.675 (1,436)	0.905 (1,436)	0.869 (1,436)	0.671 (1,436)	0.848 (1,427)	0.853 (1,423)	0.823 (952)	0.798 (1,423)
	FTSE A50		0.939 (1,436)	0.663 (1,436)	0.993 (1,436)	0.910 (1,436)	0.658 (1,436)	0.590 (1,401)	0.613 (1,397)	0.588 (946)	0.519 (1,397)
	CSI 300			0.863 (1,436)	0.948 (1,436)	0.984 (1,436)	0.856 (1,436)	0.586 (1,398)	0.605 (1,394)	0.608 (943)	0.522 (1,394)
	CSI 500				0.682 (1,436)	0.885 (1,436)	0.988 (1,436)	0.470 (1,398)	0.480 (1,394)	0.507 (943)	0.428 (1,394)
	SSE 50					0.921 (1,436)	0.676 (1,436)	0.593 (1,398)	0.615 (1,394)	0.590 (943)	0.523 (1,394)
	SSE A						0.872 (1,436)	0.580 (1,398)	0.598 (1,394)	0.602 (943)	0.518 (1,394)
	SZSE A							0.468 (1,398)	0.476 (1,394)	0.501 (943)	0.427 (1,394)
	FTSE China 50								0.986 (1,456)	0.982 (958)	0.963 (1,456)
	HSCEI									0.962 (955)	0.944 (1,456)
	MSCI China Free										0.957 (955)

(continued)

(Number of cases)

Period	Index	Mainland-listed stocks						HK-listed Chinese stocks		HK & over-seas-listed Chinese stocks	HK key index
		FTSE A50	CSI 300	CSI 500	SSE 50	SSE A	SZSE A	FTSE China 50	HSCEI	MSCI China Free	HSI
2011	CES 120	0.863 (244)	0.831 (244)	0.687 (244)	0.864 (244)	0.834 (244)	0.704 (244)	0.921 (247)	0.922 (246)	. (0)	0.902 (246)
	FTSE A50		0.963 (244)	0.800 (244)	0.995 (244)	0.955 (244)	0.825 (244)	0.622 (237)	0.629 (236)	. (0)	0.598 (236)
	CSI 300			0.916 (244)	0.965 (244)	0.990 (244)	0.934 (244)	0.579 (237)	0.587 (236)	. (0)	0.559 (236)
	CSI 500				0.803 (244)	0.922 (244)	0.990 (244)	0.460 (237)	0.470 (236)	. (0)	0.445 (236)
	SSE 50					0.957 (244)	0.825 (244)	0.624 (237)	0.631 (236)	. (0)	0.602 (236)
	SSE A						0.931 (244)	0.592 (237)	0.598 (236)	. (0)	0.573 (236)
	SZSE A							0.471 (237)	0.481 (236)	. (0)	0.454 (236)
	FTSE China 50								0.993 (246)	. (0)	0.974 (246)
	HSCEI									. (0)	0.970 (246)
	MSCI China Free										. (0)
2012	CES 120	0.894 (243)	0.874 (243)	0.741 (243)	0.899 (243)	0.872 (243)	0.743 (243)	0.900 (247)	0.903 (247)	. (0)	0.863 (247)
	FTSE A50		0.968 (243)	0.827 (243)	0.991 (243)	0.962 (243)	0.833 (243)	0.642 (237)	0.660 (237)	. (0)	0.588 (237)
	CSI 300			0.925 (243)	0.975 (243)	0.990 (243)	0.930 (243)	0.606 (237)	0.624 (237)	. (0)	0.550 (237)
	CSI 500				0.841 (243)	0.928 (243)	0.991 (243)	0.486 (237)	0.502 (237)	. (0)	0.432 (237)
	SSE 50					0.973 (243)	0.842 (243)	0.649 (237)	0.668 (237)	. (0)	0.594 (237)
	SSE A						0.924 (243)	0.614 (237)	0.631 (237)	. (0)	0.557 (237)
	SZSE A							0.487 (237)	0.500 (237)	. (0)	0.436 (237)
	FTSE China 50								0.987 (247)	. (0)	0.962 (247)
	HSCEI									. (0)	0.951 (247)
	MSCI China Free										. (0)

(continued)　　　　　　　　　　　　　　　　　　　　　　　　　　　　　(Number of cases)

Period	Index	Mainland-listed stocks						HK-listed Chinese stocks		HK & over-seas-listed Chinese stocks	HK key index
		FTSE A50	CSI 300	CSI 500	SSE 50	SSE A	SZSE A	FTSE China 50	HSCEI	MSCI China Free	HSI
2013	CES 120	0.911 (238)	0.892 (238)	0.678 (238)	0.909 (238)	0.884 (238)	0.671 (238)	0.889 (233)	0.878 (232)	0.903 (234)	0.840 (232)
	FTSE A50		0.958 (238)	0.691 (238)	0.995 (238)	0.941 (238)	0.680 (238)	0.683 (231)	0.682 (230)	0.683 (232)	0.611 (230)
	CSI 300			0.853 (238)	0.963 (238)	0.986 (238)	0.842 (238)	0.668 (231)	0.668 (230)	0.680 (232)	0.609 (230)
	CSI 500				0.705 (238)	0.862 (238)	0.986 (238)	0.513 (231)	0.516 (230)	0.546 (232)	0.485 (230)
	SSE 50					0.946 (238)	0.692 (238)	0.681 (231)	0.680 (230)	0.681 (232)	0.611 (230)
	SSE A						0.840 (238)	0.665 (231)	0.667 (230)	0.677 (232)	0.610 (230)
	SZSE A							0.512 (231)	0.513 (230)	0.543 (232)	0.485 (230)
	FTSE China 50								0.988 (244)	0.986 (237)	0.960 (244)
	HSCEI									0.977 (236)	0.946 (244)
	MSCI China Free										0.964 (236)
2014	CES 120	0.887 (245)	0.851 (245)	0.458 (245)	0.880 (245)	0.844 (245)	0.445 (245)	0.864 (238)	0.871 (238)	0.840 (245)	0.773 (238)
	FTSE A50		0.938 (245)	0.454 (245)	0.993 (245)	0.911 (245)	0.454 (245)	0.580 (238)	0.628 (238)	0.539 (245)	0.462 (238)
	CSI 300			0.707 (245)	0.947 (245)	0.973 (245)	0.700 (245)	0.563 (238)	0.599 (238)	0.533 (245)	0.446 (238)
	CSI 500				0.479 (245)	0.725 (245)	0.981 (245)	0.337 (238)	0.323 (238)	0.346 (245)	0.278 (238)
	SSE 50					0.918 (245)	0.476 (245)	0.573 (238)	0.619 (238)	0.530 (245)	0.448 (238)
	SSE A						0.707 (245)	0.585 (238)	0.619 (238)	0.557 (245)	0.474 (238)
	SZSE A							0.316 (238)	0.304 (238)	0.330 (245)	0.260 (238)
	FTSE China 50								0.969 (247)	0.980 (247)	0.928 (247)
	HSCEI									0.942 (247)	0.871 (247)
	MSCI China Free										0.954 (247)

(continued) (Number of cases)

Period	Index	Mainland-listed stocks						HK-listed Chinese stocks		HK & overseas-listed Chinese stocks	HK key index
		FTSE A50	CSI 300	CSI 500	SSE 50	SSE A	SZSE A	FTSE China 50	HSCEI	MSCI China Free	HSI
2015	CES 120	0.943 (244)	0.929 (244)	0.685 (244)	0.944 (244)	0.908 (244)	0.679 (244)	0.795 (237)	0.814 (237)	0.780 (244)	0.726 (237)
	FTSE A50		0.928 (244)	0.614 (244)	0.994 (244)	0.902 (244)	0.606 (244)	0.609 (237)	0.646 (237)	0.587 (244)	0.517 (237)
	CSI 300			0.848 (244)	0.936 (244)	0.986 (244)	0.838 (244)	0.635 (237)	0.667 (237)	0.627 (244)	0.553 (237)
	CSI 500				0.632 (244)	0.880 (244)	0.988 (244)	0.532 (237)	0.554 (237)	0.544 (244)	0.479 (237)
	SSE 50					0.909 (244)	0.624 (244)	0.612 (237)	0.650 (237)	0.591 (244)	0.522 (237)
	SSE A						0.865 (244)	0.632 (237)	0.663 (237)	0.625 (244)	0.549 (237)
	SZSE A							0.534 (237)	0.553 (237)	0.545 (244)	0.484 (237)
	FTSE China 50								0.982 (247)	0.990 (247)	0.960 (247)
	HSCEI									0.967 (247)	0.922 (247)
	MSCI China Free										0.959 (247)
2016 Jan - Nov	CES 120	0.890 (224)	0.887 (222)	0.777 (222)	0.895 (222)	0.867 (222)	0.760 (222)	0.859 (225)	0.857 (223)	0.857 (229)	0.828 (223)
	FTSE A50		0.946 (222)	0.792 (222)	0.988 (222)	0.920 (222)	0.773 (222)	0.574 (221)	0.592 (219)	0.587 (225)	0.525 (219)
	CSI 300			0.933 (222)	0.957 (222)	0.991 (222)	0.920 (222)	0.573 (218)	0.585 (216)	0.589 (222)	0.516 (216)
	CSI 500				0.804 (222)	0.959 (222)	0.994 (222)	0.515 (218)	0.517 (216)	0.530 (222)	0.456 (216)
	SSE 50					0.932 (222)	0.785 (222)	0.573 (218)	0.591 (216)	0.585 (222)	0.525 (216)
	SSE A						0.945 (222)	0.564 (218)	0.573 (216)	0.577 (222)	0.505 (216)
	SZSE A							0.500 (218)	0.502 (216)	0.514 (222)	0.443 (216)
	FTSE China 50								0.991 (225)	0.973 (227)	0.979 (225)
	HSCEI									0.960 (225)	0.965 (225)
	MSCI China Free										0.956 (225)

Appendix 2

List of constituents of CES China 120 Index
(as of 30 Nov 2016)

No.	Stock code	Stock name	Listing exchange	Stock type	Weight (%)
1	135	Kunlun Energy Company Limited	SEHK	Red chip	0.16
2	144	China Merchants Port Holdings Company Limited	SEHK	Red chip	0.22
3	151	Want Want China Holdings Limited	SEHK	Others	0.34
4	267	CITIC Limited	SEHK	Red chip	0.62
5	270	Guangdong Investment Limited	SEHK	Red chip	0.30
6	322	Tingyi (Cayman Islands) Holding Corp.	SEHK	Others	0.18
7	384	China Gas Holdings Limited	SEHK	P chip	0.18
8	386	China Petroleum & Chemical Corp.	SEHK	H shares	1.23
9	392	Beijing Enterprises Holdings Limited	SEHK	Red chip	0.16
10	656	Fosun International Limited	SEHK	P chip	0.26
11	688	China Overseas Land & Investment Limited	SEHK	Red chip	0.87
12	700	Tencent Holdings Limited	SEHK	P chip	9.72
13	728	China Telecom Corporation Limited	SEHK	H shares	0.46
14	762	China Unicom (Hong Kong) Limited	SEHK	Red chip	0.60
15	836	China Resources Power Holdings Co. Limited	SEHK	Red chip	0.22
16	857	PetroChina Company Limited	SEHK	H shares	0.99
17	883	CNOOC Limited	SEHK	Red chip	1.55
18	939	China Construction Bank Corporation	SEHK	H shares	4.94
19	941	China Mobile Limited	SEHK	Red chip	4.62
20	960	Longfor Properties Co. Limited	SEHK	P chip	0.15
21	966	China Taiping Insurance Holdings Co. Limited	SEHK	Red chip	0.28
22	992	Lenovo Group Limited	SEHK	Red chip	0.33
23	998	China CITIC Bank Corporation Limited	SEHK	H shares	0.40
24	1044	Hengan International Group Co. Limited	SEHK	P chip	0.47
25	1109	China Resources Land Limited	SEHK	Red chip	0.46
26	1288	Agricultural Bank Of China Limited	SEHK	H shares	0.71
27	1398	Industrial and Commercial Bank of China Limited	SEHK	H shares	3.66
28	1880	Belle International Holdings Limited	SEHK	Others	0.20

(continued)

No.	Stock code	Stock name	Listing exchange	Stock type	Weight (%)
29	2007	Country Garden Holdings Company Limited	SEHK	P chip	0.35
30	2318	Ping An Insurance (Group) Co. of China Limited	SEHK	H shares	1.99
31	2319	China Mengniu Dairy Company Limited	SEHK	Red chip	0.39
32	2328	PICC Property & Casualty Co., Limited	SEHK	H shares	0.53
33	2601	China Pacific Insurance (Group) Co., Limited	SEHK	H shares	0.74
34	2628	China Life Insurance Company Limited	SEHK	H shares	1.49
35	3328	Bank of Communications Co., Limited	SEHK	H shares	0.56
36	3333	China Evergrande Group	SEHK	P chip	0.19
37	3799	Dali Foods Group Company Limited	SEHK	P chip	0.08
38	3968	China Merchants Bank Co., Limited	SEHK	H shares	0.78
39	3988	Bank of China	SEHK	H shares	2.62
40	6808	Sun Art Retail Group Limited	SEHK	Others	0.17
41	600000	Shanghai Pudong Development Bank Co., Ltd.	SSE	A shares	1.49
42	600011	Huaneng Power International Inc.	SSE	A shares	0.32
43	600015	Hua Xia Bank Co., Ltd.	SSE	A shares	0.62
44	600016	China Minsheng Banking Corp., Ltd.	SSE	A shares	2.25
45	600018	Shanghai International Port (Group) Co., Ltd.	SSE	A shares	0.17
46	600019	Baoshan Iron &Steel Co., Ltd.	SSE	A shares	0.32
47	600023	Zhejiang Zheneng Electric Power Co., Ltd.	SSE	A shares	0.23
48	600028	China Petroleum & Chemical Corporation	SSE	A shares	0.54
49	600030	CITIC Securities Co., Ltd.	SSE	A shares	1.40
50	600036	China Merchants Bank Co., Ltd.	SSE	A shares	1.92
51	600048	Poly Real Estate Group Co., Ltd.	SSE	A shares	0.70
52	600050	China United Network Communications Ltd.	SSE	A shares	0.57
53	600104	SAIC Motor Corporation Ltd.	SSE	A shares	0.84
54	600276	Jiangsu Hengrui Medicine Co., Ltd.	SSE	A shares	0.66
55	600485	Beijing Xinwei Telecom Technology Group Co., Ltd.	SSE	A shares	0.20
56	600519	Kweichow Moutai Co., Ltd.	SSE	A shares	1.60
57	600585	Anhui Conch Cement Co. Ltd.	SSE	A shares	0.36
58	600606	Greenland Holdings Corporation Limited	SSE	A shares	0.04
59	600637	Shanghai Oriental Pearl Media Co., Ltd.	SSE	A shares	0.33
60	600690	Qingdao Haier Co., Ltd.	SSE	A shares	0.33
61	600795	GD Power Development Co., Ltd.	SSE	A shares	0.39
62	600837	Haitong Securities Company Limited	SSE	A shares	1.36
63	600871	Sinopec Oilfield Service Corporation	SSE	A shares	0.07
64	600887	Inner Mongolia Yili Industrial Group Co., Ltd.	SSE	A shares	1.20

(continued)

No.	Stock code	Stock name	Listing exchange	Stock type	Weight (%)
65	600893	Avic Aviation Engine Corporation Plc.	SSE	A shares	0.27
66	600900	China Yangtze Power Co., Ltd.	SSE	A shares	0.88
67	600958	Orient Securities Company Limited	SSE	A shares	0.45
68	600999	China Merchants Securities Co., Ltd.	SSE	A shares	0.55
69	601006	Daqin Railway Co., Ltd.	SSE	A shares	0.44
70	601088	China Shenhua Energy Co., Ltd.	SSE	A shares	0.34
71	601166	Industrial Bank	SSE	A shares	2.25
72	601169	Bank of Beijing Co., Ltd.	SSE	A shares	1.23
73	601186	China Railway Construction Corporation Ltd.	SSE	A shares	0.43
74	601211	Guotai Junan Securities Co., Ltd.	SSE	A shares	0.30
75	601288	Agricultural Bank of China Ltd.	SSE	A shares	1.23
76	601318	Ping An Insurance (Group) Company of China Ltd.	SSE	A shares	3.94
77	601328	Bank of Communications Co., Ltd.	SSE	A shares	1.61
78	601336	New China Life Insurance Co. Ltd.	SSE	A shares	0.39
79	601390	China Railway Group Limited	SSE	A shares	0.55
80	601398	Industrial and Commercial Bank of China Ltd.	SSE	A shares	1.10
81	601601	China Pacific Insurance (Group) Co., Ltd.	SSE	A shares	0.95
82	601628	China Life Insurance Company Limited	SSE	A shares	0.43
83	601633	Great Wall Motor Co. Ltd.	SSE	A shares	0.13
84	601668	China State Construction Engineering Corporation Ltd.	SSE	A shares	1.65
85	601669	Power Construction Corporation of China, Ltd.	SSE	A shares	0.33
86	601688	Huatai Securities Co., Ltd.	SSE	A shares	0.65
87	601727	Shanghai Electric Group Co. Ltd.	SSE	A shares	0.26
88	601766	CRRC Corporation Limited	SSE	A shares	1.10
89	601788	Everbright Securities Co. Ltd.	SSE	A shares	0.21
90	601800	China Communications Construction Co. Ltd.	SSE	A shares	0.25
91	601818	China Everbright Bank Co. Ltd.	SSE	A shares	0.65
92	601857	PetroChina Co. Ltd.	SSE	A shares	0.37
93	601898	China Coal Energy Co. Ltd.	SSE	A shares	0.11
94	601901	Founder Securities Co., Ltd.	SSE	A shares	0.35
95	601985	China National Nuclear Power Co., Ltd.	SSE	A shares	0.34
96	601988	Bank of China Ltd.	SSE	A shares	0.74
97	601989	China Shipbuilding Industry Co. Ltd.	SSE	A shares	0.66
98	601998	China Citic Bank Corporation Limited	SSE	A shares	0.21
99	1	Ping An Bank Co., Ltd.	SZSE	A shares	0.66
100	2	China Vanke Co., Ltd.	SZSE	A shares	2.10

(continued)

No.	Stock code	Stock name	Listing exchange	Stock type	Weight (%)
101	69	Shenzhen Overseas Chinese Town Co., Ltd.	SZSE	A shares	0.24
102	166	Shenwan Hongyuan Group Co., Ltd.	SZSE	A shares	0.42
103	333	Midea Group Co., Ltd.	SZSE	A shares	0.97
104	538	Yunnan Baiyao Group Co., Ltd.	SZSE	A shares	0.36
105	625	Chongqing Changan Automobile Co. Ltd.	SZSE	A shares	0.37
106	651	Gree Electric Appliances, Inc. of Zhuhai	SZSE	A shares	1.37
107	725	BOE Technology Group Co., Ltd.	SZSE	A shares	0.69
108	776	GF Securities Co., Ltd.	SZSE	A shares	0.58
109	858	Wuliangye Yibin Co., Ltd.	SZSE	A shares	0.68
110	895	Henan Shuanghui Investment & Development Co., Ltd.	SZSE	A shares	0.22
111	1979	China Merchants Shekou Industrial Zone Holdings Co., Ltd.	SZSE	A shares	0.46
112	2024	Suning Commerce Group Co., Ltd.	SZSE	A shares	0.44
113	2252	Shanghai RAAS Blood Products Co., Ltd.	SZSE	A shares	0.21
114	2304	Jiangsu Yanghe Brewery Joint-Stock Co., Ltd.	SZSE	A shares	0.43
115	2415	Hangzhou Hikvision Digital Technology Co., Ltd.	SZSE	A shares	0.46
116	2594	BYD Co. Ltd.	SZSE	A shares	0.30
117	2736	Guosen Securities Co., Ltd.	SZSE	A shares	0.44
118	2739	Wanda Cinema Line Co., Ltd	SZSE	A shares	0.31
119	300059	East Money Information Co., Ltd.	SZSE	A shares	0.46
120	300104	Leshi Internet Information & Technology Corp., Beijing	SZSE	A shares	0.38

Note: H shares are issued by companies incorporated in Mainland China and are listed on HKEX; red chips are shares issued by Mainland enterprises incorporated outside Mainland China and controlled by Mainland government entities through direct or indirect shareholding and/or representation on the company's board and are listed on HKEX; P chips are shares issued by privately controlled Mainland enterprises incorporated outside Mainland China and are listed on HKEX.

Source: CESC website

Chapter 2

Shanghai and Shenzhen Stock Connect

A "Mutual Market" for Mainland and
global investors

23 March 2017

Summary

The Shanghai-Hong Kong Stock Connect ("Shanghai Connect"), the first initiative under the Mutual Market Access pilot programme (the "Pilot Programme") between Mainland China and Hong Kong launched on 17 November 2014, offers a brand new official channel for overseas investors to invest in the Mainland stock market and for Mainland investors to invest in the Hong Kong stock market. The channel enables closed-loop Renminbi funds flow across the border in an orderly manner, thereby reducing the potential financial risk impact on the Mainland domestic market. The extended initiative — Shenzhen-Hong Kong Stock Connect ("Shenzhen Connect"), with an expanded scope of eligible securities — was already launched on 5 December 2016. The "Mutual Market" model across Shanghai, Shenzhen and Hong Kong has been basically formed. The Mutual Market model between Mainland China and Hong Kong is a symbolic breakthrough in the capital account opening process of Mainland China, under which global investment opportunities will be increasingly opened up to Mainland investors and more Mainland investment opportunities will be opened up to global investors.

Experience of the Shanghai Connect shows that Mainland investors have an increasing appetite for investment in Hong Kong stocks through Southbound trading. Their investment is not limited to large-cap blue chips but also in smaller-sized stocks in various industries. In Northbound trading, global investors also have an increasing interest in smaller-sized Mainland stocks of diversified industries. The Shenzhen Connect covers more small-sized stocks to meet the needs of both Mainland and global investors. Regulators on both sides had reached consensus on extending the Mutual Market Access scheme to cover exchange-traded funds (ETFs), for which specific schedule will be separately announced. Subject to regulatory approval, the scheme can possibly be extended to bonds and other securities, commodities and derivative products in the future. Through Southbound trading under the "Mutual Market" model, Mainland investors are open to global asset allocation opportunities for potentially better returns and an increasingly diversified scope of investment and risk management instruments than in the domestic market. Trading experience in an international market would also help nurture the maturity of Mainland investors, especially the retail investors.

With increasing cross-border investment activities, there will likely be increasing demand for the inclusion in the Mutual Market model of related cross-border portfolio hedging tools such as RMB equity derivatives, RMB interest rate and currency derivatives.

1 The Stock Connect pilot programme
— Unprecedented connectivity with the Mainland stock market towards a "Mutual Market"

The Mutual Market Access pilot programme (the "Pilot Programme"), launched in November 2014 with initially the Shanghai-Hong Kong Stock Connect (Shanghai Connect), is an unprecedented mechanism that connects stock trading, albeit within a confined scope, between the Mainland stock market and an overseas market. Prior to this, foreign participation channels in the Mainland stock market had been limited mainly to the Qualified Foreign Institutional Investor (QFII) scheme and the Renminbi Qualified Foreign Institutional Investor (RQFII) scheme and foreign retail investors could only participate through investment funds offered by QFIIs and RQFIIs[1]. In the opposite direction, the Qualified Domestic Institutional Investor (QDII) scheme and the Renminbi Qualified Domestic Institutional Investor (RQDII) scheme had been the only national official channels of Mainland participation in overseas stock markets[2]. Following successful and smooth operation of the Shanghai Connect, the Shenzhen-Hong Kong Stock Connect (Shenzhen Connect) was announced in August 2016 and subsequently launched in December that year. Hereinafter, Shanghai Connect and Shenzhen Connect are collectively referred to as the "Stock Connect" scheme.

The Stock Connect scheme is in fact a milestone step in the capital account opening of Mainland China. With daily quota imposed and a closed loop of cross-border funds flow, it allows cross-border capital investment activities to take place and develop in an orderly way with close monitoring, thereby reducing the potential financial risk impact onto the Mainland domestic stock market. The programme is scalable in size, scope and market

---

1   The B-share market (traded in foreign currency and separated from the A-share market) on the Shanghai and Shenzhen Stock Exchanges, which was launched in 1992 as the Mainland's first attempt of stock market opening to foreign investors, had become inactive in the new wave of market opening.

2   There are special pilot schemes launched by local governments, e.g. the Qualified Domestic Limited Partnership (QDLP) programme in Shanghai and the Qualified Domestic Investment Enterprise (QDIE) in Shenzhen Qianhai. However, these are limited to mainly privately offered funds or investment vehicles not widely accessible by general investors as these target mainly institutional investors and high-net-worth individuals, compared to QDII products which target general investors.

segments to match with the pace of the opening of the Mainland market as it further develops. The vision is to establish a "Mutual Market" of Mainland China and Hong Kong for Mainland and global investors.

Sub-sections below give a brief on the two Stock Connect schemes. Section 2 presents the hitherto performance since the launch of Stock Connect and Section 3 discusses the opportunities that the "Mutual Market" model offers.

1.1 Shanghai Connect

The pilot programme for the establishment of mutual stock market access between Mainland China and Hong Kong — Shanghai Connect — was jointly announced by the China Securities Regulatory Commission (CSRC) and the Hong Kong Securities and Futures Commission (SFC) in April 2014. Mutual order-routing connectivity and related technical infrastructure (Trading Links) were built by the Stock Exchange of Hong Kong Limited (SEHK), a wholly-owned subsidiary of HKEX, and the Shanghai Stock Exchange (SSE). Correspondingly the clearing and settlement infrastructure (Clearing Links) was established by the Hong Kong Securities Clearing Company Ltd (HKSCC), also a wholly-owned subsidiary of HKEX, and the securities clearing house in Mainland China — China Securities Depository and Clearing Corporation Ltd (ChinaClear). After several months' market preparation and system testing, the Shanghai Connect was formally launched on 17 November 2014. The theme of the programme is to enable Hong Kong and overseas investors to trade SSE-listed securities in the Mainland market (SH Northbound Trading) and Mainland investors to trade SEHK-listed securities in the Hong Kong market (SH Southbound Trading), within the eligible scope of the programme.

In the initial phase, eligible securities in SH Northbound Trading comprise SSE-listed A shares (the SSE-listed "Northbound stocks") which are:

- Constituent stocks of the SSE 180 Index and SSE 380 Index; or otherwise
- A shares which have corresponding H shares listed on SEHK;

except those which are not traded in Renminbi (RMB) and those under risk alert[3].

Eligible securities in SH Southbound Trading comprise SEHK-listed shares on the Main Board (the SEHK-listed "Southbound stocks") which are:

- Constituent stocks of the Hang Seng Composite LargeCap Index (HSLI); or
- Constituent stocks of the Hang Seng Composite MidCap Index (HSMI); or otherwise
- H shares which have corresponding A shares listed on the SSE;

---

3 Shares which are placed under "risk alert" by SSE, including shares of "ST companies" and "*ST companies" and shares subject to the delisting process under the SSE rules.

except those which are not traded in Hong Kong dollar (HKD) and H shares which have the corresponding SSE-listed A shares put under risk alert.

Among the Northbound stocks, the SSE 180 Index constituents are the most representative 180 A shares on the SSE while the SSE 380 Index constituents are the 380 stocks with modest scale, representing the segment of emerging blue chips outside the SSE 180 Index on the SSE[4]. The Northbound stocks in SSE 180 Index are therefore considered "large-cap" stocks in parallel with the Southbound HSLI stocks and the Northbound stocks in SSE 380 Index are considered "mid-cap" stocks in parallel with the Southbound HSMI stocks.

Within the eligible scope of securities, there were 715 SSE-listed Northbound stocks as at the end of February 2017 (including 139 stocks eligible for sell only[5]) and 317 SEHK-listed Southbound stocks[6].

In respect of investor eligibility, all Hong Kong and overseas investors are allowed to participate in Northbound trading while only Mainland institutional investors and those individual investors who hold an aggregate balance of not less than RMB 500,000 in their securities and cash accounts are allowed to participate in Southbound trading.

In Northbound trading, investors in Hong Kong will trade through Hong Kong brokers and trades are executed on the SSE platform. In Southbound trading, Mainland investors will trade through Mainland brokers and trades are executed on the SEHK platform. Northbound trading and Southbound trading follow the market practices of their respective trade execution platforms. In particular, day trading is not allowed for Mainland A shares market while being permissible in the Hong Kong market. Notably, Northbound SSE stocks are traded and settled in RMB only and Southbound SEHK stocks are traded in HKD and settled by the Mainland investors with ChinaClear or its clearing participants in RMB.

Trading under the Shanghai Connect is subject to investment quota, initially with an Aggregate Quota on the maximum cross-border investment value and a Daily Quota. The Aggregate Quota was set at RMB 300 billion for Northbound trading and RMB 250 billion for Southbound trading. This Aggregate Quota was subsequently abolished on the announcement date (16 August 2016) of the Shenzhen Connect. The Daily Quota, which is currently still applicable, limits the maximum net buy value to RMB 13 billion for Northbound stocks and RMB 10.5 billion (~HKD 11.7 billion as at end-2016) for Southbound stocks.

Since the launch of Shanghai Connect in November 2014, investor interest in Northbound and Southbound trading fluctuated over time. Since launch, Northbound trading

---

4   Source: SSE website.

5   Originally eligible SSE-listed stocks which subsequently cease to be eligible according to the set criteria.

6   Source: HKEX and SSE websites, viewed on 28 February 2017.

was more active than Southbound trading for most of the time. However, the trend since late 2015 shows that Southbound trading value has picked up and surpassed Northbound trading. Northbound trading in December 2016 increased significantly upon the launch of Shenzhen Connect. Data showed that nearly 30% of Northbound trading in the month accrued to Shenzhen Connect. (See section 2 for details.)

1.2   Shenzhen Connect

On 16 August 2016, the CSRC and the SFC jointly announced the establishment of the Shenzhen Connect — the mutual stock market access between Shenzhen and Hong Kong. This is an extended version of the Mutual Market Access pilot programme between Mainland China and Hong Kong on the foundation of the Shanghai Connect which has been running successfully since launch. The Shenzhen Connect will be established by the Shenzhen Stock Exchange (SZSE), SEHK, ChinaClear and HKSCC in a similar way as for the Shanghai Connect. This was subsequently launched on 5 December 2016. With this in place, the Pilot Programme now comprises the following Trading Links:

		Stock Connect	
		Shanghai Connect	Shenzhen Connect
Trading type	Northbound	Northbound Shanghai Trading Link	Northbound Shenzhen Trading Link
	Southbound	Southbound Hong Kong Trading Link under Shanghai Connect	Southbound Hong Kong Trading Link under Shenzhen Connect

Northbound eligible securities under the Shenzhen Connect comprise the following:
- All constituent stocks of the SZSE Component Index (SZCI) and of the SZSE Small/ Mid Cap Innovation Index (SZII) which have a market capitalisation of RMB 6 billion or above; and
- All SZSE-listed A shares of companies which have corresponding H shares listed on SEHK;

except those which are not traded in RMB and those under risk alert[7].

On top of Southbound eligible securities of the Shanghai Connect, the scope of Southbound eligible securities under the Shenzhen Connect is expanded to include the following:

- All constituent stocks of the Hang Seng Composite SmallCap Index (HSSI) which has a market capitalisation of HK$5 billion or above; and
- All SEHK-listed H shares of companies which have corresponding A shares listed on the SZSE;

except those which are not traded in HKD and H shares which have the corresponding A shares put under risk alert.

While Mainland eligible investors for Southbound trading under Shenzhen Connect are the same as that under Shanghai Connect, eligible investors for Northbound trading of shares listed on the ChiNext board of SZSE are confined, at the initial stage, to institutional professional investors as defined in the relevant Hong Kong rules and regulations[8].

The same daily quota as for the Shanghai Connect is applied to the Shenzhen Connect and aggregate quota is no longer applicable.

Table 1 below summarises the common and different key features for the Shanghai and Shenzhen Connect schemes.

Table 1. Key features of Shanghai Connect and Shenzhen Connect		
Feature	Shanghai Connect	Shenzhen Connect
Northbound eligible securities	• Constituents of the SSE 180 Index and SSE 380 Index • SSE-listed A shares which have corresponding H shares listed on SEHK	• Constituent stocks of the SZSE Component Index and of the SZSE Small/Mid Cap Innovation Index which have a market capitalisation of RMB 6 billion or above • SZSE-listed A shares of companies which have corresponding H shares listed on SEHK
	• Excluding risk alert stocks and stocks not traded in RMB	
	• 576 eligible stocks for buy and sell (as of 28 Feb 2017)	• 904 eligible stocks for buy and sell (as of 28 Feb 2017)

---

7   Shares which are placed under "risk alert" by SZSE including shares of "ST companies and "*ST companies" and shares subject to the delisting process under the SZSE rules.

8   See the definition of "Institutional Professional Investor" in the Securities and Futures (Professional Investor) Rules under the Securities and Futures Ordinance.

(continued)

Table 1. Key features of Shanghai Connect and Shenzhen Connect		
Feature	**Shanghai Connect**	**Shenzhen Connect**
Southbound eligible securities	• Constituents of the Hang Seng Composite LargeCap Index (HSLI) • Constituents of the Hang Seng Composite MidCap Index (HSMI)	
	• H shares which have corresponding A shares listed on the SSE	• Constituents of the Hang Seng Composite SmallCap Index (HSSI) which has a market capitalisation of HK$5 billion or above • H shares which have corresponding A shares listed on the SSE or the SZSE
	• Excluding H shares where the respective A shares are risk alert stocks and stocks not traded in HKD	
	• 317 stocks (as of 28 Feb 2017)	• 417 stocks (100 on top of Shanghai Connect Southbound stocks) (as of 28 Feb 2017)
Northbound eligible investors	• All Hong Kong and overseas investors (individuals and institutions)	• ChiNext eligible stocks are initially open only to institutional professional investors • All Hong Kong and overseas investors (individuals and institutions) for other eligible stocks
Southbound eligible investors	• Mainland institutional investors, and individual investors with an aggregate balance of ≥RMB 500,000 in their securities and cash accounts	
Daily quota	• Northbound: RMB 13 billion • Southbound: RMB 10.5 billion	
Aggregate quota	• Nil	
Northbound trading, clearing and settlement	• Following SSE and ChinaClear Shanghai market practice	• Following SZSE and ChinaClear Shenzhen market practice
Southbound trading, clearing and settlement	• Following SEHK and HKSCC market practice	

2 Hitherto performance of Stock Connect (up to end-2016)

2.1 Overall Northbound and Southbound trading

Both Northbound and Southbound trading value have fluctuated over time, along with changes in market sentiment. Yet, the average daily trading value (ADT) of Northbound trading has maintained a relatively steady level between 1% and 1.6% of the ADT of the Mainland total A-share market. On the other hand, after a rise and fall of Southbound trading in the first 9 months after launch, a strong upward trend was observed for Southbound trading relative to the SEHK Main Board total market trading since the fourth quarter of 2015, rising from 2.1% of the Main Board ADT in September 2015 to 10.8% in September 2016. The upward trend of Southbound trading continued despite a decrease somewhat at times. For the first time after April 2015, Southbound ADT exceeded Northbound ADT in June 2016 and time and again in subsequent months. (See Figures 1 and 2 below.)

Notably, during the 17 Northbound trading days from the launch of Shenzhen Connect on 5 December 2016 to the end of 2016, Northbound trading under Shenzhen Connect constituted 27% of Stock Connect's total Northbound trading value and 40% of Stock Connect's total buy trade value. This shows that global investors have considerable interest in Shenzhen stocks.

Figure 1. Stock Connect average daily trading value (Nov 2014–Dec 2016)

(a) Northbound

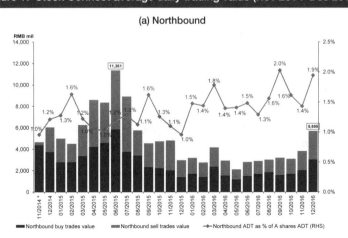

* Starting from 17 Nov 2014 when Shanghai Connect was launched.

Note: Shenzhen Connect data is included since the launch date of Shenzhen Connect (5 Dec 2016); the base reference data of Mainland A-share market includes SZSE A-share market since that date.

(continued)

Figure 1. Stock Connect average daily trading value (Nov 2014–Dec 2016)

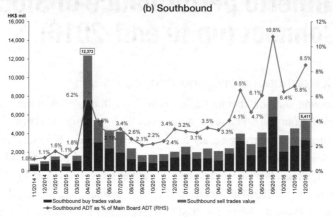

(b) Southbound

* Starting from 17 Nov 2014 when Shanghai Connect was launched.

Note: In calculating the ratio to the total market ADT, Northbound/Southbound trading values were two-sided (buy and sell values counted separately) while the total market trading values were one-sided transaction values (buy and sell counted in a single transaction value). Shenzhen Connect data is included since the launch date of Shenzhen Connect (5 Dec 2016).

Source: HKEX.

Figure 2. Average daily Stock Connect Southbound total trading (buy and sell) value in comparison with Northbound (Nov 2014–Dec 2016)

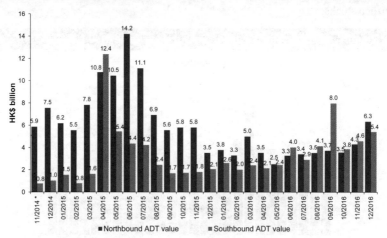

* Starting from 17 Nov 2014 when Shanghai Connect was launched.

Note: Northbound trading values are converted to HKD using month-end exchange rates from Hong Kong Monetary Authority website. Shenzhen Connect data is included since the launch date of Shenzhen Connect (5 Dec 2016).

Source: HKEX.

Moreover, Southbound trading had much higher average daily net buy values than Northbound trading since late 2015. Net sell value was recorded for Southbound trading in only two months since launch up to the end of 2016, vis-à-vis 6 months for Northbound trading (see Figure 3). Of the 485 Southbound trading days during the period, 86% had a net buy value (vs 56% for Northbound out of its 494 trading days). However, the net-buy daily quota consumption has been low for both Northbound and Southbound trading — under Shanghai Connect, only 18% of the Northbound trading days and 16% of the Southbound trading days had a daily quota usage exceeding 10%; and 6% respectively of the Northbound trading days and Southbound trading days had the usage exceeding 20%.[9] Under Shenzhen Connect, 4 out of 17 Northbound trading days (24%) had a daily usage exceeding 10% and only one day (6%) had the usage exceeding 20%; Southbound trading's daily quota usage had never exceeded 10%. (See Figure 4 and Table 2 below.)

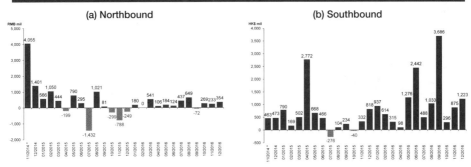

Figure 3. Average daily net buy/sell value in Stock Connect Northbound and Southbound trading (Nov 2014–Dec 2016)

* Starting from 17 Nov 2014 when Shanghai Connect was launched.

Note: Shenzhen Connect data is included since the launch date of Shenzhen Connect (5 Dec 2016).

Source: HKEX.

---

9 For Northbound trading, a daily quota usage exceeding 10% means that the net-buy value exceeds 10% of the given Northbound daily quota of RMB 13 billion, i.e. net-buy value amounts to over RMB 1.3 billion. The same applies to Southbound trading, but with reference to the Southbound net-buy daily quota of RMB 10.5 billion and currency converted to HKD based on the daily RMB/HKD exchange rate as obtained from Thomson Reuters.

Figure 4. Net-buy daily quota consumption in Stock Connect (17 Nov 2014–31 Dec 2016)

(a) Shanghai Connect — Northbound

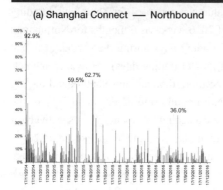

(b) Shanghai Connect — Southbound

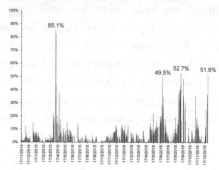

(c) Shenzhen Connect — Northbound

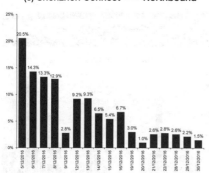

(d) Shenzhen Connect — Southbound

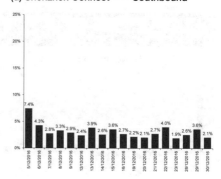

(e) Stock Connect — Total Northbound

(f) Stock Connect — Total Southbound

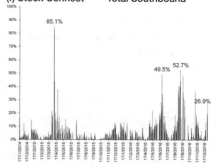

Note: Since 5 December 2016, Stock Connect total Northbound and total Southbound daily quota is the sum of those of Shanghai Connect and Shenzhen Connect.

Source: HKEX; daily trading value converted to RMB at daily exchange rates obtained from Thomson Reuters for calculating Southbound quota consumption.

Table 2. The usage of daily quota in Stock Connect (17 Nov 2014–31 Dec 2016)	Shanghai Connect		Shenzhen Connect	
	Northbound	Southbound	Northbound	Southbound
Total no. of trading days	494	485	17	18
% of trading days with net buy	55%	85%	100%	100%
Daily quota usage range	Northbound (no. of days / % of total)	Southbound (no. of days / % of total)	Northbound (no. of days / % of total)	Southbound (no. of days / % of total)
>0% - 10%	186 / 37.7%	317 / 65.4%	13 / 76.5%	18 / 100%
>10% - 20%	58 / 11.7%	66 / 13.6%	3 / 17.6%	0 / 0%
>20% - 30%	17 / 3.4%	10 / 2.1%	1 / 5.9%	0 / 0%
>30% - 40%	5 / 1.0%	11 / 2.3%	0 / 0%	0 / 0%
>40% - 50%	1 / 0.2%	6 / 1.2%	0 / 0%	0 / 0%
>50% - 60%	4 / 0.8%	2 / 0.4%	0 / 0%	0 / 0%
>60% - 70%	2 / 0.4%	0 / 0%	0 / 0%	0 / 0%
>70% - 80%	0 / 0%	0 / 0%	0 / 0%	0 / 0%
>80% - 90%	0 / 0%	2 / 0.4%	0 / 0%	0 / 0%
>90% - 100%	1 / 0.2%	0 / 0%	0 / 0%	0 / 0%
Total Stock Connect	Northbound		Southbound	
Total no. of trading days	494		485	
% of trading days with net buy	56%		86%	
Daily quota usage range	Northbound		Southbound	
>0% - 10%	191 / 38.7%		324 / 66.8%	
>10% - 20%	58 / 11.7%		67 / 13.8%	
>20% - 30%	17 / 3.4%		9 / 1.9%	
>30% - 40%	5 / 1.0%		10 / 2.1%	
>40% - 50%	1 / 0.2%		6 / 1.2%	
>50% - 60%	4 / 0.8%		1 / 0.2%	
>60% - 70%	2 / 0.4%		0 / 0%	
>70% - 80%	0 / 0%		0 / 0%	
>80% - 90%	0 / 0%		2 / 0.4%	
>90% - 100%	1 / 0.2%		0 / 0%	

2.2 Global investors' interest in Northbound stocks

At the initial stage after the launch of Shanghai Connect, global investors' trading and holding of Northbound stocks were predominantly in the large-cap SSE 180 Index constituents (94% of trading value in 2014 and 96% of holding value at the end of 2014). Northbound trading in the mid-cap SSE 380 Index constituents gradually increased from 6% during 2014 to 23% during 2016. Northbound holding of these mid-cap stocks jumped to 22% at the end of 2015 and decreased to 17% at the end of 2016, still much higher than the 4% at the end of 2014. Nevertheless, global investors' main interest was in the Mainland large-cap blue chips under Shanghai Connect. (See Figures 5a and 5b.)

For Shenzhen Connect launched on 5 December 2016, trading and holding of Northbound stocks by global investors also concentrated on the blue-chips in SZCI — 90% of trading value in 2016 and 93% of year-end holding value. (See Figures 5c and 5d.)

Figure 5. Stock Connect — Distribution of Northbound trading value and investor holding value by stock type (Nov 2014–Dec 2016)

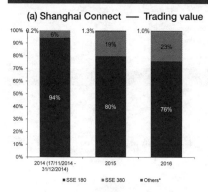

(a) Shanghai Connect — Trading value

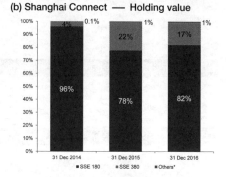

(b) Shanghai Connect — Holding value

(c) Shenzhen Connect — Trading value

(d) Shenzhen Connect — Holding value

(continued)

Figure 5. Stock Connect — Distribution of Northbound trading value and investor holding value by stock type (Nov 2014–Dec 2016)

(e) Stock Connect — Total trading value

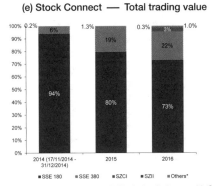

(f) Stock Connect — Total holding value

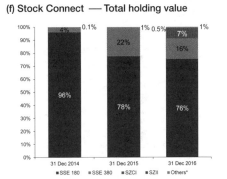

* Others include non-eligible-index A shares with SEHK-listed H shares and stocks removed from Stock Connect eligible stock list during the period (sell-only stocks).

Note: Shenzhen Connect data starts from the launch date of Shenzhen Connect (5 Dec 2016). Percentages may not add up to 100% due to rounding.

Source: HKEX.

Under Shanghai Connect, global investors maintained a steady interest in Mainland consumer stocks (Consumer Discretionary and Consumer Staples) — these stocks' trading contribution rose somewhat to 20% share of Northbound trading value in 2016 since launch and had an increasing percentage share of Northbound holding value. The share of consumer stocks was even higher under Shenzhen Connect — 47% of Northbound trading value in 2016 and 58% of year-end holding value. Shanghai and Shenzhen Connect combined, global investors' holding of Mainland consumer stocks was as high as 38% as at the end of 2016.

Mainland Industrial stocks under Shanghai Connect also attracted considerable interest — over 17% of Northbound trading value and period-end holding value in 2016. The degree of dominance of Financial stocks under Shanghai Connect (all were large-cap SSE 180 constituents) gradually reduced — from 51% of Northbound trading value and 43% of period-end Northbound holding value in 2014 to 31% and 20% respectively in 2016. Information technology (IT) stocks under Shenzhen Connect were considerably attractive to global investors, sharing 16% by trading value in 2016 and 15% by period-end holding value. The launch of Shenzhen Connect further drove up the share of IT stocks in Northbound trading. (See Figure 6.)

Figure 6. Stock Connect — Distribution of Northbound trading value and investor holding value by industry sector (Nov 2014–Dec 2016)

(a) Shanghai Connect — Trading value

(b) Shanghai Connect — Holding value

(c) Shenzhen Connect — Trading value

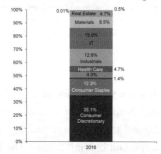

(d) Shenzhen Connect — Holding value

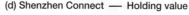
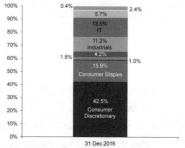

(e) Stock Connect — Total trading value

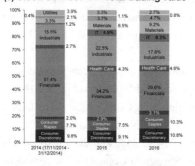

(f) Stock Connect — Total holding value

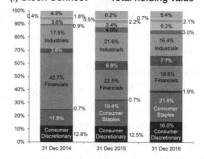

■ Consumer Discretionary   ■ Consumer Staples   ■ Energy
▨ Financials   ■ Health Care   ▨ Industrials
■ Information Technology   ■ Materials   ■ Real Estate
□ Telecommunication Services   □ Utilities

Note: Shenzhen Connect data starts from the launch date of Shenzhen Connect (5 Dec 2016). Percentages may not add up to 100% due to rounding.

Source: HKEX; stock classification is according to Global Industry Classification Standard (GICS) obtained from Bloomberg or Thomson Reuters.

While the majority share of Northbound trading in the large-cap SSE 180 stocks was contributed by Financial stocks, considerable share of Northbound trading in the mid-cap SSE 380 stocks was contributed by Industrial stocks. Consumer stocks contributed a significant share of Northbound trading in the large-cap SSE 180 stocks while the share of consumer stocks and IT stocks in Northbound trading in the mid-cap SSE 380 stocks showed an increasing trend. Apparently, although the mid-cap SSE 380 stocks had no financial stocks, IT stocks, Material stocks and Health Care stocks in the indices could attract relatively more diversified investment by industry than in the large-cap SSE 180 stocks. (See Figures 7 and 8.)

Figure 7. Shanghai Connect — Distribution of Northbound trading value and investor holding value by industry sector for SSE 180 Index constituents (Nov 2014–Dec 2016)

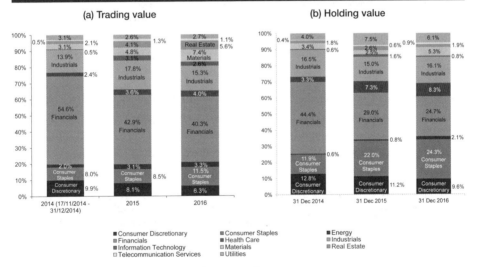

Note: Percentages may not add up to 100% due to rounding.

Source: HKEX; stock classification is according to Global Industry Classification Standard (GICS) obtained from Bloomberg or Thomson Reuters.

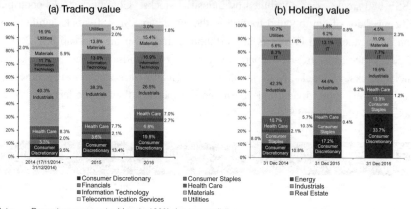

Figure 8. Shanghai Connect — Distribution of Northbound trading value and investor holding value by industry sector for SSE 380 Index constituents (Nov 2014–Dec 2016)

Note: Percentages may not add up to 100% due to rounding.

Source: HKEX; stock classification is according to Global Industry Classification Standard (GICS) obtained from Bloomberg or Thomson Reuters.

For Shenzhen Connect, in the Northbound trading of the blue-chip SZCI, the shares of consumer stocks in trading value and holding value were considerably high and the share of IT stocks were also quite high. In the Northbound trading of the SZII stocks, Industrial, IT and Material stocks had rather high percentage share in trading value, while Consumer Discretionary stocks took the lead in holding value. (See Figures 9 and 10.)

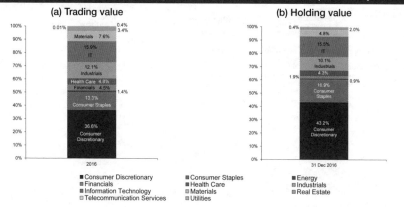

Figure 9. Shenzhen Connect — Distribution of Northbound trading value and investor holding value by industry sector for SZCI constituents (Dec 2016)

Note: Shenzhen Connect data starts from the launch date of Shenzhen Connect (5 Dec 2016). Percentages may not add up to 100% due to rounding.

Source: HKEX; stock classification is according to Global Industry Classification Standard (GICS) obtained from Bloomberg or Thomson Reuters.

Figure 10. Shenzhen Connect — Distribution of Northbound trading value and investor holding value by industry sector for SZII constituents (Dec 2016)

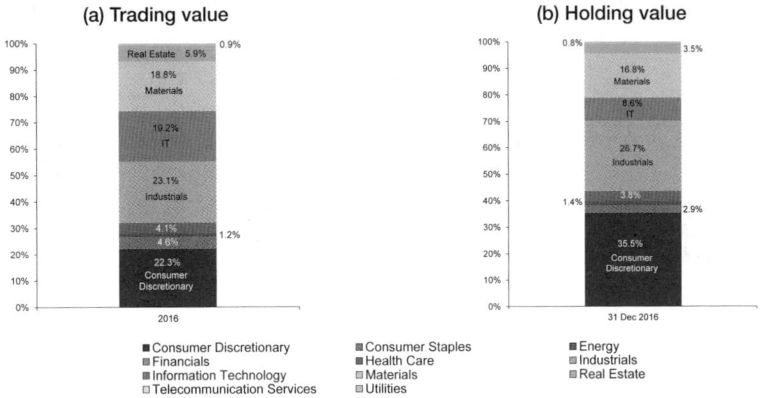

Note: Shenzhen Connect data starts from the launch date of Shenzhen Connect (5 Dec 2016). Percentages may not add up to 100% due to rounding.

Source: HKEX; stock classification is according to Global Industry Classification Standard (GICS) obtained from Bloomberg or Thomson Reuters.

2.3   Mainland investors' interest in Southbound stocks

Figure 11 shows the Southbound trading and investor holding in SEHK-listed stocks by eligible stock type. In contrast to Northbound investment, the majority of Southbound trading value and period-end holding value was in the mid-cap HSMI constituents at the launch of the scheme in 2014. This had shifted to some extent to HSLI constituents in 2016. Nevertheless, HSMI stocks still maintained a considerable share in 2016 (~40%).

Apart from the large-sized HSLI stocks and the mid-cap HSMI stocks, Southbound eligible stocks also include the small-cap HSSI stocks under Shenzhen Connect. HSSI stocks constituted considerably high percentage share in Southbound trading under Shenzhen Connect during the 18 trading days in 2016, the same as the 42% for HSMI stocks and even had a higher percentage share in holding value than HSMI stocks (46% vs 42%). However, owing to the low market value of HSSI stocks by definition, the percentage share of HSSI stocks in total holding value of Southbound stocks under Stock Connect was very small (4%).

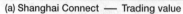

Figure 11. Stock Connect — Distribution of Southbound trading value and investor holding value by stock type (Nov 2014–Dec 2016)

(a) Shanghai Connect — Trading value

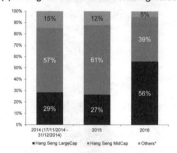

(b) Shanghai Connect — Holding value

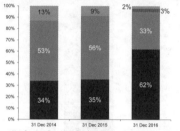

(c) Shenzhen Connect — Trading value

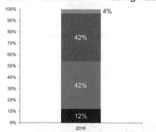

(d) Shenzhen Connect — Holding value

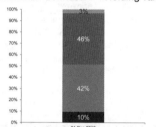

(e) Stock Connect — Total trading value

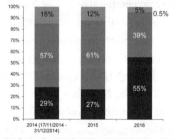

(f) Stock Connect — Total holding value

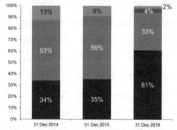

* Others include non-eligible-index H shares with SSE-listed or SZSE-listed (since 5 Dec 2016) A shares and stocks removed from Stock Connect eligible stock list ("sell-only" stocks).

Note: Shenzhen Connect data starts from the launch date of Shenzhen Connect (5 Dec 2016). The stock types in calculating holding value are the period-end status, which may have changed during the period. Due to this reason, stocks held under Shanghai Connect may include HSSI stocks while the trading of these stocks in such status is counted into "others". Percentages may not add up to 100% due to rounding.

Source: HKEX for trading value; Webb-site Who's Who Database for holding values. Stock classification according to Hang Seng Indexes Co., Ltd.

In terms of industry sector, there had been increasing investor interest in Financial stocks, which had the largest share in trading and holding values in 2016. Other more popular sector stocks were Consumer Goods, and Properties and Construction. However, Southbound trading and holding under Shenzhen Connect were not concentrated on Financial stocks but distributed across Consumer Goods, Properties and Construction, IT and Industrials sectors. (See Figure 12.)

However, the dominance of Financial stocks in Southbound investment was mainly in respect of HSLI stocks. For Southbound investment in the mid-cap HSMI stocks, Mainland investor trading and holding were very much diversified across different industrial sectors. In 2016 up to August, Southbound trading and period-end holding had considerable share in Consumer Goods stocks (~25-27%, and close to 30% when including Consumer Services stocks) and Properties and Construction Stocks (~16-18%) among HSMI constituents. Financial stocks ranked third in Southbound trading and holding of HSMI constituents, in contrast to their dominance in that of HSLI constituents. Similar different distributions by industry sector in trading and holding values for HSLI and HSMI stocks were observed under Shenzhen Connect. As for among HSLI stocks under Shenzhen Connect, IT stocks apparently attracted considerably high percentage share in trading and holding values. (See Figures 13 to 15.)

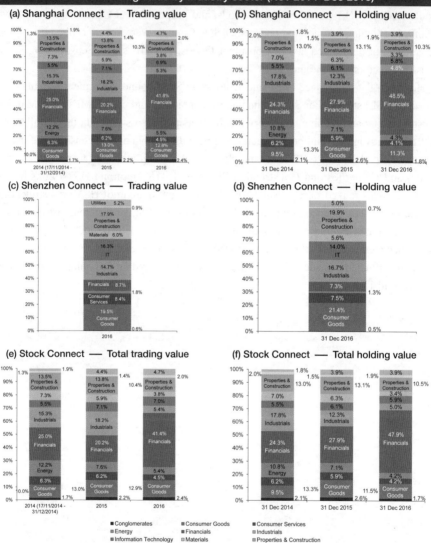

Figure 12. Stock Connect — Distribution of Southbound trading value and investor holding value by industry sector (Nov 2014–Dec 2016)

Note: Shenzhen Connect data starts from the launch date of Shenzhen Connect (5 Dec 2016). Percentages may not add up to 100% due to rounding.

Source: HKEX for trading value; Webb-site Who's Who Database for holding values. Stock classification according to Hang Seng Indexes Co., Ltd.

Figure 13. Stock Connect — Distribution of Southbound trading value and investor holding value by industry sector for HSLI constituents (Nov 2014–Dec 2016)

(a) Shanghai Connect — Trading value

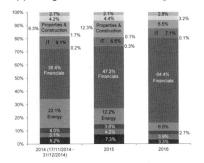

(b) Shanghai Connect — Holding value

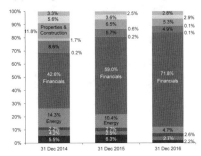

(c) Shenzhen Connect — Trading value

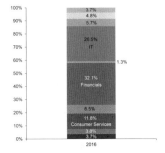

(d) Shenzhen Connect — Holding value

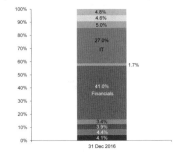

(e) Stock Connect — Total trading value

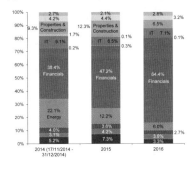

(f) Stock Connect — Total holding value

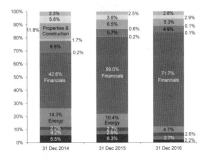

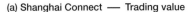

■ Conglomerates   ■ Consumer Goods   ■ Consumer Services
■ Energy   ■ Financials   ■ Industrials
■ Information Technology   ■ Materials   ■ Properties & Construction
□ Telecommunications   ▨ Utilities

Note: Shenzhen Connect data starts from the launch date of Shenzhen Connect (5 Dec 2016). The stock types in calculating holding value are the period-end status, which may have changed during the period. Percentages may not add up to 100% due to rounding.

Source: HKEX for trading value; Webb-site Who's Who Database for holding values. Stock classification according to Hang Seng Indexes Co., Ltd.

Figure 14. Stock Connect — Distribution of Southbound trading value and investor holding value by industry sector for HSMI constituents (Nov 2014–Dec 2016)

(a) Shanghai Connect — Trading value

(b) Shanghai Connect — Holding value

(c) Shenzhen Connect — Trading value

(d) Shenzhen Connect — Holding value

(e) Stock Connect — Total trading value

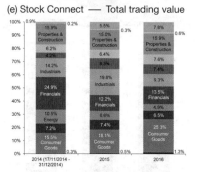

(f) Stock Connect — Total holding value

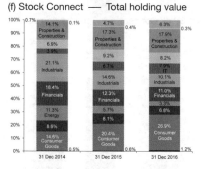

- ■ Conglomerates
- ■ Consumer Goods
- ■ Consumer Services
- ▨ Energy
- ■ Financials
- ▨ Industrials
- ■ Information Technology
- ▨ Materials
- ■ Properties & Construction
- ▢ Telecommunications
- ▨ Utilities

Note: Shenzhen Connect data starts from the launch date of Shenzhen Connect (5 Dec 2016). The stock types in calculating holding value are the period-end status, which may have changed during the period. Percentages may not add up to 100% due to rounding.

Source: HKEX for trading value; Webb-site Who's Who Database for holding values. Stock classification according to Hang Seng Indexes Co., Ltd.

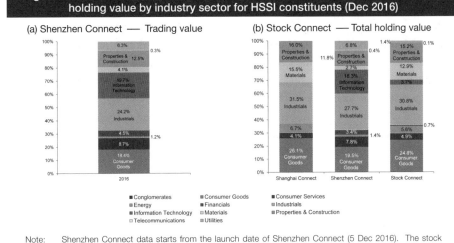

Figure 15. Stock Connect — Distribution of Southbound trading value and investor holding value by industry sector for HSSI constituents (Dec 2016)

Note: Shenzhen Connect data starts from the launch date of Shenzhen Connect (5 Dec 2016). The stock types in calculating holding value are the period-end status, which may have changed during the period. Percentages may not add up to 100% due to rounding.

Source: HKEX for trading value; Webb-site Who's Who Database for holding values. Stock classification according to Hang Seng Indexes Co., Ltd.

In other words, as far as Shanghai Connect offers, Mainland investors have considerable investment interest in Southbound trading of mid-cap stocks of a variety of industry sectors. While large-cap stocks would have high concentration in Financial stocks, **smaller-sized stocks in fact could offer a diversified scope of stocks by industry sector of interest to the Mainland investors.**

3   The "Mutual Market" model
—— Opportunities to Mainland and global investors

After the launch of Shenzhen Connect, the basic model of "Mutual Market" across Shanghai, Shenzhen and Hong Kong has been established, albeit within the limited scope of eligible securities. As the mutual stock market access scheme is scalable, this potentially opens up a Mainland-Hong Kong mutual stock market of a combined equity market value of US$10,514 billion (as of end-2016) and an average daily equity turnover of about US$84.3 billion (2016), ranking 2nd by market value (following New York Stock Exchange) and 2nd by equity market turnover among world exchanges[10]. Moreover, the "Mutual Market" model may go beyond equities in multiple dimensions. As mentioned in the joint announcement of the CSRC and the SFC on 16 August 2016 regarding the in-principle approval for the establishment of Shenzhen Connect, the two authorities have reached a consensus to include exchange-traded funds (ETFs) as eligible securities under the scheme. A launch date will be announced in due course after Shenzhen Connect has been in operation for a period of time and upon the satisfaction of relevant conditions. In addition, the CSRC and the SFC will jointly study and introduce other financial products to facilitate and meet the need of Mainland and global investors to manage price risks in each other's stock markets.

Under the "Mutual Market" model, overseas products of various kinds could be offered to the Mainland investors and vice versa for offering different Mainland products to global investors. Southbound trading opens up a regularised channel for Mainland investors, both individuals and institutions, to invest in overseas assets. The channel is a closed system with prudential monitoring of the usage of the daily quota, and yet with considerable flexibility in the absence of an aggregate quota. Without the limitation of an aggregate quota, investors may allocate their portfolio investment in cross-border assets more freely than before. This offers global asset allocation opportunities to Mainland investors. Since the system is closed in the sense that the RMB (after converted into HKD) used to purchase overseas assets under the model will be reverted back to Mainland China (after being converted back to RMB) upon the sale of the overseas assets, there is essentially no capital outflow problem in the long run. The model effectively extends the universe of investable assets for Mainland investors. Under such an environment, the connectivity channel compensates the relative shortage of investable assets in the Mainland, allowing Mainland investment monies to possibly obtain better potential returns from investing in overseas than in the domestic market. Acknowledging the advantage of this, the China Insurance Regulatory Commission (CIRC) issued a policy document[11] in early September 2016 to allow insurance funds to participate in Southbound trading under the Shanghai Connect. Eligible Mainland investors for Southbound trading under Shenzhen Connect are the same as Shanghai Connect. The expanded investment scope in the Mutual Market with

---

10 World Federation of Exchanges (WFE) statistics, from WFE website (on 20 January 2017 for market value data and on 1 March 2017 for trading value data). Average daily turnover was calculated from the combined shares turnover value for 2016 from WFE statistics using the total number of trading days (244 days) for the Mainland market. Ranking was based on 2016 up to December combined trading value.

11 《關於保險資金參與滬港通試點的監管口徑》, 9 September 2016.

Shenzhen Connect in place would offer more diversified investment choices to the Mainland investors in Southbound trading.

In addition, Southbound trading is effectively investment in a foreign currency, i.e. HKD which is pegged to the US dollar, for Mainland investors. At the time of expected depreciation of the RMB, Southbound investment could offer alternative investment options from a currency value perspective.

As the eligible instruments under the "Mutual Market" model are expandable, it is believed that the Mutual Market will offer an increasingly diversified scope of investment tools to Mainland investors, albeit in the short term the available instruments may only be cash market securities including equities and possibly ETFs. After the launch of Shenzhen Connect, Southbound eligible securities include the HSSI constituents with a market capitalisation of HK$5 billion or above, in addition to HSLI and HSMI constituents. All H shares with A shares listed in the Mainland market are included, not just those with SSE-listed A shares. HSLI and HSMI already cover up to 95% of the total market capitalisation of the Hang Seng Composite Index (HSCI), which in turn covers the top 95% of the total market capitalisation of the Hong Kong market[12]. As a result, some further 100 stocks have been added to the eligible list. More importantly, the expanded scope includes stocks in a large variety of industries, including new economy sectors of information technology and consumer goods and services.

Moreover, the trading experience offered by an international stock market like Hong Kong where the dominant participants are international professional institutional investors will be of value to Mainland domestic investors, especially the retail investors. Professional investment strategies in a mature market are usually based on stock fundamentals, and economic and industrial factors. These would help balancing the short-term speculative trading behaviour of certain Mainland investors. Southbound trading experience is therefore expected to help nurture the maturity of the Mainland investor base.

Apart from secondary market trading, connectivity in the primary market, i.e. the initial public offering (IPO) market, can be offered under the model subject to regulatory approval, allowing investors on either market to subscribe for IPOs in the other market. Products to be covered in the future (subject to regulatory approval) may also extend to bonds, commodities and risk management tools including equity derivatives, RMB interest rate and currency derivatives. In fact, in view of the successful implementation of the Stock Connect scheme, a more imminent issue is to meet the needs of investors for hedging their

---

12 The Hong Kong market universe of the HSCI refers to all stocks and real estate investment trusts ("REITs") that have their primary listings on the SEHK, excluding securities that are secondary listings, foreign compani es, preference shares, debt securities, mutual funds and other derivatives. (Source: Hang Seng Indexes Co. Ltd. website)

cross-border stock portfolios. In the Mainland, investors are allowed to trade Hong Kong stocks but there are no Hong Kong index/stock futures and options for hedging. Similarly in Hong Kong, there is also the lack of A-share hedging tools like A-share index futures and options. Related derivatives on either market may be included in the Mutual Market model in the future.

The "Mutual Market" model in fact is a symbolic breakthrough in Mainland China's capital account opening process. In the long run, a highly diversified suite of investment and risk management tools could be offered under the "Mutual Market" model before the potential full opening of the Mainland capital market. Mainland investors could thereby benefit from enhanced asset allocation and investment portfolio management, and global investors could benefit from an open channel to more Mainland investment opportunities with related risk management tools available.

Chapter 3

Primary Equity Connect

A breakthrough opportunity for
Mainland-Hong Kong mutual market
connectivity and RMB internationalisation

28 August 2017

Summary

With the launch of the Shanghai-Hong Kong Stock Connect (Shanghai Connect) in November 2014 and the Shenzhen-Hong Kong Stock Connect (Shenzhen Connect) in December 2016 (collectively referred to as the "Stock Connect scheme"), the Mainland-Hong Kong Mutual Market platform was basically formed. However, the platform is currently confined to secondary equity market trading only and investors on either side of the border are barred from the primary equity market on the other side. This has deprived investors of investment opportunities from initial public offers (IPO) across the border and has essentially hindered the pooling of liquidity in the Mutual Market to support its function in fund raising by issuers.

The Primary Equity Connect (PEC) initiative[1] put forward in HKEX's Strategic Plan 2016-2018 would provide a breakthrough opportunity to help complete the mutual market connectivity in the equity market segment. The concept of PEC is to provide a mechanism to allow Mainland investors to subscribe for IPOs in the Hong Kong market (Southbound) and global investors in Hong Kong to subscribe for IPOs in the Mainland market (Northbound). Upon listing of the IPO shares issued, trading by investors from the other market would be enabled through the existing Stock Connect mechanism. Under this connectivity model, shares subscription under PEC and shares trading under Stock Connect are effectively contained within a closed-loop system.

In respect of the Mainland market and the Hong Kong market separately, PEC is believed to be mutually beneficial to each market given the development bottleneck in the internationalisation of both markets. The current degree of internationalisation of the Mainland stock market is relatively low in multiple dimensions, including investor base, issuer base and institutional structure. The participation of Qualified Foreign Institutional Investors (QFIIs) and Renminbi QFIIs (RQFIIs) in the Mainland stock market remains relatively low (with an aggregate shareholding value less than 0.3% of the total "negotiable market capitalisation", i.e. market value of listed and marketable securities, on the Shanghai and Shenzhen stock exchanges). No foreign companies are yet allowed to list on the Mainland exchanges; and the Mainland market practices are not necessarily in line with international practices. The Hong Kong stock market, on the other hand, is highly internationalised in terms of investor participation but not in terms of listed issuers. Given these weaknesses, PEC would help enhance the international dimensions of both the Mainland market and the Hong Kong market and beyond which the Mutual Market as a whole.

---

1   Subject to regulatory approvals.

Internationalisation of the Mainland-Hong Kong Mutual Market is not a goal in itself but part of the bigger strategy for China to achieve a better balanced economy, more effective market opening and ultimately a higher degree of RMB internationalisation. The PEC initiative under the Mutual Market model would offer an opportunity to improve the current situation.

For the Mainland, PEC would (1) offer new opportunities for Mainland investors' global asset allocation and therefore an improved national balance sheet, (2) facilitate two-way market opening at lower cost, (3) help achieve one more step forward in RMB capital account convertibility, (4) support the development of the market's international investor base, (5) provide more listing opportunities for Mainland enterprises and (6) help nurture the Mainland investor base, while containing potential risks with suitable control measures. For Hong Kong, PEC would help attract the listing of international companies, further activate the market by increasing investor participation and also benefit market intermediaries through more business opportunities.

The implementation of PEC would involve additional issues, including regulatory and operational issues, beyond the Stock Connect scheme. Nevertheless, it is believed that these challenges could be largely overcome with a suitable model design that could meet the best interests of the Mainland-Hong Kong Mutual Market. This would ultimately benefit China's national accounts and the big strategy of RMB internationalisation.

1 A mutual market without primary market connectivity

1.1 Secondary market connectivity — Stock Connect

The Mainland-Hong Kong Mutual Market Access pilot programme (the "Pilot Programme") was launched on 17 November 2014, initially with the commencement of the Shanghai-Hong Kong Stock Connect (Shanghai Connect), allowing cross-border stock investments in the Mainland and Hong Kong markets. The Shenzhen-Hong Kong Stock Connect (Shenzhen Connect) was subsequently launched on 5 December 2016. The (initially applied) aggregate quota for the Pilot Programme was abolished immediately upon the official announcement of the Shenzhen Connect on 16 August 2016. (Hereinafter, the Shanghai Connect and Shenzhen Connect are collectively referred to as the "Stock Connect scheme".) By then, a mutual market platform across Shanghai, Shenzhen and Hong Kong was basically formed. This potentially opens up a Mainland-Hong Kong mutual stock market of a combined equity market value of US$10,514 billion (as of end-2016) and an average daily equity turnover of about US$84.3 billion (2016), ranking 2nd by market value (following New York Stock Exchange) and 2nd by equity market turnover among world exchanges[2]. Moreover, the "Mutual Market" model may go beyond equities as eligible trading instruments, including exchange-traded funds (ETFs)[3].

The Stock Connect scheme enables Hong Kong and overseas investors to trade securities listed on the Shanghai Stock Exchange (SSE) or Shenzhen Stock Exchange (SZSE) in the Mainland market (Shanghai (SH) or Shenzhen (SZ) Northbound Trading respectively under Shanghai Connect and Shenzhen Connect) and Mainland investors to trade securities listed on the Stock Exchange of Hong Kong (SEHK) in the Hong Kong market (SH or SZ Southbound Trading respectively under Shanghai Connect and Shenzhen Connect), within the eligible scope of the programme.

---

2 World Federation of Exchanges (WFE) statistics, from WFE website (on 20 January 2017 for market value data and on 1 March 2017 for trading value data). Average daily turnover was calculated from the combined shares turnover value for 2016 from WFE statistics using the total number of trading days (244 days) for the Mainland market. Ranking was based on 2016 up to December combined trading value.

3 As mentioned in the joint announcement of the China Securities Regulatory Commission (CSRC) and the Hong Kong Securities and Futures Commission (SFC) on 16 August 2016 regarding the in-principle approval for the establishment of Shenzhen Connect, the two authorities have reached a consensus to include ETFs as eligible securities under the scheme. A launch date will be announced in due course after Shenzhen Connect has been in operation for a period of time and upon the satisfaction of relevant conditions.

Eligible securities in SH Northbound Trading comprise SSE-listed constituent stocks of the SSE 180 Index and SSE 380 Index and otherwise the A shares which have corresponding H shares listed on SEHK, except those which are not traded in Renminbi (RMB) and those under risk alert[4]. Eligible securities in SZ Northbound Trading comprise all SZSE-listed constituent stocks of the SZSE Component Index (SZCI) and of the SZSE Small/Mid Cap Innovation Index (SZII) which have a market capitalisation of RMB 6 billion or above, and otherwise all SZSE-listed A shares of companies which have corresponding H shares listed on SEHK, except those which are not traded in RMB and those under risk alert by SZSE. Eligible securities in SH Southbound Trading comprise SEHK Main Board-listed constituent stocks of the Hang Seng Composite LargeCap Index (HSLI) and the Hang Seng Composite MidCap Index (HSMI), or otherwise H shares which have corresponding A shares listed on the SSE, except those which are not traded in Hong Kong dollars (HKD) and H shares which have the corresponding A shares put under risk alert. On top of these, eligible securities in SZ Southbound Trading also include all the constituent stocks of the Hang Seng Composite SmallCap Index (HSSI) which have a market capitalisation of HK$5 billion or above, and all SEHK-listed H shares of companies which have corresponding A shares listed on the SZSE, except those which are not traded in Hong Kong dollars (HKD) and H shares which have the corresponding A shares put under risk alert.

As of 28 June 2017, there were 574 eligible (for both buy and sell) Northbound stocks and 310 eligible Southbound stocks under Shanghai Connect; and 901 eligible (for both buy and sell) Northbound stocks and 418 eligible Southbound stocks under Shenzhen Connect. In other words, around 44% and 45% in number of listed A shares respectively on the SSE and SZSE, and around 24% that on SEHK Main Board are Stock Connect eligible securities[5]. By the end of 2016, the average daily Northbound trading value constituted about 2% of the Mainland A-share market total turnover and the average daily Southbound trading value constituted about 8% of the SEHK Main Board total turnover[6].

1.2 Primary market opportunities

The Stock Connect scheme is an unprecedented mechanism that connects the Mainland stock market with an overseas market. However, currently this is confined

---

4   Shares which are placed under "risk alert" by SSE, including shares of "ST companies" and "*ST companies" and shares subject to the delisting process under the SSE rules.

5   Source: HKEX, SSE and SZSE websites.

6   See Chapter 2 of this book, "Shanghai and Shenzhen Stock Connect — A 'Mutual Market' for Mainland and global investors".

**to secondary equity market trading only and investors on either side of the border
are barred from the primary equity market on the other side.** This has deprived
investors of the investment opportunities offered by initial public offers (IPO) of newly
listed companies in the market across the border. According to the statistics of the World
Federation of Exchanges (WFE), Hong Kong, Shanghai and Shenzhen were among the
top ten markets by IPO funds raised in the past two years (see Figure 1). Hong Kong itself
ranked first by IPO funds raised in 5 out of the past 8 years. (see Figure 2.)

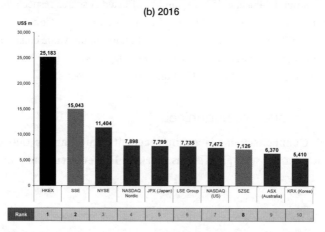

Figure 1. Top 10 exchanges by IPO funds raised (2015 & 2016)

Source: WFE website.

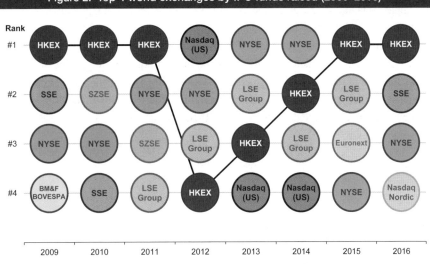

Figure 2. Top 4 world exchanges by IPO funds raised (2009–2016)

Source: WFE website.

In respect of the equity market, the Mainland-Hong Kong Mutual Market established through secondary market connectivity under Stock Connect is incomplete without primary market connectivity. In fact, the absence of primary market connectivity may harm investor interests under secondary market connectivity and bring about market unfairness. The recent case of the spin-off of BOCOM International Holdings Company Limited (BOCOM International) by Bank of Communications Co., Ltd (BOCOM Bank) was an example. BOCOM Bank has its H shares listed in Hong Kong and its A shares listed on the SSE. BOCOM Bank announced its proposed spin-off and listing of BOCOM International on the SEHK in August 2016. BOCOM Bank would provide its existing shareholders with an assured entitlement to the new shares in BOCOM International. However, due to the impediments arising from legal and policy perspectives, BOCOM Bank could only provide the assured entitlement to its existing H-share shareholders but not to its existing A-share shareholders[7]. As advocated by market participants[8], the introduction of primary market connectivity would help resolve this kind of market unfairness.

In its Strategic Plan 2016-2018, HKEX put forward the Primary Equity Connect (PEC)

---

7   An explanation was given in BOCOM Bank's announcement dated 12 September 2016. The reasons include the lack of mechanism under Shanghai Connect for A-share shareholders to subscribe newly issues shares in the Hong Kong market.

8   As reported in *Oriental Daily News* and *Hong Kong Economic Times* on 18 January 2017.

together with Shenzhen Connect, as initiatives to further expand the mutual market connectivity. After implementation of Shenzhen Connect in December 2016, the PEC initiative is expected to further open up opportunities for Mainland investors' global asset allocation. It will also support the further internationalisation of the Mainland stock market and help complete the connectivity mechanism, thereby enabling the pooling of liquidity in the Mutual Market to fulfil its function in fund raising by issuers and stock trading by investors. In a broader sense, the Mutual Market with PEC and possibly other connectivity initiatives could assist China's bigger roadmap in economic development and RMB internationalisation. These are elaborated in sections below.

2   Bottlenecks in Mainland and Hong Kong stock market development

Further market opening and internationalisation have been key policy directions of the Mainland capital market development. The 13th Five-Year Plan on Economic and Social Development (2016-2020) outlines the initiative of constructing a new pattern of all-round opening, which includes expanding two-way financial industry opening and capital market opening. In particular, it has been the central policy to develop Shanghai into an international financial centre (IFC). For this, initiatives were raised in a plan issued in 2015 by the People's Bank of China (PBOC) jointly with other government departments[9] to speed up the development of the Shanghai IFC. In March 2017, the State Council further issued a plan[10] with financial reform and market opening initiatives in the China (Shanghai) Pilot Free Trade Zone (Shanghai FTZ). These include further deepening innovative opening and orderly progress in pilot schemes on capital account opening and internationalisation of financial practices in the zone.

---

9   "Plan on Further Advancing New Financial Market Opening Pilot Schemes in the China (Shanghai) Pilot Free Trade Zone and Speeding Up the Development of the Shanghai International Financial Centre" (《進一步推進中國（上海）自由貿易試驗區金融開放創新試點加快上海國際金融中心建設方案》), 30 October 2015.

10   "Plan for Comprehensive Deepening Reform and Opening of China (Shanghai) Pilot Free Trade Zone" (《全面深化中國（上海）自由貿易試驗區改革開放方案》), 31 March 2017.

Across the border, Hong Kong has been a well-known IFC[11] with active international investor participation in its capital market which has market practices following international standards. With its strengths, the Hong Kong capital market has been supporting the opening and internationalisation of the Mainland capital market through various means including cross-border listing of Chinese enterprises and secondary market trading through Stock Connect. **While internationalisation of the Mainland market is expected to be a long-term process, the Hong Kong market itself also has its weaknesses in the international dimension.** Both sides would need to have some breakthrough in an innovative way, which would be mutually beneficial to each other. The bottlenecks of each market are examined in sub-sections below and the innovative breakthrough is discussed in Section 4.

2.1   "Internationalisation" of the Mainland stock market

On the investor side, eligible foreign investors investing in the Mainland stock market before the launch of Stock Connect were confined only to **Qualified Foreign Institutional Investors (QFIIs) and Renminbi Qualified Foreign Institutional Investors (RQFIIs)**. As of 26 April 2017, an aggregate investment quota of US$90,765 million (~RMB 626,638 million) for 281 QFIIs and of RMB 542,004 million for 183 RQFIIs were authorised by the State Administration of Foreign Exchange (SAFE)[12]. Among them, Hong Kong registered institutions got the largest number and quota value — 23% of QFII quota and 49% of RQFII quota. (see Figure 3.)

However, the total authorised QFII and RQFII quota were less than 3% of the total negotiable market capitalisation of the Shanghai and Shenzhen stock markets[13]. Moreover, not all the QFII and RQFII investment quota would be invested in the stock market.

---

11 Hong Kong is the top 4th global financial centre according to the ranking by Global Financial Centres Index, March 2017, released by Z/Yen in partnership with the China Development Institute (CDI) in Shenzhen. Shanghai, Beijing and Shenzhen ranked 13th, 16th and 22nd respectively.

12 Source: SAFE website.

13 Based on the total authorised quota as of 26 April 2016 compared to the total negotiable market capitalisation of equity securities of RMB 41,405 billion on the Mainland exchanges — SSE (RMB 25,410,107 million) and SZSE (RMB 15,995,368 million) — as at the end of March 2017 (source: SSE and SZSE monthly statistics on their respective websites). "Negotiable market capitalisation" is the market value of listed and marketable securities.

Figure 3. QFII and RQFII authorised quota by registered location (26 Apr 2017)

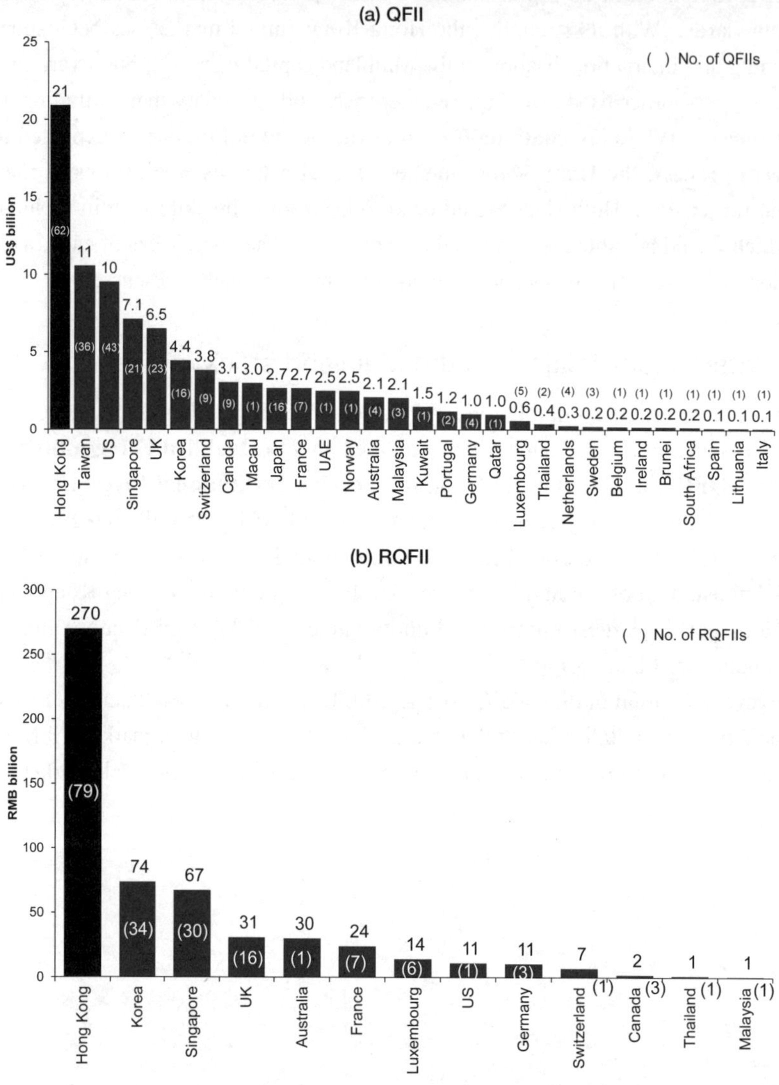

(a) QFII

(b) RQFII

Source: SAFE website.

As at the end of 2016, there were a total of 326 thousand "legal person" investor accounts with the China Securities Depository & Clearing Co., Ltd (CSDC), the central clearing house for the stock markets in Shanghai and Shenzhen.　Out of these, there were only 1,088 QFII accounts (543 in Shanghai and 545 in Shenzhen) and 1,078 RQFII accounts (534 in Shanghai and 544 in Shenzhen), i.e. **less than 1% by number of accounts in total**[14].

In terms of market value of shareholdings, the total investment of QFIIs in the Mainland stock market was RMB 114,440 million as at the end of March 2017, **less than 0.3% of the total negotiable market capitalisation on the SSE and SZSE**[15].　Stock Connect has opened another channel for foreign investors to access the Mainland market, yet the participation is still in a small scale — Northbound trading constituted 2% or less of turnover value in Mainland A shares. (see Section 1.1 above.)

On the issuer side, no foreign companies are yet allowed to list in the Mainland domestic stock market.　The initiative of an International Board on the SSE was raised in 2009 by the Shanghai Government[16] and its exploration was supported by the Central Government in its 12th Five-Year Plan for National Economic and Social Development released in 2011. However, little progress has been observed so far.

On the cash product side other than equities, the only foreign products are some ETFs with foreign assets as the underlying.　As at the end of 2016, there were 6 cross-border ETFs (8%) out of a total of 75 ETFs on the SSE and only two out of 48 ETFs on the SZSE were on foreign underlyings (Hang Seng Index and Nasdaq 100 respectively)[17].

On the market structure side, the Mainland stock market is highly dominated by retail investors.　The market practices, rules and regulations are formulated to meet the special needs of the Mainland market in its course of development, which are not necessarily in line with international practices.

In conclusion, **the degree of internationalisation of the Mainland stock market is relatively low in multiple dimensions, including investor base, issuer base and institutional structure.**　Increasing foreign participation would be conducive towards Shanghai's goal to be an international financial centre.

2.2　"Internationalisation" of the Hong Kong stock market

In terms of investor participation, the Hong Kong stock market is highly internation-

---

14　Source: CSDC Monthly Statistics, December 2016, CSDC website.　Note that different QFII and RQFII fund products would have different investor accounts.

15　Source of QFII investment: Southwest Securities research report on QFII 2017Q1 shareholding status, 1 May 2017.

16　The *Shanghai Government's Opinions on the Implementation of the State Council's Policy on Establishing "Dual Centres"* (《貫徹國務院關於推進〈兩個中心〉建設實施意見》), 11 May 2009.

17　Source: SSE and SZSE websites.

alised. Foreign investor trading in the HKEX cash market constituted a bigger proportion than local investor trading (40% and 36% respectively in 2016). (see Figure 4.) Origins of foreign investors spread across the world — there were 18 reported origins of overseas investors in Asia and 53 reported origins of overseas investors outside Asia, Europe and the US[18].

Figure 4. Distribution of HKEX cash market trading value by investor origin (2016)

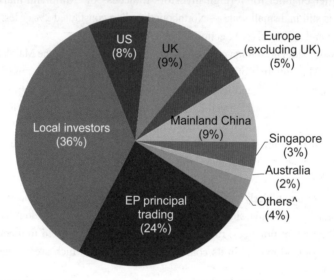

^ Others comprise investors from Japan, Taiwan, Rest of Asia and Rest of the World.

Note: Numbers may not add up to 100% due to rounding.

Source: HKEX Cash Market Transaction Survey 2016.

From 2010 to 2017Q1, almost all IPOs in Hong Kong had made international offers to global investors and over 80% of the IPO funds raised came from international offers[19] during the period (see Figure 5). In other words, **global investors are active in both the primary market subscription for IPOs and the secondary market trading in Hong Kong**.

---

18 Source: HKEX Cash Market Transaction Survey 2016.

19 Global investors participating in international offers would include Hong Kong investors, investors from Mainland China and other overseas investors; and many institutional investors in Hong Kong have an overseas origin.

Figure 5. The share of international offers in Hong Kong IPOs (2007–2017Q1)

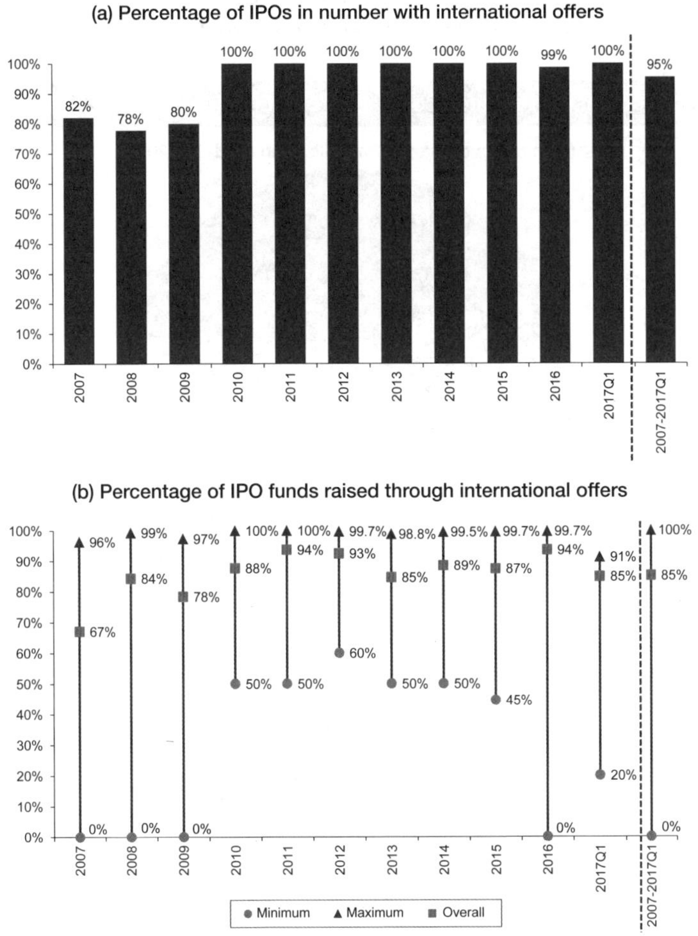

(a) Percentage of IPOs in number with international offers

(b) Percentage of IPO funds raised through international offers

● Minimum   ▲ Maximum   ■ Overall

Source: HKEX.

In terms of listed issuers, however, the overall HKEX market (Main Board and Growth Enterprise Market (GEM)) comprises, to the majority, Mainland enterprises — H shares, red chips and Mainland private enterprises (MPEs) — by market capitalisation (64% as at the end of June 2017) and turnover value (74% for January to June 2017) (see Figure 6).

Among the newly listed companies during the past decade (2008 – 2017Q1), only 8% were of origins other than Hong Kong and Mainland China, contributing 20% of total IPO funds raised. The largest share went to MPEs in terms of number (47%) and H-share companies in terms of IPO funds raised (48%). (see Figure 7.) In other words, **the Hong Kong primary market serves predominantly Hong Kong and Mainland companies and has not delivered its full potential to serve international companies.**

Figure 6. Composition of listed companies on SEHK Main Board by classification

(a) In terms of market capitalisation (End-Jun 2017)

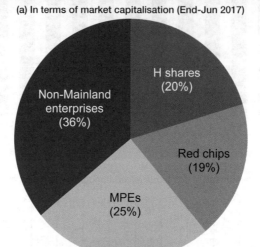

(b) In terms of turnover value (Jan-Jun 2017)

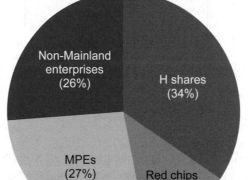

Source: HKEX.

Figure 7. Composition of newly listed companies on SEHK Main Board by classification (2008–2017Q1)

(a) In terms of number of companies

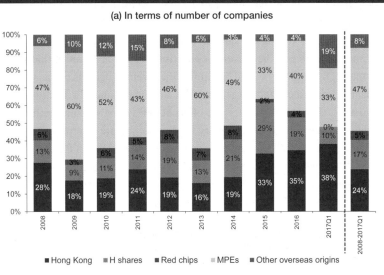

■ Hong Kong   ■ H shares   ■ Red chips   ■ MPEs   ■ Other overseas origins

(b) In terms of IPO funds raised

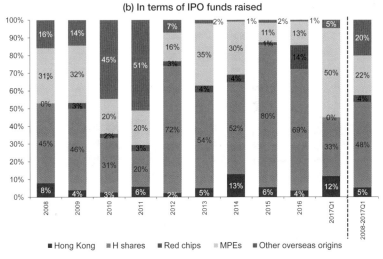

■ Hong Kong   ■ H shares   ■ Red chips   ■ MPEs   ■ Other overseas origins

Source: HKEX.

In conclusion, **the Hong Kong stock market is highly internationalised in terms of investor participation but with a relatively low degree of internationalisation in terms of listed issuers.**

2.3 "Internationalisation" of the Mainland-Hong Kong Mutual Market

The concept of the Mainland-Hong Kong Mutual Market is to open up to international and Mainland investors the access to a sizable market with combined Mainland and international elements through a specially design connectivity model. In respect of the equity market segment, the current connectivity of Stock Connect that is limited to secondary market trading only would undermine the potential benefits of the Mutual Market to participants, both issuers and investors. The major function of a stock market is fund raising by issuers. Towards this end, share offers at times of new listing are fundamental as these provide on the one hand the necessary funding for private enterprises to grow and on the other hand the diverse investment opportunities for investors. In the case of the SEHK Main Board, IPO funds raised constituted as high as 54% of total equity funds raised in 2011 and a significant share of 40% in the recent year of 2016; and share offers were the key post-listing fund raising method in the past decade. (see Figure 8.)

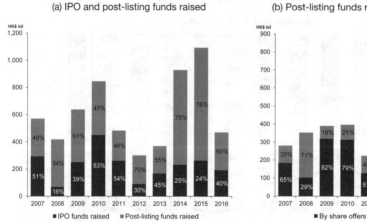

Figure 8. Equity funds raised on SEHK Main Board and percentage share of funds raised by IPO and post-listing issues and share offers (2007–2016)

(a) IPO and post-listing funds raised   (b) Post-listing funds raised by method

■ IPO funds raised   ■ Post-listing funds raised   ■ By share offers   ■ By other methods

Note: IPO fund raising methods comprise offer for subscription, offer for sale and offer for placing. For post-listing equity funds raised, share offer methods comprise placing, rights issue and open offer; other methods comprise consideration issue, warrant exercise and share option scheme.

Source: HKEX.

The internationalisation of the Mainland-Hong Kong mutual stock market would mean the market's possession of an international investor base and an international issuer base. The Hong Kong stock market, possesses an international investor base, but is weak in terms of an international issuer base. The Mainland stock market has yet to see considerable progress in both ends. With secondary market connectivity, a step forward has been made towards internationalisation of the Mutual Market. However, as discussed above, the mutual market connectivity is not complete in the absence of primary market connectivity and this would undermine the efforts towards internationalisation.

3 China's bigger roadmap: National account, market opening and RMB internationalisation

Internationalisation of the Mainland-Hong Kong Mutual Market is not a goal in itself but part of the bigger strategy for China to achieve a better balanced economy, more effective market opening and ultimately a higher degree of RMB internationalisation.

3.1 Possible improvement in the national balance sheet

China's national balance sheet[20] shows that on the assets side the nation had a growing amount of outward direct investment (ODI) and other foreign assets (excluding international reserves) during 2007 to 2013 — ODI grew from RMB 718 billion to RMB 6,147 billion; and the latter grew from RMB 3,866 billion to RMB 6,161 billion. As a percentage of total financial assets, ODI had an increasing share which rose to 1.7% in 2013 while other foreign assets had a similar share but on a decreasing trend. Comparatively, international reserves constituted a more significant percentage share of financial assets (7.5%) in the national balance sheet in 2013. The amount of international reserves maintained at a relatively high level of US$3,010.5 billion in 2016, despite a drop from the recent high of US$3,843.0 billion in 2014[21]. On the liabilities side of the balance sheet, foreign direct investment (FDI) and other foreign liabilities also had an increasing amount during 2007 to 2013, while both maintained a steady percentage share of total liabilities of about 4% and 1.5% respectively in recent years. (see Figure 9.)

---

20 Source: Wind (primary source: China Academy of Social Sciences).
21 Source: National Bureau of Statistics of the PRC.

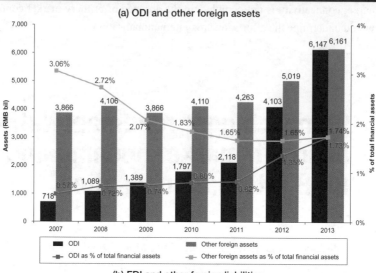

Figure 9. China's outward/foreign direct investment and other foreign assets/liabilities (2007–2013)

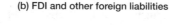

(a) ODI and other foreign assets

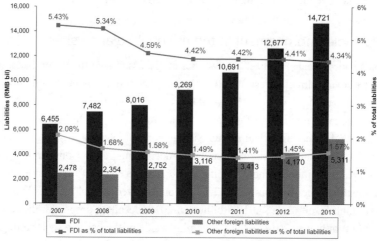

(b) FDI and other foreign liabilities

Source: Wind.

Note: All foreign liabilities are financial liabilities.

Despite growing foreign investment, current account statistics showed that China achieved negative investment income from 2010 to 2014. (see Figure 10.) This is in contrast to the achievement of the US national accounts. Figure 11 shows clearly that the

US achieved an increasing negative net international investment position and an increasing trend in positive investment income during the past decade from 2007 to 2016. The majority proportion of the US investment income came from portfolio investment. (see Figure 12.)

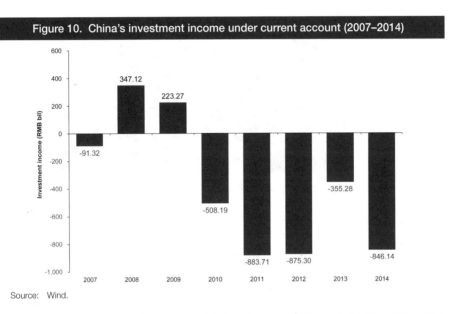

Figure 10. China's investment income under current account (2007–2014)

Source: Wind.

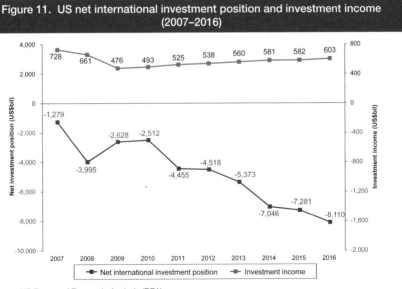

Figure 11. US net international investment position and investment income (2007–2016)

Source: US Bureau of Economic Analysis (BEA).

Figure 12. Composition of US investment income (2007–2016)

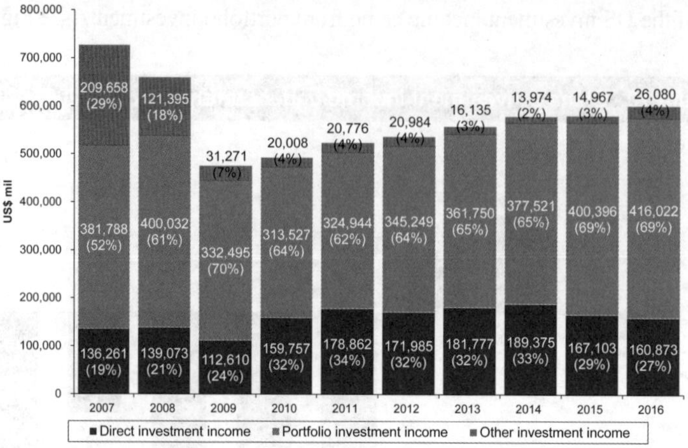

Source: US Bureau of Economic Analysis (BEA).

In fact, the US has maintained a positive net income balance for over two decades since the 1990s, while its net international investment position has been deteriorating. This aroused academic interest in solving this longstanding puzzle that the US is a net borrower from the rest of the world and yet manages to receive net income on its external position. Two major factors were identified: (1) the US has a positive net external equity balance and a negative net external debt balance, and the income yield on equity has been higher than the income yield on debt; and (2) the US earns a persistently higher income yield on its FDI assets than foreigners earn on their direct investments in the US[22]. Nobel Laureate, Paul Krugman, commented that "American assets, often taking the form of foreign subsidiaries of US corporations, earn a higher rate of return than US liabilities … when there is a lot of foreign money parked in (US) Treasuries … As a result, income from US-owned assets abroad consistently exceeds payments on foreign-owned assets in the US"[23]. A study found that the differential income earned from US-owned foreign assets relative to income paid to foreign-owned US assets could be attributed to, among other less significant factors, the differential tax treatment (US tax rate is generally higher than that of countries where the US owns foreign assets), and the higher risk in the foreign invested countries (generating

---

22 Alexandra Heath, "What explains the US net income balance?", Bank for International Settlements, BIS Working Paper No. 223, January 2007.

23 Source: "US Net Investment Income", 31 December 2011, The Opinion Pages — The Conscience of a Liberal, Paul Krugman, The New York Times (https://krugman.blogs.nytimes.com).

higher risk-adjusted returns than foreign-owned US assets)[24]. The same study also found that for portfolio equity and debt, the average yields on claims and liabilities were nearly identical (for the period 1990 to 2010).

The US case is insightful to China, which looks like a mirror image of the US — negative net investment income on positive net foreign assets. China has large international reserves which are mostly invested in US treasuries[25], of which the yield is low. On the other hand, China pays relatively high costs in attracting FDI with its preferential policies, while its ODI has been facing a certain degree of opposition and regulatory obstacles in the target countries. Moreover, China's ODI in developed countries like the US and Europe would have lower risk-adjusted returns than FDI in a developing country like China. **The US case demonstrates a way to improve China's national balance sheet, which is to increase its net international investment income by liberalising outward portfolio investment. Towards this end, developing the Mainland-Hong Kong Mutual Market platform with a high degree of internationalisation would be a promising approach** (see Section 4 below).

3.2　Further financial market opening

As discussed in Section 2 above, further financial market opening is a major policy line in the 13th Five-Year Plan and much policy support has been put in for developing Shanghai into an IFC. Back in 2009, the idea of establishing an international board on the SSE was raised with government policy support, allowing foreign companies to list on the domestic exchange. This policy initiative and the policies supporting ODI are manifestations of China's economic development policy advancement — moving from firstly letting Chinese enterprises go out for fund raising (and foreign capital going in) (since the 1990s) to secondly letting domestic capital go out through ODI and portfolio investment channels like Qualified Domestic Institutional Investor (QDII) scheme (since 2006) and Stock Connect (since 2014), and furthermore to **the future possible fund raising by foreign enterprises in the domestic market.**

The vision of market opening is expected to be two-way (i.e. both inward and outward) comprehensive financial market opening — allowing foreign capital to invest in Mainland China's domestic financial products and allowing domestic capital to invest in foreign financial products. Given the current differences between the Mainland financial

---

24　Stephanie E. Curcuru and Charles P. Thomas, "The return on US direct investment at home and abroad", Board of Governors of the Federal Reserve System, International Finance Discussion Paper No. 1057, October 2012.

25　The composition of China's foreign exchange reserves is not officially disclosed. Some sources give the estimates of about 70% in US dollar assets based on China's economic data. (Source: Wikipedia;〈揭秘：中國 3 萬億美元外匯儲備是如何配置〉, http://finance.sina.com.cn/).

market system and practices and those of the international developed markets as well as the Mainland authorities' concern of domestic financial stability, the path from the current limited opening to a comprehensive degree of market opening is not short and easy. The Mainland-Hong Kong Mutual Market platform, being a closed-loop system with a gradual approach in scope expansion, would facilitate this lengthy market opening process. (see Section 4 below.)

3.3 Towards full capital account convertibility

An orderly realisation of RMB capital account convertibility is a state objective written in the 13th Five-Year Plan and financial market opening is the key towards this objective. An analysis revealed that out of the 40 sub-items of the capital account as classified by the International Monetary Fund (IMF), there are only a few sub-items which remain unconvertible in China [26]. These outstanding sub-items relate mainly to the issuance of equity securities, money market instruments, derivatives and other instruments by non-residents. To help RMB become a convertible and freely usable currency, as stated in the PBOC's 2015 Annual Report, more financial market opening initiatives would include:

- Further liberalisation and facilitation of (1) investment in overseas financial markets by domestic residents and (2) investment in the Mainland financial market by foreign investors; and
- **Allowing qualified foreign companies to issue shares in the domestic market.**

Towards this end, **the Mainland-Hong Kong Mutual Market platform with connectivity in the primary market would help move one more step towards RMB capital account convertibility.** (see Section 4 below.)

4 Primary Equity Connect
— A breakthrough opportunity

Given the development bottleneck for internationalisation of the Mainland and Hong

---

26 The other items are either convertible or basically convertible or partially convertible. Source:〈人民幣資本項目開放的現狀評估及趨勢展望〉, 4 April 2016,《第一財經》(http://www.yicai.com/).

Kong stock markets, the mutual market connectivity model covering both the primary IPO market (Primary Equity Connect) and the secondary stock trading market (Stock Connect) would offer a possible solution. Figure 13 below shows a conceptual model.

Figure 13. Conceptual model of Mainland-Hong Kong mutual market connectivity in both the primary and secondary markets

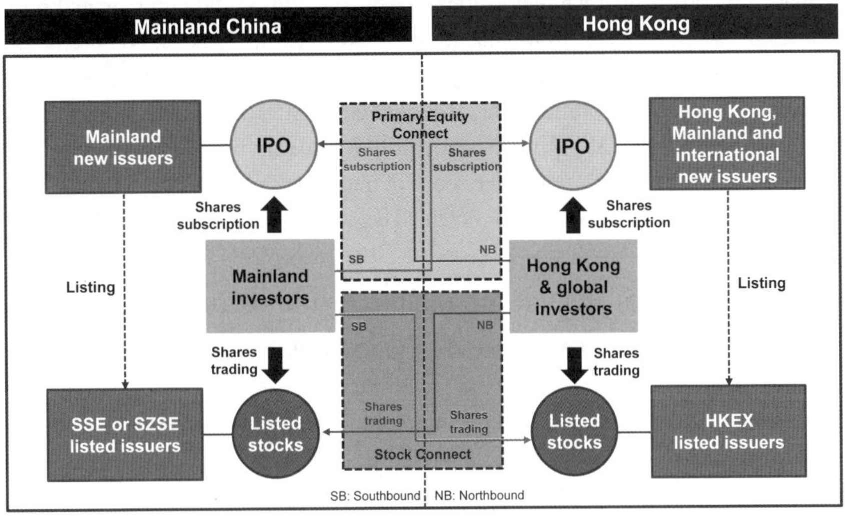

Note: Primary Equity Connect and Stock Connect form a closed system of money flow.

The concept of **Primary Equity Connect (PEC)** is to allow Mainland investors to subscribe for IPOs in the Hong Kong market (Southbound) and global investors in Hong Kong to subscribe for IPOs in the Mainland market (Northbound). Upon listing of the IPO shares issued, trading by investors from the other market would be enabled through the existing Stock Connect scheme. Under this connectivity model, **shares subscription under PEC and shares trading under Stock Connect are effectively contained within a closed-loop system**.

Compared to the QFII and RQFII schemes, foreign investors accessing the Mainland stock market through Stock Connect need not apply for special licences or investment quotas from the Mainland authorities, or open accounts with Mainland brokers or custodians. They adopt their familiar trading practices in trading Mainland stocks through Stock Connect as they do for trading in the Hong Kong stock market. Similarly, Mainland investors adopt their familiar Mainland trading practices in accessing the Hong Kong market through Stock Connect. Moreover, Stock Connect enables Mainland retail investors to directly invest in the Hong Kong market without the need to go through products of QDIIs. A seamless link of PEC for primary market shares subscription with Stock Connect for secondary market trading would be particularly beneficial to investors, owing to such trading efficiency and convenience provided by Stock Connect.

PEC would help enhance the international dimension of the Mainland market by improving international investor participation and the international dimension of the Hong Kong market by broadening the international issuer base. In addition, PEC would open up more investment opportunities to Mainland investors which the Mainland domestic market may not be able to offer in the short term. More importantly, as discussed in Section 3, the Mainland-Hong Kong Mutual Market model could support China's bigger roadmap to achieve a better national balance sheet, further two-way market opening and RMB internationalisation. The expansion of the scope of mutual market connectivity to the primary equity market would be a breakthrough. Potential benefits of PEC are discussed below.

Potential benefits of PEC to the Mainland market and investors

(1) New opportunities for global asset allocation and therefore an improved national balance sheet

During the ten years from 2006 to 2015, domestic savings had increased by a compound annual growth rate (CAGR) of 14% to about RMB 33,708 billion (~US\$5,115 billion)[27]. In the same period, overseas direct investment (ODI) had achieved a CAGR of 32%, while overseas securities investment had a negative CAGR of -0.2% despite the launch of the QDII scheme in 2006[28]. Since 2011, the ratio of overseas securities investment to total domestic savings had maintained at a relatively low level of about 5% while ODI had shot up to about 21% at the end of 2015. (see Figure 14.)

---

27 Source: Wind.
28 CAGR figures were calculated based on annual data obtained from Wind.

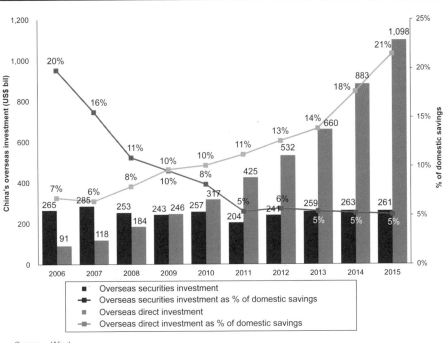

Figure 14. Mainland China's overseas direct investment and securities investment (2006–2015)

Source: Wind.

In addition, the RMB has suffered a depreciation trend since August 2015[29] (see Figure 15) and there has been fear that this may continue for a period of time. Mainland investors, including investment arms of government authorities and enterprises, are eager to seek non-RMB assets for portfolio diversification and hedging against further drop in the value of the RMB.

---

29 This was triggered by the reform of the RMB to USD exchange rate formation mechanism by the PBOC on 11 August 2015 to make it more market-driven.

Figure 15. Daily closings of RMB indices and USD/CNY exchange rate (2014–May 2017)

Note: RXY Global CNY Index is an index launched by HKEX in partnership with Thomson Reuters that measures the performance of the onshore RMB (CNY) against a basket of major international currencies. CFETS CNY Index is the official index on CNY against a basket of major international currencies published by the China Foreign Exchange Trade System (CFETS).

Source: Thomson Reuters.

Relative to the high level of domestic savings, domestic supply of investment products in the Mainland has been inadequate to meet investment demand of Mainland residents for asset value preservation and appreciation. This leads to the situation of asset shortages in Mainland China since 1990 as documented by an academic study, and the consequential phenomenon of asset bubbles, exemplified by price surge in the property market, stock market and even consumer goods [30]. The study explains that asset shortage is measured by the excess in percentage of domestic savings (i.e. the asset demand) relative to the asset supply which is constituted by domestic bonds, shares, loans, change in short-term deposits and net purchase of foreign assets. As a point of reference, the total funds raised by equity issues on the SSE and the SZSE in 2015 was about RMB 1,540 billion[31], which was less

---

30 YANG Shenggang and LIANG Can, *Asset Shortages in China* (《中國資產短缺問題研究》), 2015 (http://www.sinoss. net).

31 In 2015, total funds raised by share issuance on the SSE was RMB 871,082 million and that on the SZSE (including Main Board, SME Board and ChiNext) was RMB 668,902 million (source: SSE and SZSE monthly statistics on their respective websites). Domestic savings were RMB 33,708 billion as at the end of 2015 (source: Wind).

than 5% of the domestic savings as at the end of 2015 — similar to the 5% level for total overseas securities investment in 2015. (see Figure 14.)

Global asset allocation has therefore become an imminent need of Mainland investors. Moreover, there is a need for more income from overseas investments to improve China's national balance sheet. While the One-Belt-One-Road (OBOR) initiative would promote income from ODI by investing in the developing countries[32], increasing overseas portfolio investment could offer potentially significant income.

Currently, apart from Southbound trading under Stock Connect, the QDII scheme is the only legitimate channel for overseas securities investment. However, this is subject to quota limit and approval. As a matter of fact, there has been no increase in QDII investment quota since March 2015, which remained at US$89,993 million by June 2017. (see Figure 16.) This was less than 2% of the domestic savings as of end-2015.

Figure 16. Month-end approved investment quota of QDII (Aug 2012–Jun 2017)

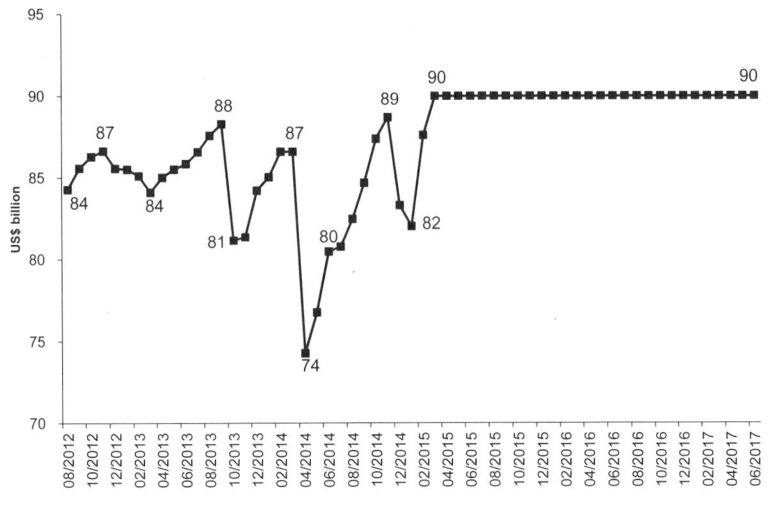

Source: SAFE website.

Besides QDII, there is known to be a "green channel" offered by SAFE under which special authorisation is obtained from SAFE for certain domestic investors (mainly

---

32 OBOR consists of the Silk Road Economic Belt (SREB) and 21st Century Maritime Silk Road (MSR) initiatives. The SREB runs through Central Asia, West Asia, the Middle East to Europe, with extension to South Asia and Southeast Asia. The MSR runs through Southeast Asia, Oceania and North Africa. The report, *Industrial cooperation between countries along the Belt and Road* (《「一帶一路」沿線國家產業合作報告》), released by the China International Trade Institute in August 2015 identified 65 countries along OBOR that will be participating in the initiative.

Mainland cornerstone investors) to subscribe for IPO shares offered by Mainland enterprises to be listed in Hong Kong. This practice started from the IPO of the Postal Savings Bank of China Co., Ltd and has been applied to IPOs of several Mainland enterprises since then. However, this is a special practice under special authorisation applied to special enterprises and is subject to special requirements imposed by SAFE[33].

Southbound PEC would open up one more official global asset allocation channel for Mainland investors, possibly in a wider scale than existing channels, by enabling them to subscribe new shares of international companies to be listed in Hong Kong. This would enhance the overseas portfolio investment of Mainland capital to a larger degree than the current only way of secondary market connectivity.

(2) Facilitating two-way market opening

In an attempt to open up the Mainland domestic market for fund raising by international companies, the success factors would lie in the attractiveness of the Mainland market to potential issuers on the considerations of funding needs and the cost of capital. Some big international companies seeking business expansion in China may be interested due to the branding effect offered by a domestic listing status. Nevertheless, there may not be many of these companies to provide a continuous supply of issuers. As the regulatory framework of the Mainland stock market is very different from that of international developed markets, potentially high compliance costs may deter the majority of potential foreign issuers to list in China, even if an international board is available in the domestic market.

In comparison, with PEC under the Mainland-Hong Kong Mutual Market model, foreign issuers would abide by the more familiar, internationalised rules and standards of the Hong Kong stock market for share offerings to the Mainland investors. As illustrated in Figure 17 below, the connectivity platform of the Mutual Market[34] enables Mainland domestic capital and financial products to go out and international capital and financial products to go into China, without the need for investors and issuers on either side of the Mutual Market to adapt to practices on the other side.

---

33 As informed by market participants, under "green channel" permission, SAFE requires remittance back to the Mainland of the funds raised by the issuer from the IPO and part, if not all, of the proceeds received by the investors upon sale of the subscribed shares. This is considered to be a harsh "double-remittance" requirement.
34 Certain initiatives in developing the platform are subject to regulatory approvals.

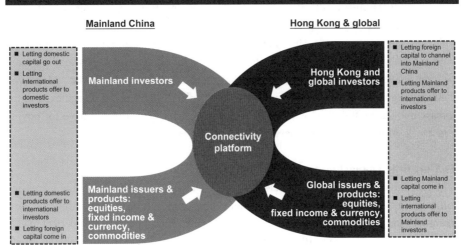

Figure 17. The connectivity platform of the Mainland-Hong Kong mutual market

(3) One more step in RMB capital account convertibility

Southbound PEC under the Mutual Market would effectively implement RMB convertibility in one more item under the capital account, i.e. the offering of shares or equity-type securities by non-residents to the Mainland domestic investors. In the future, other suitable connectivity initiatives (e.g. the offering of financial derivatives) could be introduced under the Mutual Market model for achieving RMB convertibility under the remaining capital account items.

Although in the long run China could achieve comprehensive RMB convertibility under the capital account by some other means, PEC and other connectivity initiatives under the Mutual Market model could help speed up the process in a controlled manner. (see point (7) below.)

(4) Development of international investor base

As noted in Section 2.1 above, foreign holding in the Mainland stock market was less than 0.3% of the total negotiable market capitalisation on the SSE and SZSE, compared to 40% of foreign investor trading in the Hong Kong stock market. **Northbound PEC would provide more opportunities to the Mainland for developing the international investor base in the domestic market**, which is vital to achieving its stock market internationalisation.

IPOs with share offerings through Northbound PEC would be open to all international investors. Compared to domestic IPOs that are open only to qualified foreign investors (QFIIs and RQFIIs), IPOs under Northbound PEC are expected to involve more international marketing efforts from the issuers and the securities industry (both Mainland and international). International investors would be provided with intensive issuer information for their better understanding of the investment value, possibly in a more comprehensive way than domestic offers that focus on domestic investors. Moreover, relatively large holdings by foreign investors in Mainland shares could be achieved in the primary market through PEC without the share price impact as it would have for buying large block of shares in the secondary market.

(5) More listing opportunities to Mainland enterprises

The relatively fast economic development in Mainland China has resulted in increasing domestic savings on the one hand and increasing funding needs of existing and new enterprises for growth and expansion on the other hand. This appears to be a perfect match. However, the current pace of IPOs in the Mainland stock market may not be able to satisfy the long queue of enterprise applications on a timely basis. As reported by Bloomberg[35], as of September 2016 there were over 830 IPO applicants waiting for the approval of the CSRC for listing on the SSE or the SZSE. The CSRC had put in efforts to shorten the queue[36], which was reduced to about 700 by February 2017. However, the typical waiting time was still 18 months or longer[37]. Local media reports indicated that there were more than 700 companies still waiting for approval at the provincial level before they could join the IPO queue[38]. Official exchange statistics show that the SSE and the SZSE recorded a total number of new listings of 103 and 124 respectively in 2016 and 103 and 108 respectively in 2017 up to May. Based on the increased speed of IPO processing in 2017 (about 500 per year), the existing queue, including the provincial waiting queue, could be processed in 3 years' time. However, this number does not take into account the continuous new entrants to the queue as the Chinese economy develops, not to say there could be policy suspension of IPOs at times of bearish markets (as in past experiences).

An IPO listing in Hong Kong would be an alternative to the Mainland enterprises waiting in the Mainland IPO queue. Currently, a Hong Kong listing would enable the enterprises to

---

35 "Few in China's IPO queue likely to benefit from fast-track reform", *Bloomberg*, 12 September 2016.

36 In September 2016, the CSRC allowed companies registered.in any one of 592 impoverished regions nationwide to jump the queue but the objective appeared to be wealth redistribution rather than easing the IPO bottleneck (*Bloomberg*, 12 September 2016); and in February 2017 the CSRC was reported to be considering offering a shortcut for some of the country's largest technology companies to jump the queue.

37 "China to let big tech firms jump IPO queue", *The Straits Times*, 25 February 2017.

38 "Few in China's IPO queue likely to benefit from fast-track reform", *Bloomberg*, 12 September 2016.

build an investor base mainly of Hong Kong and global investors, with Mainland investors to a much lesser extent. With Southbound PEC in place, Mainland enterprises listed in Hong Kong would also be able to offer shares to Mainland investors. In this way, an IPO listing in Hong Kong would become a practical alternative to Mainland enterprises targeting a Mainland investor base, while opening up the opportunity to reach out to global investors. **PEC thus would help relieve the IPO constraints in the Mainland at the current stage of domestic market development.**

(6) Nurturing the Mainland investor base

The international regulatory framework and market practices in Hong Kong make the Hong Kong market a valuable training ground for Mainland investors. The QDII scheme launched in 2006 has benefited the Mainland institutional investors and indirectly the Mainland retail investors through QDII products. Stock Connect launched in November 2014 further opens up the training ground directly to Mainland retail investors in respect of secondary market trading. In respect of IPO subscription in the primary market, the Mainland market practice is much different from international practice. Partly due to supply-demand imbalance, Mainland investors may have been used to price surge of IPOs at listing in the Mainland. **Southbound PEC would offer opportunities to Mainland investors in gaining international experience of IPO shares subscription and price movements (both upside and downside are possible) upon listing, thereby help nurture the Mainland investor base towards maturity.**

(7) Risks of capital outflow within control

Same as the model design of Stock Connect in the secondary market, money flows involved in PEC would be done in a closed-loop system such that on sale of the subscribed shares (presumably through Stock Connect), the money would flow back to the buying investors in the origin market. This would alleviate the concern of capital outflow from the Mainland market.

Potential benefits of PEC to the Hong Kong market and global investors

(1) Attracting the listing of international companies

Southbound PEC would open up the funding pool from Mainland investors and this would attract potential global issuers to list in Hong Kong. Given the sheer size of Mainland domestic savings and business opportunities in China, a listing in Hong Kong under Southbound PEC is believed to be attractive to international companies in consideration of sizable IPOs and companies with business development strategies in China. Moreover, with Mainland investors as potential subscribers, international companies would have greater confidence in a successful IPO in Hong Kong.

(2) Activating the market by increasing investor participation

To complete the shares offering and trading cycle in the Mutual Market, PEC in the primary market is expected to be operationally linked with Stock Connect in the secondary market so that shares subscribed under PEC by investors on either market could be traded through Stock Connect. The Hong Kong market fuelled by PEC would be further activated by expectedly increased international listings and the enlarged liquidity pool with Mainland investor participation. Increased market liquidity in both the primary market and the secondary market would be welcomed by global investors.

(3) Benefiting market intermediaries with vibrant primary and secondary market activities

The increased primary and secondary market activities would provide more business opportunities to market intermediaries in Hong Kong including investment banks, law firms, accounting and auditing firms, and securities brokers.

Potential benefits of PEC to the Mainland-Hong Kong Mutual Market

The implementation of PEC would help fulfil the Mainland-Hong Kong Mutual Market ecosystem in the equity market segment. This would be a breakthrough opportunity for linking up international investors with Mainland companies and Mainland investors with international companies, thereby helping achieve the vision of the Mutual Market in the international dimension. **Being scalable, the Mainland-Hong Kong Mutual Market could ultimately become a financial supermarket for Mainland and international investors, offering to them different kinds of financial products from all over the world.**

5   Issues for consideration in implementing PEC

Like Stock Connect, PEC is a brand new concept which would require careful consideration of practical issues and market implications for its implementation. Nevertheless, with a suitable model design, the challenges could be largely overcome and the benefits would outweigh the costs. These challenges are discussed below.

(1) Potential increase in market competition

The increased foreign participation in Mainland IPOs under Northbound PEC may be regarded as a booster on top of the strong domestic demand for Mainland IPOs. There may therefore be the concern of further boosting the considerably active IPO subscription market in the Mainland. Quite the contrary, it is believed that, with proper model design, **PEC could contribute to a fairer pricing in, and a more healthy development of, the Mainland primary market**. Under the current market practice in the shares allocation process, QFIIs and RQFIIs could only get a minute portion of the Mainland IPO shares subscribed. The shares allocation mechanism in the Mainland market is quite different from international market practice and constrains foreign participation in the primary market. Northbound PEC could be designed to allow greater foreign participation, in both new shares subscription and allotment, as well as in the pricing process. This would help bring the Mainland primary market to gradually align with international practice.

Secondly, there may be the concern from the Mainland markets about increased competition from Hong Kong for the listing of Mainland issuers under PEC, as the Mainland issuers may have higher intention to opt to list in Hong Kong instead of in the domestic markets. This, however, would not be a major issue since the existing regulatory requirements for overseas listing of Chinese enterprises, particularly for H-share companies, would not be affected under PEC. The additional attraction under PEC would be the availability of Mainland investor demand. **PEC would in fact help alleviate the overwhelming pressure on IPO processing in the Mainland**.

Thirdly, there may be concern about the potential competition from Hong Kong with the planned International Board in the Mainland for the listing of international companies. In this respect, **PEC in its conceptual model is believed not to have any impact on the International Board** for which the issuer eligibility requirements and operational details are yet to be ascertained. On the contrary, market views consider that it would probably be the overseas exchanges that might face the competition from Hong Kong under Southbound PEC for the listing of international companies. The internationalisation of the Mainland stock market in fact would need multiple initiatives for trial implementation, to be modulated as appropriate as the market develops in time. **The Mutual Market with PEC would effectively act as a kind of China's offshore international board, that could run in parallel with its onshore international board.**

For the Hong Kong market, the increased liquidity pool under Southbound PEC is generally considered beneficial rather than competitive for IPO subscriptions in Hong Kong.

(2) Regulation and investor protection

Market regulation for PEC would be a major area that requires careful consideration. This would include eligibility criteria for issuers and investors, and the obligations and liabilities of exchanges, market regulators, intermediaries, issuers and investors on both sides of the Mutual Market. **The ultimate objective would be to provide a level-playing field with adequate investor protection and risk control.**

Mainland and Hong Kong stock markets have different primary market regimes, including IPO eligibility criteria, listing requirements and disclosure requirements in IPO prospectus. For IPOs under Southbound PEC targeting Mainland investors, additional issuer eligibility and disclosure requirements may need to be imposed[39]. Conversely, additional disclosure requirements may also be needed for IPOs under Northbound PEC to be offered to local and global investors in Hong Kong. Moreover, there may be certain state-level restrictions on foreign ownership of Mainland enterprises in strategic industries. As a result, certain specific IPO eligibility criteria or restrictions on foreign subscription for Northbound PEC may apply.

In case of any incident relating to regulatory matters of issuers under PEC that might impact investors' interests, the application of which market's investor protection regime, the roles of the market regulators on either side, the liabilities of the issuers, exchanges and intermediaries would need to be clarified. In addition, cross-border investors' credit risk would be a concern to market intermediaries. Nevertheless, these could be tackled by a clear and sound regulatory framework and suitable operational design.

Moreover, **extensive investor education** would need to be done in order to equip

---

39 As reference, cross-border offerings of shares to investors in markets like the US and the UK would need to follow certain regulatory requirements, including information disclosure requirements, of the market where the cross-border offering is made.

investors on each side of the Mutual Market with knowledge of the primary market practices and investment behaviour on the other side for self-protection.

(3) Operational issues

Whether it is Northbound or Southbound, PEC would involve an extension of IPO shares subscription to investors on the other side of the Mutual Market. In contrast to secondary market trading under Stock Connect where there are no changes in the listed stocks' home market trading, clearing and settlement practices, the shares subscription and allotment practices under PEC in the IPO home market would need to be carefully designed to accommodate cross-border investors.

Questions to be answered would include but are not limited to: (1) Would cross-border retail investors be allowed to subscribe for PEC shares or would PEC be opened only to cross-border institutional investors? (2) How would PEC shares be allotted to cross-border investors? (3) Would there be a separate subscription pool for cross-border subscription or a combined pool with domestic market subscription? (4) Would cross-border subscription be subject to different market rules or follow the IPO home market rules? (5) Would cross-border investors be served by intermediaries in the IPO home market or in the investor's market? (6) Would intermediaries which serve cross-border investors be subject to different regulatory requirements, e.g. Know-Your-Client (KYC) rules and placement guidelines?

Since IPO procedures and market practices are very different between Hong Kong and the Mainland, **considerable efforts are expected to formulate the PEC operating model design in order to meet the interests of both issuers and investors, as well as to maintain the basic principles of market fairness, openness and integrity**.

Besides, the post-IPO shares trading support for PEC by **a seamless link with Stock Connect** is necessary. A proper regulatory and operational framework will have to be prepared for this arrangement. Given the usual market practice of funding offered by brokers to investors for IPO shares subscription in Hong Kong, there is advocacy from the market for enabling block trades in Stock Connect to facilitate such funding service support. The reason is that, without block trade facility, there would be market impact on the stock price for investors to sell a large volume of their subscribed shares in the secondary market in order to repay the brokers for the funding. Implementing PEC without a corresponding block trade facility in Stock Connect would be a deterrent to IPO subscriptions by cross-border investors who need funding support from brokers. These kinds of operational detail would need to be catered for in linking PEC with Stock Connect.

6   Conclusion

The Mainland-Hong Kong Mutual Market established upon the launch of the Stock Connect scheme in respect of the secondary market trading dimension is incomplete in the absence of primary market connectivity. The introduction of PEC would fulfil the fundamental functions of the Mutual Market in fund raising by issuers in addition to stock trading by investors. It could also improve the current deficiencies in the international dimension of the Mainland and Hong Kong stock markets and therefore of the Mutual Market. With PEC, the Mutual Market could offer more effective support to China's bigger roadmap to achieve a better balanced economy, further market opening and ultimately a higher degree of RMB internationalisation.

PEC would offer mutual benefits to both sides of the Mutual Market. To the Mainland, PEC would (1) offer new opportunities for Mainland investors' global asset allocation and therefore an improved national balance sheet, (2) facilitate two-way market opening at lower costs, (3) help achieve one more step in RMB capital account convertibility, (4) support the development of the market's international investor base, (5) provide more listing opportunities for Mainland enterprises and (6) help nurture the Mainland investor base, while containing potential risks with suitable control measures. To Hong Kong, PEC would help attract the listing of international companies, further activate the market by increasing investor participation and also benefit market intermediaries with more business opportunities.

The implementation of PEC would involve additional issues, including regulatory and operational issues, beyond the Stock Connect scheme. Nevertheless, it is believed that these challenges could be largely overcome with a suitable model design that could meet the best interests of the Mainland-Hong Kong Mutual Market. This would ultimately benefit China's national accounts and the big strategy of RMB internationalisation.

Abbreviations

CSRC	China Securities Regulatory Commission
ETF	Exchange-traded fund
FDI	Foreign direct investment
IFC	International financial centre
IPO	Initial public offer
ODI	Overseas direct investment
PBOC	People's Bank of China
PEC	Primary Equity Connect
QDII	Qualified Domestic Institutional Investor
QFII	Qualified Foreign Institutional Investor
RQFII	Renminbi Qualified Foreign Institutional Investor
SAFE	State Administration of Foreign Exchange
SEHK	Stock Exchange of Hong Kong
SSE	Shanghai Stock Exchange
SZSE	Shenzhen Stock Exchange
WFE	World Federation of Exchanges

Remark

This research report has made reference to views and feedback on the PEC concept sought from primary market participants in Hong Kong, including investment banks, broker firms, law firms, accounting firms and fund managers.

Chapter 4

CES Stock Connect Hong Kong Select 100 Index

A key measure of Hong Kong market
investment under Stock Connect

24 November 2017

Summary

The Mainland-Hong Kong Mutual Market was basically established after the launch of the Shanghai-Hong Kong Stock Connect in November 2014 and the Shenzhen-Hong Kong Stock Connect in December 2016 (collectively referred to as the Stock Connect scheme). This has opened up more opportunities for Mainland investors, with fewer restrictions than previous channels like Qualified Domestic Institutional Investors (QDII), for investing in overseas markets. Statistics show that there is a strong and growing interest from the Mainland investors in trading stocks listed in Hong Kong through Southbound Trading under the scheme.

As the Mutual Market attracts increasing investor interest, related index services to support the continuous growth and development of the market have become inevitable. Indices not only serve as benchmark measures to track the performance of domestic/regional/global markets, segments of a market or cross-markets, they are also increasingly used as the underlying benchmarks for passive investment instruments such as exchange-traded funds (ETFs) or derivatives such as index futures and options to gain exposure to, or to hedge against investment in, a market.

CES Stock Connect Hong Kong Select 100 Index (CES SCHK100) is a unique Stock Connect-related index that tracks the Hong Kong stocks available for Southbound Trading (the Southbound Stocks), with the following special characteristics:

(1) Having a considerably high coverage of Southbound stocks in terms of market capitalisation and turnover value;

(2) Tracking investment of a pure Hong Kong concept, with a high representativeness of stocks listed in Hong Kong and not in the Mainland market at the same time, therefore representing pure investment opportunities outside Mainland China, with only moderate correlation with movements in the Mainland domestic stock market;

(3) Having a relatively high coverage of stocks of the growth sectors among Southbound stocks, e.g. Mainland private enterprises and stocks in the New Economy industries;

(4) Owing to its stock composition, having historically higher PE ratio but lower dividend yield than HSI and HSCEI, and also lower return volatility than the key Hong Kong and Mainland indices, for most of the time since its launch.

Given the index's high representativeness of Southbound eligible stocks and growth-sector stocks in the New Economy, the potential of investment opportunities in its constituent stocks is yet to be exploited further through Southbound Trading. Towards this end, CES SCHK100 is potentially a useful benchmark for developing passive investment instruments like ETFs for Southbound Trading.

1 Investment opportunities to Mainland investors in the Mainland-Hong Kong Mutual Market

1.1   Establishment of the Mainland-Hong Kong Mutual Market[1]

The Mutual Market Access pilot programme (the "Pilot Programme"), launched in November 2014 with initially the **Shanghai-Hong Kong Stock Connect (Shanghai Connect)**, opens up a brand new channel for overseas investors to invest in the Mainland stock market and Mainland investors to invest in overseas assets. The Shanghai Connect enables Hong Kong and overseas investors to trade securities listed on the Shanghai Stock Exchange (SSE) in the Mainland (**Northbound Trading**) and Mainland investors to trade securities listed on the Stock Exchange of Hong Kong Limited (SEHK) in Hong Kong (**Southbound Trading**), within the eligible scope of the programme through the trading and clearing infrastructure connections between the two exchanges and between the corresponding clearing houses in the two markets.

Following successful and smooth operation of Shanghai Connect, the **Shenzhen-Hong Kong Stock Connect (Shenzhen Connect)** was subsequently launched in December 2016. Similar to Shanghai Connect, Shenzhen Connect enables Northbound Trading of securities listed on the Shenzhen Stock Exchange (SZSE) by Hong Kong and overseas investors and Southbound Trading of securities listed on the SEHK by Mainland investors through infrastructural connections between the SEHK and SZSE and between their clearing houses. The scope of eligible securities under Shenzhen Connect covers also the smaller sized ("small-cap") stocks in addition to the "large-cap" and "mid-cap" stocks under the Shanghai Connect[2]. (Hereinafter, Shanghai Connect and Shenzhen Connect are collectively referred to as the "Stock Connect" scheme.)

After the launch of Shenzhen Connect, the basic model of "**Mutual Market**" across Shanghai, Shenzhen and Hong Kong has been established, albeit within the scope of eligible securities. As the mutual stock market access scheme is scalable, this potentially

---

1   See also Chapter 2 of this book, "Shanghai and Shenzhen Stock Connect — A 'mutual market' for Mainland and global investors".
2   See websites of the HKEX, SSE and SZSE for the eligible securities under the Stock Connect scheme.

opens up a Mainland-Hong Kong mutual stock market of a combined equity market value of US$10,514.0 billion (as of end-2016) and an average daily equity turnover of about US$84.3 billion (2016), ranking 2nd by market value (following New York Stock Exchange) and 2nd by equity market turnover among world exchanges[3].

1.2   Value to the Mainland investors

Prior to the launch of the Stock Connect scheme, the Qualified Domestic Institutional Investor (QDII) scheme was the only official channel for Mainland investors to invest in overseas markets, subject to investment quota approved by the State Administration of Foreign Exchange (SAFE). While Mainland qualified institutional investors could directly invest overseas after obtaining QDII licences and the authorised quota, Mainland retail investors who would like to invest in overseas markets could do so through available QDII products. However, there has been no increase in the aggregate QDII investment quota since March 2015 up to October 2017, which remained unchanged at US$89,993 million during the period[4].

Southbound Trading under Stock Connect opens up a new regularised channel for Mainland investors, both individuals and institutions, to directly invest in overseas assets. In contrary to the QDII scheme, no aggregate quota is applied to trading under Stock Connect, which is subject only to a Daily Quota which is applied on a net-buy basis[5]. The channel is a closed-loop system with prudential real-time monitoring of the usage of the Daily Quota, and yet with considerable flexibility in the absence of an aggregate quota. As the Daily Quota applies on a net-buy basis and is reset every day (once it is reached during the day, only cross-border sell orders of eligible securities are accepted), there is virtually no limitation of stock turnover during the day. The Mutual Market under Stock Connect therefore effectively extends the universe of investable assets for Mainland investors which could possibly offer investment opportunities of good potential returns to them.

Southbound Trading under Stock Connect has been on a strong rising trend in the recent year, demonstrating **a strong and growing interest from the Mainland investors in trading stocks listed in Hong Kong through this channel**. This would give rise to increasing

---

3   World Federation of Exchanges (WFE) statistics, from WFE website (on 20 January 2017 for market value data and on 1 March 2017 for trading value data). Average daily turnover was calculated from the combined shares turnover value for 2016 from WFE statistics using the total number of trading days (244 days) for the Mainland market. Ranking was based on 2016 up to December combined trading value.

4   Source: SAFE website.

5   The Daily Quota limits the maximum net buy value to RMB 13 billion for Northbound stocks and RMB 10.5 billion for Southbound stocks for each of Shanghai Connect and Shenzhen Connect.

demand for associated services, including index services and related investment vehicles. (See Appendix 1 for details of Southbound Trading activities under Stock Connect.)

2 Index services for Mutual Market investment

2.1   Indexing for the Mainland-Hong Kong Mutual Market

To better serve the Mainland-Hong Kong Mutual Market with related market services, Hong Kong Exchanges and Clearing Ltd (HKEX), SSE and SZSE jointly established **China Exchanges Services Company Ltd (CESC)** in 2012. CESC's business started with developing cross-border indices covering the Hong Kong, Shanghai and Shenzhen markets. These indices provide the foundations for tradable index products for the benefits of Mainland and global investors for investing in the Mutual Market. Figure 1 below presents the current CESC family of indices.

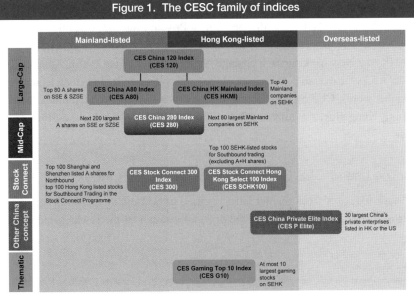

Figure 1. The CESC family of indices

Source: CESC website.

The CESC indices serving the Mainland-Hong Kong Mutual Market have two main categories:

(1) CES China indices — Indices comprising Mainland company stocks listed in Mainland China and/or those listed in Hong Kong

- **CES China 120 Index (CES 120)** — Comprising 120 large-cap companies, with 80 A-share companies on the Mainland exchanges and 40 Mainland companies on SEHK

- **CES China A80 Index (CES A80)** — Comprising the 80 A-share companies constituting the CES 120

- **CES China HK Mainland Index (CES HKMI)** — Comprising the 40 Mainland companies constituting the CES 120

- **CES China 280 Index (CES 280)** — Comprising 280 mid-cap companies with size following the 120 large-cap companies in CES 120, with 200 A shares on the Mainland exchanges and 80 Mainland companies on SEHK

(2) CES Stock Connect indices — Indices comprising eligible stocks under Stock Connect that are listed in Mainland China and/or those listed in Hong Kong

- **CES Stock Connect 300 Index (CES 300)** — Comprising the top 100 A shares on each of SSE and SZSE for Northbound Trading and top 100 Hong Kong listed stocks for Southbound Trading

- **CES Stock Connect Hong Kong Select 100 Index (CES SCHK100)** — Comprising the largest 100 stocks on SEHK for Southbound Trading, excluding H shares which have A shares listed in the Mainland (A+H shares)

Other CESC indices include the CES China Private Elite Index and the CES Gaming Top 10 Index. The former tracks the 30 largest China's private-owned enterprises listed on the SEHK, the New York Stock Exchange, the NASDAQ or the NYSE American[6] while the latter tracks the overall performance of the gaming stocks listed in Hong Kong.

In other words, **CES SCHK100 is a unique Stock Connect-related index that tracks the Hong Kong stocks available for Southbound Trading under the Stock Connect scheme**.

2.2 The use of CES SCHK100 as a benchmark index

A stock index (or stock market index) is used to measure the performance of a stock market or a segment of it, based on the price movements of the stocks listed and traded

---

6   Formerly known as the American Stock Exchange (AMEX) in the early days and more recently as NYSE MKT.

on the market. Stock indices are widely used by investors as benchmarks to compare the performance of their own investment portfolios or those of financial managers against the performance of the overall market or the market segment that the stock index measures.

A stock index is created, formulated, produced, maintained and disseminated in response to market needs. Wherever there is investor interest in investing in a market or market segment or across different markets, there will be a need for stock indices to benchmark the performance of their investments. The Dow Jones Industrial Average (DJIA) is known to be the first stock index, which was created in the 19th century to measure the performance of the New York stock market. Nowadays, there are different types of stock indices covering different market scopes. (see illustration in Table 1.)

Table 1. Major types of stock index		
Type of stock index	Nature	Examples
"National" indices	Measure the performance of the key stock markets (or major stock exchanges) of a given nation or locality	**US:** DJIA and NASDAQ 100; **UK:** FTSE 100; **Japan:** Nikkei 225; **Hong Kong:** Hang Seng Index (HSI); **Shanghai:** SSE Composite; **Shenzhen:** SZSE Composite
"Segment" indices	Measure the performance of a specific segment of an exchange market	**Hang Seng Composite Size Indices:** HSLI, HSMI, HSSI [7]; **SZSE:** SME Composite Index, ChiNext Composite Index
"Regional" indices	Measure the performance of stock markets in a specific region defined geographically or by the level of industrialisation or income	EURO STOXX 50 [8], MSCI Emerging Markets Index [9]
"Sectoral" indices	Measure the performance of a specific industry sector of stocks, which may be within the same market or across different markets	Hang Seng Industry Sub-indexes[10], STOXX Asia/Pacific 600 industry indices[11]

---

7   HSLI is the Hang Seng Composite LargeCap Index; HSMI is the Hong Kong Composite MidCap Index; HSSI is the Hang Seng Composite SmallCap Index.

8   EURO STOXX 50 Index is a blue-chip index for the Eurozone, covering 50 stocks from 11 Eurozone countries (source: STOXX website, https://www.stoxx.com/).

9   The MSCI Emerging Markets Index captures large- and mid-cap representation across 24 emerging market countries, with 843 constituents (source: MSCI website, https://www.msci.com/).

10 The Hang Seng Industry Sub-indexes consist of Finance, Utilities, Properties, and Commerce and Industry sub-indices.

11 The STOXX Asia/Pacific 600 industry index series consists of indices covering respectively 19 industry sectors across multiple Asia/Pacific markets including Japan, Hong Kong, Australia and Singapore (source: STOXX website, https://www.stoxx.com/).

(continued)

Table 1. Major types of stock Index		
Type of stock index	Nature	Examples
"World" / "Global" indices	Measure the performance of the stock markets from multiple regions across the world	MSCI World Index[12], S&P Global 100[13]
"Thematic" and "Strategy" indices	Indices created on certain investment themes or strategies	Hang Seng China Enterprises Index for listed Chinese companies in Hong Kong (HSCEI)[14], HSI Volatility Index

Indices can have multiple versions which differ in the calculation methodology. They can differ based on how the index constituents are weighted and on how dividends are accounted for. The common versions include the **price return indices** which only consider the price of the constituents, and the **total return indices** which account for dividend reinvestment. Another dimension of difference lies in the weighting method — price only, full capitalisation or free float-adjusted. DJIA is the best example of a **price-weighted index**[15] while HSI is **free float-adjusted market capitalisation weighted** index with a 10% cap on individual securities[16], which is the most common index methodology in the stock market.

The CES SCHK100 is considered a "thematic" benchmark index for measuring the performance of Southbound investment under Stock Connect.

Apart from being a performance measure, stock indices (or indices in general that apply to asset markets other than stocks, such as property and commodities) are **increasingly used as the underlying assets of passive investment instruments such as exchange-traded funds (ETFs) or derivatives such as index futures and options** to gain exposure to, or to hedge against investment in, a market. ETFs, in particular, have become increasingly popular in the recent years — according to ETFGI[17], US$3.548 trillion was invested in the 6,630 ETFs or exchange-traded products (ETPs) listed globally at the end of 2016, with

---

12 The MSCI World Index captures large- and mid-cap representation across 23 developed market countries, with 1,654 constituents (source: MSCI website, https://www.msci.com).

13 The S&P Global 100 Index measures the performance of multi-national, blue-chip companies of major importance in the global equity markets, with 100 highly liquid constituents (source: S&P Dow Jones Indices website, https://us.spindices.com).

14 The HSCEI was initially designed to track H-share companies listed in Hong Kong. The index compiler, Hang Seng Indexes Company Ltd, announced in August 2017 the addition of red chips (which are state-controlled enterprises incorporated outside Mainland China) and Mainland private enterprises (MPEs or "P-chips") to the index in five phases from March 2018 to March 2019.

15 This owes very much to legacy reasons as the DJIA was introduced at a time when automation of index calculation by computer means was not available.

16 Source: Hang Seng Indexes website, http://www.hsi.com.hk.

17 ETFGI LLP is a wholly independent research and consultancy firm providing services to leading global institutional and professional investors, the global ETF and ETP industry, its regulators and advisers (http://etfgi.com).

a net inflow of US$389.34 billion in the year, recording 35 consecutive months of net inflows[18]. According to BlackRock, a net inflow of US$189.1 billion into the global ETF market was recorded in the first quarter of 2017, the highest record so far; of this, US$109.1 billion flowed into stock ETFs[19].

An ETF is a marketable security that tracks an index, a commodity, bonds, or a basket of assets. ETFs trade like common stocks on a stock exchange. Investment in an ETF replicates an investment in its underlying asset(s). For a stock index ETF, the underlying asset composition will be the same as the relative weightings of the constituents in the index. The increasing popularity of ETFs lies in the simple way it offers to investors to gain investment exposure to virtually any asset class, geography or sector, at relatively low costs vis-à-vis active fund management[20].

Therefore, the availability of a suitable index for a market or an investable segment of an asset group is not only important for market performance measurement, but also helps promoting or facilitating investment, thereby driving up liquidity, in the underlying market or asset group by enabling the creation of ETFs based on the index. For achieving the latter successfully through ETFs, two conditions must be met: (1) the index must be investable, i.e. the components of the index must be tradable in a free and open market; (2) it must be possible to buy all the components of the index in accordance with their respective weightings in the index without incurring very high transaction costs or having market impact.

For the Mainland-Hong Kong Mutual Market that attracts increasing investor interest, related index services and index-related investment products have become inevitable as a support to the continuous growth and development of the market.

2.3   Mutual Market index-linked investment vehicles

The first cross-border index-linked investment product [21] developed specially for the Mainland-Hong Kong Mutual Market was introduced in 2013 — the CES 120 index futures launched on 12 August 2013 on HKEX's derivatives market. In the same year, three related index ETFs were also listed on HKEX's securities market (see Table 2). These are considered "Mutual Market concept" index-linked products as their underlying indices are members of the cross-border index family, which are different from ETFs based on indices of Hong Kong stocks or Mainland A shares or Mainland companies listed in Hong Kong or overseas that are built independently of the Mutual Market concept.

---

18 Source: "ETF industry grew faster than hedge funds in 2016 — ETFGI", *International Adviser*, 6 March 2017.
19 Source: 〈貝萊德：首季全球 ETF 吸 1.4 萬億新高〉, *Hong Kong Economic Times*, 4 May 2017.
20 For reference, see "The evolution of the ETF industry", *Pensions & Investments*, 31 January 2017.
21 "Cross-border" refers specifically to Mainland-Hong Kong cross border. Cross-border investment products mean products that enable Mainland investors to invest in Hong Kong assets and/or vice versa.

Table 2. Mutual Market concept index-linked products in Hong Kong (as of end-Aug 2017)			
Futures product			Launch date
CES China 120 Index Futures			12/08/2013
ETFs (dual-counter stocks)	HKD counter	RMB counter	Listing date
ChinaAMC CES China A80 Index ETF*	3180	83180	26/08/2013
CSOP CES China A80 ETF	3137	83137	23/09/2013
E Fund CES China 120 Index ETF	3120	83120	21/10/2013

Note: The ETF was delisted on 10 November 2017.
Source: HKEX.

In Hong Kong, cross-border index-linked products have been available for Hong Kong and overseas investors to invest in the Mainland stock market long before the launch of Stock Connect. The first ETF listed on SEHK with Mainland-listed shares included as the underlyings was the iShares MSCI China Index ETF[22] listed on 28 November 2001. This was followed by the iShares FTSE A50 China Index ETF on 18 November 2004. As of end-August 2017, out of a total of 93 physical stock index ETFs listed on the SEHK, there were about 30% ETFs which have Mainland-listed shares included as underlyings[23]. The underlying indices include MSCI China Index, FTSE China A50 Index, SSE 50 Index, CSI indices, CES indices and others. All these ETFs on SEHK constituted 28% of total turnover in ETFs in 2017 up to August[24].

In the Mainland, there were a total of 85 and 53 ETFs listed respectively on the SSE and the SZSE as at the end of August 2017. Of these, there was only 3 cross-border ETFs on the SSE, which are on the HSI and HSCEI, and only one cross-border ETF on the SZSE, which is on the HSI. Turnover value in the single ETF on HSI ranked 5th among all ETFs listed on the SZSE for the period January to August 2017, constituting 1.6% of the total turnover of ETFs on the SZSE during the period (while the top 3 ETFs by turnover value already contributed 87% of total ETF turnover on the SZSE during the period)[25].

---

22 The MSCI China Index comprises companies incorporated in the People's Republic of China (PRC) and listed either on the SSE, the SZSE or the SEHK. Constituents include Mainland-listed B shares, Hong Kong-listed H shares, P-chips and foreign listings (e.g. American Depositary Receipts). MSCI announced in June 2017 the inclusion of China A shares in the MSCI Emerging Markets Index and MSCI ACWI Index series beginning in June 2018. This would include MSCI China Index.
23 The figures excluded dual counting of dual-counters. Source: HKEX website.
24 Source: Calculation based on statistics in Trading Summary for ETFs, August 2017, on HKEX website.
25 Source: Monthly Statistics on the SZSE website.

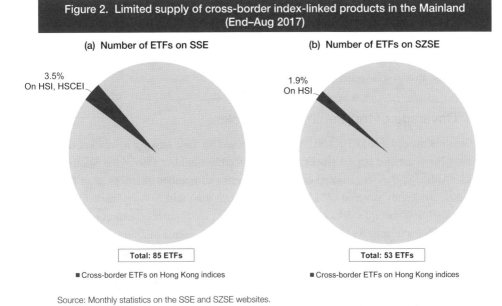

Figure 2. Limited supply of cross-border index-linked products in the Mainland (End–Aug 2017)

(a) Number of ETFs on SSE

(b) Number of ETFs on SZSE

3.5%
On HSI, HSCEI

1.9%
On HSI

Total: 85 ETFs

Total: 53 ETFs

■ Cross-border ETFs on Hong Kong indices

■ Cross-border ETFs on Hong Kong indices

Source: Monthly statistics on the SSE and SZSE websites.

It is apparent that **there is only a limited supply of cross-border index-linked products in the Mainland to serve investment in the Mutual Market while the potential demand is expected to be strong**.

3   The CES SCHK100 Index
— A tradable index for Southbound investment

CES SCHK100 was launched on 15 December 2014. It has a base value of 2000 points on the base date of 31 December 2008. The index constituents consist of the top 100 SEHK-listed stocks by market value that are available for Southbound Trading under the Stock Connect scheme, excluding the dual-listed A+H shares. It adopts the common index compilation method of free float-adjusted market capitalisation weighted with a 10% cap. Real-time data of the index is disseminated at 5-second intervals.

The special feature of the index is that it is a unique Stock Connect-related index that tracks the Hong Kong stocks available for Southbound Trading under the Stock Connect scheme. The usefulness of the index in Southbound Trading lies in **its representativeness of Southbound eligible stocks and its different composition vis-à-vis other existing Hong Kong and Mainland market indices for alternative investment opportunities**. This is illustrated in sub-sections below.

3.1 Coverage of top Southbound stocks

Constituents of CES SCHK100 are selected based on their ranking by market capitalisation. These are the largest stocks with considerable liquidity among the Southbound stocks. Although they constituted only 24% in terms of number of stocks as at the end of August 2017, they covered 68% by market capitalisation and 53% by market turnover value[26] of all Southbound stocks (see Figure 3). Except for H shares (bearing in mind that CES SCHK100 excludes A+H stocks), the index has a high degree of representativeness for each other stock type (see Section 3.2).

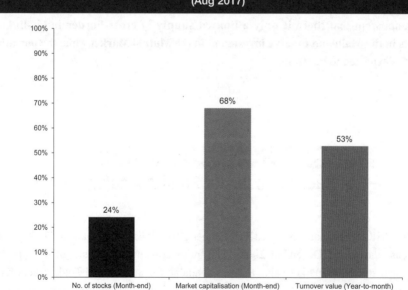

Figure 3. Percentage share of CES SCHK100 stocks in all Southbound stocks (Aug 2017)

Source: Websites of HKEX, SSE, SZSE and CESC for stock lists; HKEX for market data.

---

26 The "market turnover value" refers to the stocks' total turnover in the market, not just Southbound Trading.

3.2   Pure Hong Kong concept, with high representativeness

CES SCHK100 consists of stocks listed only in Hong Kong, excluding A+H stocks which already have A shares listed in the Mainland market. The index therefore measures pure exposure to Hong Kong listed stocks, which is a specialised opportunity to the Mainland investors for investing through Stock Connect on top of their domestic stock investments.

Compared to the collection of all Southbound stocks, CES SCHK100 represents more for non-H shares in terms of market capitalisation — 33% vs 25% for Hong Kong stocks, 31% vs 26% for Mainland private enterprises (MPEs) and 25% vs 20% for red chips; and in terms of turnover value (for 2017 up to August) — 25% vs 16% for Hong Kong stocks; 38% vs 27% for MPEs and 21% vs 13% for Red Chips (see Figure 4). In particular, MPEs represent the key growth sector in China's current economic transformation stage into the New Economy.

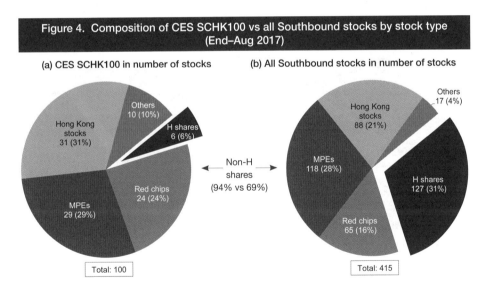

Figure 4. Composition of CES SCHK100 vs all Southbound stocks by stock type (End–Aug 2017)

(a) CES SCHK100 in number of stocks

Others 10 (10%)
Hong Kong stocks 31 (31%)
H shares 6 (6%)
Red chips 24 (24%)
MPEs 29 (29%)
Total: 100

Non-H shares (94% vs 69%)

(b) All Southbound stocks in number of stocks

Others 17 (4%)
Hong Kong stocks 88 (21%)
MPEs 118 (28%)
H shares 127 (31%)
Red chips 65 (16%)
Total: 415

(continued)

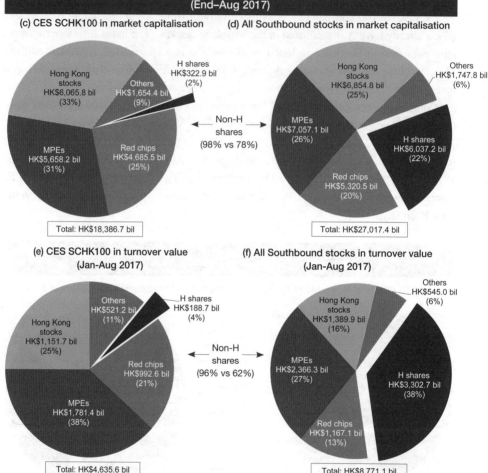

Figure 4. Composition of CES SCHK100 vs all Southbound stocks by stock type (End–Aug 2017)

(c) CES SCHK100 in market capitalisation

(d) All Southbound stocks in market capitalisation

Total: HK$18,386.7 bil

Total: HK$27,017.4 bil

(e) CES SCHK100 in turnover value (Jan-Aug 2017)

(f) All Southbound stocks in turnover value (Jan-Aug 2017)

Total: HK$4,635.6 bil

Total: HK$8,771.1 bil

Note: Stock classification by type is done by HKEX based on the listed company's China/non-China origin and in consideration of the origin of establishment and place of incorporation for companies other than H shares, red chips and MPEs.

Source: Websites of HKEX, SSE, SZSE and CESC for stock lists; HKEX for market data.

Excluding H shares, CES SCHK100's coverage of Southbound stocks by each other stock type is very high in terms of market capitalisation (80%-95%) and market turnover value (75%-96%), albeit to a somewhat lesser degree in terms of Southbound turnover value (62%-91%) (see Figure 5).

Figure 5. CESC SCHK100 stocks' coverage of all Southbound stocks by stock type (Aug 2017)

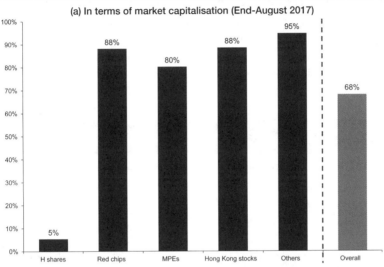

(a) In terms of market capitalisation (End-August 2017)

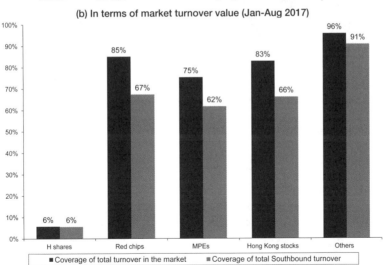

(b) In terms of market turnover value (Jan-Aug 2017)

■ Coverage of total turnover in the market   ■ Coverage of total Southbound turnover

Note: Stock classification by type is done by HKEX based on the listed company's China/non-China origin and in consideration of the origin of establishment and place of incorporation for companies other than H shares, red chips and MPEs.

Source: Websites of HKEX, SSE, SZSE and CESC for stock lists; HKEX for market data.

3.3 Resemblance by industry sector, with more exposure to the New Economy

CES SCHK100 stocks have a composition by industry sector as diverse as all South-bound stocks, but with a lesser degree of contribution from Financials (22% vs 31% in terms of market capitalisation and 20% vs 31% in terms of market turnover value). Instead, it has heavier weightings by market capitalisation and turnover value on stocks in the Information Technology (IT) sector, the perceived high-growth sector in the New Economy (see Figure 6).

Figure 6. Composition of CES SCHK100 vs all Southbound stocks by industry sector (End–Aug 2017)

(a) CES SCHK100 in number of stocks

(b) All Southbound stocks in number of stocks

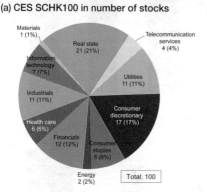

(c) CES SCHK100 in market capitalisation

(d) All Southbound stocks in market capitalisation

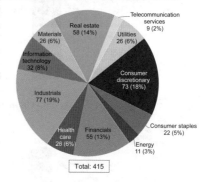

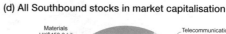

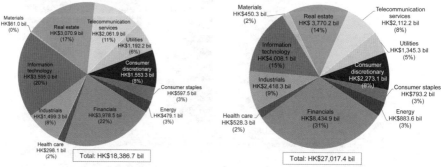

(continued)

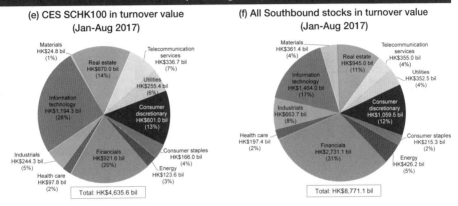

Figure 6. Composition of CES SCHK100 vs all Southbound stocks by industry sector (End–Aug 2017)

Source: Websites of HKEX, SSE, SZSE and CESC for stock lists; HKEX for market data.

As shown in Figure 7, CES SCHK100 has a relatively high coverage of Southbound stocks in the New Economy sectors[27]:

- IT — 90% by market capitalisation, 82% by market turnover value;
- Consumer Staples — 75% by market capitalisation, 77% by market turnover value;
- Consumer Discretionary — 68% by market capitalisation, 57% by market turnover value;
- Health Care — 56% by market capitalisation, 50% by market turnover value.

Figure 7. CES SCHK100 stocks' coverage of all Southbound stocks by industry sector (August 2017)

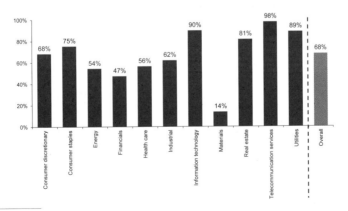

(a) In terms of market capitalisation (End-August 2017)

27 The New Economy sectors refer to the sectors included in the CSI MarketGrader China New Economy Index (source: China Securities Index Co., Ltd. (CSI) website).

(continued)

Figure 7. CES SCHK100 stocks' coverage of all Southbound stocks by industry sector (August 2017)

(b) In terms of market turnover value (Jan-Aug 2017)

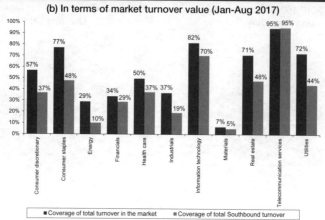

Source: Websites of HKEX, SSE, SZSE and CESC for stock lists; HKEX for market data.

3.4 Performance versus key Hong Kong and Mainland indices

Figure 8 shows the daily movements of CES SCHK100 in comparison with the key stock indices in Hong Kong, Shanghai and Shenzhen stock markets. For the past 8+ years since the base date (31 December 2008) of CES SCHK100 up to end-August 2017, the index achieved a cumulative return of 102%. This was unmatched by the key indices of HSI (66%) and HSCEI (36%) in Hong Kong and surpassed only by the SSE 380 (118%), the SZSE A Share Index (125%) and the SZSE SME Composite Index (144%) (see Table 3).

Figure 8. Daily closings of CES SCHK100 and key Hong Kong and Mainland indices (Rebased on 31 Dec 2008) (31 Dec 2008–31 Aug 2017)

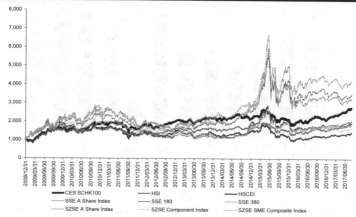

Source: CESC for CES SCHK100, Thomson Reuters for other indices.

Table 3. Cumulative returns of CES SCHK100 and key Hong Kong and Mainland indices (31 Dec 2008–31 Aug 2017)

Index	Cumulative return
CES SCHK100	102.23%
HSI	66.48%
HSCEI	35.86%
SSE A Share Index	61.03%
SSE 180	72.52%
SSE 380	118.58%
SZSE A share Index	125.23%
SZSE Component Index	51.15%
SZSE SME Composite Index	144.08%
ChiNext Composite Index	74.67%

Note: Returns are natural logarithmic returns.

Source: Calculated from daily closings of indices — CESC for CES SCHK100, Thomson Reuters for other indices.

CES SCHK100 was found to have better historical return performance both in the short term and in the long term than the Hong Kong key indices. On the other hand, Mainland indices showed bigger fluctuation in their return performance over time, especially for the SZSE indices. Their returns fluctuated a lot in the past years and were at a much lower (or negative) level in the recent year compared to the positive return achieved by CES SCHK100. In fact, CES SCHK100 has an annualised volatility of daily returns lower than the key Hong Kong and Mainland indices for most of the years since its launch. (See Table 4 and Figure 9.)

Table 4. Period returns of CES SCHK100 and key Hong Kong and Mainland indices up to 31 Aug 2017

Index	1-year	3-year	5-year	7-year
CES SCHK100	21.41%	15.26%	46.11%	54.55%
HSI	19.67%	12.26%	36.16%	30.89%
HSCEI	16.87%	2.98%	19.65%	-0.95%
SSE A Share Index	8.59%	41.64%	49.56%	24.15%
SSE 180	15.79%	49.27%	56.52%	31.00%
SSE 380	0.01%	41.02%	68.42%	36.73%
SZSE A share Index	-4.44%	47.44%	84.14%	50.93%

(continued)

Table 4. Period returns of CES SCHK100 and key Hong Kong and Mainland indices up to 31 Aug 2017				
Index	1-year	3-year	5-year	7-year
SZSE Component Index	0.54%	32.16%	27.56%	-4.71%
SZSE SME Composite Index	-2.35%	48.21%	88.19%	54.82%
ChiNext Composite Index	-16.09%	45.57%	121.22%	83.82%

Note: Returns are natural logarithmic returns.

Source: Calculated from daily closings of indices — CESC for CES SCHK100, Thomson Reuters for other indices.

Figure 9. Return and volatility of CES SCHK100 and key Hong Kong and Mainland indices (2009–Aug 2017)

(a) Annual/period returns

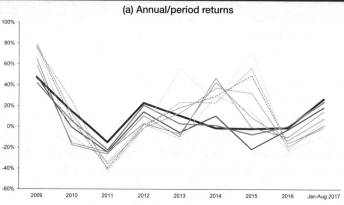

(b) Annualised volatility of daily returns

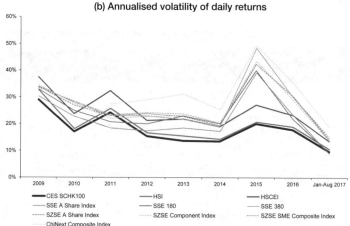

Note: Returns are natural logarithmic returns. Annualised volatility is the annualised standard deviation of daily returns during the period.

Source: Calculated from daily closings of indices — CESC for CES SCHK100, Thomson Reuters for other indices.

Despite the different return performance, CES SCHK100 had a relatively high correlation of daily returns with the key Hong Kong indices of HSI and HSCEI — correlation coefficients of 0.978 and 0.899 respectively with HSI and HSCEI during the period from January 2009 to August 2017. On the contrary, the correlation of the daily returns of CES SCHK100 with the Mainland key indices was at a moderate, lower level — a correlation coefficient of 0.5 or below (see Table 5).

Table 5. Correlation coefficients of daily returns of CES SCHK100 with key Hong Kong and Mainland indices (Jan 2009–Aug 2017)	
Correlation with index	Correlation coefficient
HSI	0.978
HSCEI	0.899
SSE A Share Index	0.507
SSE 180	0.509
SSE 380	0.435
SZSE A share Index	0.437
SZSE Component Index	0.461
SZSE SME Composite Index	0.407
ChiNext Composite Index	0.341

Note: Correlation coefficients are Pearson correlation coefficients; all are statistically significant at 0.1% level.

Source: Calculated from daily closings of indices — CESC for CES SCHK100, Thomson Reuters for other indices.

Among indices tracking Hong Kong stocks only, the month-end price-earnings (PE) ratios of CES SCHK100 was higher than those of HSCEI all the time and also higher than those of HSI for most of the time during December 2014 to August 2017, albeit it had a lower dividend yield than the two indices for most of the time during the same period[28] (see Figure 10). It might reflect that the constituents of CES SCHK100 come more from the growth sector at the current stage of economic development (see Section 3.3 above), generating capital gains from profit re-investment rather than distributing profits as dividends, while many constituents of HSI and HSCEI would be companies in the traditional economy which are at a relatively mature stage of development.

---

28 PE ratios and dividend yields for the indices are weighted average of those of the respective index constituents.

Moreover, compared to the Shenzhen stock market which is generally perceived to comprise of growth stocks, CES SCHK100 has a higher dividend yield (3.08%) than the SZSE Component Index (1.04%), the SME Index (0.84%) and the ChiNext Index (0.69%) as of end-August 2017[29].

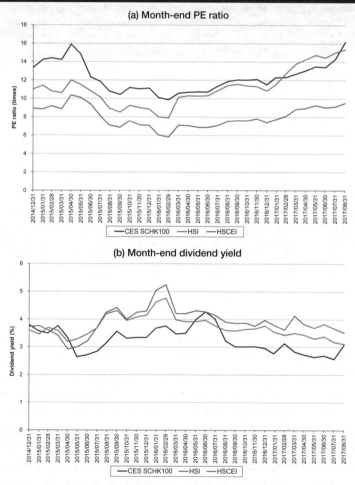

Figure 10. PE ratios and dividend yields of CES SCHK100 and key Hong Kong indices (Dec 2014–Aug 2017)

(a) Month-end PE ratio

(b) Month-end dividend yield

Note: The figures are weighted average of those of the respective index constituents.

Source: CESC for CES SCHK100, Hang Seng Indexes website for HSI and HSCEI.

29 Source of dividend yields of SZSE indices: CNINDEX monthly index report, August 2017 (http://index.cninfo.com.cn/).

3.5   Opportunities offered by CES SCHK100 for Southbound investment

In summary, CES SCHK100 has the following characteristics for tracking investment through Southbound Trading under Stock Connect:

(1)   Having a relatively high coverage of Southbound stocks in terms of market capitalisation and turnover value;

(2)   Tracking investment of a pure Hong Kong concept, with a high representativeness of stocks listed in Hong Kong and not in the Mainland market at the same time, therefore representing pure investment opportunities outside Mainland China, with only moderate correlation with movements in the Mainland domestic stock market;

(3)   Having a relatively high coverage of stocks of the growth sectors among Southbound stocks, e.g. MPEs and stocks in the New Economy industries;

(4)   Owing to its stock composition, having historically higher PE ratio but lower dividend yield than HSI and HSCEI, and also lower return volatility than the key Hong Kong and Mainland indices, for most of the time since its launch.

Interestingly, despite the above facts, it was found that during 2017 up to August the share of Southbound Trading in the total market turnover of constituent stocks of CES SCHK100 (10%) was to a lesser degree than that in the overall market turnover of all Southbound stocks (14%). The same observation was obtained whether in terms of different stock types or in terms of different industry sectors (see Figure 11).

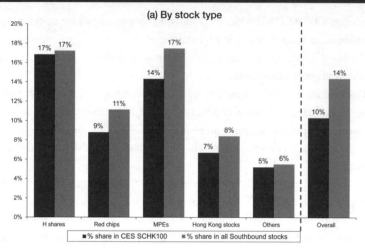

Figure 11. Percentage share of Southbound Trading in total market turnover of CES SCHK100 and of all Southbound stocks (Jan–Aug 2017)

(a) By stock type

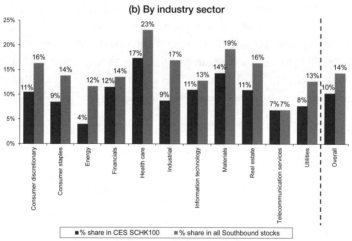

(b) By industry sector

Source: Websites of HKEX, SSE, SZSE and CESC for stock lists; HKEX for market data.

In other words, **the potential of investment opportunities in CES SCHK100 stocks is yet to be exploited further through Southbound Trading**. The index fulfills the pre-requisites to support the development of passive investment instruments like ETFs for Southbound Trading — the index constituents are tradable in a free and open market and trading of the constituents is expected to have little market impact as they are the top 100 Southbound stocks by size. (see Section 2.2 above on the conditions for the success of

ETFs.) CES SCHK100 therefore would be a useful benchmark for developing passive investment instruments like ETFs for Southbound Trading.

4   Conclusion

The establishment of the Mainland-Hong Kong Mutual Market after the launch of the Stock Connect scheme has opened up more opportunities for Mainland investors, with fewer restrictions than previous channels like QDII, for investing in overseas markets. Statistics show that there is a strong and growing interest from the Mainland investors in trading stocks listed in Hong Kong through Southbound Trading under the scheme.

CESC SCHK100 is a unique Stock Connect-related index that tracks the Hong Kong stocks available for Southbound Trading, representing pure investment opportunities outside Mainland China. Given its high representativeness of Southbound eligible stocks and growth-sector stocks in the New Economy, the potential of investment opportunities in its constituent stocks is yet to be exploited further through Southbound Trading. Towards this end, CES SCHK100 is potentially a useful benchmark for developing passive investment instruments like ETFs for Southbound Trading.

Appendix 1

Southbound Trading activities under Stock Connect

Since the launch of Shanghai Connect, Northbound Trading had been much more than Southbound Trading in terms of average daily trading value (ADT) for most of the time before 2016. A rising trend of Southbound Trading was observed since late 2015, slowly at the beginning and in a rapid pace after the launch of Shenzhen Connect in December 2016 (see Figure A1). The ratio of Southbound Trading relative to the SEHK Main Board total market trading rose from 1.0% of the Main Board ADT[30] in September 2015 to the highest level of 6.1% in September 2017. This compared to about 1% for the ADT of Northbound Trading as a percentage of the Mainland total A-share market. (see Figure A2.) Moreover, Southbound ADT exceeded Northbound ADT time and again in a number of months during the period.

Moreover, Southbound Trading had much higher average daily net buy trade values than Northbound Trading for most of the time since late 2015. Net sell trade value was recorded for Southbound Trading in only two months since launch up to September 2017, vis-à-vis 6 months for Northbound Trading. (see Figure A3.) Up to September 2017, the cumulative net buy-in value of Hong Kong stocks by Mainland investors through Southbound Trading was HK$604.0 billion, compared to the cumulative net buy-in value of RMB 314.5 billion (~HK$372.0 billion) of Mainland stocks by global investors through Northbound Trading.

---

30 For more appropriate comparison, total Southbound Trading on 2-sided basis (buy and sell) was divided by two to give a figure on one-sided basis for calculating the ratio relative to the SEHK Main Board total turnover value (also on a one-sided basis). The same approach is applied in calculating the ratio of Northbound Trading to the Mainland total A-share market turnover value.

Figure A1. Average daily value of Stock Connect total trading (buy and sell)
—— Northbound and Southbound (Nov 2014–Sep 2017)

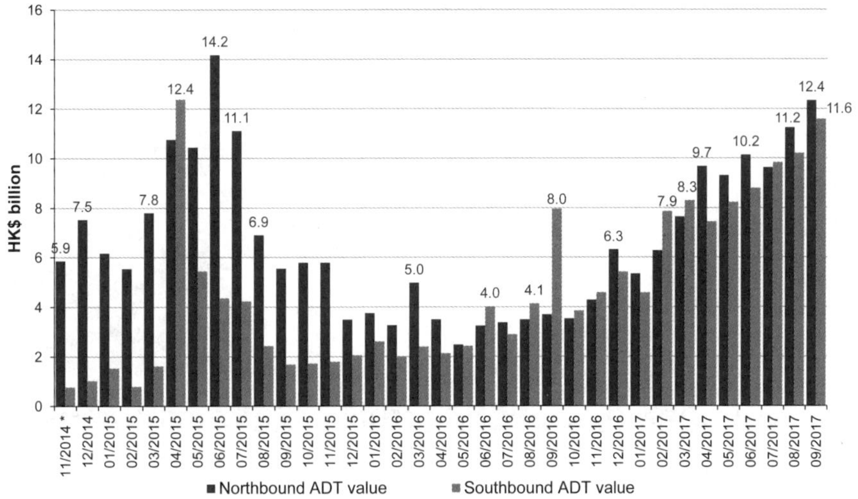

■ Northbound ADT value   ■ Southbound ADT value

* Starting from 17 Nov 2014 when Shanghai Connect was launched.

Note: Stock Connect total trading values include buy and sell trades. Northbound Trading values are
        converted to HKD using month-end exchange rates from Hong Kong Monetary Authority website.
        Shenzhen Connect data is included since the launch date of Shenzhen Connect (5 Dec 2016).

Source: HKEX.

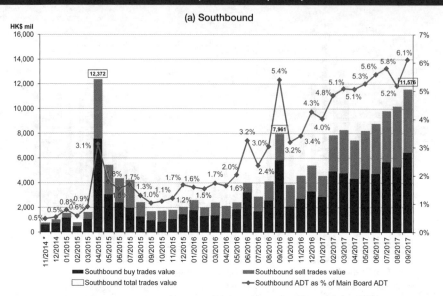

Figure A2. Stock Connect average daily trading value and percentage share of market total (Nov 2014–Sep 2017)

(a) Southbound

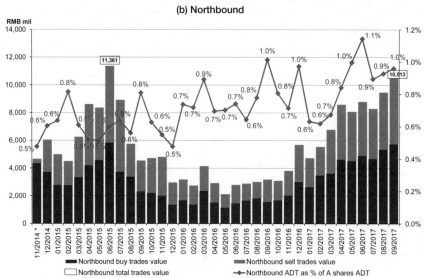

(b) Northbound

* Starting from 17 Nov 2014 when Shanghai Connect was launched.

Note: In calculating the ratio to the total market ADT, Northbound/Southbound Trading total value (buy and sell) was divided by two to give one-sided figures before calculating its percentage share of the one-sided total market trading value (buy and sell counted in a single transaction value). Shenzhen Connect data is included since the launch date of Shenzhen Connect (5 Dec 2016). The base reference data of Mainland A-share market includes SZSE A-share market since that date.

Source: HKFX.

Figure A3. Average daily net buy/sell trade value in Stock Connect Southbound and Northbound Trading (Nov 2014–Sep 2017)

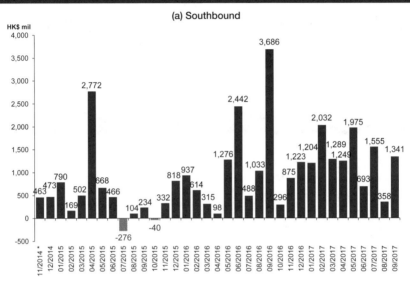

(a) Southbound

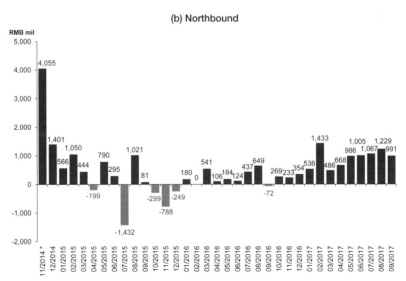

(b) Northbound

* Starting from 17 Nov 2014 when Shanghai Connect was launched.

Note: Shenzhen Connect data is included since the launch date of Shenzhen Connect (5 Dec 2016).

Source: HKEX.

Appendix 2

List of constituent stocks of CES SCHK100

(End-September 2017)

Stock code	Stock name	Weight (%)
700	Tencent Holdings Limited	11.27
5	HSBC Holdings Plc. (Hong Kong)	10.21
1299	AIA Group Limited	7.89
941	China Mobile Limited	5.52
1	CK Hutchison Holdings Limited	3.06
388	Hong Kong Exchanges and Clearing Limited	2.93
2888	Standard Chartered Plc.	2.87
16	Sun Hung Kai Properties Limited	2.09
883	CNOOC Limited	2.04
1113	CK Asset Holdings Limited	1.91
2388	BOC Hong Kong (Holdings) Limited	1.82
11	Hang Seng Bank	1.65
2	CLP Holdings Limited	1.61
27	Galaxy Entertainment Group Limited	1.60
2007	Country Garden Holdings Company Limited	1.50
3	The Hong Kong and China Gas Company Limited	1.40
175	Geely Automobile Holdings Limited	1.34
688	China Overseas Land & Investment Limited	1.26
3333	China Evergrande Group	1.21
6	Power Assets Holdings Ltd.	1.15
1928	Sands China Limited	1.12
2018	AAC Technologies Holdings Inc.	1.10
2382	Sunny Optical Technology (Group) Company Limited	1.08
1658	Postal Savings Bank of China Company Limited	1.01
4	Wharf Holdings Limited	0.96
66	MTR Corporation Limited	0.92
762	China Unicom (Hong Kong) Limited	0.89
267	CITIC Limited	0.76
17	New World Development Company Limited	0.75

(continued)

Stock code	Stock name	Weight (%)
1109	China Resources Land Limited	0.75
2328	PICC Property & Casualty Company Limited	0.72
12	Henderson Land Development Company Limited	0.71
1114	Brilliance China Automotive Holding Limited	0.71
669	Techtronic Industries Company Limited	0.69
2319	China Mengniu Dairy Company Limited	0.68
728	China Telecom Corporation Limited	0.63
1093	CSPC Pharma Pharmaceutical Group Limited	0.63
1038	CKI Infrastructure Holdings Limited	0.61
1044	Hengan International Group Company Limited	0.61
288	WH Group Limited	0.55
384	China Gas Holdings Limited	0.53
23	Bank of East Asia Limited	0.52
2313	Shenzhou International Group Holdings Limited	0.52
20	Wheelock and Company Limited	0.51
83	Sino Land Company Limited	0.49
1177	Sino Biopharmaceutical Limited	0.49
2688	ENN Energy Holdings Limited	0.49
656	Fosun International Limited	0.48
19	Swire Pacific Limited	0.47
101	Hang Lung Properties Limited	0.47
151	Want Want China Holdings Limited	0.47
1099	Sinopharm Holdings Company Limited	0.47
1359	China Cinda Asset Management Company Limited	0.44
144	China Merchants Port Holdings Company Limited	0.43
966	China Taiping Insurance Holdings Company Limited	0.43
270	Guangdong Investment Limited	0.41
291	China Resources Beer (Holdings) Company Limited	0.39
960	Longfor Properties Company Limited	0.39
992	Lenovo Group Limited	0.38
371	Beijing Enterprises Water Group Limited	0.37
522	ASM Pacific Technology Limited	0.37
1128	Wynn Macau Limited	0.37
1357	Meitu Inc.	0.37
1972	Swire Properties Limited	0.35
2689	Nine Dragons Paper Holdings Limited	0.33
425	Minth Group Limited	0.32
2282	MGM China Holdings Limited	0.32

(continued)

Stock code	Stock name	Weight (%)
586	China Conch Venture Holdings Limited	0.31
836	China Resources Power Holdings Company Limited	0.31
6808	Sun Art Retail Group Limited	0.31
10	Hang Lung Group Limited	0.30
257	China Everbright International Limited	0.30
322	Tingyi (Cayman Islands) Holding Corporation	0.30
607	Fullshare Holdings Limited	0.30
1169	Haier Electronics Group Company Limited	0.30
2020	ANTA Sports Products Limited	0.30
69	Shangri-La Asia Limited	0.29
135	KunLun Energy Company Limited	0.28
551	Yue Yuen Industrial (Holdings) Limited	0.28
981	Semiconductor Manufacturing International	0.28
1193	China Resources Gas Group Limited	0.28
659	NWS Holdings Limited	0.27
683	Kerry Properties Limited	0.27
1816	CGN Power Company Limited	0.27
14	Hysan Development Company Limited	0.26
494	Li & Fung Limited	0.26
813	Shimao Property Holdings Limited	0.26
3311	China State Construction International Holdings	0.26
392	Beijing Enterprises Holdings Limited	0.24
867	China Medical System Holdings	0.23
8	PCCW Limited	0.22
1060	Alibaba Pictures Group Limited	0.22
3377	Sino-Ocean Group Holding Limited	0.22
165	China Everbright Limited	0.21
3320	China Resources Pharmaceutical Group Limited	0.20
880	SJM Holdings Limited	0.18
293	Cathay Pac Airways Limited	0.16
241	Alibaba Health Information Technology Limited	0.13
3799	Dali Foods Group Company Limited	0.13
1929	Chow Tai Fook Jewellery Group Limited	0.12

Chapter 5

The rise of China's new-economy sector

Financing needs and Hong Kong's new role

19 December 2017

Summary

As traditional growth drivers such as investment and exports recede, the new-economy sector, which is typically characterised by new industries, new industry forms and new business models, has developed rapidly and gradually becomes the driving force for the structural transformation of the Chinese economy and its new growth impetus. However, the lack of a diversified and multi-layer financial market has restricted the growth of China's emerging industries. On one hand, the traditional model of bank lending and private equity venture funds has not been able to fully satisfy the financing needs of new-economy companies. On the other hand, the path of listing with initial public offering may not be able to satisfactorily resolve the issues about the founder(s) shareholding structure. How to enhance the financing capabilities of the capital market with the funds raising suitable for innovative technology companies will be crucial for promoting the development of these companies and for the rapid growth of the new-economy sector.

Setting up an open, diversified and innovative financial market would be conducive to enhancing the international competitiveness of China's new-economy sector. China's economic transformation and the rise of its innovation industries will provide impetus to Hong Kong to assume a new role and positioning in the process of China's economic liberalisation and Chinese companies' internationalisation.

1 Rise of new-economy industries adds new impetus to China's economy

The current change in China's industrial structure is unprecedented. The growth engine is gradually shifting away from the traditional model of exports and investment to a new direction characterised by economic transformation and business innovation.

1.1   As traditional growth engines weakened, the old economy of investment and exports began to stagnate

China has undergone the world's longest period of the highest and the most extensive economic growth in the past 40 years. This was driven by investment and exports, relying heavily on capital-intensive, energy-intensive and labour-intensive economic structure. Between 1978 and 2016, Mainland China had an average annual growth rate of 9.6% in gross domestic product (GDP) and has become the world's second largest economy, with the GDP per capita rising from RMB 385 to RMB 53,980. (see Figure 1.)

Figure 1. China's GDP per capita and GDP growth rate (1978–2016)

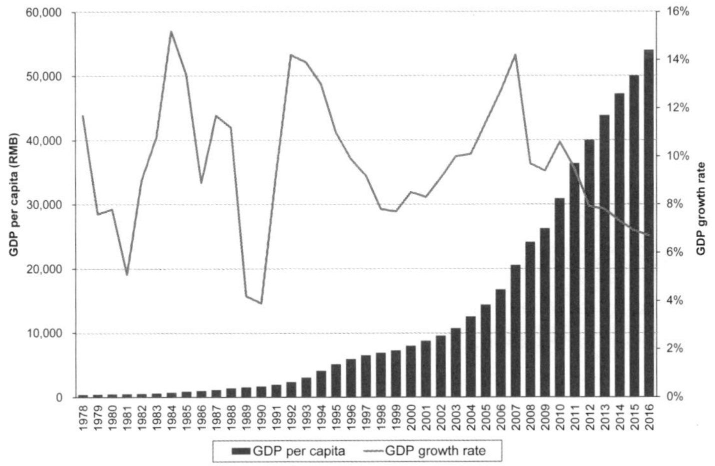

Source: Wind.

However, the global financial crisis of 2008 led the world economy into stagnation and precipitated tremendous changes in the international economic landscape. China's economy also saw signs of a cyclical decline under the superimposition of three economic adjustment periods and growth hit a bottleneck. The economic growth impetus offered by investment has been decreasing as the marginal return on investment continues to decline. Fixed-asset investment growth per annum has subsided from 24.33% in 2006 to 8.1% in 2016. As a result, the contribution of investments to total GDP growth had reduced from the peak of 8.0% to 2.8% in 2016. As for trade, global trade volume has also shrunk substantially to below 3% for the past 5 years. China's growth in trade declined nearly 8% in 2015 and further declined in 2016[1]. At the current time of de-globalisation, China's trade growth is no longer expected to outpace global trade's average growth. The contribution of trade to China's GDP growth has been negative since 2009. (see Figure 2.)

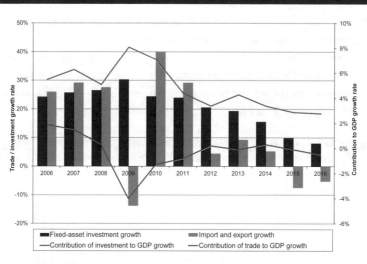

Figure 2. China's trade and investment growth (1996–2016)

Source: Wind.

1.2 Technology and policies drive new-economy sector

Characterised by new industries, new industry forms and new business models, the new-economy sector is developing rapidly in China and gradually becomes a major driver

---

1   Source: National Bureau of Statistics.

of the development and structural transformation of the Chinese economy.

The rise of the new-economy sector is attributable to two key factors:

Firstly, new technologies have enabled revolutionary changes in business models. This is particularly apparent in five major areas:

(1) **The internet** has evolved from an instrument that improves communication to a core infrastructure that supports the entire businesses. Digitalisation of business operations have led to new business models and changes in consumer behaviour.

(2) **Data** has become a new kind of resources. There is a new wave of revolution driven by technologies such as knowledge innovation, cloud computing, artificial intelligence and big data.

(3) **Sharing economy** is a model to match supply and demand of idle resources at lower cost and with higher efficiency. This model is applied in multiple sectors ranging from accommodation (e.g. AirBnB) to transport (e.g. Didi Dache) and have facilitated brand new business models such as collaborative consumption, cooperative economy and point-to-point economy.

(4) **Financial technology (Fintech)** covers payment and clearing, digital currency, online lending, blockchain, robo adviser, smart contracts and many other fields. The combination of finance and technology creates new business models and are revolutionising traditional services in the financial market. A research by McKinsey[2] indicates that an average of 40% of traditional banking revenue would be affected by the extensive adoption of digital banking.

(5) **An industry-wide upgrade of manufacturing.** The adoption of new energy, new materials and high-end equipment manufacturing has become a new direction for the world's industrial growth and is also providing a solid support for the structural recovery of the global economy.

Secondly, a series of policies and strategies have been introduced by various countries after the 2008 Global Financial Crisis, providing guidance and institutional support for new-economy industries.

On the global front, developed countries shifted their focus back to the real economy following the financial crisis. Apart from formulating re-industrialisation strategies that focus on reinvigorating the manufacturing sector, incentive-based policies were introduced in biopharmaceuticals, electronic equipment, intelligent technology, materials technology, clean energy and other new-economy industries as an attempt to cultivate new-economic growth drivers.

---

2   Source: *Where is the future of Chinese banking industry*, McKinsey research, 2016

The US government was the first to actively support emerging industries including advanced manufacturing technologies, intelligence manufacturing, new energy, biotechnology and information-driven businesses. The aim is to encourage the return of high-end manufacturing to the country to create more high value-added and high-tech domestic jobs, so as to re-establish a new and highly competitive industrial framework in the US. In December 2009, the US announced a framework to reinvigorate the country's manufacturing sector, followed by the implementation of the Advanced Manufacturing Partnership and a National Strategic Plan for Advanced Manufacturing in June 2011 and February 2012 respectively. Under the "re-industrialisation" strategy, the US' share of the world's manufacturing industry rose steadily to 16.6% in 2014 and by the end of 2015, the manufacturing sector had added a total of 700,000 people to its workforce, showing some early positive results in the rebuilding of the industry in the US[3].

European governments have also implemented policies that reinvigorate industries or reshape future industries. Germany introduced the "Industry 4.0" national strategy which targets emerging industries and aim for an upgrade of the entire manufacturing value chains to improve competitive advantage. This induces great repercussions throughout the world. The strategy signifies the integration of internet technology (e.g. cloud computing, big data, 3D printing, network security) with the manufacturing sector. The key purpose is to enhance the intelligence level of the manufacturing industry and to create adaptive and resource-efficient intelligent manufacturing factories. Since the introduction of "Industry 4.0", 47% of companies in Germany have participated in the scheme, and 12% have implemented the strategy[4] and have become the key drivers of economic transition and growth in the local market and in Europe.

In China, the Decision of the State Council on Accelerating the Cultivation and Development of Strategic Emerging Industries was announced in 2010, followed by the "Made-in-China 2025" strategic plan in 2015. The strategic objective of these is the creation of a strong manufacturing country, with nine strategic missions and focuses[5] specified. It is hoped that with global vision and strategic thinking, technology revolution can penetrate the entire manufacturing industry to give the country's manufacturing sector a new competitive edge. Such policy initiative not only provides an important strategic guideline for China to capture

---

3   Source: U.S. Bureau of Labor Statistics.

4   Source: Findings of surveys by Germany's three major associations (German Machine Tool Builders' Association, German Association for Information Technology, Telecommunications and New Media, and German Electrical and Electronic Manufacturers' Association).

5   The nine strategic missions and focuses are the new generation IT industry, high-end CNC machine tools and robots, aerospace equipment, marine engineering equipment and high-tech vessels, advanced rail transportation equipment, energy saving and new energy vehicles, power equipment, agricultural equipment, new materials, biomedical and high-performance medical equipment and production services ancillary to these industries.

a commanding position in the future development of the new-economy and high-tech industries, it also has important strategic significance in the transformation of China's economic growth dynamics that will move China forward as a manufacturing superpower.

Table 1. Comparison of major policies that support high-tech manufacturing in US, Germany, China and other countries		
Country	Time	Major policies
US	2009	A Framework for Revitalizing American Manufacturing
	2011	Advanced Manufacturing Partnership
	2012	National Strategic Plan for Advanced Manufacturing
	2013	A Roadmap for U.S. Robotics — From Internet to Robotics
	2015	A Strategy for American Innovation
Germany	2010	High-Tech Strategy 2020 for Germany
European Union (EU)	2006	Creating an Innovative Europe
	2010	Europe 2020
	2014	EU's Horizon 2020
Japan	2007	Japan's Innovation Strategy 2025
Korea	2009	Planning and Development Strategies for New Growth Drivers
China	2010	Decision of the State Council on Accelerating the Cultivation and Development of Strategic Emerging Industries
	2015	Made-in-China 2025

Source: Government websites of the respective countries and related news.

1.3 China's new-economy industries enter a phase of critical strategic development

The restructuring of the global economy and technological revolution have brought about major opportunities for the Chinese new-economy sector and yet the Chinese economy also needs innovation to improve the quality, efficiency and sustainability of its development. Once the new economy replaces the old economy as the crucial driver of future economic and productivity growth, this will signify China's departure from the "L-shaped" economic trend and the success of its economic transformation.

According to China's 13th Five-Year Plan, the output value of strategic emerging industries would exceed RMB 60 trillion by 2020, contributing more than 15% of GDP. According to Mckinsey's projections, the 12 major technological breakthroughs including mobile internet, cloud computing, advanced robots, new-generation genomes, etc. would directly produce an annual economic value of US$14-33 trillion by 2025[6]. It is apparent that the next 10 years will be a critical strategic development phase for the sectors of high-end equipment manufacturing, new-generation information technology (IT) and information services, new materials, new energy, energy saving and environmental protection and other emerging industries in China.

The growth of emerging industries[7] has outpaced the overall economic growth as the Chinese economy's structural enhancement accelerates. By the end of 2016, there were 1,152 A-share listed companies in strategic emerging industries, representing 38% of the total number of listed companies. Their operating revenue in 2016 was RMB 3.25 trillion, up 17.7% year-on-year, and has been growing for four consecutive years. Their profits in the same year grew by 22.3% — the growth rate was up 8.5%-point from that in the previous year and was 16.2%-point above the average growth rate of all listed companies, showing further elevation in their relative performance. The average gross profit margin was 25.1%, which was 7.5%-point above the A-share market average. The sector continued to be the market leader in term of effectiveness. According to Caixin China New Economy Index, the input in new-economy sector accounted for 31.8% of the input in total economy as of April 2017[8]. It is apparent that there is a rebalancing between the old and new economies in China — the new-economy sector, characterised by new industries, new industry forms and new business models has preliminarily become the major force supporting the performance of Chinese enterprises.

---

6   Source: 〈麥肯錫發佈 12 大顛覆技術物聯網、雲、機器人、自動汽車在列〉, Xinhua website, 29 September 2015.
7   China does not have specific definition for new-economy companies. In this paper, the concept of "emerging industries" falls within the scope of new economy.
8   Source: Caixin (財新智庫) website (http://pmi.caixin.com/2017-05-02/101085054.html).

Figure 3. Comparison of annual growth in operating revenue between listed companies in strategic emerging industries and all A-share companies (2013–2016)

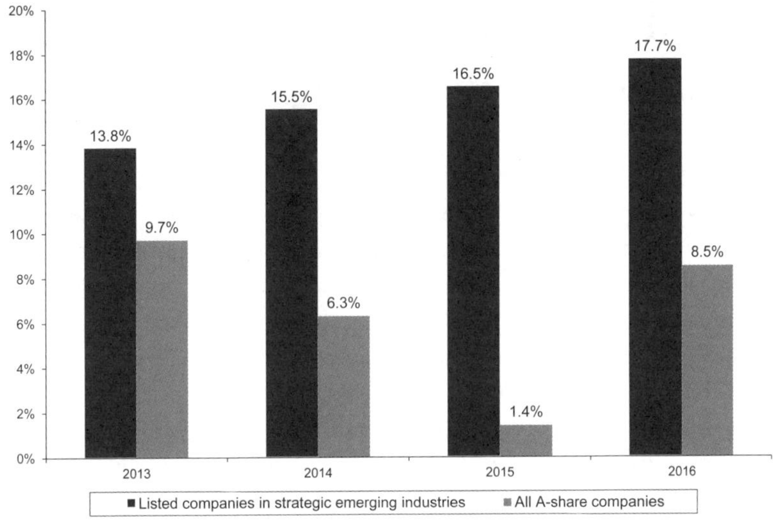

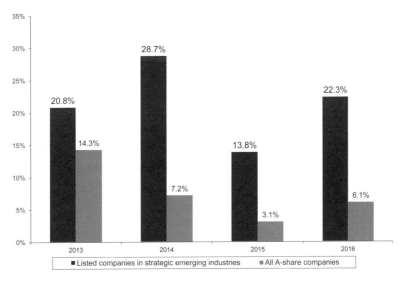

Source: State Information Centre.

Figure 4. Comparison of average profit growth rates between listed companies in strategic emerging industries and all A-share companies (2013–2016)

Source: State Information Centre.

2 Existing channels fail to fully satisfy the financing needs of new-economy sector

Technology companies carry different risk characteristics and financing needs at different stages of development. Start-ups of innovative enterprises in general would have core technologies and intellectual property rights, but suffer from the lack of capital, long establishment period and high risks and uncertainties. They are typically in need of a greater array of financing tools and a nurturing environment that can support innovation and industrialisation of new technologies.

Noteworthily, China's diversified and multi-layer investment and financing mechanism is not yet well developed. This has constrained the growth of new-economy sector in the country. Capital raising on the Shenzhen Stock Exchange (SZSE)'s Small and Medium-Sized Enterprises Board (SME Board) and ChiNext, where Mainland private enterprises dominate, accounted for only a small portion of the entire financing system in China. In 2016, the SME Board and ChiNext raised RMB 706.8 billion in total, compared to the total social financing of RMB 17.8 trillion in China. Of all social financing, RMB 2.9 trillion was raised through corporate bonds[9] and RMB 1.7 trillion was equity financing in the stock market by non-financial enterprises. (see Figure 5.) Furthermore, a research indicated that more than 85% of medium to large non-state-owned enterprises relied on self-owned funds, while the remaining companies sought capital from outside parties (of which, two-thirds relied on bank loans and one-third raised funds via the capital market). (see Figure 6.) It is clear that bank lending, bond and equity financing constituted a small proportion of financing for innovative technology companies, indicating that these existing channels are inadequate to satisfy the financing needs of new-economy industries. **The enhancement of the fund-raising capabilities of the capital market will be crucial for promoting the development of innovative technology companies and for the rapid growth of the new-economy sector in China.**

---

9   Source: Wind.

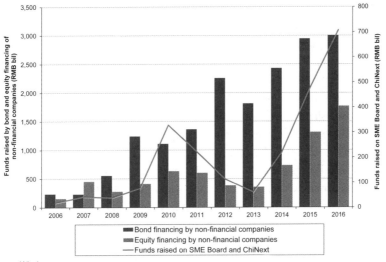

Figure 5.　Funding through SME Board and ChiNext, bond financing and equity financing (2006–2016)

Source:　Wind.

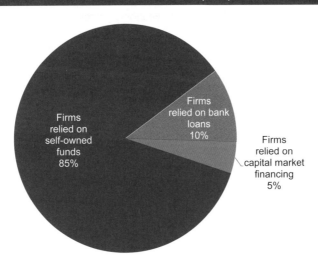

Figure 6.　Medium to large non-state-owned enterprises' reliance on different sources of fund (2016)

Source:　A research report on the development of private enterprises in China (中國民營企業發展研究報告) published in a think-tank magazine《民銀智庫研究》, Issue 57.

2.1 Bank lending

New-economy companies usually find it difficult to obtain bank loans at their early stage of development for a number of reasons owing to their special characteristics.

Firstly, new technologies and their innovative models involve a considerable degree of uncertainty and specialties which do not conform to banks' conventional way of assessing businesses. Secondly, unlike traditional industries, new-economy companies (especially those in the internet business) are typically asset-light with low barriers of entry. Therefore, these companies usually deploy significant amount of resources (before making any profits) to expand market share in order to deter competition in their early days of development. From the banks' perspective, these companies typically lack profit track records or a profit model and a stable cash-flow history; nor do they have a sound credit record and suitable guarantors. These factors make it difficult for banks to extend credit to them due to risk considerations. Thirdly, financial institutions can only charge technology innovative enterprises a fixed interest rate under existing laws in China such that the return cannot reflect the high risk premium. This reduces the incentive of financial institutions to extend credit to technology innovative enterprises.

As a result, the financing support provided by China's existing commercial banks to new-economy and high-tech companies is significantly insufficient. As of the end of 2015, outstanding bank loans offered to new-economy and high-tech companies[10] amounted to only RMB 562.7 billion, or less than 1% of total bank loans provided in the country. It is also significantly lower than the share of new-economy sector's contribution to China's GDP. (see Figure 7.)

---

10 These refer to companies in the industry sectors of "information transmission, computer services and software", and "scientific research, technical services and geological exploration".

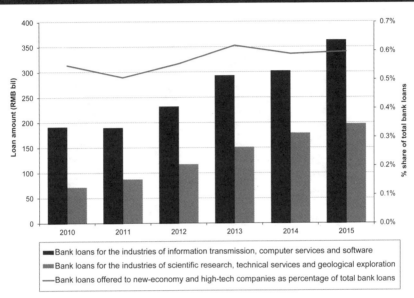

Figure 7. Funding obtained by new-economy and high-tech enterprises from banks — Amount and percentage share (2010–2015)

■ Bank loans for the industries of information transmission, computer services and software
■ Bank loans for the industries of scientific research, technical services and geological exploration
— Bank loans offered to new-economy and high-tech companies as percentage of total bank loans

Note: New-economy and high-tech companies refer to those in the industry sectors of "information transmission, computer services and software", and "scientific research, technical services and geological exploration".

Source: Wind.

2.2   Venture capital and private equity funds

Venture capital (VC) and private equity (PE) funds are the major sources of financing for new-economy and technology enterprises, the latter is more important for the development of new-economy companies with lager fund size. Investment in the new-economy sector was on the rise globally. Worldwide PE investment increased from US$45.2 billion in 2012 to US$128.5 billion in 2015. (see Figure 8.)

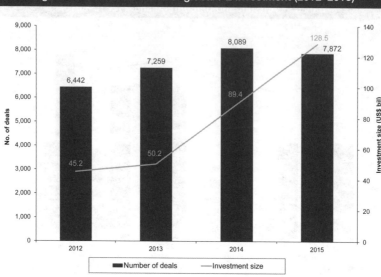

Figure 8. The annual trend of global PE investment (2012–2015)

Source: Venture Plus Q4 2015, KPMG and CB Insights.

Generally speaking, early-stage technology companies benefit more from PE than from going public.

(1) From a strategic point of view, PE support enables companies to focus on long-term strategies unhindered by short-term performance, minimising the time and resources spent by management on communication with external shareholders. This allows the companies to maintain their competitive edge without disclosing their business secret.

(2) PE funds also provide a series of value-added services in addition to financial capital. Out of a desire to add value to their investment, PE institutions usually assist these enterprises to improve their management and governance and deploy resources in areas such as human capital, marketing and business strategy to nurture enterprise development. In addition, they may also provide industry knowledge, supplier and client information that could assist these enterprises to grasp a timely understanding of the industry and enhance their competitiveness with integrated internal and external resources.

(3) As a result of continuous global monetary easing, PE funds have been able to provide technology companies with sufficient long-term funding. There are an increasing number of companies which have obtained valuations exceeding US$1 billion through funding in the primary market.

This explains why a large number of technology companies procrastinates their listings. In 2014, the average age for a technology company going public is 11 years, compared to around 4 years in 1999. The extension of time-to-listing has resulted in much longer investment periods before most venture capitalists can realise their investment returns in the public market[11].

However, from the investor's perspective, general investors would demand a higher degree of transparency and information disclosure from the technology companies than what could be offered by companies with private equity funding. PE investors, on the other hand, often ask for a commitment of earnings within 7 to 10 years from the partnership with new-economy companies. Going for a listing or doing mergers and acquisitions are, therefore, the major ways to realise returns on their investment. This is the main reason that PE investors would urge their invested companies to go for a listing.

2.3　Initial public offer (IPO) and listing

IPO is not only the primary channel for PE funds to exit their investment, but also a means whereby a company can obtain substantial capital from the stock market, enhance company reputation, and achieve business objectives and strategic breakthrough for the company's long-term development.

Due to the unique nature of new-economy companies in terms of business model and technological innovation, such companies will not only consider the basic criteria of a listing venue such as its funding capacity and liquidity, but also other factors relevant to the strategic development of a technology company. For example, they will consider whether the listing venue choice of industry peers favours the development of an industry conglomeration effect, whether there are enough analysts to conduct in-depth analysis of the unique business models of new-economy companies and whether after listing they can foster a greater momentum in new innovation with the technological capabilities in the local market.

Another factor not to be overlooked is the fact that the equity interest of founders would gradually be diluted after listing along with continuous expansion of equity financing, affecting their control over the company's strategic direction. Such development is especially unfavourable to technology companies where the business model is centred on the founders. Although tapping into new capital will allow the listed company to realise new opportunities and value creation, the transition from controlling management by the

---

11　Source: McKinsey research, "Grow fast or die slow: Why unicorns are staying private", May 2016.

founders to the democratic and transparent governance mechanism of a listed company may lead to conflicts. Striking a balance between growth and control is therefore a key consideration for technology entrepreneurs whose companies have gone public or are preparing to go public.

The prevailing practice to satisfy the needs of enterprises, especially those of private investors or founders of technology companies, is the adoption of a dual-class share structure (weighted voting rights). This is a flexible design in governance structure which allows founders of technology companies, after an IPO, to maintain control of the company even if they no longer hold the majority of its shares. Take the example of Baidu, which is listed on NASDAQ. Baidu adopted a dual-class share structure similar to that of Google — the newly issued and publicly offered A shares in the US and the original B shares. B shares have 10 voting rights per share while A shares have only one voting right per share. B shares can be converted into A shares at the ratio of 1:1, but A shares cannot be converted into B shares. This weighted voting rights structure enables Baidu's founding shareholder to have controlling voting rights by holding only a small percentage of shares, thereby maintaining effective control over the enterprise after its listing. In this way, not only Baidu successfully raised the capital it required, its founders are also able to maintain control at the same time.

3   Hong Kong will assume a new role in the internationalisation of China's new-economy companies

Hong Kong, with its unique geographical advantages and highly professional and international investment landscape, has always been a market connector between Mainland China and the rest of the world. Against the backdrop of economic transformation and the rise of innovation industries in the Mainland, how should Hong Kong co-exist and integrate with new-economy companies in the Mainland and provide them with a suitable international platform? And how would such integration drive the development of Hong Kong's science and innovation ecosystem, thereby turning it into a new impetus for the city's economic transformation?

3.1   Having the "home advantage" in adopting international practices on Chinese soil, Hong Kong can become the preferred capital market of new-economy companies

Looking beyond major factors such as market size and turnover, regulatory transparency, legal landscape, market valuation and fund-raising efficiency, the familiarity with Chinese companies and the ability to attract international investors are also crucial factors that Chinese companies would take into account when choosing the listing and fund-raising venue (referred to as "home country effect"). These are the reasons why Hong Kong has been selected by many Mainland companies as their first stop towards internationalisation.

In the past 30 years of growth in Mainland enterprises, Hong Kong has demonstrated itself as a hub connecting international capital and Mainland enterprises, and performed multiple functions for Mainland enterprises in their quest for financing, reform and transformation. Since the first Mainland-incorporated company listed in Hong Kong in 1993, Hong Kong has become the primary overseas capital market for Mainland companies to raise capital. Overseas-incorporated red-chip companies were already seen in the Hong Kong market back in the 1980s. As of end-November 2017, 1,041 Mainland companies (including H shares, red chips and private enterprises) were listed in Hong Kong, which accounted for 50% of the total number of listed companies and 66% of the total market capitalisation on the Hong Kong stock market[12]. These statistics reflect that Hong Kong has a unique home advantage when it comes to the overseas listing of Mainland companies. With these advantages, Hong Kong could offer better value discovery and pricing capabilities for Mainland's new-economy companies than its counterparts in the US and other markets. It therefore can provide the most suitable international platform for these companies' capital deployment, restructuring, mergers and acquisitions (M&A) and expansion during their different stages of strategic development and growth. The Hong Kong capital market can be transformed to help cultivate quality technology companies with transformative and innovative traits for China.

---

12 Source: HKEX.

3.2 Hong Kong can make use of its open and international environment to help new-economy companies set up regional headquarters and internationalise

Hong Kong is a well-recognised international financial centre (IFC) with the key success factors of an international fund-raising platform — an open market, the rule of law, international regulatory and market structures, professional talent pool and an English/ Chinese bilingual environment. Amid the rapid development of the Mainland financial market in recent years, Hong Kong still possesses these unique advantages as an IFC that provides important functional support for Mainland companies' overseas investment and their quest for business partners.

In recent years, an increasing number of Mainland-based high-tech and new-economy companies have ventured out of the Mainland to expand their outbound investment. They acquire overseas advanced technologies, research and development (R&D) capabilities and dominant projects, and set up regional headquarters overseas. In supporting the "out-going" strategy of Mainland enterprises, apart from providing a low-cost financing platform, Hong Kong can utilise its advantages of management resources and communication convenience to help Mainland high-tech enterprises break through trade barriers and regulatory obstacles, and formulate effective strategies for their global expansion. For example, China Everbright leveraged on its Hong Kong's arm to acquire Lampmaster, a global leader in sophisticated industrial equipment, in 2015. Based on its Hong Kong platform, the company not only set up a global M&A fund to raise capital, but also benefited from Hong Kong's legal environment and pool of talents. Another example is Gold Wind Science & Technology which acquired the direct-drive permanent magnet technology from Germany's wind turbine manufacturer, Vensys. The acquisition expanded the company's technological know-how and facilitated further advancement based on the US technology, making it China's leading wind turbine manufacturer today. Gold Wind has operations in Europe, Australia, South Africa and Latin America, and hires Hong Kong employees for managerial roles in its overseas operations to stay connected with overseas markets and to better align with client needs.

Going forward, when more high-tech and innovative companies engage in investment and financing businesses in the international market, they can make use of Hong Kong's flexible range of financial tools and structures, and enjoy the convenience ensued from it. Hong Kong could connect these companies with different overseas markets and would become the cradle for their growth. It would establish a platform for Mainland companies to acquire international counterparts and advanced technology, to seek overseas technology partners, to arrange for technological cooperation and to formulate global plans.

3.3   Hong Kong can connect global investors with China's new-economy assets and become a key platform for their asset allocation to China's new-economy sector

A significant feature of China's new-economy companies is that precedents of technology or innovation are adopted from developed countries, cultivated in China and transformed into new manifestations with global significance. For instance, CB Insights database shows that there are 183 unicorn companies[13] in the world, of which 43 are from China. Among those with a valuation of US$10-60 billion are Ant Financial, Xiaomi, Didi Chuxin, Lu.com, Zhongan Online Insurance[14], DJI and Meituan. All of these are operating in the most active fields of the new economy sector, including internet finance, e-commerce and media. Each of these companies has the potential to become a mega offering when it gets listed. However, the growth of such companies could be a complicated and lengthy process, and so their investors are bound to face tremendous risks and uncertainties.

Hong Kong's strength lies in its position not only as the primary market for Mainland companies to venture overseas, but also as the platform that international investors are most familiar with for investing in Mainland assets. New-economy assets with various forms have been taking shapes in China. Taking the list of unicorn companies announced by Great Wall Enterprise Institute and the Torch Centre under China's Ministry of Science and Technology, China's 131 unicorn companies[15] have an overall valuation amounting to US$487.6 billion or an average valuation of US$3.72 billion per company. China is now the world's second country after the US with the largest number of unicorn companies. Hong Kong is more familiar with the operations of Mainland enterprises and their profit models compared to other markets. It is also easier for investors in Hong Kong to communicate with the management of Mainland technology companies and to exert external governance. The gathering of new-economy companies in Hong Kong will shape an ecosystem for further development of new-economy sectors and the associated investor base. Through its wide array of products and its well-established pool of professionals and market functions, Hong Kong can establish an effective link between China's new-economy assets and global capital, achieving connection and integration of global financial resources with Mainland assets, and convert the achievements of new-economy technology companies into growth drivers of the Mainland economy. At the same time, international new-economy companies which have a favour for China's market can use Hong Kong as

---

13 Unicorn companies refer to companies that were set up within 10 years' time, have a valuation exceeding US$1 billion, had obtained private equity investment but are not yet listed.
14 ZhongAn Online Insurance was listed on HKEX on 28 September 2017.
15 The definition of unicorn enterprise in this survey is somewhat different from the one used by CB Insights.

a springboard — getting a listing in Hong Kong as the first step to tap into China's market, and then expanding their business in China afterwards.

3.4 Convergence of technology companies and their capital in Hong Kong will accelerate the development of Hong Kong's technology ecosystem, and promote the co-existence, integration and collaborative development of Hong Kong and Mainland China new-economy companies

Technological innovation is growing rapidly in the Hong Kong local market. A series of developments including the establishment of the Innovation and Technology Bureau and the Academy of Sciences of Hong Kong, innovative technology measures and plans for future technology infrastructure mentioned in the 2016-17 Financial Budget, and the technology and innovation policies mentioned in the Policy Addresses in 2015 and 2016 that involve more than HK$18.2 billion in government spending have instilled great confidence into Hong Kong's technology and innovation industry development[16]. **The entry of Mainland and global new-economy companies will enrich the landscape of upstream, midstream and downstream companies in the technology industry, leading to the formation of regional innovation core system, and strengthening the competitive advantage of companies in the industry agglomeration. Combined with the Hong Kong government's development plan for "a smart city, healthy aging and robotics", attracting the entry of new-economy companies will speed up the formation of a "smart city ecosystem" and promote Hong Kong's sustainable development as an international metropolis and green eco-city.**

Through the creation of an open, diversified and innovative multi-layer financial market, new energy of technological innovation and urban transformation can be injected into Hong Kong. Hong Kong can also help enhance the international competitiveness of Mainland's new-economy companies, and assume a new role and positioning in the process of China's economic liberalisation and Mainland companies' internationalisation.

---

16 Source: Hong Kong's Innovation and Technology Bureau website.

Part 2

Fixed Income and Currency (FIC)

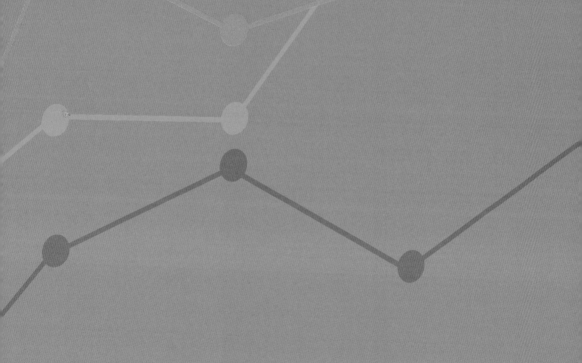

Chapter 6

TR / HKEX RMB
Currency Indices (RXY)

October 2016

Summary

The TR/HKEX RMB Currency Indices (RXY Indices) provide an independent, transparent and unbiased valuation of the RMB against the currencies of China's most important trading partners.

The RXY Indices adopt a reliable unambiguous calculation methodology, based on WM/Reuters[1] Intraday Spot Rates and strictly follow the International Organisation of Securities Commissions (IOSCO) principles for financial benchmarks. This gives the RXY Indices the advantage of becoming commonly used RMB benchmarks in comparison with any in-house RMB valuation models developed by market players which are mostly not disclosed to the public.

The RXY Indices complements the Mainland Central Bank's CFETS RMB Index — the most regarded RMB Index developed for policy purpose — by being highly correlated with the latter and at the same time delivering hourly valuations, transparency and accessibility to all market participants. The RXY Indices are probably the only currently RMB tradable indices that are publicly available to the market. They are suitable for serving as references for financial instruments including futures, options and exchange traded funds (ETFs). This would help market participants who are looking for more RMB investment and hedging tools in the course of increasing internationalisation of the RMB and liberalisation of the Mainland financial market.

---

1   WM/Reuters is "World Markets Company/Reuters".

1   Serving the need for a tradable RMB index

1.1   The increasing internationalisation of the Renminbi (RMB)

On 11 December 2015, the Mainland Central Bank — the People's Bank of China (PBOC) — launched three new RMB currency policy indices on the China Foreign Exchange Trade System (CFETS) to benchmark the RMB performance against baskets of major international currencies. These comprise the CFETS RMB Index, the BIS Currency Basket RMB Index and the SDR Currency Basket RMB Index[2].

The new policy addresses the importance of shifting global emphasis of the RMB exchange rate from against the US dollar (USD) towards against multiple currencies. Such a move aims to decrease the perceived volatility of the RMB on the dual exchange rate screen, where large swings in the currency rate sometimes are caused merely by economic events in the USA, and may have little relationship with the international value of the RMB.

The move to a multicurrency basket in measuring the RMB lies in the fact that China undoubtedly had grown into a major international trading country and the RMB is increasingly used in international trade and financial activities. A bilateral exchange rate of USD to onshore RMB (USD/CNY) could not reflect the trade and financial relationships of China with multiple countries across the globe.

In recent years, China has traded with an increased number of countries around the world, and become the major trading partner for many of them. The World Trade Organisation (WTO) trading statistics[3] show that China overtook Japan as the leading exporter in 2004, surpassed the USA in 2007 and Germany in 2009, and became the world's leading exporter. In 2015, China's merchandise exports were US\$2.27 trillion, maintaining the first place in the world ranking. The range and diversity of China's trading partners boosts demand for the RMB and highlights the limitations of using a bilateral exchange rate for measuring the RMB. (See Figure 1.)

---

2   CFETS RMB Index mainly refers to CFETS currency basket, including CNY versus FX currency pair listed on CFETS. BIS Currency Basket RMB Index refers to the Bank for International Settlements (BIS) currency basket. The SDR Currency Basket RMB Index refers to the International Monetary Fund (IMF)'s Special Drawing Right (SDR) currency basket.

3   WTO, "World trade in 2015-2016", April 2016.

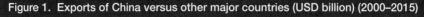

Figure 1. Exports of China versus other major countries (USD billion) (2000–2015)

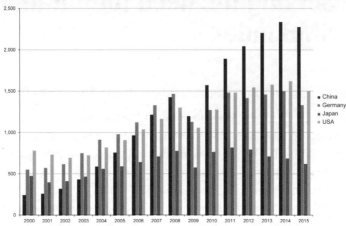

Source: UN Commodities trading database.

In addition to China's strong position in international trade, the use of the RMB as a settlement and investment currency also increases over time. According to SWIFT[4], the market share of the RMB as an international payment currency ranked fifth in July 2016, moving up two notches in two years' time. (See Figure 2.)

Figure 2. RMB's share as an international payment currency (Jan 2014 & Jul 2016)

Customer initiated and institutional payments. Messages exchanged on SWIFT. Based on value.

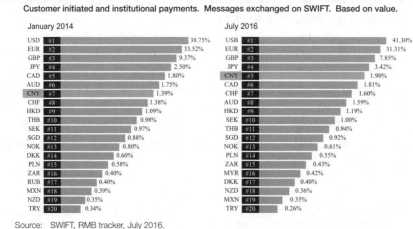

Source: SWIFT, RMB tracker, July 2016.

---

4   SWIFT, "RMB tracker", August 2016.

The demand for RMB will likely continue to grow, to be supported by the massive "One Belt One Road" initiative launched by the Chinese President Xi Jinping in 2013. Implementation of this development initiative will connect China with Asia, East Asia, Europe and Africa by land (the "Silk Road Economic Belt") and by sea (the "21th Century Maritime Silk Road") by enhancing existing trade infrastructure and building up new infrastructure. China has committed US$40 billion to China's Silk Road infrastructure fund[5], US$100 billion to the Asian Infrastructure Investment Bank (AIIB)[6], and US$50 billion to the New Development Bank[7] to support projects under the initiative. This will likely extend the span of RMB internationalisation and boost demand in RMB for trade and financial transactions relating to "One Belt One Road".

1.2   Liberalisation of the RMB market

Over the last two decades, China has demonstrated tremendous economic growth, supported by development of its exchange rate infrastructure. At the initial stage of China's globalisation, the RMB was pegged to the USD. In 2005, the pegged regime was switched to a managed floating rate system, which allowed the RMB to trade in a daily band of ±0.3% against the USD[8]. The band was widened to ±2% in March 2014[9]. A milestone exchange rate system reform in fact occurred in 2015, which is a cornerstone in RMB internationalisation. On 11 August 2015, the PBOC introduced a major change in the USD/CNY fixing regime, under which the exchange rate is to be fixed with market makers' submitted rates that make "reference to the closing rate of the interbank foreign exchange market on the previous day, in conjunction with demand and supply conditions in the foreign exchange market and exchange rate movement of the major currencies in the international market"[10].

The shift to the new market-based fixing regime shows the readiness of China to allow the RMB, to a considerable extent, to be driven by market forces with reference to major currencies. The steps taken by the Chinese government towards RMB internationalisation were applauded by the International Monetary Fund (IMF), which announced in November 2015 the inclusion of the RMB in SDR, to be effective in October 2016.

The marketisation reform of the foreign exchange regime laid a solid foundation for

---

5   "China's Silk Road dream falls into place with US$40b fund", *South China Morning Post*, 17 February 2015.

6   "China ratifies Asian Infrastructure Investment Bank agreement", *Economic Times*, 4 November 2015.

7   New Development Bank website, http://www.ndb.int/brics-bank-to-begin-funding-of-projects-from-april-kamath.php.

8   "A Managed Floating Exchange Rate Regime is an Established Policy", PBOC, 15 July 2016.

9   "Public Announcement No. 5", PBOC, 17 March 2014.

10   "China defends new currency regime", *Financial Times*, 13 August 2015.

more liberalised cross-border financial activities using the RMB. Further liberalisation measures were subsequently introduced. These include the relaxation of quotas and streamlining of the application process for Qualified Foreign Institutional Investors (QFII), and the opening of direct access of foreign investors to the China Interbank Bond Market (CIBM). Further market liberalisation creates an increasing demand from global market participants and policy makers for a clear RMB benchmark for analysing movements of the RMB, to better observe trends and directions for gauging their global RMB exposure and to better balance policy mandates relating to RMB.

Shortly after implementation of the new USD/CNY fixing regime, the PBOC launched the CFETS RMB Index in December 2015, which includes 13 currencies of China's key trading partners. The index was launched together with the BIS Currency Basket RMB Index and the SDR Currency Basket RMB Index. The indices provide new benchmarks of the RMB exchange rate movements in world economic activities.

Table 1. Recent policy path on the internationalisation of RMB towards a global reserve currency	
Date	Policy
11 Aug 2015	PBOC USD/CNY fixing regime reform
30 Nov 2015	Announcement of RMB inclusion in Special Drawing Rights (SDR) by IMF
11 Dec 2015	CFETS RMB Index launch
4 Feb 2016	State Administration of Foreign Exchange (SAFE) relaxation on Qualified Foreign Institutional Investors (QFII) quota and simplification of application process
24 Feb 2016	PBOC announcement on allowing direct access of foreign institutional investors to the China Interbank Bond Market (CIBM)
27 May 2016	PBOC & SAFE announcement of implementation rules for CIBM direct access scheme
1 Oct 2016	Inclusion of RMB in SDR by IMF

Apart from the need for monitoring RMB movements, there will be increasing demand from global market participants for hedging their RMB exchange rate risk when managing their RMB investment portfolios or for monetising their views on the RMB exchange rate. To serve such needs, a transparent and tradable RMB index and exchange rate tools based on it are desirable. It is on this background that the TR/HKEX RMB Currency Index series (RXY Indices or RXY Index series) is designed and introduced. Jointly developed by HKEX and Thomson Reuters, RXY Index series was officially launched on 23 June 2016. The index series is believed to be able to meet market needs that otherwise could not be met by existing RMB indices. (see elaboration in sections 2 and 3 below.)

2 International experience in exchange rate indices

2.1 Central Banks' indices

The modern history of foreign exchange (FX) markets shows that the Central Banks of many countries have developed currency indices as economic indicators for comparing the exchange rates of their country's currencies against their major trading partners. One example is the Trade-Weighted Dollar Index introduced by the US Federal Reserve in 1973. Currently the Index includes 26 currencies and their weightings are revised on an annual basis. The index's primary function is to serve as a policy macroeconomic indicator. Despite its supremacy, the index has not developed into a financial tradable instrument.

In the Euro area, starting 1999 the European Central Bank (ECB) publishes two effective exchange rates (EERs) of the Euro. One of the Euro EERs is calculated against 19 currencies of major trading partners of the Euro zone and the other one reflects the trade relations with a broader currency set of 38 countries. In a similar way as the Trade-Weighted Dollar Index, the broad-based Euro EER has the currency weightings revised on an annual basis and is a strong indicator of the currency value. Both Euro EERs are predominantly used as important indicators for assessing the external economic conditions and international price and cost competitiveness. The EERs' behavior is also an important element of the ECB's evaluation of the monetary situation in the Euro area and setting up strategies for the EU monetary policy.

The most recent notable development in Central Banks' FX indices is the CFETS RMB Index, launched in November 2015. Although the methodology of the index composition is not fully transparent to the public, the index reflects China's currency value against currencies of its important trade partners. The index is calculated with reference to a basket of 13 currencies directly traded against the RMB on CFETS. The weight of each currency in the index is calculated by international trade weight with adjustments of re-export trade factors.[11] Since its launch, the CFETS RMB Index has undoubtedly played an important role in guiding the FX market participants' attention away from the bilateral USD/CNY rate to the reference to a basket of currencies when measuring RMB performance.

---

11 Source: CFETS website (http://www.chinamoney.com.cn).

2.2 Tradable indices

In contrast to the policy currency indices run by the Central Banks, financial market players have developed other currency benchmarks, which are adapted to the financial markets' requirements and are more suitable for trading. One of the most successful examples is the USD Index (USDX) created by the Intercontinental Exchange (ICE) Futures U.S., which includes only six currencies (CAD, CHF, EUR, GBP, JPY, SEK). USDX futures contracts were subsequently listed in November 1985 on the ICE Futures U.S. Since the inception, the currency weights in the index were revised only once in 1999 for replacing a few European currencies with the Euro (EUR). Despite the fact that the current basket of the USDX does not entirely reflect the latest US economic relationships due to the shift in the economic landscape towards China over the last decade, the USDX futures have become the world's most widely-recognised traded currency index futures, with 12 million contracts traded in 2015[12]. A number of multimillion exchange traded funds (ETFs) are linked to the index (e.g. Powershares DB Bullish and Bearish funds, WisdomTree Bloomberg USDX fund).

Another example is the Euro Index (EURX or EXY) launched in January 2006 by the New York Board of Trade (NYBOT) which was later acquired by the ICE. The ICE Euro Index measured the value of the Euro against a basket of five currencies (USD, GBP, JPY, CHF, and SEK) and initially reflected the weightings calculated by the ECB for deriving the EERs. However, in May 2011 the ICE Futures U.S. ended the trading of futures and options on the index and shortly after that discontinued the calculation of the ICE Euro Index.

While USD and EUR are major international currencies inducing demand for development of their corresponding tradable indices, there are rarely indices developed on currencies of developing countries. The RXY Index series is the latest example of tradable currency indices, for the first time on RMB, in the course of the currency's internationalisation process. The primary index of this series — TR/HKEX Global CNH Index (Global RXY Index) — is particularly of relevance.

---

12 The ICE, US Dollar Index Futures, Historical Monthly Volumes.

Table 2. Comparison of Central Banks' indices and their tradable peers			
Central Bank's index	Federal Reserve's Trade-Weighted Dollar Index	European Central Bank's Euro Effective Exchange Rates (EERs)	CFETS RMB Index
Launch	1973	1999	2015
Publisher	Federal Reserve	ECB	PBOC
Number of constituents	26	19/38	13
Rebalancing	Annually	Annually	Annually
Calculation	Geometric average	Geometric average	Geometric average
Weighting	Trade weighted	Trade weighted	Trade weighted
Tradability	nil	nil	nil
Tradable index	USDX / DXY	EURO / EXY	Global RXY CNH
Launch	1985	2006	2016
Publisher	ICE Futures U.S.	ICE Futures U.S.	Thomson Reuters/HKEX
Number of constituents	6	5	14
Rebalancing	Fixed	Fixed	Annually Adjusted
Tradable products	Futures, Options, ETFs	Futures, Options	Not yet available*

*   Products including futures, options and ETFs may be introduced based on RXY Indices. HKEX is considering introducing RXY Index futures, subject to regulatory approval.

Source: Board of Governors of the Federal Reserve System, European Central Bank, CFETS, HKEX, ICE Futures U.S.

2.3   Other indices

In the RMB market, some market participants initiated their own in-house indices to measure RMB performance. Banks often develop their internal benchmarks for conducting macroeconomic analysis and look at the FX market's trends. For such purposes real effective exchange rate (REER) indices, or in other words, indices adjusted for domestic inflation rate, are the best indicators as they demonstrate relative strength or weakness of the domestic currency in comparison to other currencies. In addition to REERs, banks also create RMB models focusing on some specific areas (e.g. exports or certain industry sectors) or on some aspects of RMB development, such as the theme of globalisation. Some banks use their internal RMB indices as benchmarks or reference rates for developing FX derivatives strategy.

On the asset management side, in-house RMB indices are often used for reference by market institutions' trading desks as indicators of directions of RMB movement. Such internal instruments are also often used to analyse performance of RMB-denominated assets in investment portfolios or to identify hedging strategy of the RMB exposure. However, due to competition among investment managers and their desire to produce higher returns, in most cases the internal models are not disclosed to the public.

The in-house indices described above of various market institutions including banks, securities firms or asset management companies, whilst may be good benchmarks for in-house valuation purposes, cannot become the financial industry's leading RMB measuring instruments. Part of the reason is that these indices are built by the institutions to meet their internal targets which would be different from one another, for which independence would be an issue. Secondly, in-house RMB indices are rarely in compliance with international standards like the International Organisation of Securities Commissions (IOSCO) principles, partly due to the costs and labour involved in aligning, maintaining and certifying the internal benchmarks in accordance with the international principles. Therefore, in-house instruments cannot become industry benchmarks due to their non-independence, non-compliance with the international rules for financial benchmarks and often non-transparency in their calculation methodologies.

3.   The RXY Index series

The RXY Indices are developed by independent parties — HKEX and Thomson Reuters — with the objective of offering market participants a RMB performance indicator of the highest industry standard.

The RXY Indices comprise a Primary Index — the **TR/HKEX Global CNH Index** — and three variant indices — **TR/HKEX Global CNY, TR/HKEX Reference CNH** and **TR/HKEX Reference CNY**. The indices differ by their base currency basket — a global basket which includes 14 currencies, and a reference basket which consists of 13 currencies (more information on the baskets' constituent currencies is provided in 3.1 below) — and the RMB measure — onshore RMB (CNY) and offshore RMB (CNH). They have a base date of 31 December 2014 with a base value of 100, same as the CFETS RMB Index, and the historical data of the RXY Indices is available back to 31 Dec 2010. Figure 3 gives a summary of the four indices and Figure 4 shows their historical performance.

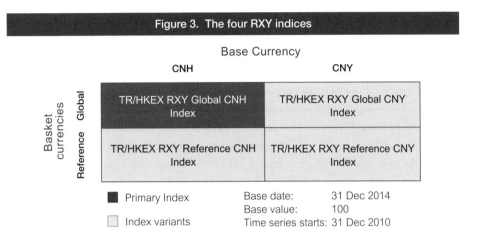

Figure 3. The four RXY indices

Base Currency

		CNH	CNY
Basket currencies	**Global**	TR/HKEX RXY Global CNH Index	TR/HKEX RXY Global CNY Index
	Reference	TR/HKEX RXY Reference CNH Index	TR/HKEX RXY Reference CNY Index

■ Primary Index

☐ Index variants

Base date: 31 Dec 2014
Base value: 100
Time series starts: 31 Dec 2010

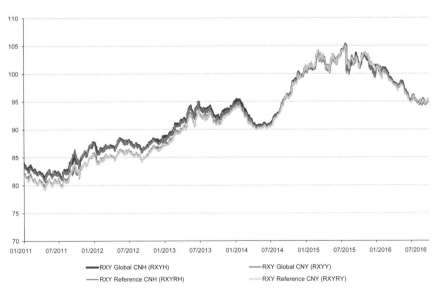

**Figure 4. Historical performance of the four TR/HKEX RXY indices
(31 Dec 2010–30 Sep 2016)**

━━RXY Global CNH (RXYH)   ━━RXY Global CNY (RXYY)
━━RXY Reference CNH (RXYRH)   ━━RXY Reference CNY (RXYRY)

Source: HKEX.

523

The RXY Indices are managed in accordance with the IOSCO principles for financial benchmarks[13], which ensure that governance of the indices is objective and rules-based, and that the indices are transparently calculated from underlying rates supplied by WM/Reuters. Being in compliance with the IOSCO principles, the RXY Indices have the advantage of being qualified not only as benchmarks for RMB valuation, but also as underlying references for financial instruments such as futures, options and ETFs. Furthermore, market regulations imposed on investment products have become more stringent after the 2008 Global Financial Crisis and require higher product transparency and integrity, including those of the products' references or underlying assets. Under such regulatory environment, being IOSCO-complaint enables the RXY Indices to be used for issuing investment products to meet the needs of different types of institutional and retail investors.

Thomson Reuters administers the RXY Indices. Through a framework that includes a committee of subject matter experts and a dedicated index manager, Thomson Reuters is responsible for maintaining the integrity and quality of the RXY Indices, and for carrying out regular work and duties, including the following:

- To interpret the index methodology and implement the annual rebalance procedure;
- To review feedback received from index stakeholders;
- To develop and implement changes to the index methodology if required by feedback from stakeholders or by market events;
- To manage interaction with the **Index Advisory Group** ("IAG") and **Index Action Committee** ("IAC") in respect of rebalances and index methodology changes; and
- To report to the **Thomson Reuters Benchmarks Oversight Committee** ("TRBOC").

Following interaction with the IAC and, where required, the IAG, the Index Manager is responsible for determining any changes to the index methodology.

The IAC is an internal Thomson Reuters group of subject matter experts (indices as well as asset classes) that support the Index Manager with additional advice related to methodology interpretation or changes to the methodology. Specifically, the Index Manager may communicate the feedback obtained from the IAG, which includes a representative from HKEX, and/or index stakeholders to the IAC and solicit its advice. The IAC in turn reports to the TRBOC.

3.1 Product design

The RXY Indices are designed in such a way that the index will rise when the base currency (CNY or CNH) appreciates in value against the base basket of currencies, and will

---

13 Source: Document on IOSCO website (www.iosco.org/library/pubdocs/pdf/IOSCOPD415.pdf).

decline when the base currency depreciates in value against the base basket of currencies. The indices are calculated on an hourly basis and have their close values at 4 pm Hong Kong time on the trading day.

Each RXY index, I_t, is computed at any point of time, t, in accordance with the formula below:

$$I_t = I_0 \cdot \prod_i \left(\frac{FX_{i,t}}{FX_{i,0}}\right)^{\omega_i}$$

Where $FX_{i,t}$ is the spot FX rate of currency i in the basket against the base currency, at time t. I_0 and $FX_{i,0}$ are the index and spot FX rates at the last rebalance time respectively, and w_i is the weight of currency i (such that $\Sigma w_i = 1$). The indices are calculated using geometric averaging algorithm and spot FX rates conversion (all FX rates used in the above formula are derived from spot FX rates against USD).

The RXY Indices are based on two baskets: the global basket which consists of 14 currencies, and the reference basket which consists of 13 currencies. Each currency basket is trade-weighted, where the weighting of each currency is determined by the actual trade volume between China and the corresponding country of the currency.

The 14 currencies included in the global basket are:
• AUD — Australian dollar
• CAD — Canadian dollar
• CHF — Swiss franc
• EUR — Euro
• GBP — British pound
• HKD — Hong Kong dollar
• JPY — Japanese yen
• KRW — Korean won
• MYR — Malaysian ringgit
• NZD — New Zealand dollar
• RUB — Russian ruble
• SGD — Singaporean dollar
• THB — Thailand baht
• USD — US dollar

The reference basket exclude KRW, i.e. having the same set of 13 currencies as currently in the CFETS RMB Index basket, with similar weightings as well (see Figure 5 below). The currency weightings in the RXY Indices are rebalanced on an annual basis in order to reflect the most recently available trading data. The weightings are derived from the United Nations Commodities Trade Statistics (UN Comtrade)[14], which provides annual trade volumes between China and other countries. The trade data from the Hong Kong Census and Statistics Department[15] is used to adjust the annual bilateral exports from Mainland China to Hong Kong as reported by the UN Comtrade. Such adjustment reflects the fact that a substantial amount of the exports from Mainland China to Hong Kong is not for domestic use and requires recalculation to obtain the actual amount of exports from Mainland China that are absorbed by Hong Kong. The reference to the international source of trade statistics and transparent mechanism of the weighting composition of the RXY Indices make the changes in the indices' constituents highly predictable.

Figure 5. Currency weights in RXY and CFETS RMB Indices

Currency weightings in RXY Global Currency Indices[1]

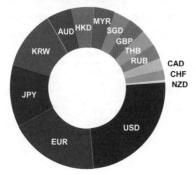

| Currency | Currency Weights | | |
	RXY Global Currency Indices[1]	RXY Reference Currency Indices[1]	CFETS RMB Index[2]
USD	24.69%	28.09%	26.40%
EUR	18.47%	21.03%	21.39%
JPY	12.27%	13.97%	14.68%
KRW	12.14%	0.00%	0.00%
AUD	5.02%	5.72%	6.27%
HKD	4.89%	5.56%	6.55%
MYR	4.29%	4.88%	4.67%
SGD	3.55%	4.04%	3.82%
GBP	3.46%	3.93%	3.86%
THB	3.32%	3.78%	3.33%
RUB	2.99%	3.41%	4.36%
CAD	2.45%	2.79%	2.53%
CHF	1.95%	2.22%	1.51%
NZD	0.51%	0.58%	0.65%

Notes:
(1) Weights in RXY Indices are effective from 3 October 2016 until 29 Sep 2017.
(2) CFETS RMB Index was introduced on 11 December 2015. The currency weights in the index were obtained from CFETS's announcement on its website on that date.

Source: HKEX, CFETS.

The rebalancing of the RXY Indices involves updating the weightings of constituents in each index basket on an annual basis, with the rebalancing cycle starting in June of each year, after the UN Comtrade and Hong Kong Census and Statistics Department's annual trade statistics become available. The IAG studies the preliminary weighting calculations and announces the updated weightings on the last business day of June. The new weightings become effective from the first trading day of October.

---

14 United Nations Commodities trade statistics, www.comtrade.un.org.
15 Hong Kong Census and Statistics Department, http://www.censtatd.gov.hk/home/.

The RXY Index series is designed to possess the characteristics of a tradable index — regulatory compliance, reliable data source, reputable compiler, transparent methodology and frequent publishing (hourly). These may not be fulfilled by the only other publicly available official CFETS RMB Index.

3.2   Usage and benefits

The RXY Indices are based on FX rates provided by WM/Reuters, which is regulated by the UK Financial Conduct Authority. WM/Reuters spot FX rates are continuously monitored to ensure that they meet the highest standards of industry best practice, and therefore minimise the chance of price manipulation or control.

From a policy perspective, the RXY Indices offer market analysts and economists a transparent, hourly benchmark for analysing the behavior of the RMB and for building their predictions on the RMB trends. The RXY Indices would be reliable proxies for the CFETS RMB Index due to their high degree of correlation with the latter, and they therefore also provide convenient tools to the market for analysing Chinese authorities' FX policy. By monitoring movements of the USD/CNY fixing rate announced by the PBOC on a daily basis and comparing it with the RXY Indices, market participants can see the direction of the Chinese FX policy.

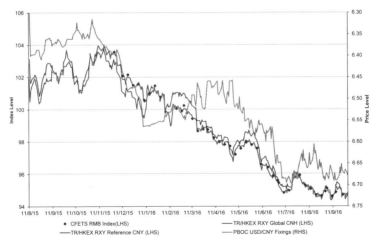

Figure 6. RXY and CFETS RMB Indices performance versus USD/CNY fixing (11 Aug 2015–30 Sep 2016)

Note: For comparison purpose, the CFETS RMB Index is rebased on 30 November 2015 to the same level as the TR/HKEX RXY Reference CNY Index.

Source: HKEX, CFETS, Thomson Reuters.

From a trading perspective, the RXY Indices could serve as good indicators of the market-driven direction of RMB movements and market participants could capitalise on taking their views on the RMB directions. International players with RMB exposure may use the RXY Indices for better management of their currency risks.

Table 3. Risk and return profile of the RXY indices											
	Return						Risk			Correlation with CFETS RMB Index (30/11/2015 – 30/09/2016)	Beta vs CFETS RMB Index (30/11/2015 – 30/09/2016)
Index	Jul 2016		Aug 2016		Sep 2016		30-day realised volatility (ending)				
	M-o-M	Y-o-Y	M-o-M	Y-o-Y	M-o-M	Y-o-Y	29/7/ 2016	31/8/ 2016	30/9/ 2016		
TR/HKEX RXY Global CNH (RXYH)	0.09%	-8.30%	-0.74%	-6.43%	-0.46%	-8.48%	3.44%	3.38%	3.95%	0.9963	0.6432
TR/HKEX RXY Global CNY (RXYY)	-0.05%	-8.34%	-0.68%	-7.28%	-0.46%	-8.02%	3.40%	2.97%	3.43%	0.9817	0.6703
TR/HKEX RXY Reference CNH (RXYRH)	0.48%	-7.99%	-0.72%	-6.06%	-0.38%	-8.30%	3.31%	3.22%	3.90%	0.9933	0.6570
TR/HKEX RXY Reference CNY (RXYRY)	0.34%	-8.02%	-0.65%	-6.90%	-0.38%	-7.72%	3.55%	3.09%	3.57%	0.9860	0.6533

Note: "M-o-M" return is the month-on-month return of the index as of the month-end date relative to the previous month-end index;
"Y-o-Y" return is the year-on-year return of the index as of the month-end date relative to the corresponding month-end of the previous year.

Source: HKEX, Thomson Reuters, CFETS.

In addition, the RXY Indices can also become useful tools for the Chinese authorities. The open dialog between the PBOC and the major FX market players about the mechanism of setting the exchange rates may suggest that, driven by market forces, the RXY Indices could possibly be used as a price discovery tool by the relevant Chinese authorities.

The RXY Indices are designed to **provide references for financial instruments including futures, options and ETFs**. The design of the indices aims at ensuring that any derivatives referenced to the RXY Indices can be priced fairly by permitting arbitrage trades. To meet the fast development of RMB internationalisation and the consequent growing demand for RMB financial products, RMB hedging tools such as futures and options contracts on RXY Indices would be useful. RXY Indices would have the potential to develop into an extremely important tool in the chain of FX products creation and valuation.

Chapter 7

HKEX's Five-Year China Ministry of Finance Treasury Bond Futures

The world's first RMB bond derivatives
accessible to offshore investors

24 April 2017

Summary

China's debt capital market has now become the third largest in the world at RMB 56.3 trillion, or about US$8.1 trillion[1], after a rapid expansion over years. China also made significant strides to advance RMB internationalisation and the openness of the domestic financial market. Although the current share of foreign holdings of Chinese bonds is still at a low level, foreign capital shows a strong appetite for Chinese sovereign bonds, and foreign holdings in the sovereign bond segment has significantly increased after the formal inclusion of RMB into the International Monetary Fund's Special Drawing Right (SDR) basket in October 2016. If China implements the pilot Bond Connect scheme between Hong Kong and the Mainland in the near future, the increased foreign investment in Chinese bonds would result in a surging demand for related risk management.

Developing effective hedging support and providing foreign exchange (FX) access are important for foreign investors to increase their exposure to RMB assets. To date, there are a number of interest rate risk management products in the onshore market, which provides supportive tools to hedge RMB interest rate risks. Along with further opening up of the domestic foreign exchange market to foreign investors recently, some eligible foreign investors can also directly access mainland derivatives. HKEX's T-Bond Futures utilizes the product strength of offshore market and is carefully designed with a few distinguishing features, in order to ensure that the trading of this product would unlikely have an adverse impact on the onshore market.

Based on the experience of developed countries, the introduction of treasury bond futures plays an important role in improving the pricing function of the underlying bond market, promoting the liquidity of spot market and enriching the means of interest rate risk management of bond investors. A majority of empirical studies finds either no significant effect, or else a decrease in volatility, of the spot market following the introduction of treasury bond futures. HKEX's T-Bond Futures provides a solid tool for foreign investors to hedge against interest rate volatility of RMB assets, and could be regarded as a quickening step to support the development of the onshore fixed income market and facilitate foreign capital flows into China's domestic bond market. Banks, asset management companies, brokerage firms and insurance companies are the main target users of this product.

---

1   Source: Wind (data as at end-2016).

1   The openness of China's domestic bond market

China's debt capital market has now become the third largest in the world at RMB 56.3 trillion, or about US$8.1 trillion, after a rapid expansion at a simple average annual growth rate of 21% in outstanding value over the past five years. (see Figure 1.)

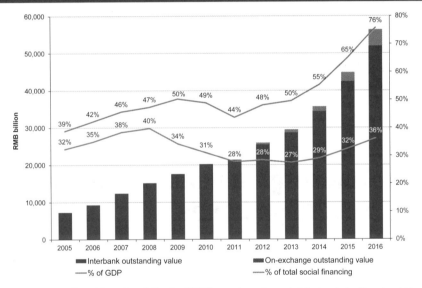

Figure 1. Year-end outstanding value of China's debt securities and its share of GDP and total social financing (TSF) (2005–2016)

Source: China Central Depository & Clearing (CCDC) annual reports on statistical analysis of bond market (2011 — 2016), excluding data for on-exchange market and bonds settled in Shanghai Clearing House (SCH) (2005 — 2011) due to data unavailability. GDP and TSF data are from Wind.

China has promoted its debt capital market to play a more prominent role in financial resources allocation. Meanwhile, it also made significant strides to advance RMB internationalisation and the openness of the domestic financial market. One major direction is to invite more foreign participation to tap into the domestic bond market in order to boost the diversity and variety of the bond sector and further increase the scale and depth of the domestic financial market. In 2010, China took the first step to allow qualified institutions to use offshore RMB to invest in the interbank bond market, and then introduced Renminbi Qualified Foreign International Investor (RQFII) program in 2011 and relaxed the invest-ment restrictions of Qualified Foreign International Investor (QFII) in 2013 as another steps towards opening domestic bond market.

Since 2015, a number of notable liberalisation measures were launched to further facilitate foreign investors to access China's interbank bond market (CIBM). Specifically, in late-May 2015, the People's Bank of China (PBOC) allows offshore RMB clearing and participating banks to conduct repurchase (repo) financing by using their onshore bond holdings. In mid-July 2015, the PBOC further eased the scope of eligible bond transactions by allowing eligible entities to engage in bond trading, bond repo, bond lending, bond futures, interest rate swaps and other trades permitted by the PBOC in the interbank market. In February 2016, the PBOC released new regulations which relaxed the rules applicable to foreign institutional investors accessing the interbank bond market. In May 2016, China further published the detailed rules to clarify the investment procedure of foreign institutional investors in the interbank bond market.

Those policy moves, to some extent, pass on a message to the market that China is on the way to further open its capital account and encourage more foreign capital inflows.

2   The surging demand for Chinese sovereign bonds

Given the formal inclusion of RMB into the SDR basket in 2016, the demand for RMB assets, especially RMB bond assets, would steadily grow when central banks and global investors started to consider reallocating funds into RMB-denominated assets. The achievement of SDR status increases the global acceptance of RMB as a global investment and reserve currency, which would most likely trigger an increasing demand for RMB-denominated assets from both public- and private-sectors internationally. According to our estimation, if the holding share of RMB assets by global institutions or individuals could reach to 10% of total domestic bond market, it could be expected that over RMB 9.5 trillion would flow into relevant RMB bond assets in the coming years.

Typically, debt securities, especially the sovereign bonds are a top asset class for central banks and global fund managers in assets management. Although the share of foreign holdings of Chinese bonds are still at a low level, foreign capital shows a strong appetite for Chinese sovereign bonds, and foreign holdings in the sovereign bond segment have significantly increased recently. In 2016, the foreign holdings of Chinese government and policy-bank bonds increased by RMB 233 billion, a six-fold rise compared with RMB 35 billion in 2015. The foreign ownership in China's sovereign bond market rose to 3.93% from 2.62% at end-2015. (see Figure 2.) Due to the current low (or negative) yield environment among sovereign bonds of major developed countries, the movement of capital reallocation from other financial segments to Chinese sovereign bond segment would likely be stronger after China takes more welcoming steps to foreign participants.

Figure 2. The share in China's sovereign bond market by investor type (End–2016 vs End–2015)

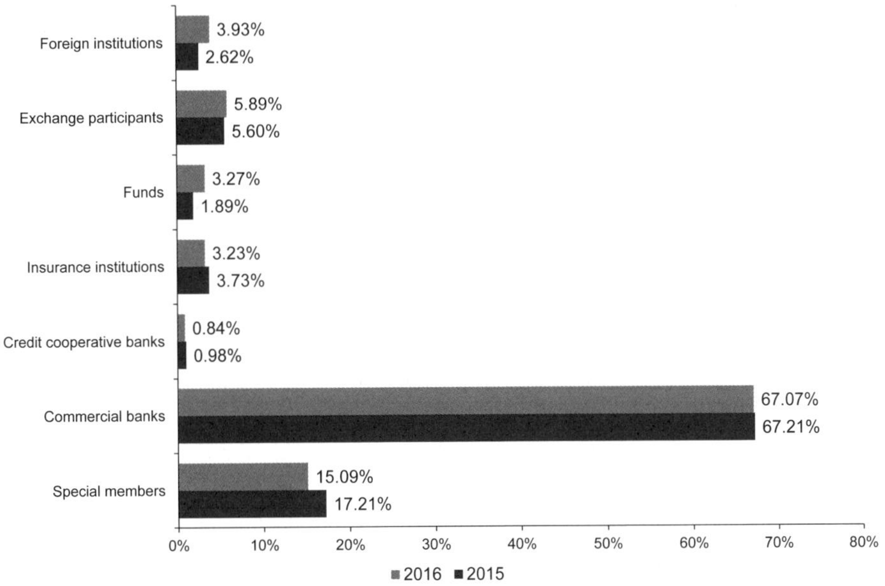

Source: Wind.

3   The advantage of offshore market in risk hedging and accessibility

Developing an effective hedging support and providing the FX access are important for foreign investors to increase their exposure to RMB assets. Currently, China's domestic FIC derivatives market is relatively deep and liquid, with a range of FX products available (including spot, forwards, swaps, and options) and also treasury bond futures products available. However, foreign institutions are not yet to be allowed access to the domestic treasury bond futures for risk hedging. Moreover, domestic insurance companies and banks, the main holders of Treasury cash bonds, are not yet to be allowed to participate in the trading of treasury bond futures. Such segmentation of the domestic bond market could split liquidity and market depth.

The availability of bond futures with better liquidity can help foreign investors improve their ability to hedge against interest rate risks via risk transfer and channeling, and increase their willingness to hold a larger portion of Chinese bond assets. On 15 March 2017, Li Ke-qiang, the Premier of the State Council of the People's Republic of China, publicly stated the plan to set up bond market links between Hong Kong and Mainland China[2]. Under such pilot Bond Connect scheme to be implemented in future, the increased foreign investment in Chinese bonds could result in a surging demand for related risk management.

It is on this backdrop that HKEX's 5-Year China Ministry of Finance Treasury Bond Futures (T-Bond Futures) is designed and introduced. To date, there are a number of interest rate risk management products in the onshore market, which provides supportive tools to hedge RMB interest rate. Along with further opening up of the domestic foreign exchange market for foreign investors recently, certain eligible foreign investors can also directly access mainland derivatives. HKEX's T-Bond Futures utilizes the product strength of offshore market to provide such differentiation. Its introduction in the offshore market provides a solid tool to help foreign investors hedge against interest rate volatility of RMB assets, could be regarded as a quickening step to facilitate foreign capital flows into China's domestic bond market.

---

2   Mr Li answered questions from domestic and foreign journalists at a news conference after the national legislature's annual session concluded in Beijing on 15 March 2017 and stated that China are preparing to implement for this year a pilot bond market scheme connecting between Hong Kong and the Mainland allowing for the first time overseas capital to buy Mainland RMB bonds.

4   Onshore and offshore hedging tools for Chinese bond assets

Sovereign bond futures are an important section in the exchange-traded interest rate derivatives market. They are designed to allow price convergence to the most liquid sovereign bonds at the stated maturity (e.g. 2-, 5-,10- or 30-year). This makes sovereign bond futures a valuable instrument for hedging interest rate exposure represented by sovereign bond yields. For example, sovereign bond derivatives can be used for hedging by a corporate borrowing at a fixed spread above the government treasuries, or a fund manager investing in this corporate's bonds.

Currently, the sovereign bond futures available in the China's domestic market are the 5-year and 10-year MOF T-Bond Futures contracts listed on the China Financial Futures Exchange (CFFEX). The 5-year contract was introduced on 6 September 2013, followed by the 10-year contract on 20 March 2015. As of March 2017, the average daily turnover (ADT) of these bond futures amounted to RMB 67.73 billion with an open interest (OI) of RMB 84.57 billion. (see Figure 3.) However, these products have not been available for foreign investors to hedge interest rate risks of RMB assets, and the liquidity is limited due to the absence of major participants, such as domestic insurance companies and banks.

The offshore market also lacks efficient RMB rates hedging tools for mid- to long-term yield curve before the launch of T-Bond Futures. Previously, the management tools for hedging RMB interest rate risks in the offshore market are the non-deliverable interest rate swap ("NDIRS") and offshore RMB (CNH) interest rate swap ("IRS"). The pricing of NDIRS is more influenced by speculative factors instead of fundamental capital flows, and NDIRS is therefore not generally regarded as an efficient tool to hedge RMB interest rate exposure. Along with the growth of CNH money market, CNH IRS has developed further and more market transactions have shifted from NDIRS to CNH IRS. However, the key issue in CNH IRS pricing is that the offshore RMB deposit rates differ from the rates onshore due to the relatively low market liquidity and lack of demand for lending. This contributes to a different pricing of the CNH IRS from the rates in the domestic market. (see Figure 4.) HKEX's MOF T-Bond Futures can serve as a benchmark tool of long-term interest rate of Chinese domestic assets for offshore investors, supplementing the existing CNY NDIRS yield curve.

Figure 3. The turnover of the 5-year and 10-year MOF T-Bond Futures contracts listed on CFFEX (Sep 2013 – March 2017)

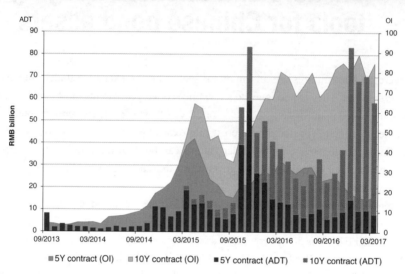

Source: Bloomberg.

Figure 4. The performance of NDIRS and CNY IRS (2013 – 2016)

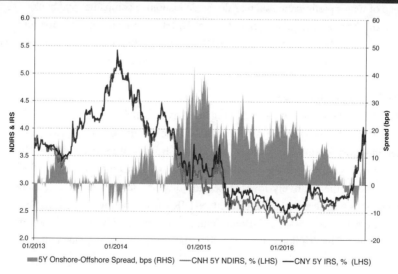

Source: Bloomberg.

5   Product design: methodology and applications

HKEX's T-Bond Futures is designed in a similar way as CFFEX's bond futures contract in that the underlying is the onshore China Ministry of Finance treasury bonds and the coupon rate is 3% per annum. The difference is that the domestic bond futures adopt a physical delivery design, known as "cheapest-to-deliver", which allows the short-position holder to deliver the cheapest among the eligible bond securities to the long position at contract expiry. Moreover, the product design of HKEX's T-Bond Futures has similarities[3] with the government bond futures listed on the Australian Securities Exchange (ASX) and Korea Exchange (KRX).

5.1   Principles for construction of the bond basket

Construction of the bond basket of HKEX's T-Bond Futures is based on the principles including transparency, predictability, liquidity, ease of replication and reliability.

(a)   **Transparency and predictability:** The bond basket and reference price is based on a rule-based design, with its methodology made publicly available, including pricing and valuation process, formula and models. HKEX reserves the right to exercise discretion, when necessary, due to the substantial changes in China's treasury bond issuance policy.

(b)   **Constituents liquidity:** The bond basket constituents should exhibit good liquidity in general for hedging purposes. Therefore, the bonds selected as the bond basket constituents must be in the top 3 most liquid issues based on ChinaBond's[4] relative liquidity measure as of the date of basket construction.

(c)   **Ease of replication:** Based on historical performance, the bond basket's total trading volume should represent at least 50% of that of the bond universe. Futures based on the bond basket should track the 5-year MoF T-Bond performance closely. Therefore, investors can easily replicate the underlying bond basket for hedging purposes.

---

3   Similar to the government bond futures listed on ASX and KRX, the HKEX's T-Bond Futures contract is based on a cash settlement methodology based on an underlying basket of bonds, Both ASX and KRX will announce the underlying basket of bonds ahead of the contract's first trading day. On the last trading day, the contract will be cash settled based on the average yield of the bond constituents in the bond basket.

4   ChinaBond is China Central Depository & Clearing Co. Ltd, the domestic central depository of Chinese bonds.

(d) **Price reliability:** The reference price of the bond basket (5-year MOF T-Bond) is provided by ChinaBond on a daily basis. ChinaBond, is a fully state-owned non-bank official financial institution authorised by the China MOF to develop and operate the national treasury bond depository system.

5.2 Determination of daily reference price for each futures contract

ChinaBond shall, in accordance with the procedures and methodologies provided by Hong Kong Futures Exchange, a fully-owned subsidiary of HKEX, determine the bond basket and calculate the daily reference price of the bond basket for each Futures Contract.

The bond basket before the listing of each futures contract (quarterly) is determined according to the below arrangements:

(a) The date of basket determination is defined as 5 working days before the listing date of the futures contract;

(b) The bonds must be in the top 3 most liquid issuances based on ChinaBond's relative liquidity measure; and

(c) The liquidity measure is based on the trading data for the last 22 working days dating back from the date of basket determination.

Similar to the design of ASX and KRX Government Bond Futures, daily reference price for each futures contract is calculated based on the formulas as below:

(a) Collect the yield from ChinaBond for the constituent bonds in the Basket of Bonds, denoting as r_1, r_2, r_3

(b) Calculate the simple average yield to maturity of the Basket of Bonds with formula:

$$r = \frac{\sum_{i=1}^{3} r_i}{3}$$

c) Calculate daily reference price of the bond basket: Nominal 5-year term bond with coupon rate of 3% paid on an annual basis, with formula:

$$\sum_{i=1}^{5} \frac{3\% \times 100}{(1+r)^i} + \frac{100}{(1+r)^5}$$

Where r is the average yield to maturity calculated in (b).

5.3　Hypothetical examples for illustration[5]

Example 1 — Hedging against interest rate movement

Assume a fund manager, concerned about a potential tightening of monetary conditions in China, wants to hedge against the interest rate risk. On 31 Oct 2016, the fund manager holds RMB 100 million nominal value of treasury bond 160014.IB at price 101.813 with a duration of 5.901. HKEX's T-Bond Futures Mar-17 contract is traded at 102.282, with a duration of 4.80. With the objective of neutralising the dollar duration, he hedges his holding by selling 245 contracts of HKEX's T-Bond Futures Mar-17. By 26 Jan 2017, the yield has gone up and the value of the bond has decreased to 98.439 (-3.374), recording a loss of RMB 3.4 million. The price of HKEX's T-Bond Futures Mar-17 drops to 99.480 (-2.802). The portfolio manager closes the position, gaining RMB 3.4 million. The loss of RMB 3.4 million from the cash bond holding is covered by RMB 3.4 million profit from T-Bond futures positions.

Example 2 — Duration management

Assume a portfolio manager has a diversified bond portfolio of RMB 300 million market value with duration of 7.00. She has the flexibility to adjust duration either up or down by 10% under the fund's stated investment objectives. The portfolio manager expects rates to fall. She therefore intends to increase duration to 7.70. HKEX's T-Bond Futures contract is currently traded at 102.282, with a duration of 4.80. She can buy 86 contracts of HKEX's T-Bond Futures.

Example 3 — Synthetic bond

Assume a foreign institutional investor does not have access to China's onshore bond market, but wishes to create a synthetic cash bond position in order to gain proxy bond exposure due to the China market's yield differential. He can buy 100 contracts of HKEX's T-Bond Futures, creating a proxy bond position with notional value of RMB 50 million.

Example 4 — Credit spread trade

Assume an investor expects the yield of a corporate bond to diverge from the yield of the HKEX's T-Bond Futures. If the investor expects the credit spread (the yield of a corporate bond minus the yield of the HKEX's T-Bond Futures) to narrow, he can consider buying the corporate bond and selling the HKEX's T-Bond Futures. Alternatively, if the investor expects the credit spread to widen, he can consider selling the corporate bond and buying the HKEX's T-Bond Futures.

5　These examples do not constitute investment advice and independent advice should be sought where appropriate. In the case of risky strategies, investors may lose the entirety of their investment.

5.4　Pro-forma performance analysis

The reference price of the HKEX's pro-forma futures has an annualised correlation of 92.1% against CFFEX's futures (September 2013 to December 2016), which is based on a physically-delivered design. (see Figure 5.) Therefore, it could facilitate international investors to effectively address the growing interest rate risk management demand. Banks, asset management companies, brokerage firms and insurance companies will be the main target users of this product.

Figure 5.　Correlation between HKEX's pro-forma futures against CFFEX's futures (Sep 2013 to End–2016)

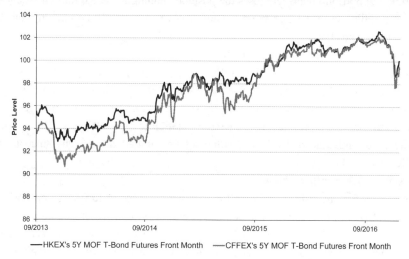

HKEX's 5Y MOF T-Bond Futures Front Month　　CFFEX's 5Y MOF T-Bond Futures Front Month

Source:　Bloomberg, HKEX.

In addition, HKEX's T-Bond Futures could be regarded as a proxy for the RMB bond yield index due to the high degree of correlation between the two. HKEX's pro-forma futures bond basket yield tracks closely the Sovereign Bond Yield (5Y) published by ChinaBond. (see Figure 6.) The yield-to-maturity (YTM) of the two series has an annualised correlation of 98.3% over the past six years (2011-2016). Therefore, HKEX's T-Bond Futures provide a relatively convenient tool for the market to evaluate Chinese bond assets.

Figure 6. Correlation between HKEX's pro-forma futures against ChinaBond's Government Bond Yield (Jun 2008 to End–2016)

———HKEX's MOF T-Bond Futures Front Month Basket YTM   ———China T-Bond 5Y YTM

Source: Wind, HKEX.

6   Interaction and effectiveness

Based on the experience of developed countries, the introduction of treasury bond futures plays an important role in improving the pricing function of the underlying bond market, promoting the liquidity of spot market and enriching the means of interest rate risk management of bond investors. Across the market literature, a majority of empirical studies finds either no significant effect, or else a decrease in volatility, of the spot market following the introduction of treasury bond futures[6].

HKEX's T-Bond Futures is carefully designed with a few distinguishing features to ensure that the trading of this product would unlikely have an adverse impact on the onshore market. In fact, this product serves the function of supporting the development of the onshore fixed income market. These features are as below:

---

6   See "The Impact of Futures Trading on the Spot Market for Treasury Bonds" (Shantaram Hegde, 1994) and "The Impact of Derivatives on Cash Markets: What Have We Learned?" (Stewart Mayhew, 2000).

(a) HKEX's T-Bond Futures contract is cash settled for difference in RMB cash in the offshore market. At each futures contract expiry, the amount of transactions to be exchanged between market participants for settlement purpose happen in the offshore market, and only represent a fraction of the full contract notional amount. The impact of settlement process on liquidity is therefore considerably less compared to a physically-delivered futures contract.

(b) HKEX's T-Bond Futures contract is settled to the price based on the average yield of three constituent bonds in the bond basket, which represents the top three most liquid onshore T-Bonds within the bond universe. This final settlement price design reduces the risk of manipulation on any individual underlying bond. Under such design, HKEX's T-Bond Futures in effect provide investors the exposure to a part of the bond yield curve, rather than the exposure to individual bond (please refer to Section 5.1 and 5.2 for further details of the final settlement price).

In addition, as the HKEX's T-Bond Futures contract will converge to the final settlement price at expiry, any significant price deviation between HKEX's T-Bond Futures and the similar onshore product would be costly and for offshore market participants to take on more positions. A case in point was the CNH IRS market where the pricing differential between the onshore and offshore rates contributed to a lack of liquidity. Based on the pro-forma analysis, the average yield of HKEX's T-Bond Futures contract dummy bond basket has a high correlation with the onshore 5-year treasury bond yield (98.3% from 2011 to 2016), and the daily reference price of HKEX's T-Bond futures contract is also highly correlated with CFFEX's T-bond futures price (92.1% from September 2013 to December 2016) (please refer to Section 5.4 for further details). Taking a one-way position in HKEX's T-Bond Futures contract that is of sufficient magnitude to affect the onshore market stability would be very difficult, if not impossible, in practice.

(c) HKEX's T-Bond Futures contract is traded in a regulated, centralised and transparent exchange platform. This improves the market transparency and offers useful information to participants on price expectations and open interest levels.

(d) As is the case with other HKEX listed futures products, there are several measures in HKEX's trading and clearing rules and in relevant Securities and Futures Commission regulations, which can be used to deter the accumulation of large open positions of the T-Bond Futures contract, and thus may minimise the risk of unwanted volatility in the market, such as:

- Requiring additional concentration collateral from clearing participants with a large share of the outstanding open interest, thus effectively lowering the leverage on large open positions;

- Requiring exchange participants (either acting for their own account or on behalf of any client) to report large open positions (LOPs) in the contract to the HKEX. HKEX also has the power to require additional contextual information from any LOP holders to justify their large positions;
- Imposing position limits to cap the position that can be held by a single beneficial owner. Position limits are taken seriously, and breaching them might constitute a breach of relevant HKEX rules and the Securities and Futures Ordinance, including potential criminal liability. HKEX and the SFC both can take remedial action against any breaches, including forcing a participant to reduce their positions in a timely and orderly manner where appropriate.

China is now the fastest growing bond market in the world and is the third largest after the United States and Japan. International participation in China's bond market continues to increase, driven by the further opening up of China interbank bond market, the broadening of international acceptance of the RMB, the inclusion in the emerging markets bond indices and the yield differential compared to developed markets. **HKEX's 5-Year China Ministry of Finance T-Bond Futures contract is the world's first bond derivatives accessible to offshore investors. It is an efficient, transparent and easy-to-access tool which may help investors to manage against China interest rate risk exposure.**

Chapter 8

Tapping into China's domestic bond market

An international perspective

16 May 2017

Summary

A well-developed RMB bond market with a high level of foreign participation is an essential attribute that underpins RMB as an international reserve currency. The growth potential of foreign holdings of RMB bonds would be considerably large, given the size of China's economy and the RMB bond market. However, due to the restrictions under the current market opening programmes for foreign investors, the degree of foreign participation in China's bond market is significantly lower than those in the countries with international currencies and even some emerging markets. This reveals the needs to enhance market infrastructure, trading rules and financial products with innovative measures in order to further advance RMB internationalisation.

At present, China runs three main programs that allow foreign investors to access the domestic bond market, namely the Qualified Foreign Institutional Investor (QFII) scheme, the Renminbi Qualified Foreign Institutional Investor (RQFII) scheme and eligible institutions in the Mainland's interbank bond market (PBOC Eligible-Institutions scheme)[1]. Although related regulations have been gradually relaxed, the rules on quota administration, account management, or fund remittance are still major hurdles in effective investment strategies and funds allocations of foreign participants. Moreover, some institutional features of the domestic market are of key concerns that need to be addressed in order to promote more active foreign participation. These include issues like market fragmentation, less diversified market structure, under-developed credit rating system, and potential credit risk.

Domestic bond market development has been one of the state policy priorities for both the Mainland capital market development and RMB internationalisation. To further promote foreign participation in China's domestic bond market, the following potential improvements can be considered: (1) Further integrating trading platforms and foreign participation schemes; (2) Accelerating the pace of cross-border product innovation to bridge the offshore foreign exchange (FX) market strength with the domestic bond market; and (3) Linking up onshore and offshore bond markets, as the Bond Connect scheme jointly announced by the People's Bank of China (PBOC) and Hong Kong Monetary Authority (HKMA), to diffuse international practices and standards to the domestic market. A cross-border Bond Connect platform will offer a well-developed financial infrastructure and market practices in line with international legal and regulatory standards. This would reduce regulatory burdens and offer a more convenient trading environment for both foreign participants and domestic investors. This measure could be regarded as part of a wider effort to further open China's capital markets and to make RMB-denominated assets more accessible to foreign participants, and strengthen the role of Hong Kong as a gateway between the Mainland and international markets.

---

1   See Section 4 for details of the schemes.

1 The potential for foreign participation in China's domestic bond market

Over the past decade, China has made significant progress in developing its bond market, with measures ranging from steadily liberalising interest rates to gradually easing capital controls. As a result, China's bond market experienced a rapid expansion at a simple average annual growth rate of 21% in outstanding value over the past five years, and has become the third largest in the world at RMB 56.3 trillion[2], or about US$8.1 trillion (see Figure 1 in Chapter 7). However, China's bond market is still modest as a percentage of gross domestic products (GDP) compared to countries with international currencies. Foreign participation in China's bond market has remained minimal at around 2.52% of the whole market, and 3.93% of the sovereign debt market[3], significantly lower than that in Japan, U.S. and even some emerging markets (see Figure 1), indicating a large room to advance the foreign participation in China's domestic bond market.

The inclusion of RMB in the Special Drawing Right (SDR) basket of the International Momentary Fund (IMF) opens a special window for global participants to tap into China's bond market. Entering SDR basket is regarded as an official endorsement of RMB as part of the international financial system and is an important milestone for China to integrate into the global financial system. The importance of SDR status is more than symbolic. From an investment perspective, SDR inclusion itself will not directly spur significant investment needs, as the SDR basket is a supplementary international reserve asset of around US$288 billion, within which RMB accounts only for 10.92% weighting[4]. However, the attainment of a SDR status will increase the global acceptance of RMB as a global investment and reserve currency, which would most likely trigger an increasing demand for RMB-denominated assets from both public- and private-sectors internationally, and hence would lead to a steady global asset diversification from other financial segments into Chinese assets, especially into RMB-denominated bonds and relevant financial products.

---

2   Source: BIS, Wind, as of end-2016.
3   Source: CCDC, as of end-2016.
4   Source: IMF website.

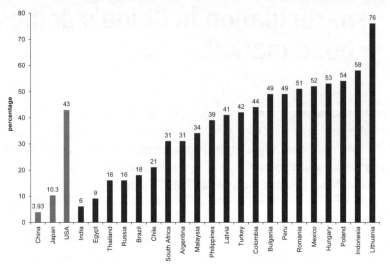

Figure 1. The shares of foreign ownership in sovereign debt market in China, Japan, US and major emerging economies

Note: The grey bars refer to the shares of foreign ownerships in China, Japan and USA, and the blue bars refer to those in major emerging market economies.

Source: Emerging markets data are from BIS and IMF report (2015); China data is from Wind, as of end-2016; Japanese data is from Asian bonds online, as of end-2015; US data is from Federal Reserve, U.S. Treasury, as of end-2016.

In respect of the public sector, current holdings of RMB assets (including bonds, stocks, loans and deposits) by foreign governments and quasi-official sectors amounted to RMB 666.7 billion[5], equivalent to around 1% of total official foreign exchange reserves world-wide[6]. This is much smaller than the share of Australia dollar (AUD) or Japanese Yen (JPY) in official global foreign exchange reserves, which are 1.94% and 4.48% respectively as of end of 2016Q3. (see Figure 2.) If the foreign holdings of RMB by public sector could roughly reach the level of AUD, US$110 billion of global reserve would be shifted into RMB assets. A further rise to a level comparable to that of JPY in global FX reserves could result in a US$400 billion capital inflow into RMB assets.

In respect of the private sector, China's bond assets are not well represented in international benchmarks at present. If Chinese assets were included in some international indices, such as J.P. Morgan Emerging Markets Bond Index (EMBI Global Index) which

---

5   See *RMB Internationalisation Report* (2015)《人民幣國際化報告（2015）》, the PBOC.
6   Similarly, the RMB constituted 1.1% of total official foreign currency assets in 2015, according to IMF statistics.

is widely-used as the central reference point in international fixed income market, China's weight in the index would be about one-third, according to an IMF report. (see Figure 3.) Furthermore, if supporting policies are in place to facilitate bigger access by institutional investors and private investors to the domestic bond market, foreign holdings of Chinese bonds could increase to a level comparable to those in other international currencies, e.g. about 10% of total bond market.

Assuming that the growth rate of China's bond market in the next few years is same as compound annual growth rate of TSF in the past five years, i.e. 14%, and that foreign holdings of Chinese bonds reach the level of 10% of the whole market, then the foreign holdings of Chinese bonds could reach RMB 9.51 trillion, or 9.93% of GDP, by 2020. (see Table 1.) The growth potential of foreign participation in China's domestic bond market would then be considerably large.

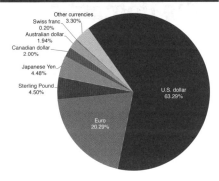

Figure 2. Currency composition of official foreign exchange reserves (End of 2016Q3)

Note: The renminbi was included in the category of "other currencies."

Source: International Financial Statistics (IFS).

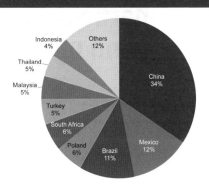

Figure 3. EMBI Global Index, if Chinese bonds included

Source: IMF Global Financial Stability Report, April 2016.

Table 1. The projection of foreign participation in China's domestic bond market (by 2020)		
	2016	2020
GDP (RMB billion)	74,413	95,730
Total domestic bond market (RMB billion)	56,305	95,100
Foreign holdings in domestic bond market (RMB billion)	853	9,510
As % of GDP	1.15%	9.93%

Note: Calculations are based on the following assumptions — (1) The annual growth rates of GDP and China's bond market are 6.5% and 14% respectively; (2) Foreign holdings account for 10% of total debt outstanding value.

Source: Foreign holdings data in 2016 is from the PBOC; Wind for 2016 data; author's calculations for 2020 estimation.

2 The benefit of increased foreign participation in the domestic bond market

Firstly, having investors with different investment objectives will spur a wider range of investment strategies, and help to channel capitals towards the most productive industries. Therefore, encouraging various types of foreign investors to enter into the domestic bond market would help build a diversified investor structure, activate trading and contribute to a more competitive market. This would further increase the scale and depth of the domestic financial market.

Secondly, facilitating foreign holdings of Chinese bonds is a key to increase the international use of the RMB. Increasing foreign holdings would be one of the important factors in the assessment of a currency to be widely usable or not. However, current foreign holdings of China's bonds are much lower than that of the economies with international currencies. Take the case of the US treasury market where the investors broadly consist of financial institutions, private individual investors and foreign entities. As of end-2016, government entities (Federal Reserve and local governments) accounted for 23% of total holdings of US treasures. Apart from them, mutual funds and foreign investors are also major participants. In particular, the share held by foreign participants was over 40% of total outstanding value. The rest was owned by banks (less than 5%), insurance companies, and an assortment of trusts and other types of investors. (see Figure 4.) Since debt securities are typically a top asset class for central banks and global fund managers, the tradability and usability of China's bond assets for foreign investors are crucial to advance RMB internationalisation and support RMB as a meaningful reserve currency.

Figure 4. Percentage of outstanding value held by the diversified investor base in the US sovereign bond market (End–2016)

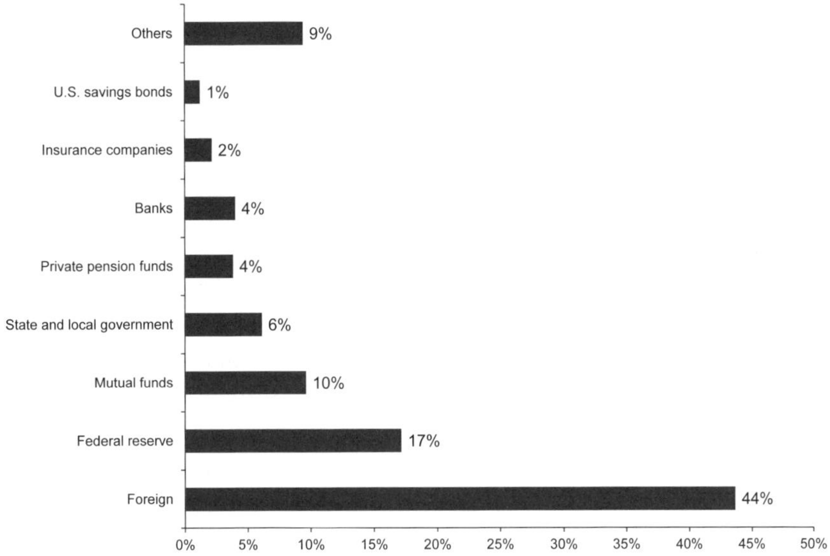

Note: The debts include treasury bills, notes, bonds, and special State and Local Government Series securities.

Source: Federal Reserve, U.S. Treasury.

Thirdly, a deep bond market with a wide variety of instruments and long-term investors would help to absorb the impact of fluctuations of foreign capital flows and enhance global investors' confidence in holding RMB-denominated assets. In 2016, foreign holdings of China's domestic bonds continued to increase, in spite of the weak RMB exchange rate. (see Section 3 below.) Even foreign participation in other asset types fell, foreign capitals showed a steady preference for bond assets. Foreign participation in China's bond market is expected to increase, which would compensate the capital outflow and back up the exchange rate of the RMB in the medium term.

3 Current structure of China's domestic bonds held by foreign participants

China is on its way of setting up an enlarged regime for promoting foreign participation in its domestic bond market. Along with the broad reach of offshore Renminbi centers around the world, and bilateral currency swap lines with a wide range of countries, China's approvals of qualified investors and investment quota under the RQFII and QFII schemes have been accelerating over the past few years. Meanwhile, the PBOC has also provided faster approvals for foreign institutions to gain access to the interbank bond market. Therefore, foreign capital inflows to China's onshore bond market have steadily increased. As of end-2016, foreign holdings of China's domestic bonds have reached a new high of RMB 852.6 billion, 13% higher than the previous year[7].

Figure 5 shows that the overall foreign holdings of Chinese domestic assets, including bonds, equities (stocks), loans and deposits, amounted to RMB 3.03 trillion at the end of 2016. Among them, foreign holdings in bonds and equities continued to increase, while those in deposits and loans fell significantly. Notably, the share of bond assets in overall foreign holdings rose to 28% from 20% at end-2015, versus a decline in the share of deposits from 41% to 30% during the period (see Figure 5), reflecting a significant shift in foreign capital's allocation to bond assets.

---

7 Source: Wind.

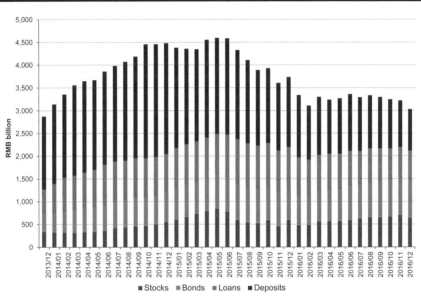

Figure 5. Total foreign holdings in China's domestic market by asset type (Dec 2013–Dec 2016)

Source: PBOC.

Within bond allocations, most of the foreign capital flowed into rates rather than credit bonds. Foreign participants increased their holdings by RMB 233 billion of government and policy-bank bonds in 2016, a six-fold rise compared with RMB 35 billion in 2015[8]. Foreign participation in China's sovereign bond market rose to 3.93% from 2.62% at end-2015. (see Figure 6.)

Among the investor types with increased holding value of sovereign bonds in 2016, foreign investors contributed 13% of the total increased value, behind only nationwide commercial banks (38%) and city-level commercial banks (19%), becoming the third biggest buyer of sovereign bonds in 2016. In contrast, foreign holdings in credit bonds fell to a record low of RMB 49.4 billion, accounting for only 6% of total foreign holdings in bond assets at the end of 2016. (see Figures 7 and 8.)

---

8   Source: Wind.

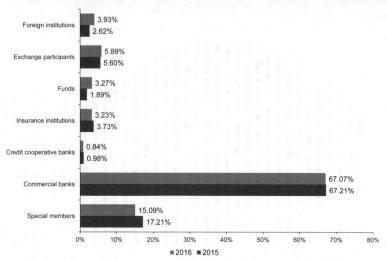

Figure 6. The share in China's sovereign bond market by investor type (End–2016 vs End–2015)

Source: Wind.

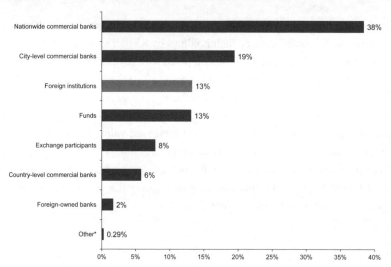

Figure 7. The share in increased value of sovereign bonds by investor type in 2016

Note: Excluding the following investor types which had their sovereign bond holding value decreased in 2016: special members, rural cooperative banks, credit cooperative banks, securities companies and insurance institutions.

Others include rural banks, other commercial banks, non-bank financial institutions, non-financial institutions, individuals and other institutes.

Source: ChinaBond website.

Figure 8. Foreign holdings by bond type in value terms (End–2016)

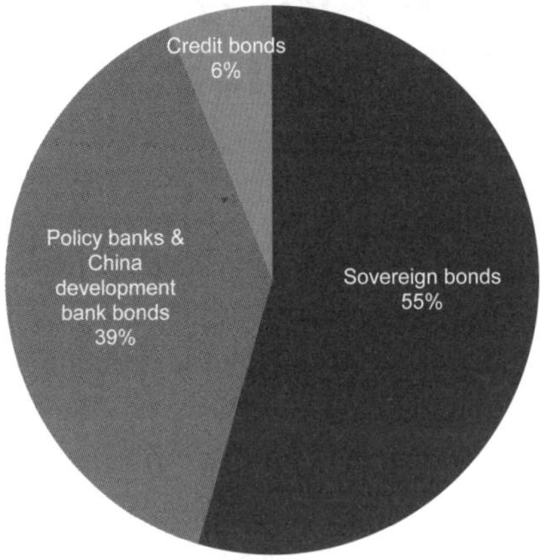

Notes: Foreign holdings of credit bonds include enterprises bonds recorded at CCDC, medium-term notes (MTNs) recorded at CCDC and SCH, commercial papers (CPs) and super commercial papers (SCPs) recorded at Shanghai Clearing House (SCH).

Source: Wind.

The increased proportion of sovereign bonds in foreign holdings may reflect the fact that foreign investors are prone to be more cautious to Chinese assets amid increasing credit defaults in China's bond market recently. Due to China's weak market infrastructure, particularly the lack of creditable rating agencies, foreign institutions tend to hold sovereign and high-rating bonds as part of their foreign exchange reserves. However, the incentive of diversifying to higher-yield assets and credit bonds would likely be stronger in the near future, given the currently low (or negative) yield environment in major developed markets. In this spirit, the credit bond sector could expand faster than government bond sector, once the market infrastructure and credit issues in China's bond market are considerably improved.

Overall, at the end of 2016, foreign holdings of Chinese bonds increased to 2.52% of total outstanding value from 2.03% as of end-2015[9], with 411 foreign institutions registered in China's bond market[10]. However, the foreign participation in China's bond market is still at an early stage due to the restrictions under the current market opening programs. This reveals the need to enhance market infrastructure, trading rules and financial products with innovative measures in order to further broaden and deepen China's bond market.

---

9   The calculation is based on the data from ChinaBond website.
10 Source: ChinaBond website.

4   The latest policy changes in current schemes

In the past, China had remained prudent towards opening up its financial market and sought to restrict the movement of capital in-and-out of the country that might impose potential threats to the stability of the domestic financial system. Hence, China's bond market has been largely closed to foreign investors, resulting in the low share of foreign ownership in the domestic bond market. In recent years, China has taken steps to open up its bond market to catch up with the rapid pace of capital account liberalisation and RMB internationalisation.

At present, China runs three main schemes that allow foreign investors to access the domestic bond market, namely the QFII, the RQFII and the PBOC Eligible-Institutions schemes, as explained below.

4.1   The Qualified Foreign Institutional Investor (QFII) scheme

The QFII scheme was launched in 2002 which initially allowed foreign investors to access the exchange market including bonds traded on the exchange market. Subsequently, there were substantial changes in the QFII regime, lowering entry barriers for foreign institutions and expanding the investment scope. In March 2013, the authorities relaxed the QFII investment restrictions and QFIIs are allowed to access the interbank bond market. In 2016, the Mainland authorities further relaxed the controls by simplifying the administration of investment quotas, the capital remittance and repatriation arrangements, and shortening the lock-up period. (see details in Table 2.) By the end of 2016, US$87.3 billion of investment quota were granted to 276 QFIIs[11].

4.2   The Renminbi Qualified Foreign Institutional Investor (RQFII) scheme

The RQFII scheme was introduced as an extension of the QFII scheme in December 2011. Under this scheme, foreign investors can deploy offshore RMB funds to invest in onshore assets. In the first phase, Hong Kong-based subsidiaries of Chinese fund management

---

11 Source: SAFE website.

and securities companies could apply for a RQFII license and investment quota to invest in China's capital market. The RQFII regime was subsequently expanded to more countries and regions, including developing and developed countries. As of end-2016, the aggregate quota was increased to RMB 1,510 billion from RMB 270 billion initially, and a total of RMB 528 billion quota was granted to 175 RQFIIs[12].

Similar to the QFII scheme, the rules for RQFII scheme have been gradually relaxed. In 2013, major policy changes of RQFII were made under which the "20% equities/80% bonds" restriction on asset allocation was removed and the scope of permitted investment was expanded to include stock index futures and fixed income products traded on the inter-bank bond market.

In 2016, the RQFII and QFII schemes were significantly liberalised. In February, the State Administration of Foreign Exchange (SAFE) issued the *Foreign Exchange Administrative Rules on Domestic Securities Investment by QFIIs*. In September, the PBOC and SAFE jointly issued *the Circular Concerning the Relevant Matters on Domestic Securities Investment by RQFIIs*. The changes under this "New Regime" are mainly in respect of quota administration, account management, and fund remittance of QFIIs and RQFIIs, which have been the major hurdles in effective investment strategies and fund allocations of foreign participants. The major changes are:

(1) **New administration on investment quota**: Under the New Regime, an investment quota is calculated according to a certain percentage of asset scale rather than the original investment quota limit. Moreover, investment quotas for foreign sovereign wealth funds, central banks and monetary authorities are unlimited based on their actual needs.

(2) **Relaxation on inward remittance of principal amount and relevant lock-up period**: The New Regime removes the restriction on remitting the principal amount into China within 6 months after their investment quota being approved, and shortens the lock-up period of the principal amount for QFIIs/RQFIIs to 3 months upon the aggregate principal amount remitted into China reaching RMB 100 million for RQFIIs, or US$20 million for QFIIs.

(3) **Relaxation on outward remittance of funds**: The New Regime permits outward remittance of funds by RQFIIs after the expiry of the relevant lock-up period. In terms of QFIIs, outward remittance of principal is no longer subject to prior SAFE approval. However, outward remittance of funds by QFIIs is still subject to threshold limitations.

---

12 Source: SAFE website.

(4) **Improved account management**: The New Regime liberalises the quantity limitation on opening bank accounts on each QFII. Account management on QFIIs and RQFIIs is also unified.

4.3 Eligible institutions in the Mainland's interbank bond market (PBOC Eligible-Institutions scheme)

This pilot scheme was launched by the PBOC in 2010 to allow qualified foreign institutions to use offshore RMB to invest in the interbank bond market. At launch, three types of institutions were eligible, including foreign central banks or monetary authorities, offshore RMB clearing and participating banks. Meanwhile, sovereign wealth funds and international organizations may also access the domestic interbank bond market under this scheme.

Since 2015, a number of notable liberalisation measures under this scheme were introduced to further facilitate foreign investors to enter China's interbank bond market. Specifically, in late-May 2015, the PBOC allows offshore RMB clearing and participating banks to conduct repurchase (repo) financing by using their onshore bond holdings. In mid-July 2015, the PBOC further eased the scope of eligible bond transactions by allowing eligible entities to participate in onshore interbank bond market to engage in bond trading, bond repo, bond lending, bond futures, interest rate swaps and other trades permitted by the PBOC, without any prior approval by the PBOC or any quota restrictions.

In February 2016, the PBOC released No. 3 Announcement[13] which further relaxed the rules applicable to foreign institutional investors accessing the interbank bond market. Firstly, the categories of eligible foreign institutional participants were extended to all qualified foreign institutional investors, including commercial banks, insurance companies, securities companies, fund management companies, as well as other types of financial institutions and medium-to-long-term institutional investors recognised by the PBOC. Secondly, the No. 3 Announcement further relaxed FX limitations which have been imposed on the foreign institutional investors. And thirdly, the No. 3 Announcement abides by the macro-prudential administration regime, and hence does not impose quota limit on specific investors. In May 2016, China further published the detailed rules to clarify the investment procedure of foreign institutional investors in the interbank bond market to facilitate the implementation of the No.3 Announcement.

The measures in 2016 moved a further step to open up the domestic bond market.

---

13 See〈中國人民銀行公告 2016 年第 3 號〉.

However, further enhancements are advisable. For example, the current scope of qualified investors is still limited to financial institutions; some restrictions exist on bond products and quote limit, etc.; and the access procedure of the domestic interbank bond market could be further simplified and clarified in order to attract more foreign participation.

Table 2. Current framework for QFII, RQFII and PBOC Eligible-Institutions schemes			
	QFII	RQFII	PBOC Eligible-Institutions scheme
Regulatory approvals	• CSRC: QFII/RQFII license • SAFE: QFII quota • PBOC: Pre-filing for CIBM access		Pre-filing with PBOC
Investment quota	• Only needs to pre-file with SAFE if requested quota is within the base quota or obtain approval if the requested quota exceeds base quota. • The base quota is calculated according to a certain percentage of asset scale.		• Employment of the macro-prudential administration regime to foreign investors • No specific investment quota requirements. Applicant may pre-file with PBOC the anticipated investment value
Eligible fixed income products	• On-exchange market: government bonds, enterprise bonds, corporate bonds, convertible bonds, etc. • Interbank market: cash bonds		• Foreign reserves institutions: all cash bonds, repos, bond borrowing and lending, bond forwards, IRS, FRA, etc. • Other foreign institutions: all cash bonds and other products permitted by the PBOC, offshore RMB clearing / participating banks can also trade repos.
Foreign exchange management	Onshore with the local custodian	Has to remit in off-shore RMB (obtain from offshore)	Onshore/offshore
Lock-up period on principal repatriation	3 months	3 months, no restriction on open-ended fund clients	Nil
Frequency of repatriation and restrictions	Daily for open-ended funds, with threshold limitations	Daily for open-ended funds	The ratio of accumulated outward remittance need to meet some basic requirements

Abbreviations:

CSRC   China Securities Regulatory Commission
IRS   Interest Rate Swap
FRA   Forward Rate Agreement
CIBM   China's interbank bond market

Source: As of end-2016. Please refer to the website of PBOC, CSRC and SAFE for the most updated rules and policies.

5 Institutional features that may limit foreign participation

In addition to restrictive access to the domestic market, certain institutional features are of key concerns that need to be addressed in order to promote more active foreign participation.

5.1 Fragmented framework across both trading platforms and instruments

The Chinese domestic bond market remains fragmented in terms of its regulatory framework across both instruments and trading platforms. There are multiple regulators supervising various debt instruments traded on different markets, mainly the stock exchanges and the interbank bond market. Depending on the product and the market, foreign participants need approvals from different regulators.

Table 3. Two major domestic bond markets in China		
	Interbank bond market	On-exchange market
Regulator	PBOC	CSRC
Trading platform	China Foreign Exchange Trade System	Shanghai / Shenzhen Stock Exchanges
Central securities depository	China Central Depository & Clearing (CCDC)/ Shanghai Clearing House (SCH)	China Securities Depository and Clearing Corporation (CSDCC)
Available instruments	Central government bonds, local government bonds, policy bank bonds, central bank bills, enterprise bonds, medium-term Notes (MTNs), commercial papers (CPs), commercial bank bonds, financial institution bonds, interbank negotiable certificates of deposit, asset-backed securities, repos, bond lending, bond forwards, interest rate swap, etc.	Central government bonds, local government bonds, municipal bonds, enterprise bonds, corporate bonds, convertible bonds, asset-backed securities, private placement bond issued by small and medium-sized enterprise
Key investors	Institutional investors (banks, securities companies, insurance companies, funds, financial companies, enterprises, offshore institutions, etc.)	Securities companies, insurance companies, funds, financial companies, individual investors, enterprises

Source: PBOC, CSRC.

Domestic institutional investors mainly trade in the interbank bond market, leading to over 90% of total bond turnover taking place in the interbank market, and less than 10% on the Shanghai and Shenzhen stock exchanges in 2016[14]. Each trading platform has their own set of restrictions, and not all products can be traded in both markets. Basically, several types of bond (government bonds, enterprise bonds and corporate bonds) can be traded on both the interbank and exchange markets, while most of the rest (such as policy bank bonds, financial bonds, central bank bills, MTNs, CPs, repos, bond lending, etc.) are traded only in the interbank market. Convertible bonds and private placement bonds are traded in the exchange market.

As the bond market is divided into multiple segments with different regulators, liquidity is split and the market depth is depressed. Moreover, most hedging products are traded only in the interbank bond market. This raises the risk issue for most foreign participants, especially funds and securities companies, as most of them mainly access exchange market through QFII and RQFII schemes.

5.2   The concentrated investor base in the domestic bond market

Another key factor which hampers the liquidity of China's bond market would be the high concentration of investor structure. As of end-2016, commercial banks held 58.5% of the overall bond outstanding value. If special institutions (mostly the PBOC and policy banks) are included, the combined holdings by the banking sector account for over 60% of the total market. The holdings are even more concentrated in the government bond sector where the domestic banks held around 80% of outstanding value. In comparison, the proportion held by other non-bank financials, including insurance companies, funds and exchange participants, who tend to trade more actively, was 32% of the overall bond outstanding value at the end of 2016[15].

Such a lopsided investor base is hardly to nurture bond market liquidity. China's bond turnover ratio, at 2.79 in 2016, is much lower than in the US, and is also lower than the levels of Japan and the Republic of Korea when they started to internationalise their currencies and when commercial banks dominated the bond markets in the 1990s[16]. The lower turnover ratio in China can be explained by the less diversified investor profile, as well as the dominance of commercial banks in the bond markets. A well-diversified investor base and the resulting higher market liquidity are two essential attributes for an international RMB as well as for a well-developed bond market in China.

---

14 Source: Wind.
15 Source: Wind.
16 See details in the People's Republic of China's financial market: are they deep and liquid enough for RMB internationalization?, ADBI working paper, April 2014.

5.3 Lack of differentiation and transparency in credit ratings

Currently, nearly 90% of the domestic bonds are given AA or above grading by the domestic rating agencies[17]. The credit spreads of Chinese bonds, especially credit bonds, are often not wide enough to compensate for the underlying credit risk. Compared to international standards, there exists a huge credit rating gap and difference in assessment metrics between domestic and international rating agencies, making it quite difficult for foreign investors to identify credit differentials of Chinese corporate bonds. It is necessary to align the domestic market more with international rating standards and practices and to allow the entry of international rating agencies as well, so that foreign investors can easily track China's credit quality and to derive more sound differentiation in credit risks.

6 Possible improvements

To further promote foreign participation in China's domestic bond market, the following potential improvement could be considered:

6.1 Integrate trading platforms and current foreign participation schemes

Market size and liquidity are the key attributes to the trading and pricing efficiency in the bond market. As mentioned in section 5, most of China's domestic bonds are still issued and traded separately on the interbank and exchange markets, with only a small portion of instruments available on both markets. Moreover, the trading volumes of these two markets are extremely imbalanced. The trading volume on interbank market is far exceeding 90%, while the liquidity on exchange traded bonds is relatively low.

Most of the foreign participants under QFII and RQFII schemes, such as securities companies, funds, or small and medium-sized institutional investors, gain access only to the exchange market, as the entry requirements and transaction cost on the interbank market are relatively high. The exchange market is less liquid and much smaller, leading to a higher credit spread and weak hedging capability. A more integrated trading platform could help to

---

17 Source: Wind.

build up a favorable critical mass to improve pricing capability.

Furthermore, the recent policy changes under QFII and RQFI schemes have made their respective policies on investment quota, fund remittance and account management closer to each other.   There would likely be further unification or integration of the QFII and the RQFII schemes, so as to reduce transaction cost and better support a more diversified investor base.

6.2   Accelerate the pace of cross-border product innovation to effectively bridge the offshore FX product strength with the domestic bond market

Increased foreign investment in Chinese bonds would result in a surging demand in related risk management. To facilitate risk diversification in RMB bond investment, it is necessary to introduce more instruments in the domestic bond market. Besides, FX instruments are also essential to hedge RMB exchange rate risk of RMB bond investment for foreign investors. Despite the further opening up of the domestic FX market recently to foreign investors, leveraging on the strength and abundant supply of hedging tools in the offshore market to hedge the domestic bond assets could be another effective way. On 10 April 2017, Hong Kong Exchanges and Clearing Limited launched 5-Year China Ministry of Finance Treasury Bond Futures contract. As the world's first onshore interest rates product accessible to off shore players, the new contract is an efficient, transparent and easy-to-access tool to manage against China interest rate risk exposure[18].

The advantage of offshore RMB market lies in that it is freely accessible to anyone, including private-sector entities. The liquidity of the offshore FX market has also improved considerably, with the FX turnover in RMB reaching a significant proportion of the onshore turnover volume. The Hong Kong offshore RMB market provides a solid foundation for the sustained development of RMB derivatives and hedging tools to facilitate foreign participants' risk management in FX volatility for holding Chinese bond assets.

---

18   See Chapter 7 of this book, "HKEX's Five-Year China Ministry of Finance Treasury Bond Futures — The world's first RMB bond derivatives accessible to offshore investors".

6.3 Link up onshore and offshore bond markets to diffuse international practices and standards to the domestic market comprehensively

Similar to the Shanghai/Shenzhen-Hong Kong Stock Connect schemes, setting up a cross-border platform and developing a mutual market access to both the onshore and offshore bond markets (Bond Connect) can be a promising solution to further improve trading convenience and pricing efficiency in RMB bonds. Investors in one market would be able to trade bonds in the other market through the bridging of the Hong Kong and Mainland financial infrastructure institutions.

From the trading perspective, Bond Connect scheme would enable bond market integration across the border and between on-exchange market and interbank bond market, thereby improving liquidity. Furthermore, this integrated market can make available more standardised instruments for developing effective benchmarks for RMB-denominated assets and improve the pricing efficiency of Chinese bond assets.

Although international investors can now directly participate in the domestic RMB market, including FX and bond markets, the offshore market still serves as one central pillar supporting the RMB as a global currency. Given the well-developed offshore financial environment and infrastructure in Hong Kong, a cross-border Bond Connect scheme could reduce regulatory burdens and offer more convenient institutional conditions for foreign investors, such as credit rating with international standards and better investor protection.

Through Bond Connect, a great variety of international bonds will also be made available to the Mainland investors for their global asset allocation strategies. Through participating in an international trading platform together with professional international investors, Mainland investors could also gain experience in international market practice and regulation. In this way, Bond Connect could help develop a mature and professional investor base in, and the breadth and depth of, the Mainland domestic bond market.

Chapter 9

The HKEX USD/CNH Options Contract

An RMB currency risk management tool

17 August 2017

Summary

The introduction of the HKEX's Renminbi (RMB) currency options is driven by the growing demand from market participants for diversified tools for trading and hedging the offshore RMB (CNH) exchange rate.

HKEX's RMB currency options contracts are complementary to the family of HKEX's RMB currency futures contracts. These options products could serve as risk management tools against non-linear sensitivities and offers volatility trading opportunities on RMB exchange rates, addressing market demand not previously satisfied by HKEX's RMB currency futures [1]. In the course of ongoing RMB exchange rates liberalisation process and the associated policy developments, the exchange rates are in transition from policy rates to market-driven rates. This is expected to create higher volatility in the USD/CNH exchange rate. The one-month implied volatility of USD/CNH was around 1% to 2% in the month before the RMB exchange rate reform in August 2015[2]. It has increased sharply to a range of 4% to 10% in the year subsequent to the reform. The increase in spot USD/CNH volatility could provide an opportunity for the introduction of RMB currency options contracts to enable volatility trading and facilitate hedging for market participants.

Furthermore, the global over-the-counter ("OTC") RMB options market already has a sizeable average daily turnover of approximately US$18 billion[3] with an average transaction size of US$150 million[4] as of 2016. Unlike two to three years ago when CNH structured forward positions dominated the market risk profile in the OTC CNH derivatives market, almost all new volatility risks are now hedged by vanilla options which are standardised calls/puts with no special features.

In light of the relative lack of transparency in the OTC market, the associated margin requirements by new regulations and the counterparty risks, there is an increasing demand for bringing RMB currency options trading onto the exchange market. The USD/CNH Options contracts listed on the Hong Kong Futures Exchange ("HKFE"), and centrally cleared through the HKFE Clearing Corporation Limited ("HKCC"), provide price transparency and reduce counterparty risks in this important and growing CNH options market.

---

1   Compared to currency futures which offer linear exposure of the underlying currency rate, currency options offer exposure of non-linear risk sensitivities such as second-order derivative of the underlying (i.e. gamma), volatility (i.e. vega) and time (i.e. theta).

2   Source: Bloomberg.

3   Triennial Central Bank Survey of foreign exchange and OTC derivatives markets in 2016, Bank of International Settlement.

4   "Emerging Markets Currency Guide 2016", HSBC.

1 Macro environment: market demand and current support

1.1 Two-way volatility drives demand for RMB risk management tools

On 11 August 2015 the People's Bank of China (PBOC) introduced a new market-based managed floating framework for RMB exchange rate fixing, which is based on the previous close of the onshore RMB (CNY) rate, together with market supply and demand factors with reference to a basket of currencies (the reform). The 1-month implied volatility of the USD/CNH rate increased from 1%-2% in the month before the August 2015 reform to 4%-10% in the year subsequent to the reform.

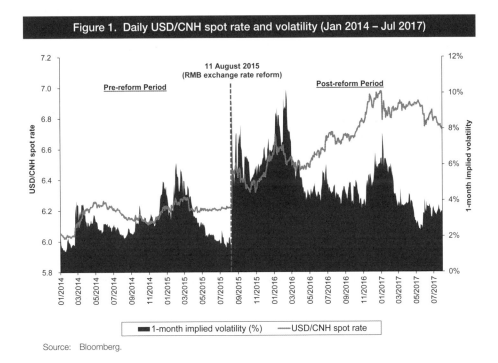

Figure 1. Daily USD/CNH spot rate and volatility (Jan 2014 – Jul 2017)

Source: Bloomberg.

The internationalisation of the RMB has entered a new stage. Hong Kong's role as the most critical offshore RMB hub connecting the Mainland and the world has become increasingly important. Hong Kong should continue to build on its own strength and, based on its new role as a "mutual market", open up bigger room for innovation for the better of its long-term development. Such a new role may be perceived from three different angles: (1) an enhanced offshore RMB market; (2) a RMB risk management centre; and (3) a gateway market. These are elaborated below.

(1) For the offshore RMB market, the impetus of market growth used to heavily rely on RMB appreciation expectations and arbitrage trading between onshore and offshore markets. Now, the market's depth and breadth have been building up as more and more relevant financial products are introduced and more risk management tools are provided for the use of more and more portfolio management tactics. All these serve appropriately the demand for global allocation of RMB assets and the related cross-border capital flows. In developed countries such as Japan, the UK and the US, the size of their credit, equity and bond markets is more than 5 times of their GDP[5], whereas the corresponding figure is around 2.1 times in China. This indicates substantial room for further development in China in terms of financial market deepening and financial product diversification. Together with the inclusion of RMB in the International Monetary Fund's Special Drawing Right (SDR) basket of currencies, two-way volatility is expected to become the new norm. More and more investors have become increasingly aware of this market change and have started to manage their exchange rate risk exposure.

Through further developing and enriching its range of multi-facet financial products and related financial services, **Hong Kong can continue to well position its offshore RMB market** as a major venue for cross-border investments (especially overseas investments from the Mainland) and related risk management. Now is a good time for Hong Kong to further build up the market depth and effectiveness of its offshore RMB market. Through more effective and sensible pricing benchmarks for onshore and offshore RMB markets, segregation in the RMB pricing structure could be improved and a reasonable price difference between the onshore and offshore RMB markets could be maintained. This would require further connectivity between the onshore and offshore RMB bond markets, foreign exchange (FX) markets and derivatives markets, enhanced market liquidity and an increased number and diversity of market participants.

---

5   Source: International Monetary Fund (IMF) database.

(2) **Hong Kong is well positioned to become a risk management centre** to facilitate the current transformation and adjustment process of the Mainland economic and financial system. For example, greater flexibility in the RMB exchange rate is broadly expected in the next stage of the currency's path towards internationalisation, and in such process the demand for exchange rate risk management is bound to be substantial. Meanwhile, Mainland companies are expanding their global reach and their participation in projects in countries along the Belt and Road[6]. In this process, Hong Kong is also well positioned to serve the demand from these companies for overseas investment risk management and their global presence.

(3) Following the gradual opening of China's financial market, Hong Kong will not only be an active destination market for investment purposes by Mainland entities, but is also becoming **the gateway market** for Mainland entities to invest in other markets. This is becoming more evident after the launch of the Mainland-Hong Kong Mutual Market Access pilot programme. The launch of Shanghai-Hong Kong Stock Connect (Shanghai Connect) in November 2014, the Shenzhen-Hong Kong Stock Connect (Shenzhen Connect) in December 2016, and the mutual bond market access on bond markets (Bond Connect) in July 2017, has linked up Hong Kong, Shenzhen and Shanghai into a sizeable mutual market. With Hong Kong as the gateway, such a mutual market will support the global asset allocation of Mainland funds and provide sound infrastructure and platforms for international funds to invest in the Mainland capital market. Foreseeably, if such a framework of connectivity is expanded to other product types, Hong Kong's key role as a gateway market will be further strengthened. With the increasing cross-border investment activities, the demand for risk management is expected to increase, possibly multi-fold.

1.2   The support offered by HKEX's RMB products and platforms

Against the macro backdrop analysed above, the overseas market saw growing interest in China's fixed income and currency (FIC) market and greater demand for risk management and investment. HKEX has been devoting efforts in multiple dimensions with an aim to be an offshore RMB product trading and risk management centre. HKEX's platforms now have a variety of RMB products including bonds, exchange traded funds (ETFs), real estate investment trust (REIT), equities, RMB fixed income and currency (FIC) derivatives

---

6   A development proposal proposed by Chinese President Xi Jinping in September 2013 that focuses on connectivity and cooperation between Eurasian countries.

and commodity derivatives. The HKEX RMB product suite is provided with an aim to match market demands.

HKEX launched its USD/CNH Futures contract in 2012, which has seen turnover take off since 2015 and is now one of the most actively traded RMB futures contract in the world[7]. HKEX then moved to diversify its product offerings by launching new CNH currency pairs against the Japanese yen, Euro, and Australian dollar, which began trading on 30 May 2016 to facilitate cross-currency hedging. In addition to RMB currency risk management tools, RMB interest rate risk management tools on HKEX were also enriched upon the introduction of HKEX's Five-Year China Ministry of Finance Treasury Bond Futures (Bond Futures) on 10 April 2017. The Bond Futures would be useful tools for interest rate hedging, especially upon the launch of the connectivity scheme between the bond markets in Mainland and Hong Kong (Bond Connect) on 3 July 2017. Bond Connect is a pilot scheme that connects China's interbank bond market with the world, giving international investors "Northbound" access to trade bonds directly on the China Foreign Exchange Trading System (CFETS), the Mainland interbank bond market trading platform, for the first time.

There is also tremendous potential market demand for an RMB Currency Index benchmark as the RMB becomes a reserve currency and the market focuses on the relationship between the RMB and global currencies. In June 2016, HKEX launched the TR/HKEX RMB Currency Index series (RXY Indices or RXY Index series) which is jointly developed with Thomson Reuters, allowing market participants to conveniently monitor the RMB's movements. HKEX also plan to introduce futures and options on the index in the future to provide the market with effective RMB risk management tools.

In addition, HKEX plans to launch a full suite of different RMB products. As a start in the commodities segment, HKEX launched dual-currency (USD and RMB pricing and settlement) physical delivery Gold Futures Contracts on 10 July 2017. This new product offers gold producers, users and investors a practical solution to manage risks arising from the gaps between the gold spot and futures markets, as well as the price difference between the RMB and USD.

Furthermore, enhancements in HKEX's infrastructural platforms add to a solid foundation for further development of RMB derivatives in Hong Kong. HKEX's subsidiary, OTC Clearing Hong Kong Ltd (OTC Clear), commenced business in 2013. This is a key piece of infrastructure to serve clearing service needs of FIC market participants, especially in regionally-traded products and in particular RMB-based derivatives.

---

7   See Chapter 14 of this book, "HKEX research report, "HKEX towards an offshore RMB product trading and risk management centre".

1.3   HKEX USD/CNH Futures: one of the world's most liquid USD/CNH contracts

There is an increased market awareness of the merits or even necessity of hedging RMB exchange rates. The first RMB derivative product traded on HKEX was the USD/CNH Futures launched in September 2012. Investors, both individuals and institutions, have started to realise how the RMB FX volatility can have an impact on their investment portfolios in terms of RMB assets, liabilities and cash flow. The RMB's two-way movement has become accepted as a metric in investors' risk management framework.

In 2016, the HKEX USD/CNH Futures recorded historical highs of annual contract volume and year-end open interest[8]. The record-breaking total trading volume of the product in 2016 was 538,594 contracts, an annual increase of 105%; and the record high year-end open interest was 45,635 contracts, a year-on-year increase of 98%. Its average daily volume climbed to 4,325 contracts in December 2016. The cash-settled CNH/USD futures also showed a growing contract volume in 2016H2 and its open interest has been continuously building up since launch. Its average daily volume achieved 95 contracts in December 2016; and its open interest reached the highest of 1,494 contracts at year-end.

Entering into 2017, new records of performance were seen in this product:

- Record single-day turnover of 20,338 contracts (notional value of US$2 billion) on 5 January 2017; followed by the second and third record turnovers of over 8,600 contracts (notional value of more than US$860 million) on 31 May and 1 June 2017.
- Open interest record of 46,711 contracts (notional value of US$4.7 billion) on 4 January 2017.
- Night-session record of 3,642 contracts (notional value of US$360 million) on 4 January 2017.
- Increased market participation, bringing the total number of exchange participants (EPs) having traded the product to 112.

---

8   Source: HKEX.

Figure 2. HKEX USD/CNH Futures trading performance (2012–2017 1H)

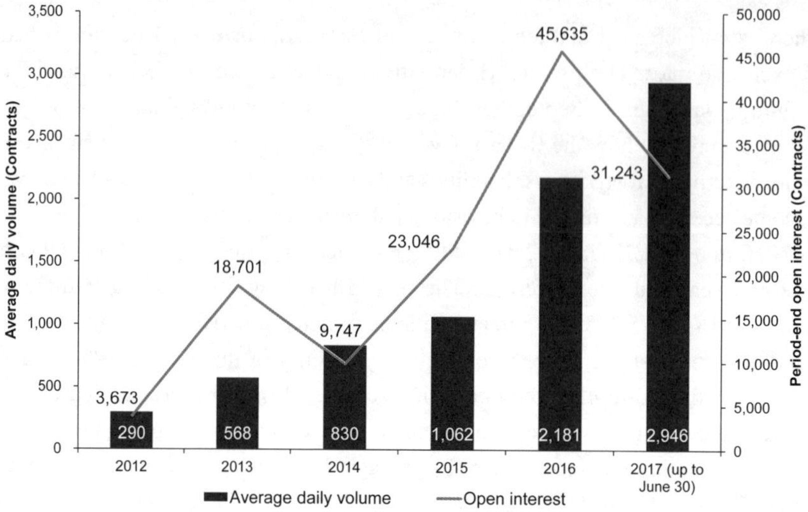

Source: HKEX.

2   HKEX USD/CNH Options: the risk management tool made available on exchange market

Currently, a sizeable OTC RMB options market already exists, with an average daily trading volume of US$18 billion[9] and an average transaction size US$150 million[10] (see Figure 3). Unlike two to three years ago when CNH structured forward positions dominated the market risk profile in the OTC CNH derivatives market, almost all new volatility risks are now hedged by vanilla options which are standardised calls / puts with no special

---

9   *Triennial Central Bank Survey of foreign exchange and OTC derivatives markets in 2016*, Bank of International Settlement.
10 *Emerging Markets Currency Guide 2016*, HSBC.

features. This demonstrates an increasing market demand to hedge currency risk using vanilla options instead of exotic-style options.

Figure 3. Average daily turnover OTC RMB currency options globally

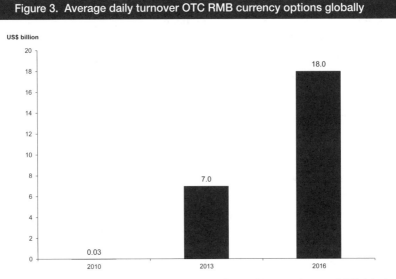

Source: Bank of International Settlement, Triennial Central Bank Survey of foreign exchange and OTC derivatives markets.

Compared to the OTC RMB currency options market, HKEX's currency options market has several characteristics, explained in sub-sections below.

2.1   Continuous quotation

Traditionally, OTC RMB currency options market is operated on a bilateral dealing, request-for-quote (RFQ) basis. Investors have to source, contact and negotiate with price providers individually to get a quote on options price and compare the prices among themselves, which might not be the most efficient way for price discovery.

In this regard, HKEX's USD/CNH Options offer a different role on trade execution and price discovery. Typically, continuous quotations on around 150 option series are available provided by dedicated liquidity providers, with the average spreads of 12-40 pips for short tenors and 80-160 pips for long tenors[11]. Such streaming of bid/ask quotations allows

---

11 Source: HKEX, July 2017; "pip" refers to percentage in point (0.0001), which is the minimum price fluctuation for a currency pair.

investors to freely execute trades at their desired strike and tenor, facilitating liquidity development. In addition to continuous quotation, investors can also submit quote requests to dedicated liquidity providers on specific strike and tenor.

One of the special characteristics of an exchange-traded market for RMB derivatives traditionally traded OTC is that it acts as a liquidity aggregator offering continuous and tight bid/ask liquidity. The regulatory and capital benefits (see Section 2.2 below) of trading listed products have become more pronounced. The exchange market offers a marketplace for orderly and transparent trading on an equal basis.

2.2 Capital efficiency

New rules in Europe (EMIR[12]) and the United States (CFTC[13]) are affecting existing OTC participants. From 1 March 2017, all in-scope counterparties (primarily financial entities and systemically important non-financial entities) with uncleared OTC portfolios must exchange variation margin daily. This requirement is relatively new for many users of OTC products, and the requirement to exchange initial margin is being made mandatory in stages over time towards full implementation by September 2020.

RMB exchange-traded derivatives provide capital efficiency to investors as a result of their comparative advantages in various aspects vis-à-vis the OTC market. Table 1 presents a table of comparison between exchange-traded RMB derivative products and OTC products.

Table 1. Comparison between RMB exchange-traded derivatives products and RMB OTC derivatives products		
Item	RMB OTC derivatives	RMB exchange-traded derivatives
Price transparency	Relatively less transparent — Need to contact each counterparty to get the price	Highly transparent — option prices are available on the HKEX website, and through information vendors and trading platforms of brokers
Central clearing	Bilateral and no central clearing	Central clearing counterparty for both sides of transactions
Credit and collateral	Need to negotiate credit lines and collateral arrangements with banks	Margin-based and cash collateral is accepted

---

12 European Market Infrastructure Regulation ("EMIR") is the regulatory technical standards pursuant to Article 11 European Market Infrastructure Regulation (EU) No. 648/2012 of the European Parliament and of the Council.
13 Commodity Exchange Act of the U.S. Commodity Futures Trading Commission ("CFTC").

(continued)

Table 1. Comparison between RMB exchange-traded derivatives products and RMB OTC derivatives products		
Item	RMB OTC derivatives	RMB exchange-traded derivatives
Settlement risk	RMB is not an eligible currency in CLS[14], therefore cannot utilise the CLS system for position netting	Position netting is available for exchange traded derivatives

Source: HKEX analysis.

2.3   Versatile hedging tool due to unique risk and reward profile

HKEX's USD/CNH Options contract is designed to mirror the characteristics of the USD/CNH futures contract to allow for cross-product hedging and cross-margining benefits, as well as to provide alternative product payoff structures for the same notional contract size.

(1)   Cross-product hedging

USD/CNH options are directly complementary to the existing HKEX USD/CNH Futures, which together could allow investors to deploy trading and hedging strategies under various market conditions with relatively low counterparty risk in comparison to OTC derivatives. They provide investors a hedging tool against RMB volatility amid the ongoing RMB liberalisation process and policy development towards a market-driven framework. (See Table 2 for a comparison between options and futures)

(2)   Cross-margining benefits

HKEX's USD/CNH Options contract is traded on a margin-basis under the SPAN methodology[15] adopted by HKCC, where net delta is a key determinant of margin requirement for futures and options of the same underlying. A client can therefore enjoy cross-margining benefits when holding USD/CNH Futures and Options positions at the same time, paying less margin compared to those required for separate outrights.

From a risk management perspective, options contract is a versatile tool due to options' unique risk and reward profile. With a variety of options/futures strategies deployable, options contracts provide exposure to multiple market parameters, e.g. spot rate, volatility and time.

Options contract is suitable for various RMB market conditions, providing flexibility of strategies to cater for various market conditions — they can be utilised in bullish, bearish, range-bound or volatile markets. (See section 3.2 on the basic applications of the product.)

---

14 Continuous Linked Settlement System — a global clearing and settlement system for cross-border foreign exchange transactions.

15 SPAN — Standard Portfolio Analysis of Risk. Please refer to the margining methodology document at www.hkex.com.hk/eng/market/rm/rm_dcrm/rm_dcrm_clearing/dmrm_clearing_settlement.htm .

Table 2. Options versus futures	
Options	Futures
• Gives the buyer the right, but not the obligation, on or before a pre-determined date, to buy (or sell) an underlying asset at a pre-determined price (the "strike price"); and the seller the obligation to sell (or buy) the asset at the strike price if the buyer exercise the right • The option price has a non-linear relationship with the underlying asset, and has a unique risk and reward structure • The buyer pays upfront price as the option's premium	• Gives the buyer the obligation to buy an asset at a pre-determined price and the seller the obligation to sell at the pre-determined price, at a specified time in the future • Futures prices have linear relationship with the underlying assets • No upfront cost (apart from margin and other fees related to trading)
Illustrations	
• Options: An investor buys a (European-style) call option on the USD/CNH rate with the strike price at 7.0 expiring in three months. Three months' later on the option's expiry date, the investor will have the right, but not the obligation, to buy USD at the exchange rate of RMB 7.0 per USD. • Futures: An investor enters into a long position of USD/CNH futures at 7.0 expiring in three months. Three months' later on the futures' expiry date, the investor will have the obligation to buy USD at the exchange rate of RMB 7.0 per USD. • Currency options are more complex than options on other asset classes, due to the fact that a call on one currency is also a put on the other currency.	

Source: HKEX analysis.

2.4 Other characteristics of exchange trading

- **Cost effectiveness**: In general options contract traded on exchange provides leverage[16] and cost effectiveness as it is traded on an option premium and margin basis, and requires upfront payment of only a fraction of the notional value. For the HKEX USD/CNH Options, related transaction cost is further reduced as the trading fee is waived for the first six months (20 March 2017 — 29 September 2017) and the SFC levy is not applied.

- **Transparency:** Being exchange-traded options, options contract is standardised, trading of which is orderly and transparent. Investors can access real-time exchange-traded options prices via, information vendors (see Appendix 3 for list of source of market information for USD/CNH options) and the trading platforms of brokers.

- **Ease-of-Access:** In general, exchange is available to different investor types, including but not limited to retail investors, corporate users, asset managers and hedge

---

16 Currency options and leverage are of high risk and not suitable for inexperienced investors or people who are less risk tolerant. For further information, please refer to HKEX website.

fund managers. For example, in HKEX, investors can access the product through the existing distribution with over 120 RMB-enabled EPs. In comparison, OTC RMB currency options market is restricted to institutional users only.

Supported by the aforementioned characteristics, the trading volume and open interest of HKEX's USD/CNH Options continue to accumulate. Up to 31 July 2017, the total trading volume since launch was 4,914 contracts (i.e. US$491 million in notional amount) and the open interest continues to hit record high. As of 31 July 2017, the open interest across all contract months reached 1,727 contracts (i.e. US$173 million in notional amount).

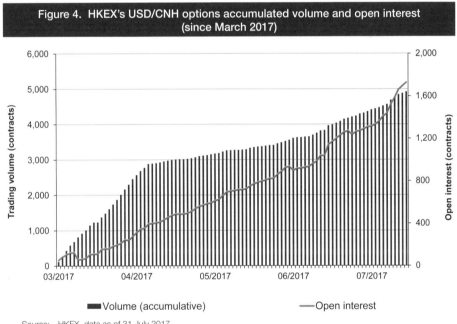

Figure 4. HKEX's USD/CNH options accumulated volume and open interest (since March 2017)

Source: HKEX, data as of 31 July 2017.

3 HKEX USD/CNH Options: product design and applications

HKEX's USD/CNH Options contract is a European-style option on spot, exercisable only at expiration date but not before. This is designed based on the prevailing OTC market practice (where the majority of FX options exist in European-style forms). Exercise at expiry results in the delivery of USD against RMB with full principal amount at the strike price, which meets the demand for principal exchange from options users.

3.1 Pricing behaviours and risk factors to monitor

The option premium, i.e. the option price, is a function of a certain number of factors including the underlying asset's strike price and spot price, interest rates (of the two underlying currencies), tenor and volatility. The Black-Scholes pricing model, commonly used for equity options, was extended by Garman and Kohlhagen[17] to price currency options.

Currency options are multi-dimensional instruments and their prices in the secondary market respond to various market parameters. Due to the diversity in the terms of options, i.e. tenor, strike, etc., it is impractical for a market participant to hedge with exactly the same options in the market. As such, option traders act as risk managers and manage their risks by closely monitoring various market parameters. There are specific measurements of the sensitivity of the option value with respect to different market parameters. These measurements are collectively known as the "Greeks". Analysis of Greeks is crucial to option valuation and risk management.

The Greeks decompose the risks contained in an option price or a portfolio of options into their various constituent parts, which in turn allows traders to decide which risks to retain and which to hedge. The different risk measures in the Greeks include:

- **Delta:** Change in price of the option for a change in the spot price of the underlying
- **Gamma:** Change in delta for a change in the spot price of the underlying
- **Theta:** Time decay of the option, i.e. change in price of the option for the passage of time
- **Vega:** Change in price of the option for a change in volatility of the underlying

---

17 M.B. Garman and S.W. Kohlhagen, "Foreign currency option values", *Journal of International Money and Finance*, 1983, Vol. 2.

- **Phi:** Change in the base currency's risk-free interest rate[18]
- **Rho:** Change in the pricing currency's risk-free interest rate[18]

3.2 Product applications of HKEX USD/CNH options

Main users of RMB currency options and futures would include corporates, asset management firms and fund houses, proprietary trading firms, brokerage firms and professional investors. These different user types could use the RMB currency products for various purposes.

Listed below are hypothetical illustrations of product applications of the RMB currency options (analysis do not include transaction costs and past performance is not an indicator for future performance).

3.2.1 Basic applications

(a) Protection on RMB depreciation

Situation

An investor worries about RMB depreciation. He needs to sell his RMB assets and convert back to USD after 3 months.

Possible application

Buying a 3-month call options (i.e. buy USD, sell RMB) with strike 6.8500 as an example.

Scenario

At expiry, if USD/CNH fixing rate appreciates to 6.7000, the option expires and is not exercised. The investor can sell his RMB assets and covert RMB to USD at a better level at 6.7000. If USD/CNH fixing rate depreciates to 7.0000, the option is exercised and the investor can sell his RMB assets and covert RMB to USD at the original strike rate (i.e. 6.8500).

Potential Risks and Returns

Potential returns: The investor is able to covert RMB to USD at a better price if the RMB depreciates.

Potential risks: The investor has to pay options premium to buy the protection.

---

18 In a currency pair, the currency used as a reference to quote is the "pricing currency" (bottom), and the currency that is quoted in relation is called the "base currency" (top).For example, for EUR/USD, EUR is the base currency and USD is the pricing currency.

(b) Protection on RMB appreciation

Situation

An investor worries about RMB appreciation. He needs to sell his USD assets and convert back to RMB after 3 months.

Possible application

Buying a 3-month put options (i.e. sell USD, buy RMB) with strike 6.8500 as an example.

Scenario

At expiry, if USD/CNH fixing rate depreciates to 7.0000, the option expires and is not exercised. The investor can sell his USD assets and covert USD to RMB at a better level at 7.0000. If USD/CNH fixing rate appreciates to 6.7000, the option is exercised and the investor can sell his USD assets and convert USD to RMB at the original strike rate (i.e. 6.8500).

Potential Risks and Returns

Potential Returns: The investor is able to covert USD to RMB at a better price if the RMB appreciates.

Potential risks: The investor has to pay options premium to buy the protection.

3.2.2 Advanced applications

(a) Yield enhancement — sell covered call

Situation

An exporter has USD receivables due in three months and targets to convert the USD into the CNH at a better rate than futures price. The exporter does not need to sell the USD upon receipt so he would rather sell it at a better rate. The exporter also wants to make some extra yield from this expected cash inflow.

Possible application

The exporter sells a Mar 2017 expiry USD/CNH call option with strike 7.1000 and receives the CNH premium of 775 pips. USD/CNH spot rate: 6.9300, Mar 2017 futures price: 7.0450, Volatility: 7.40 bid.

Scenario

At expiry, if USD/CNH fixing rate is less than 7.1000, the option expires out-of-the-money and is not exercised. The exporter keeps the CNH premium as an extra return from taking the position. If USD/CNH fixing rate is greater than 7.1000, the option is exercised and the exporter sells USD against CNH at 7.1000 which is still better than the futures price if he had hedged three months ago. The exporter keeps the CNH premium which makes his effective selling rate 7.1775.

Potential Risks and Returns

Potential returns: The exporter makes an extra return on the usage of idle cash by selling options. The exporter sells at an effectively better rate even if the option is exercised by the buyer.

Potential risks: Should the USD appreciate significantly against the CNH, the exporter may incur opportunity cost compared to selling at the prevailing market rate.

(b) Cost Reduction — Buy call spread

Situation

A portfolio manager, who has exposure to RMB-denominated assets, plans to hedge CNH depreciation by buying USD/CNH call options. The portfolio manager's hedging horizon is one year. However the long-tenor USD/CNH call option is very costly due to the time value, upward sloping volatility curve and futures curve. For example, Dec 2017 expiry USD/CNH call option with strike 7.2500 is priced at 2,515 pips. (USD/CNH spot rate: 6.9300, Dec 2017 futures price: 7.2650, volatility: 8.85 offer)

Possible application

The portfolio manager can sell Dec 2017 expiry USD/CNH call option with strike 7.5000 and receives a premium of 1,585 pips with the view that USD/CNH may depreciate, but not to the level of 7.5000. (7.5000 strike volatility 9.06 bid). The premium from the 7.5000 strike call reduces the net cost of the hedging strategy of buying USD/CNH call options with a lower strike of 7.2500. The portfolio now pays a net premium of 930 pips.

Scenario

At expiry, if USD/CNH fixing rate is less than 7.2500, both options expire out-of-the-money and are not exercised. The portfolio manager bears the net option premium as hedging cost, but it is cheaper than not having adopting the strategy. If USD/CNH fixing rate is greater than 7.2500, but less than 7.5000, the portfolio manager exercises the option he has bought and let the option he sells to expire. This is the best scenario because he keeps the hedge and has reduced the hedging cost. If USD/CNH fixing rate is greater than 7.5000, both options are exercised. The portfolio manager loses the hedge, however he makes a net cash flow of 2500 pips in the CNH, which helps compensate his hedge in the spot market.

Potential Risks and Returns

Potential returns: The strategy reduces the hedging cost by taking a view on the movement of the USD/CNH exchange rate.

Potential risks: The strategy may become a partial hedge under some circumstances.

(c) Risk reversal

Scenario

A trader has the view that USD/CNH spot rate will go higher in the next three months. He buys a Mar 2017 expiry USD/CNH call with strike 7.1500 and pays option premium 715 pips (volatility 8.35 offer). However he does not want to bear the full amount of option premium nor to be too aggressive in taking positions. He chooses to sell a Mar 2017 expiry USD/CNH put with strike 6.9500 and receives option premium 525 pips (volatility 6.70 bid). His net cost is 190 pips.

Result

USD/CNH spot rate and forward/futures curve both move up. Assuming there is a parallel shift in the spot rate and futures prices by 600 pips, the call option is worth 955 pips and put option is worth 355 pips. Net value of the strategy is priced at 600 pips. The trader has realised 200% profit by taking the right view. Alternatively the trader can choose to wait till expiry date for the call option to be exercised and the put option to expire.

Potential returns

There are several ways to capture potential return from risk reversal:

- If USD/CNH spot rate and forward rates/futures prices go higher, the call option will be worth more than the put option and the trader can choose to take profit by closing the positions.
- If market has more demand for USD/CNH call options than put options, the call option will be worth more than the put option in terms of implied volatility. (This is called the volatility skew.)

Potential risks

If USD/CNH moves to the unfavourable direction, the trader not only loses the option premium he pays for the call option, but also incurs losses from his short position on the put option. In this case, the losses have been enlarged although the initial cost is lower.

(d) Volatility play — Straddle (two options with the same strike)

Scenario

A trader has the view that USD/CNH spot rate will remain volatile trading in the near future so the volatility curve may shift higher. He buys a Dec 2017 expiry USD/CNH call with strike 7.2500 together with a Dec 2017 expiry USD/CNH put with strike 7.2500. The call option is priced at 2,490 pips (volatility 8.85 offer) and put option is priced at 2,350 pips. The total premium is 4,840 pips.

Result

The vega position from the straddle is 550 pips (275 pips for each option).

Assuming implied volatility for Dec 2017 with strike 7.2500 increases to 10.00, the call option is worth 2,810 pips and the put option is worth 2,670 pips. The strategy is now priced at 5,480 pips. The value change is 640 pips (approximately 1.15 vega).

Potential returns

A long-dated straddle gives traders the largest exposure to volatility movement. It is a direct way to trade and realise traders' view on the volatility curve. (Straddle is usually delta-neutral at the inception of the transaction.) Short-dated straddles can be used to trade volatile movement of underlyings (gamma trading).

Potential risks

Traders are exposed to volatility risk by trading straddles. In a situation where the underlying moves but the volatility does not change much, traders have to manage delta while not making money from volatility.

Other possible applications

Trades can trade strangle (long one call and one put with different strikes) for higher volatility movement. This is called "trade the wings". Traders can trade butterfly (long straddle and short strangle) if they expect there will be volatility in certain price range but not too much. So they are "financing" straddle by selling strangles.

3.3   Physical delivery on exercise

3.3.1   Call options

Assumptions:

Strike price (k) = 6.90; Official settlement price (s) = 6.95

If the settlement price > strike price, the option is exercised, if the settlement price ≤ strike price, the option expires worthless.

Physical delivery process (see Figure 5):

If the call option is exercised, on physical delivery, the buyer pays the final settlement value, i.e. contract size (100,000 USD) x k (6.90) = 690,000 CNH, to the clearing house and receives the underlying currency value equalled to the contract size (100,000 USD) from the clearing house.

On the other hand, the seller delivers the underlying currency value equalled to the contract size (100,000 USD) to the clearing house and receives the final settlement value, i.e. contract size (100,000 USD) x k (6.90) = 690,000 CNH, from the clearing house.

Figure 5. Physical delivery on exercise for call options

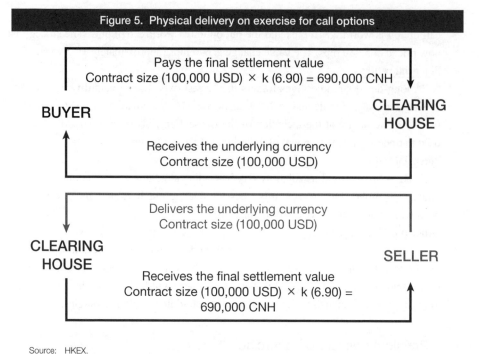

Pays the final settlement value
Contract size (100,000 USD) × k (6.90) = 690,000 CNH

BUYER

CLEARING HOUSE

Receives the underlying currency
Contract size (100,000 USD)

Delivers the underlying currency
Contract size (100,000 USD)

CLEARING HOUSE

SELLER

Receives the final settlement value
Contract size (100,000 USD) × k (6.90) =
690,000 CNH

Source: HKEX.

3.3.2 Put options

Assumptions:

Strike price (k) = 6.90; Official settlement price (s) = 6.85

If the settlement price < strike price, the option is exercised, if the settlement price ≥ strike price, the option expires worthless.

Physical delivery process (see Figure 6):

If the put option is exercised, on physical delivery, the buyer receives the final settlement value, i.e. contract size (100,000 USD) x k (6.90) = 690,000 CNH, from the clearing house and delivers the underlying currency value equalled to the contract size (100,000 USD) to the clearing house.

On the other hand, the seller receives the underlying currency value equalled to the contract size (100,000 USD) from the clearing house and pays the final settlement value, i.e. contract size (100,000 USD) x k (6.90) = 690,000 CNH, to the clearing house.

Figure 6. Physical delivery on exercise for put options

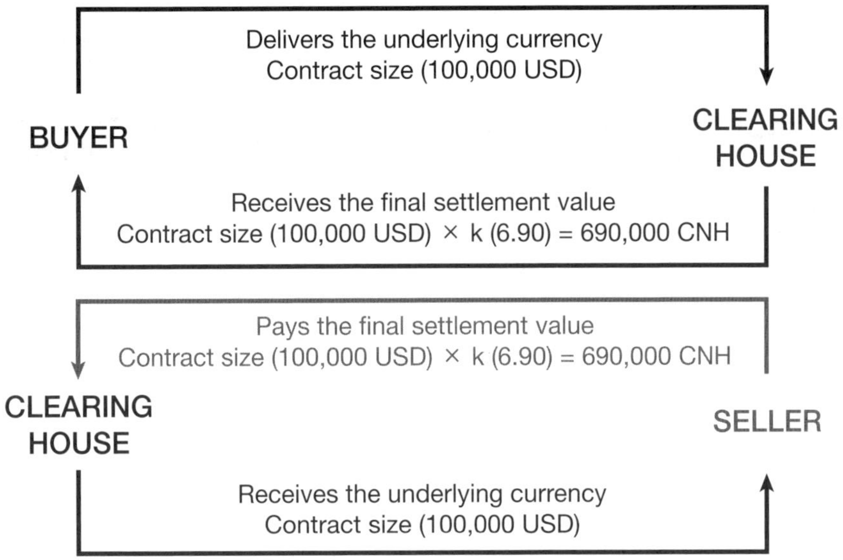

Delivers the underlying currency
Contract size (100,000 USD)

BUYER

**CLEARING
HOUSE**

Receives the final settlement value
Contract size (100,000 USD) × k (6.90) = 690,000 CNH

Pays the final settlement value
Contract size (100,000 USD) × k (6.90) = 690,000 CNH

**CLEARING
HOUSE**

SELLER

Receives the underlying currency
Contract size (100,000 USD)

Source: HKEX.

Appendix 1

HKEX USD/CNH Options contract specifications

Feature	HKEX USD/CNH Options contract features			Remark
Underlying	USD/CNH currency pair			
Contract size	US$100,000			Same as Futures
Options premium quotation	Quoted in 4 decimal places (i.e. 0.0001) on amount of RMB per USD.			Follows current OTC market quotation method
Strike prices	Strike intervals will be set at 0.05			Allows liquidity aggregation at specific strikes
Official settlement price	USD/CNY (HK) Spot Rate published by Hong Kong Treasury Markets Association (TMA) at or around 11:30 a.m. on the Expiry Day			Market benchmark in the CNH spot market
Settlement on exercise	Physical delivery on exercise			To meet the demand for principal exchange from options users
		Holder	Writer	
	Call options	Payment of the final settlement value in RMB	Delivery of US dollars	
	Put options	Delivery of US dollars	Payment of the final settlement value in RMB	
Exercise style	European style			Most popular in OTC market
Contract months	Spot month, the next three calendar months and the next four calendar quarter months			Same as Futures (except the furthest fifth calendar quarter month)
Final settlement day	Third Wednesday of the contract month			Same as Futures
Expiry day	Two Hong Kong business days prior to the final settlement day			Same as Futures

(continued)

Feature	HKEX USD/CNH Options contract features	Remark
Position limit	For the USD/CNH Futures contract, CNH/USD Futures contract and USD/CNH option contract combined, a position delta of 8,000 long or short in all contract months combined provided that: • Position delta for the spot month USD/CNH Futures contract and the spot month USD/CNH Option contract combined during the five Hong Kong business day up to and including the expiry day shall not exceed 2,000 long or short; and • The position for CNH/USD Futures contract shall not at any time exceed 16,000 net long or short contracts in all contract months combined	Combined position delta is used across RMB currency futures and options
Large open positions	500 open contracts in any one series	

Appendix 2

Trading and clearing arrangements of HKEX USD/CNH Options contracts

Maximum order size

Maximum order size is 1,000 contracts. Exchange Participants are required to submit their requests to HKEX for setting up their order size limits based on their business needs and risk management requirements.

Block trade

Block trade facilities are supported by the exchange's derivatives trading system. The volume threshold for block trades is 50 contracts (notional of US$5 million). The permissible price range is 10% for prices over or equal to 0.4, and 0.0400 for prices below 0.4.

Price makers

Some Liquidity Providers will provide continuous quotes on common strikes on screen, whereas some Liquidity Providers will quote prices upon requests for quote (RFQ).

Clearing arrangements

For clearing, Clearing Participants (CPs) have to arrange for RMB and USD settlement capability. They need to set up RMB and USD account with the Settlement Banks appointed by the HKCC and to maintain relevant mandates. Furthermore, CPs have to ensure these bank accounts are in active status and ready for physical delivery. Meanwhile, non-CPs should contact their General CPs to ascertain eligibility of clearing.

For details, please refer to the HKEX website.

Appendix 3

Accessing market information on HKEX USD/CNH Options

(1)   Information vendor access codes

Vendor	Access code
AAStocks	340900
Activ Financial	CUS/1701/9999P.HF
AFE Solution	873181-7
Bloomberg	CSX Curncy OMON <GO>
CQG	C/P.CUS
DBPower	CUS
Eastmoney	CUS
Esunny	CUS
ETNet	CUS
Fidessa	CUS_Osmy.HF
FIS Global	CUS+<STRIKE PRICE>+<MONTH CODE>+<LAST DIGIT OF THE YEAR>
Hexin Flush Financial Information Network Ltd	CUS
Infocast	CUS (Menu > Derivatives > Options > Select ""CUS"")
Interactive Data	O:CUS\MYYDD\[Strike Price]
Market Prizm	CUS <Strikes> my
QPI	P11370-P11375
SIX Financial	CUSmy
Shanghai DZH	CUS[mmyy][C/P][Strike]
Shanghai Pobo	CUSyymm-C/P-SSSSS

(continued)

Vendor	Access code
Telequote	CUSOmy
Tele-Trend	Open->Options->CUS
Thomson Reuters	0#HCUS*.HF
Wind	Quant -> CUSO.HK

(2)   Real-time prices in HKEX website
http://www.hkex.com.hk/eng/ddp/Contract_RT_Details.asp?PId=388

(3)   List of Exchange Participants offering trading services for USD/CNH Options
http://www.hkex.com.hk/eng/prod/drprod/rmb/ep-fxo.htm

(4)   List of Exchange Participants Enabled for RMB derivatives trading
http://www.hkex.com.hk/eng/prod/drprod/rmb/brokerlist.htm

Chapter 10

Innovations and implications of Bond Connect

Supporting the opening up of the Mainland financial market

15 November 2017

Summary

Bond Connect is an arrangement that enables Mainland and overseas investors to trade bonds on the Mainland and Hong Kong bond markets through the connectivity established between the financial infrastructure institutions in the Mainland and Hong Kong. Bond Connect is a major milestone of deepening mutual market access between the Mainland and Hong Kong. As a more efficient market opening channel that runs in parallel to the existing ones, Bond Connect is an innovative and explorative initiative in many ways, one which can attract a broader group of overseas investors to participate in the China Interbank Bond Market and an arrangement that international investors can better adapt to and more familiar with.

The innovations under Bond Connect are manifested in the areas of market admission in pre-trade, price discovery and information communication in trading, and custody and settlement arrangements in post-trade. It effectively connects the Mainland bond market with international practices, at lower access costs and higher market efficiency. On 3 July 2017, the Northbound Trading Link of Bond Connect was officially launched. A trading volume of more than RMB 7 billion was recorded that day. Within three months after launch, foreign holdings of the domestic debt securities increased significantly from RMB 842.5 billion to RMB 1,061.0 billion[1], which might be attributable to the launch of Bond Connect. This reflects to some extent the positive impact of Bond Connect on overseas participation in the Mainland bond market.

Bond Connect further opens up the Mainland bond market in a controlled manner, thereby providing new impetus to the market's international participation, the continued market open-up and reform, as well as to RMB internationalisation. Through Bond Connect, Hong Kong could become a convenient window for overseas investors to gain access to the Mainland bond market. This would further reinforce Hong Kong's position as an offshore RMB centre, foster the building up of an ecosystem of onshore and offshore RMB products around Bond Connect, and strengthen Hong Kong's role as an international financial centre and its intermediary function for capital flows into and out of the Mainland.

---

1   Source: China Central Depository & Clearing Co., Ltd. (CCDC) and Shanghai Clearing House (SCH) websites.

1 A new breakthrough in the opening up of the Mainland financial market

Bond Connect is an arrangement that enables Mainland and overseas investors to trade bonds on the Mainland and Hong Kong bond markets through the connectivity established between the institutional financial infrastructure in the Mainland and Hong Kong. On 16 May 2017, the People's Bank of China (PBOC) and the Hong Kong Monetary Authority (HKMA) jointly announced their approval of the establishment of Bond Connect. On 3 July, the Northbound Trading Link of Bond Connect was officially launched[2], further opening up the Mainland financial market. With the continuous development of the mutual market access between the Mainland and Hong Kong in recent years and the successive launch of the Shanghai and Shenzhen Stock Connect schemes[3] (collectively referred to as the "Stock Connect"), mutual market access is basically achieved between the Mainland and Hong Kong stock markets. As bond market is another key component of the capital market, Bond Connect is considered to be another innovative breakthrough in the opening up of the Mainland financial market in view of the tremendous room for development in the Mainland bond market and the growing international demand for RMB assets.

1.1   Tremendous room for development in the Mainland bond market

As the Mainland financial market continues to transform, the opening up of the bond market has become a critical force driving the opening up of the Mainland financial market and the internationalisation of the Renminbi (RMB). At the end of March 2017, the Mainland bond market was the world's third largest after the US and Japan, with a total outstanding value of RMB 66 trillion and with the total outstanding value of corporate credit bonds

---

2   According to current arrangements, the initial phase of Bond Connect is confined to Northbound Trading. Overseas investors from Hong Kong and other countries and areas (hereafter called "overseas investors") are allowed to invest in the China Interbank Bond Market (CIBM) through the mutual access arrangements established between the Hong Kong and Mainland financial infrastructure operating institutions in respect of trading, custody, settlement etc.

3   Stock Connect is a Mainland-Hong Kong Mutual Market Access pilot programme, under which investors in the Mainland and Hong Kong are, for the first time, able to gain direct access to each other's stock markets. Shanghai-Hong Kong Stock Connect (Shanghai Connect) was launched in November 2014 and Shenzhen-Hong Kong Stock Connect (Shenzhen Connect) was launched in December 2016.

being Asia's largest and the world's second largest[4]. Nevertheless, foreign participation in the Mainland bond market remains relatively low. If appropriate opening-up measures are adopted to attract more foreign participation in the Mainland bond market, not only will the reform of international balance of payments inflow be facilitated and the ability to reduce fluctuations in the international balance of payments be improved in the short-term, the liquidity of the Mainland bond market will also be enhanced in the medium to long term.

1.2   The pace of opening up the Mainland bond market to overseas investors has accelerated in recent years

The Mainland opened up its interbank bond market for the first time to qualified foreign institutions in 2010 and launched the RMB Qualified Foreign Institutional Investor (RQFII) scheme in 2011. Two years later, in 2013, Qualified Foreign Institutional Investors (QFIIs) were allowed to participate in the China Interbank Bond Market (CIBM). In 2015, various initiatives that substantively facilitated overseas investors' access to the CIBM were implemented. These include: in June 2015, the PBOC allowed overseas RMB clearing banks and participating banks that had entered CIBM to conduct repurchase (repo) transactions in Mainland bonds; in July 2015, the PBOC adopted policies to further facilitate investments on CIBM by overseas central banks and institutions of a similar nature (including foreign central banks or monetary authorities, sovereign wealth funds and international financial institutions), allowing them to expand their investments to cash bonds, bond repos, bond lending, bond forwards, interest rate swaps (IRS), and forward interest rate agreements (FRA), etc.; in February 2016, the PBOC announced new rules that further relaxed the eligibility of foreign institutional investors for entering CIBM, with detailed rules released in May 2016 to expand the scope of eligible foreign institutional investor types and eligible trading instruments, to abolish the investment quota and to simplify the investment procedures. Prior to the launch of Bond Connect, there were 473 overseas investors participating in CIBM, with bond holdings exceeding RMB 800 billion[5].

1.3   Bond Connect has positive impact in attracting international capital

The above measures in exploring the opening up of the Mainland bond market have laid the foundation for the launch of Bond Connect. However, at the end of 2016, foreign

---

4   Source: PBOC website.
5   Source: PBOC website.

holdings in Mainland bonds remained under 2%, which was notably below the average level of opened-up bond markets in developed economies. (see Figure 1.) The existing access channels of the Mainland bond market mentioned above mainly suit overseas central banks and large institutions that are relatively more familiar with the Mainland bond market and can afford higher operational costs to participate in it. For a large number of small to medium-sized overseas investors, new channels are needed to attract their participation and to address the challenges they encounter when participating in the Mainland bond market. It is against such a background that Bond Connect was launched.

Figure 1. Percentage of foreign holdings in the bond markets of developed countries (by value) (End–2016)

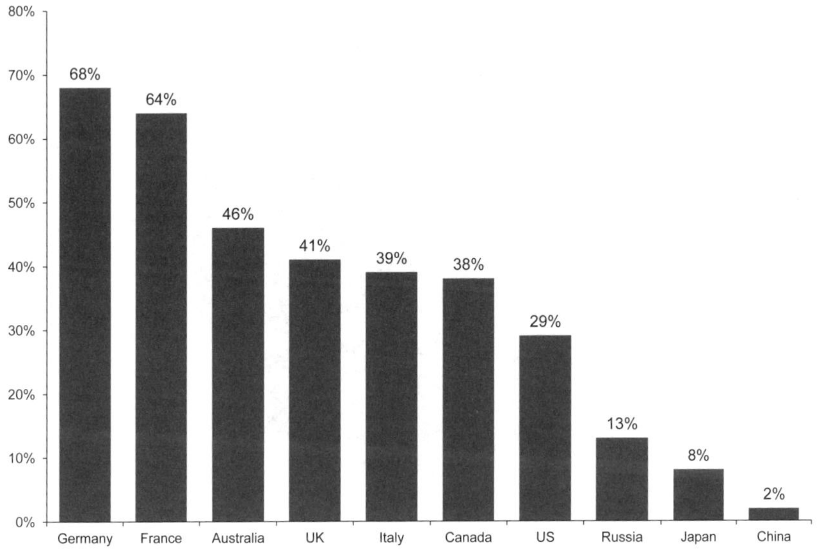

Source:　Bloomberg, Bank for International Settlements, PBOC.

In October 2016, the RMB was officially included in the Special Drawing Right (SDR) currency basket of the International Monetary Fund (IMF) with a weighting of 10.92%. This would bring new participants and capital flow into bond assets denominated in RMB and enhance the global acceptance of the RMB currency as a global investment and reserve currency, thereby boosting international institutional demand for RMB assets in both the public and private sectors. However, the proportion of RMB in official foreign exchange (FX) reserves and FX transactions are currently far below 10.92%. This suggests that the driver for RMB internationalisation in the next stage will come from the proliferation of a diverse suite of investable offshore and onshore RMB-denominated financial assets for international investors. In this context, the opening of the Mainland bond market is a crucial factor. Opening up of the bond market to a larger extent will also foster closer ties between the Mainland regulators and the international market, and promote further internationalisation of participants in the onshore financial infrastructure. Onshore Mainland financial institutions can also develop closer business connections with foreign institutional investors through Bond Connect, paving the way for greater participation by Mainland financial institutions in overseas markets.

Figure 2. The share of RMB in IMF's SDR currency basket (End–2016)

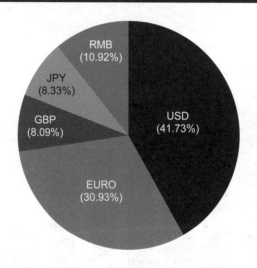

Source: IMF.

2 Bond connect effectively connects the Mainland bond market

Before the launch of Bond Connect, overseas investors can participate in the Mainland bond market through three main channels — QFII scheme, RQFII scheme and foreign institutions' direct access to CIBM (CIBM scheme). (see section 1.2 above.) Compared to these existing channels, Bond Connect is an innovative breakthrough in the "pre-trade", "trading" and "post-trade" processes. It has addressed expectations and demands of international investors seeking to participate in the Mainland bond market.

2.1   Pre-trade: Market admission and parallel channels

Same as the scope of eligible investors trading in the Mainland bond market through existing channels as regulated by the PBOC's Public Notice No. 3 [2016], overseas investors trading under Bond Connect are mainly central-bank-type institutions and medium- to long-term investors with a focus on asset allocation. This reflects that the Mainland is steadily implementing its strategy to open up the RMB capital and financial accounts and provide new alternative and convenient channels in terms of market admission, filing procedures, eligibility approval and other areas for the entry of long-term capital into the Mainland bond market. While existing channels might satisfy the needs of overseas central banks and large institutional investors to invest in the Mainland bond market, there are certain investors who are interested in investing in the Mainland bond market but are reluctant to bear the high participation cost. That is to say, under Bond Connect, there is no need for overseas investors to have in-depth understanding of the onshore trading and settlement systems of the Mainland bond market and the related Mainland laws and regulations. They only need to use the current trading and settlement practices they are familiar with. Such an arrangement considerably lowers the entry barriers and costs for overseas investors to participate in the Mainland bond market. Bond Connect is therefore a more user-friendly bond investment channel for overseas investors, as illustrated below.

Firstly, before the launch of Bond Connect, overseas investors mainly accessed CIBM through settlement agents, i.e. as "Type C" participants. In this way, foreign institutions are required to complete the necessary filing and account opening procedures through a domestic CIBM settlement agent for market entry. These procedures, to a certain extent, created obstacles for certain institutional investors to tap into the Mainland bond market. On the contrary, under the more liberalised Bond Connect mechanism, foreign institutions can gain access to the Mainland bond market through a single entry point via offshore infrastructures. Overseas investors do not need to open Mainland settlement and custody accounts and are not required to directly deal with Mainland authorities for market admission, trading qualifications and other related issues. They can make use of their existing accounts in Hong Kong to directly access the Mainland bond market. They are assured at the outset of trading that they can adhere to their familiar international laws and trading practices and at the same time fulfill market admission and filing requirements through offshore financial infrastructures. There is no need for them to get familiarised with the Mainland market practices that differ from their long-established trading and settlement practices.

In operation, the Bond Connect Company Limited (BCCL), established offshore jointly by Hong Kong Exchanges and Clearing Ltd (HKEX) and China Foreign Exchange Trade System (CFETS), provides professional guidance on market access, application review and other market-access preparation support. The time taken to process an admission filing with the PBOC is substantially shortened. In terms of operational procedures, accessing the Mainland's onshore bond market via Bond Connect aligns more with the trading practices of international investors, particularly for those institutional investors who would like to access the Mainland bond market but are unfamiliar with its rules. Bond Connect essentially enhances the pace of market entry and participation efficiency of these investors in the Mainland bond market.

Secondly, under the current regulatory requirements, overseas investors investing in the Mainland bond market through the QFII, RQFII and CIBM schemes are required to comply with certain requirements upon market entry, including those on fund remittance and lock-up period. They are also required to specify at the outset their planned investment amounts that have to be fulfilled in subsequent transactions. (see Table 1.) This, at times, may not align with the investment strategies of certain foreign institutions for flexible utilisation of funds. This is also an attribute that has deterred foreign institutions from entering the onshore bond market. Bond Connect, on the contrary, has no such restrictions at market admission, such that overseas institutions face far fewer obstacles at market entry and can directly manage their onshore transactions, allowing greater flexibility in their RMB asset allocation. This will undoubtedly increase the incentive for overseas institutions, particularly the small to medium-sized institutional investors, to participate in the Mainland bond market.

Table 1. Key framework of the current QFII, RQFII and CIBM schemes			
	QFII scheme	RQFII scheme	CIBM scheme
Regulatory approval	• China Securities Regulatory Commission (CSRC): QFII/RQFII licence • State Administration of Foreign Exchange (SAFE): QFII/RQFII investment quota • PBOC: Pre-filing for CIBM access		Pre-filing with PBOC
Investment quota	• Needs to pre-file with SAFE if the quota applied for is within the base quota; to obtain approval if the quota applied for exceeds the base quota. • The base quota is calculated as a certain percentage of asset size.		• Subject to macro-prudential administration of foreign investors • No specific investment quota requirements; need to pre-file with PBOC the planned investment value

(continued)

Table 1. Key framework of the current QFII, RQFII and CIBM schemes			
Qualified fixed-income products	• On-exchange market: government bonds, enterprise bonds, corporate bonds, convertible bonds, etc. • Interbank market: cash bonds		• Foreign reserves institutions: all cash bonds, bond repos, bond borrowing and lending, bond forwards, IRS, FRA, etc. • Other financial institutions: all cash bonds and other products permitted by the PBOC, offshore RMB clearing/participating banks can also trade repos.
FX management	Conduct FX onshore with the local custodian	Remit in offshore RMB	Onshore / offshore
	QFII scheme	**RQFII scheme**	**CIBM scheme**
Lock-up period of principal remitted in	3 months	3 months, but no such restriction for open-ended funds	No
Outward remittance frequency and restrictions	Daily limit (for open-ended funds only), and monthly limit	Daily limit (for open-ended funds only)	• Certain requirements are imposed on the ratio of accumulated outward remittance.

Source: As of end-2016. Please refer to the websites of PBOC, CSRC and SAFE for the most updated rules and policies.

Thirdly, the market access channel of Bond Connect co-exists with the existing QFII, RQFII and CIBM schemes. Overseas investors can now choose among the multiple channels with flexibility. After the launch of Bond Connect, overseas investors are able to choose an investment channel that better suits their strategies. This enables diversified and effective asset allocation and product development in the Mainland's onshore financial market. Similar adjustments in the choice of investment channels were observed after the launch of Stock Connect. This shows that existing market opening channels are complementary to each other, rather than a replacement or substitution, to serve different investment needs of a diverse investor base. Bond Connect advances the opening up of the Mainland bond market, and helps drive forward RMB internationalisation and the Mainland's capital account liberalisation.

2.2   Trading: price discovery and information efficiency

The Mainland bond market adopted three trading modes: voice trading, click-to-trade and Request-for-Quote (RFQ). As voice trading is done offline, bond trading in the Mainland market could be difficult for overseas institutions to understand in depth. Under Bond Connect, overseas investors can conduct interbank cash bond trading with the Mainland market makers by way of submitting RFQ via offshore platform, after which market makers can provide tradable price quotes for overseas investors to choose and confirm transactions. To overseas institutional investors who are not so familiar with the Mainland bond market, such trading arrangement is simpler and the price and transaction data is more transparent, symmetric and more conducive for price discovery.

Moreover, under the agent bank model, overseas investors cannot directly trade with a Mainland counterpart. They can only entrust agent banks in the Mainland to do trades on their behalf. Under Bond Connect, overseas investors can freely select a market maker for quotes and decide the timing of a transaction via an overseas electronic trading platform using an interface and trading mode they are familiar with. As a result, these overseas investors who invest in the Mainland bond market through Bond Connect do not have significant switching costs during actual operations. To those small to medium-sized overseas institutional investors who are sensitive to transaction costs, this factor is particularly important. At present, Tradeweb is the first overseas electronic trading platform available to investors for Bond Connect. Bloomberg and other electronic platforms are actively progressing on connecting with the Bond Connect platform and will launch their systems once they are ready. The ability of overseas institutions to directly request for price quotes and trade with Mainland institutions without the need to change their own trading practices has enhanced the transparency and efficiency of the entire trading process.

From the perspective of market operation, Bond Connect provides overseas investors with a channel of direct trading alternative to the existing model of agency trading as "Type C" participants. To overseas investors, particularly the institutional investors who are not so familiar with the Mainland market, Bond Connect has, to some extent, reduced agency and communication costs and increased trading efficiency, thereby contributing to the enhancement of market liquidity.

2.3 Post-trade: custody and settlement

The Mainland bond market currently adopts a single-level depository system, which is an important arrangement after due consideration following a long period of practical experience to conform to the characteristics of the Mainland bond market. In the offshore market, however, a multi-level depository arrangement with a nominee system has been the long-standing practice. Such difference in practices has created obstacles for overseas institutions to participate in the Mainland bond market. As the multi-level depository arrangement and nominee system in the international bond market have evolved over years to its present form, the business operations of overseas institutional investors have become strongly adapted to it. Should there be any major changes in this operating model, the compliance and back-end operation departments of overseas institutional investors, and the relevant regulators in the local jurisdictions, will face significant adjustment difficulties. This will restrain certain institutional investors from participating in the Mainland bond market.

Bond Connect effectively connects the "single-level depository system" as required under the see-through custodian model of the Mainland and the "multi-level depository system" under the nominee model adopted in the international bond market. For the registration, depository, clearing and settlement of bonds for overseas investors, a settlement link is established between CCDC and SCH as the onshore central depository institutions, and the Central Moneymarkets Unit of the HKMA (HKMA-CMU) as the offshore central depository institution. In this way, overseas institutions can observe Mainland market rules while adopting their long-established international practices at the same time, thereby effectively reducing the access costs involved in the operation under different market structures. The arrangement is also conducive to the development of related financial products and business models after the launch of Bond Connect. (See Figure 3 for the operating system set-up of Northbound Trading under Bond Connect.)

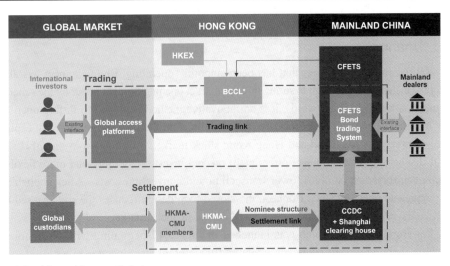

Figure 3. Northbound operating model of Bond Connect

* Bond Connect Company Limited.

Source: HKEX.

In terms of legal framework compatibility, the trading and settlement activities under the Northbound Link of Bond Connect shall comply with the regulations and business rules of the place where these activities take place. Under the nominee holding structure, overseas investors shall exercise their rights over the bond issuers through HKMA-CMU as the nominee holder. If there is a default in bond redemption, HKMA-CMU as the nominee holder of the overseas investor and registered as the bondholder, may exercise the rights of bondholder and take legal actions. Meanwhile, the overseas investor as the beneficial owner of bonds, upon provision of relevant evidence, may also take legal actions in its own name in Mainland China.

3 Mainland financial market opening and development of the offshore RMB centre in Hong Kong

3.1 Further opening up the Mainland bond market in a controlled manner

Similar to Stock Connect that has been smoothly operating, Bond Connect has also adopted a closed-loop design ensuring that the market opening brought about by the scheme is under control. In other words, it is an innovative method deployed to further open up the Mainland bond market. Stock Connect is a breakthrough achieving two-way capital flow between the Mainland and Hong Kong stock markets. Compared to the QFII and RQFII schemes, Stock Connect offers more relaxed investor eligibility and more flexible quota controls, and involves lower transaction and conversion costs. After abolishing the aggregate quota in Stock Connect, the closed-loop design reduces the risks caused by substantial capital flow into and out of the Mainland financial market.

For Bond Connect which was launched in July 2017 beginning with Northbound trading, there is also no aggregate quota and the system also operates in a closed-loop structure. By connecting cash bond markets in the Mainland and Hong Kong, the scheme is expected to help funnel international capital into the Mainland bond market, facilitating the internationalisation of the Mainland capital market.

3.2   Further strengthening and reinforcing Hong Kong's status as an offshore RMB centre

Hong Kong has been a global offshore RMB business hub[6]. The use of RMB is no longer limited to cross-border trade settlement as in the initial stage, but has also extended to a considerable level to the areas of investment, financing, hedging and foreign reserves management. Offshore RMB FX trading volume has also continued to grow[7]. The sequential launch of Shanghai Connect and Shenzhen Connect, together with the abolition of their aggregate quota, provide new channels for the opening of the Mainland capital market. Bond Connect serves a bigger purpose of complementing Hong Kong's capability in its bond market development as an international financial centre (IFC). The Southbound Bond Connect will be explored at a later stage, which is expected to attract even more Mainland capital into Hong Kong, bringing greater momentum to Hong Kong's bond market and its overall financial system.

3.3   Building an ecosystem of onshore and offshore Renminbi products associated with Bond Connect

Given the specific characteristics of the Mainland bond markets, the BCCL is expected to perform a prominent role in investor education and market coordination. The launch of Bond Connect and the smooth operation of BCCL would have a profound effect on the Mainland's onshore and offshore bond markets, in particular, in shaping the onshore and offshore financial markets towards an ecosystem associated with bond asset allocation. As bond markets are dominated by institutional investors, bond trading is closely intertwined with demands for financial derivatives trading and risk management. Bond Connect is expected to bring more international bond investors to Hong Kong to trade offshore debt products on Mainland assets. This could bring impetus to Hong Kong's bond market and drive demand for professional services in financial derivatives and risk management in RMB, and thereby benefiting the overall industry.

Currently, the Mainland domestic derivatives market offers a variety of tradable products (including forwards, swaps, options and treasury bond futures), with considerable depth and liquidity, to support hedging against RMB-related risks. Lately, the further opening up of the domestic FX market has made it possible for certain qualified overseas investors to directly make use of the domestic derivative products. Concurrently, the Hong

---

6   See *HONG KONG: The Global Offshore Renminbi Business Hub*, HKMA website, 2016.
7   See "Triennial Central Bank Survey of Foreign Exchange and OTC Derivatives Markets", Bank for International Settlements, 2016.

Kong OTC market offers a series of RMB products, including RMB FX spots, forwards, swaps and options. Similar exchange-traded products, including RMB currency futures, options and treasury bond futures[8], are also available on HKEX. All these products enable overseas participants to better hedge against their exposure in Mainland bond assets and FX risks. Upon the launch of Bond Connect, professional services in Hong Kong that are ancillary to the scheme, such as risk management and innovative RMB-denominated financial instruments, are expected to gain further traction.

3.4 Smooth operation drives overseas participation in the primary market of domestic bond issuance

The Northbound Trading Links of Bond Connect was launched on 3 July 2017 with active turnover. A total of 19 participating dealers and 70 overseas institutions executed an aggregate of 142 trades, amounting to RMB 7,048 million, on its debut. By the end of September, 184 overseas institutions have tapped into Mainland bond market via Bond Connect. Foreign holdings of Mainland domestic bonds also significantly increased from RMB 842.5 billion as of end-June before launch to RMB 1,061.0 billion[9] as of end-September. Such an increase may be attributable to the open channel and innovative regime of Bond Connect.

Direct subscription of Mainland bond issues by overseas investors was also allowed at the inception of Bond Connect. Five debt financing instruments with an aggregate value of RMB 7 billion were issued by non-financial enterprises on the scheme's debut. During the first month after launch, 4 financial bonds and 14 commercial papers with a total issue size of RMB 60.68 billion and RMB 15.5 billion respectively were issued to Mainland domestic investors and overseas investors[10]. On 26 July, the Hungarian Government was the first to issue a three-year Panda Bond of RMB 1 billion[11], which was available for subscription via Bond Connect. By the end of August 2017, CIBM bond issuers using Bond Connect included state-owned enterprises, local government enterprises and overseas governments, covering the sectors of power, telecommunications, transportation, metals, agriculture and forestry, etc. This shows that Bond Connect is gradually becoming a key channel through which overseas institutions participate in the Mainland primary bond market and has facilitated the diversification of the investor base in the market.

---

8 HKEX's pilot scheme on RMB treasury bond futures will be suspended on the maturity of the December 2017 contracts. HKEX is now working on a series of risk management instruments that support Bond Connect. New RMB interest rate products will be launched in due course.

9 Source: CCDC and SCH website.

10 Source: Wind.

11 Panda bonds are RMB-denominated bonds issued in China's domestic bond market by foreign institutions.

4   Conclusion

In conclusion, Bond Connect allows, for the first-time in history, overseas funds to trade onshore Mainland bonds via offshore infrastructure in Hong Kong for trading and settlement. Drawing on the success of Stock Connect, Bond Connect expands the channels through which overseas investors can trade in Mainland onshore bonds, and further opens up the market while retaining the long-established trading and settlement practices in overseas markets. The innovations of Bond Connect are manifested in pre-trade market admission, price discovery and information communication during trading, and post-trade custody and settlement. This effectively connects international practices and the Mainland bond market at lower access costs and higher market efficiency.

As a major IFC in the region, Hong Kong provides Bond Connect with a trading and settlement platform that conforms to international practices, bridging the Mainland and global capital markets. More entities would be attracted to participate in the Hong Kong financial market, bringing in more capital and widening the range of RMB financial products in the city. All these would facilitate the development of Hong Kong as a RMB asset allocation centre and solidify Hong Kong's status as an offshore RMB centre.

In the long term, Bond Connect would significantly improve the efficiency of cross-border investment capital flow and enhance the internationalisation of the Mainland market. The implementation of Bond Connect is expected to foster the further development of financial products and services associated with onshore RMB bonds, and contribute to the development of a more diversified investor base and a more open bond market for the Mainland.

Abbreviations

BCCL	Bond Connect Company Limited
CCDC	China Central Depository and Clearing Co., Ltd
CFETS	China Foreign Exchange Trade System
CIBM	China Interbank Bond Market
FRA	Forward interest rate agreement
FX	Foreign exchange
HKMA	Hong Kong Monetary Authority
HKMA-CMU	Central Moneymarkets Unit of the Hong Kong Monetary Authority
IFC	International financial centre
IMF	International Monetary Fund
IRS	Interest rate swap
PBOC	People's Bank of China
QFII	Qualified Foreign Institutional Investor
RFQ	Request for Quote
RQFII	RMB Qualified Foreign Institutional Investor
SCH	Shanghai Clearing House

Part 3
Commodities

Chapter 11

Hong Kong is building up itself to be an Asian gold pricing centre

10 July 2017

Summary

Hong Kong Exchanges and Clearing Limited (HKEX) launched dual-currency (USD and RMB pricing and settlement) physical delivery Gold Futures Contracts ("Gold Contracts") through its subsidiary, the Hong Kong Futures Exchange (HKFE), on 10 July 2017.

HKEX acquired the London Metal Exchange (LME) in 2012 and, consistent with the vision drawn in its 2016-2018 Group Strategic Plan, aims to transform itself into a global vertically-integrated multi-asset class exchange. Commodities are one of the core pillars of HKEX's four-pronged multi-asset strategy. The launch of the HKEX Gold Contracts is a clear demonstration of HKEX's commitment to offer attractive commodity products in Asia.

Although the Hong Kong gold market has over 100 years of history, it still lags behind other global gold trading centres such as New York and London in term of benchmarking, liquidity, and completeness of product and service offerings. However, as a global financial centre located at the gateway of the world's second largest economy and the biggest gold consumer — China, Hong Kong has the right ingredients to become an Asian gold pricing centre.

Given Hong Kong's advantages as a free market and an entrepot trade centre, Hong Kong has an active physical gold trading market and enjoys the status as one of the major bullion markets in the world. In addition, being the largest offshore RMB centre, Hong Kong has a unique position in facilitating China's RMB internationalisation. The introduction of physically delivered gold futures in Hong Kong would be a stepping stone to achieving this objective.

The fundamental market demand for gold trading in China together with the trading demand from the rest of the world, and the related risk management demand well position Hong Kong to build itself up to be an Asian gold pricing centre. In order for Hong Kong to cultivate a new gold pricing benchmark, it is necessary to form a well-functioning marketplace to link up the spot market and the futures market and to provide efficient channels to serve these markets in Hong Kong, and to complete the gold ecosystem to include other gold-related financial products and services such as gold leasing and related derivatives. The formation of the new Hong Kong Asian benchmark will be established naturally as the liquidity grows via these channels within the ecosystem.

1   The nature and uses of gold

Gold is a dense, bright, orange-yellow precious metal, which is soft, malleable and ductile. Due to its relative rarity and chemical nobility, gold is highly valued for jewellery and other decorative purposes, as an investment, and historically as a form of money. It is still a major component of central bank reserves as shown in Table 1.

Table 1. Top 20 reported official gold holdings (as at March 2017)			
Rank	Economy / Multilateral organisation	Tonnes	% of central bank reserves
1	United States	8,133.5	75%
2	Germany	3,377.9	69%
3	International Monetary Fund (IMF)	2,814.0	—
4	Italy	2,451.8	68%
5	France	2,435.9	64%
6	Mainland China	1,842.6	2%
7	Russia	1,680.1	17%
8	Switzerland	1,040.0	6%
9	Japan	765.2	2%
10	Netherlands	612.5	64%
11	India	557.8	6%
12	European Central Bank (ECB)	504.8	27%
13	Turkey	427.8	16%
14	Taiwan China	423.6	4%
15	Portugal	382.5	55%
16	Saudi Arabia	322.9	2%
17	United Kingdom	310.3	9%
18	Lebanon	286.8	21%
19	Spain	281.6	17%
20	Austria	280.0	46%

Source: IMF International Financial Statistics Database, World Gold Council.

The history of human understanding of gold can be traced back to the ancient Egyptian era more than five thousand years ago. Since then, gold and human development have been inseparably intertwined.

Chemically, gold is an element with the symbol Au (from Latin aurum) and atomic number 79 — one of the highest of the naturally-occurring elements. Gold occurs most often in free elemental form as nuggets or grains in rocks and alluvial deposits; there are also significant quantities of gold in the sea. Since the 1880s, South Africa has been the major source of world gold supply, having produced about 50%[1] of cumulative production to date. However, South Africa has recently been eclipsed by other producers, especially China.

The world's first gold coins were struck in Lydia Asia Minor around 600 BC, and gold was the base for monetary systems for most of subsequent human history. The world gold standard was abandoned only in 1971, and Switzerland continued using gold to back up 40% of its currency until 1999. Today, many central banks still keep a portion of their reserves in gold. Gold remains an important investment instrument — in the form of bullion, paper gold, derivatives and exchange traded funds (ETFs) — and in turbulent times gold is viewed as a safe haven investment. Surges in the gold price tend to come at times of war, and more recently, the 2008 Global Financial Crisis. However, as the investment universe has grown, gold's relative importance has declined.

Gold resists most acids and most alkalis, while conducting electricity well. These properties lead to its continuing use for corrosion-resistant conductors in computerised devices and electrical devices, its main industrial application. A typical mobile phone may include 50mg of gold, worth around US$2.00 in today's market price. Gold is also used in infrared shielding, coloured glass production and gold leaf. Gold is harmless when ingested, and is sometimes used for food decoration; gold salts are still used medically as anti-inflammatories.

As stated above, gold has been used as money throughout human history. The global development of trading and exchange of gold has been formed since 200 years ago. Figure 1 below shows some key events of this development in modern history — most of these happened after the abandonment of the gold standard when the Bretton Woods system collapsed in 1971.

---

1   Source: World Gold Council.

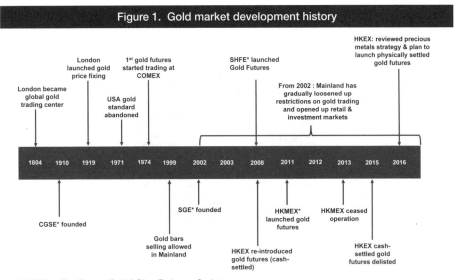

Figure 1. Gold market development history

* CGSE — The Chinese Gold & Silver Exchange Society.
  HKMEX — Hong Kong Mercantile Exchange.
  SGE — Shanghai Gold Exchange.
  SHFE — Shanghai Futures Exchange.

Source: HKEX, analysis on public sources of information.

2   Gold fundamentals

2.1   Gold supply and demand

About 53% of the world's 2016 consumption of gold is for jewellery, 37% for investments and 10% for industrial uses. (see Figure 2.) Since gold does not decay or easily react with other substances, most of the gold that have been mined by mankind over the millennia still exist, although a great part is likely lost, buried in graves, or (in the case of gold used for industrial purposes) embedded in junk in landfills. Nonetheless, gold scrap from existing privately-held stocks of jewellery and bullion, together with gold scrap from recycled industrial products are a significant source of annual supply alongside newly mined gold.

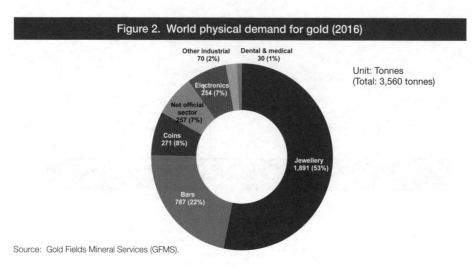

Figure 2. World physical demand for gold (2016)

Unit: Tonnes
(Total: 3,560 tonnes)

Other industrial 70 (2%)
Dental & medical 30 (1%)
Electronics 254 (7%)
Net official sector 257 (7%)
Coins 271 (8%)
Bars 787 (22%)
Jewellery 1,891 (53%)

Source: Gold Fields Mineral Services (GFMS).

The price of gold, like that of other commodities, is driven by the balance between demand for the metal for jewellery and other purposes, and physical supply from mines and scrap. However, given the magnitude of useable above-ground stocks, physical surpluses or deficits are less important than in the case of other commodities in determining price (although they may impact lead times, premia and margins along the value chain). Because gold is still an investment instrument, monetary conditions and confidence in the economy are important drivers of its price. Throughout the history, gold has acted as a safe haven during time of uncertainties and market turmoil. As shown in Figure 3, gold prices spiked up to US$870 per troy oz during hyperinflation and energy crisis in 1980, and reached US$1,895 per tory oz in 2011 after the 2008 Global Financial Crisis.

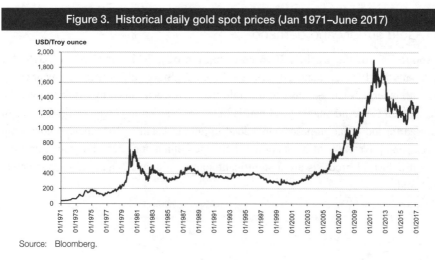

Figure 3. Historical daily gold spot prices (Jan 1971–June 2017)

USD/Troy ounce

Source: Bloomberg.

Gold mine production has been increasing over the past decade to exceed 3,000 tonnes annually as shown in Table 2 below. Gold scrap supply increased in 2016 by 8% year-on-year to 1,268 tonnes, in line with the recent increase in the gold price. And this is reversing the recent decline trend as it hit the lowest point of 1,158 of scrap supply in 2014 as shown in Table 2 below.

Table 2. World gold supply and demand (tonnes)									
	2007	2008	2009	2010	2011	2012	2013	2014	2015	2016
Supply										
Mine production	2,538	2,467	2,651	2,775	2,868	2,883	3,077	3,172	3,209	3,222
Scrap	1,029	1,388	1,765	1,743	1,704	1,700	1,303	1,158	1,172	1,268
Net hedging supply	-432	-357	-234	-106	18	-40	-39	108	21	21
Total supply	3,135	3,498	4,182	4,412	4,590	4,543	4,341	4,438	4,402	4,511
Demand										
Jewellery	2,474	2,355	1,866	2,083	2,091	2,061	2,610	2,469	2,395	1,891
Industrial fabrication	492	479	427	480	471	429	421	403	366	354
Electronics	345	334	295	346	343	307	300	290	258	254
Dental & medical	58	56	53	48	43	39	36	34	32	30
Other industrial	89	89	79	86	85	83	85	79	76	70
Net official sector	-484	-235	-34	77	457	544	409	466	436	257
Retail investment	449	937	866	1263	1616	1407	1873	1164	1162	1058
Bars	238	667	562	946	1,247	1,056	1,444	886	876	787
Coins	211	270	304	317	369	351	429	278	286	271
Total physical demand	2,931	3,536	3,125	3,903	4,635	4,441	5,313	4,502	4,359	3,560
Physical surplus/ deficit	204	-38	1,057	509	-45	102	-972	-64	43	951
ETF inventory build	253	321	623	382	185	279	-880	-155	-125	524
Exchange inventory build	-10	34	39	54	-6	-10	-98	1	-48	86
Net balance	-39	-393	395	73	-224	-167	6	90	216	341

Source: GFMS.

As shown in Table 2 above, the total physical demand for gold continued to fall by 18% in 2016 to a three-year low of 3,560 tonnes, with declines in all demand areas. Jewellery remained the biggest source of demand, followed by retail investment. However, jewellery demand fell by 21%, largely because of a sharp fall in Indian and Chinese consumption[2]. Industrial fabrication continued to decline, falling by 3% to 354 tonnes, the lowest level in a decade, because of the weak demand in all major sectors, particularly in electronics (along with ongoing substitution of gold usage in the industry) and in dental and decorative uses. However, the total non-physical investment in gold increased to 610 tonnes (largely due to ETF buying in the year compared with the net sale recorded in the previous year), as investors allocating more funds to the ETF and futures trading.

According to GFMS, the total above-ground gold stocks (defined as the cumulative historical total of mine production) increased by 1% year-on-year to 187,200 tonnes in 2016. This is equivalent to a value of about US$7.6 trillion as of June 21, 2017. As shown in Figure 4, jewellery stocks were the largest component which accounted for about 48%, followed by private investment and official holdings which accounted for about 38%.

Figure 4. Total Global above-ground gold stocks (2016)

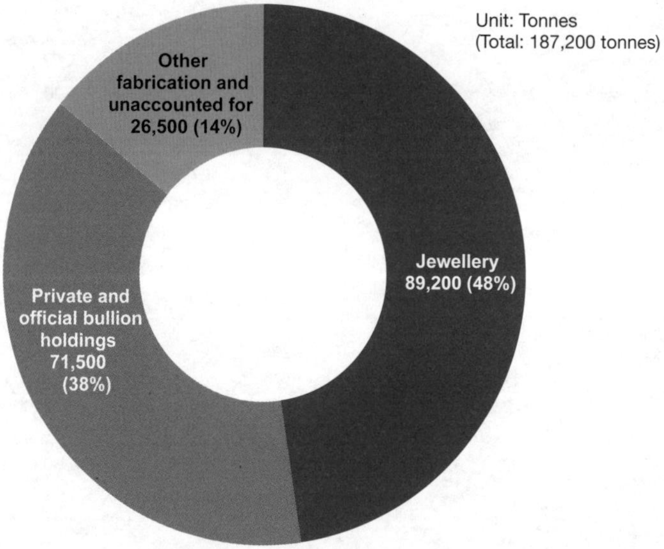

Unit: Tonnes
(Total: 187,200 tonnes)

Other fabrication and unaccounted for 26,500 (14%)

Private and official bullion holdings 71,500 (38%)

Jewellery 89,200 (48%)

Source: GFMS.

_____

2 Source: World Gold Council.

2.2   Main players in the gold value chain

There is an ecosystem of different players in the gold value chain as shown in Figure 5 below. Miners extract and process ore to deliver raw gold to processors which further refine and distribute it to consumers which in turn fabricate the gold into products that can be distributed to end-users, namely retail consumers, investors, industrial users and central banks. The value chain is supported by numerous service providers such as assayers who certify the quality and weight of gold bars, custodians who keep gold safe in their vaults, information vendors who disseminate gold prices and exchanges providing marketplaces for their members to trade gold contracts.

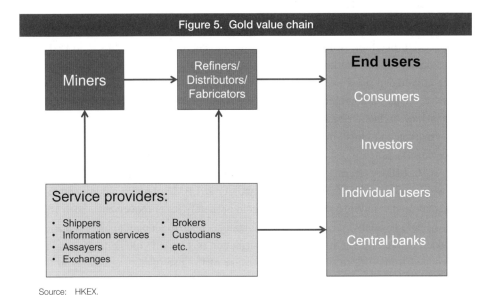

Figure 5. Gold value chain

Source: HKEX.

3   Global gold markets

The global gold market is largely distributed but dominated by London-centric over-the-counter (OTC) trading. Nevertheless, there are significant volumes as recorded in a small number of futures exchanges, principally COMEX of the CME Group and the Shanghai

Futures Exchange (SHFE). Information that affects the price of gold can be generated in many places, including mines, jewellery demand, central bank transactions, and macro-economic developments. The gold market ecosystem has been well established in the western world as they are ranging from spot trading to forward trading, gold leasing, and financing and other derivatives products. Currently, the global gold prices are set in spot fixing from London and futures trading from New York. These two western markets and the eastern markets, Mainland China and Hong Kong, are described below.

3.1　Western markets

3.1.1 London-based OTC market

London gold market development history can be traced back to over 200 years ago when London overtook Amsterdam to become the centre of the world's gold trading in 1804. London Bullion Market Association ("LBMA") — a wholesale OTC market for the trading of gold and silver — began operation in 1919, and the daily market prices were set by the five major bullion merchants and banks. The London price also influenced the gold prices in New York and Hong Kong. Currently, London is still the world's largest gold spot trading market, and global bullion trades still make reference to the London gold price as the benchmark.

In 2016, LBMA clearing members transferred and cleared around 157,828[3] tonnes of gold with a value of US$6.3 trillion. The activity at LBMA, largely among commercial banks, may lead to physical movements of gold or merely paper movements. Such 157,828 tonnes of gold transfers by LBMA clearing members in 2016 amount to around one-eighth of London market trading volume. The contribution of London volume to the world total has dropped from 90% to 65% in 2016 because of the rise of trading centres in China, Thailand and Singapore. Total world turnover for 2016 is estimated at 1,867,000 tonnes with a value of US$75 trillion. This is equivalent to 580 times mine production. Because of such ample liquidity, gold (like most currencies) trades at full carry.

3.1.2 Zurich gold market

The gold market of Zurich is not in any formal establishment as in London. Instead, the market is mainly supported by the three major Swiss banks providing liquidity and clearing

---

3　The transferred and cleared figure by the LBMA clearing members is obtained from the LBMA Website, and the figures in the rest of the paragraph are estimated by HKEX based on various of channels including gold news articles and consulting with major bullion players in the gold industry.

services in the OTC market. Switzerland is the world's largest gold transit hub as it hosts some of the world's most recognised gold refiners, such as PAMP and Metalor. In addition, it is also the world's largest private gold storage centre mainly due to its special legal framework which provides additional protection to gold owners.

3.1.3 CME Group's COMEX futures market

COMEX (currently under CME Group), formerly known as the Commodity Exchange Inc., was established in 1974 after the US abandoned the gold standard in favour of a flexible mechanism for the pricing of gold against the US dollar and as a result of increasing arbitrage or investment demand from the majority of legal entities in the US.

Prompted by the sharp volatility of the US dollar and other factors, the US gold futures market expanded rapidly in the years from 1978 to 1980. Today, COMEX is the world's largest gold futures trading centre by trading volumes, with tremendous influence on spot gold prices. The COMEX gold futures are monthly contracts with daily delivery mechanism during the delivery month. All of the delivery points are in New York City and the nearby State of Delaware. In 2016, COMEX had a total notional trading volume of 179,000 tonnes in gold futures[4].

3.2   Eastern markets

3.2.1 Mainland China gold market

In 1950, China prohibited private ownership of bullion and put the gold industry under state control. During the initial phase of economic reform from 1978 onwards, the gold market cautiously opened, mainly in the form of jewellery manufacturing in the Shenzhen Special Economic Zone. The central role of the People's Bank of China (PBOC) in the regulation, supervision and control of the purchase and distribution of gold and silver in China was confirmed by the 1983 Regulations on the Administration of Gold and Silver. The PBOC was also given the responsibility for managing the country's gold reserves. Following the 2001 abandonment by the PBOC of its controls on the purchase, allocation and pricing of gold, private sector demand for jewellery and, more recently, investment gold, expanded rapidly.

With the PBOC relinquishing direct control, prices were set on the Shanghai Gold Exchange (SGE), of which the PBOC was the key founder and key stakeholder. The SGE commenced trading on 20 October 2002. All refined gold was sold on the SGE, which was

---

4   Source: FIA.

the only market for the purchase of gold by industry and financial institutions. All imported bullion was made available through the SGE. Further liberalisation took place in 2003 when the licensing system for running businesses in gold and silver products was abolished, and in 2004 when private persons were allowed to own and trade bullion.

The Chinese gold market remains under indirect state control. While private trade in gold has largely been liberalised, the interaction between China and international markets remains restricted which is a key element of capital account controls. The SGE's launch of an International Board for gold in the Shanghai Free Trade Zone (FTZ) in 2014, which admits international trading participants, is a further cautious liberalisation measure. However, there has been little trading to date.

Key milestones in China's gold policy and market development are shown in Table 3 below.

Table 3. Milestones in China's gold policy and key market developments	
Year	Description
1950	• Gold industry under state control • Private holding of bullion prohibited
1983	• PBOC Regulations on administration of gold
1995	• Consumption tax on gold jewellery halved from 10% to 5%
1996	• New jewellery pricing structure — raw material cost separated from labour cost
1998	• PBOC Shenzhen Branch commenced gold imports from UBS, HSBC, Investco
2001	• China Gold Association established • Retail price control abolished by State Price Bureau
2002	• SGE started formal trading, trading exempt from value-added tax ("VAT")
2004	• Prohibition on gold bullion lifted
2007	• China became world's biggest gold producer
2008	• Foreign bank members admitted to SGE: HSBC, Scotia Mocatta, ANA, UBS, Standard Chartered • SHFE gold futures launched
2010	• ICBC launched Gold Accumulation Plan • Four more banks given licences to import gold
2011	• First foreign banks allowed to trade gold on SHFE: ANZ, HSBC
2012	• OTC interbank trading permitted, cleared through SGE

(continued)

Table 3. Milestones in China's gold policy and key market developments	
Year	Description
2013	• China world's biggest gold consumer • First China gold ETF launched in July • Foreign banks granted licences to import gold: ANZ, HSBC
2014	• Launch of SGE International Board in Shanghai FTZ
2016	• Launch of SGE Gold Fixing

Source: World Gold Council.

In 1978, China's gold production was less than 10 tonnes per annum. By 2016, production had increased to 453 tonnes. Measures taken to boost output included setting up the forerunner of China National Gold Group Corporation and the establishment of a special gold mining unit of the People's Liberation Army to prospect for gold and develop gold mines. Facilitative policies and investments were introduced in the 1981-85 and 1986-90 Five-Year Plans, bringing China's annual gold production to 100 tonnes in the early 1990s, with growth continuing strongly thereafter. China became the world's largest gold producer in 2007, and accounted for 14% of production in 2016.

Although gold supply from domestic production and recycling of gold has risen in recent years, it has been outstripped by domestic demand, with the result that China has swung from a gold surplus to a large gold deficit. Even though China does not publish figures on its gold imports, and China's gold imports have increased considerably since 2010, of which most was imported via Hong Kong[5].

China's gold trading is mainly concentrated on the SGE and the SHFE. The SGE is China's only legitimate physical spot gold trading venue, which connects gold production and consumption demands. The SHFE was launched in 2008 and is only the gold futures market place in Mainland China.

The SGE trades spot, and more recently spot-deferred contracts. Retail investors can trade gold by opening accounts with banks that are members of the exchange. Initially this was via a pilot scheme operated by ICBC, but the contract size of 1 kg was too large for retail investors. In July 2007, retail investors were allowed to start trading Au9999 and Au100g contracts via banks. According to SGE, by 2010, some 1.8 million retail investors accounted for 19% of the exchange's trading[6]. That year the SGE traded 5,715 tonnes, and

---

5   Source: Metals Focus.
6   Source: Shanghai Gold Exchange.

in 2016 it traded over 23,000 tonnes[7]. All gold contracts on the SGE can now be traded by retail investors. The contracts are settled by physical delivery, but most are traded for speculative purposes with positions closed before settlement.

In September 2014, the SGE opened an International Board in the Shanghai FTZ with RMB-denominated futures contracts. The contract traded a modest 78 tonnes of notional gold in the period from launch to the end of 2014, rising somewhat to 50 tonnes in the first two months of 2015. In April 2016, the SGE launched a RMB gold price fixing for the first time in history.

According to Futures Industry Association (FIA), the SHFE gold futures trading volume has been ranked second by number of contracts behind the COMEX gold globally since 2013, and the volume comparison in the past few years between the two exchange is shown in Figure 6 below.

Figure 6. Trading volume of COMEX and SHFE gold futures

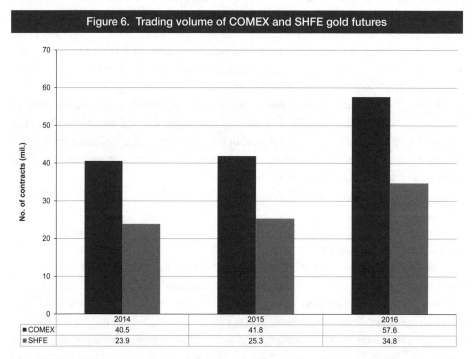

	2014	2015	2016
COMEX	40.5	41.8	57.6
SHFE	23.9	25.3	34.8

Source: FIA and SHFE.

---

7   Source: Shanghai Gold Exchange.

3.2.2 Hong Kong gold market

Hong Kong, with its proximity to the national gold fabrication centre of Shenzhen, is the main source of China's gold imports, recording 867[8] tonnes in 2016, accounting for about 86% of the China's total gold imports.

Hong Kong has been a significant centre for gold trading for over a century. As the Mainland China market opens up, Hong Kong also has a significant role as a centre for physical gold trading between China and the world.

The formation of the gold market in Hong Kong emerged upon the establishment of the Chinese Gold and Silver Exchange Society (CGSE) in 1910. Since the Hong Kong Government revoked its controls over gold imports and exports in 1974, the Hong Kong gold market has boomed. The Hong Kong gold market has now become an important gold investment and trading hub in Asia, connecting with the other time zones in Europe and the Americas. Given Hong Kong's significance in global gold trading, the five major gold merchants from London and the three major banks from Switzerland have set up trading desks and branches in Hong Kong. With the strong participation from the foreign major bullion players, gold pricing in the Hong Kong market has made reference to the London benchmark ever since.

4　Feasibility of building a Hong Kong benchmark

4.1　Is Hong Kong ready for physical delivery gold futures?

Although the Hong Kong gold market has over 100 years of history, it still lags behind other global gold trading centres such as New York and London in terms of benchmarking, liquidity, and completeness of offering in related products and services. However, as a global financial centre located at the gateway of China, the world's second largest economy and the biggest gold consumer, Hong Kong has the right ingredients to become the Asian gold pricing centre. Backed by the city's well established financial market, this

---

8　Source: Metal Focus.

special mission can be accomplished by HKEX and the major financial institutions as well as relevant regulators working together to meet the needs of the market, to ride on the international gold market trends, and gradually increase the Hong Kong gold market's pricing power outside the European and the American trading hours. With reference to the history development of the world's major commodity markets' pricing, the formation of a new pricing centre should meet two basic conditions as explained below.

(1) High fundamental market demand for trading and risk management

One of the key natural factors in forming a price benchmark is the need for large-scale physical trading activities. As shown in Figure 7, China imports about one-quarter of the world's supply, and there is about 70% of such imports go through Hong Kong.

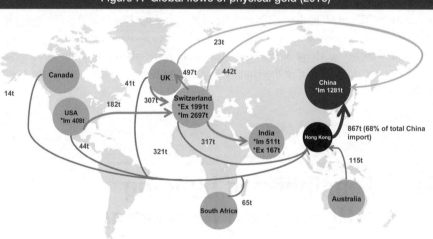

Figure 7. Global flows of physical gold (2016)

*Im: Imports;   *Ex: Exports
Note: Unit in tonnes (t).
Source: HKEX, analysis based on Metals Focus data.

With such high volume, it is fundamentally sound to develop related derivatives market to serve the risk management needs as these markets developed in New York and London previously.

(2) A sound marketplace with effective usage of spot and derivatives market

In order for Hong Kong to cultivate a new gold pricing benchmark, it is necessary to form a well-functioning marketplace to link up the spot market and the futures market and to provide efficient channels to serve these markets in Hong Kong, needless to say the offering of other gold-related financial products and services such as gold leasing and related derivatives as the London and New York markets provide currently.

4.2   The golden opportunity for Hong Kong

During the past century, Hong Kong's spot gold market has been very active, from refining, processing, inspection, to wholesale and retail, to trading and hedging. These constitute an impetus to Hong Kong's gold import and re-export trade. However, the Hong Kong gold market has been opaque and retroactive, in which the physical trading volume has been traditionally linked to London gold as the basis for pricing benchmark due to the dominance of the London benchmark globally.

Even though the SGE and SHFE have dominated activities in the Mainland Chinese gold onshore market, due to differences in the laws and regulations between the onshore market and the offshore market, and the capital and gold import/export control, and the Mainland gold benchmarks cannot be referenced commercially from Hong Kong. Therefore, the HKEX has a unique timeframe and geographical advantage to develop the right products to serve the market needs.

It is the time for Hong Kong to act to gain the pricing power in one way or another. One of the effective ways is the launch of the dual-currency (USD and RMB pricing and physical settlement) physical-delivery gold futures products by HKEX. With the launch of the HKEX gold futures market, the picture of spot and futures trading in gold will be completed in Hong Kong. The ecosystem built up can be extended to connect Mainland China with the developed western markets in both spot and derivatives trading in gold by the major bullion players via various channels as they see trading opportunities across the market. The result of which would be the formation of the new Hong Kong Asian benchmark as the liquidity grows and as world recognition is being established.

In addition, LME, the London entity of the HKEX Group, also launched gold futures (along with silver futures contracts) on London on 10 July 2017. This will provide round-the-clock, dual-location coverage for gold futures trading to meet the commercial needs of the HKEX Group clients. This can enhance London gold trading via "financialising" and "futurising" the spot trading activities into the LME liquidity pool.

As the gold futures (priced in both USD and RMB) market grows, more effective interactions between other related areas supporting the gold-trading ecosystem can be promoted. These areas include the interest rates, foreign exchange (FX) rates and gold-leasing markets. As the gold market ecosystem matures, along with an active gold futures market, this will offer fundamental support to the offshore RMB interest rate market, and ultimately RMB internationalisation. In the end, a new Asian gold benchmark can be formed naturally.

5 HKEX gold futures: product design and key technical points

5.1 New market landscape

HKEX acquired LME in 2012 and, as the vision drawn in its 2016-2018 Group Strategic Plan, aims to transform itself into a global vertically-integrated multi-asset class exchange. Commodities is one of the core pillars of HKEX's four-pronged multi-asset strategy. The launch of the HKEX Gold Contracts is a clear demonstration of HKEX's commitment to offer attractive commodity products in Asia.

As stated above, given Hong Kong's advantages as a free market and an entrepot trade centre, Hong Kong has an active physical gold trading market and enjoys the status as one of the major bullion markets in the world.

In addition, being the largest offshore RMB centre, Hong Kong has a unique position in facilitating the Mainland's RMB internationalisation. The introduction of the physically delivered gold futures in Hong Kong is a stepping stone to achieving this objective as physical gold can be used as a "backing" for a fiat currency via an interest rate of RMB and leasing interest rate of gold mechanism similar to the current USD and gold relationship.

5.2 Product design and key factors for consideration

In order to meet customers' needs, the following key factors were considered when the contracts were designed:

(a) Underlying and contract size — Gold not less than 0.9999 fineness and 1 kilogram is commonly traded by Asian customers, especially in Greater China; the physical delivery mechanism ensures market price converging to the real physical spot market to establish the new benchmark in Hong Kong, providing a robust risk management tool for the end users.

(b) Trading and settlement currency — The trading and settlement currency in USD and CNH will attract both USD- and CNH-based investors. The dual-currency Gold Contracts will generate an implied USD/CNH FX rate as the Gold Contracts have the same underlying. There will be arbitrage opportunities between this rate and other rates in the FX market, hence increasing the overall liquidity of USD/CNH while improving and smoothing out the forward curves among these markets.

(c) Contract months — The spot plus 11 following consecutive calendar months will cover the most liquid trading months in both international and Mainland futures markets and will provide the physical market with more hedging tools.

5.3   Product applications and users of HKEX gold futures

The value proposition of introducing the Gold Contracts for investors include but are not limited to the following:

(a)   To facilitate the Mainland and international investors' access to the gold market via a robust trading hub at HKEX in an Asian time zone;

(b)   To provide hedging and risk management options for investors and end users;

(c)   To provide more investment options for the growing offshore RMB deposits; and

(d)   To attract investors who prefer gold exposure in their portfolio.

The potential users and customers of the Gold Contracts are:

(a)   Physical players such as gold refiners, fabricators and jewellers who need to hedge gold price risk;

(b)   Financial players such as banks and funds who utilise the futures market to link with their gold-related investment products, and arbitrageurs who trade price disparity between onshore and offshore markets i.e. between New York, London, Shanghai and Hong Kong markets and deploy other trading strategies for FX and interest rate disparity; and

(c)   Other investors and traders who have an appetite for gold exposure.

Appendix 1

HKEX gold futures contract specifications[9]

Feature	USD gold futures	CNH gold futures
Underlying	1 kilogram gold of not less than 0.9999 fineness, bearing a serial number and identifying stamp of a Recognised Refiner	
Contract size	1 kilogram	
Trading currency	US dollars	RMB
Contract months	Spot month and the next eleven calendar months	
Minimum fluctuation / tick size	USD0.01 per gram	RMB 0.05 per gram
Trading hours (Hong Kong time)	8:30 am to 4:30 pm (day trading session), and 5:15 pm to 1:00 am the next morning (after-hours trading session)	
Last trading day	The third Monday of the contract month (postponed to the next business day if it is a Hong Kong public holiday)	
Final settlement day	The second Hong Kong business day after the last trading day	
Settlement type	Physical settlement	
Exchange fees[10]	Trading fee: USD1.00 per contract per side Settlement fee: USD2.00 per contract per side	Trading fee: RMB 6.00 per contract per side Settlement fee: RMB 12.00 per contract per side

---

9   All first-letter capitalized terms are defined in the trading and clearing rules amendments for the gold products, accessible via the following links:
    http://www.hkex.com.hk/eng/rulesreg/traderules/traderuleupdate-hkfe/Documents/49-17-HKFE-Star_e.pdf
    http://www.hkex.com.hk/eng/rulesreg/clearrules/clrruleupdate_hkcc/Documents/50-17-HKCC-Star_e.pdf
10 The exchange fees are subject to change by the Exchange from time to time.

Appendix 2

Trading and settlement arrangement and requirements[11]

1.   Trading and settlement arrangement

The trading and settlement arrangement is shown as in the chart below.

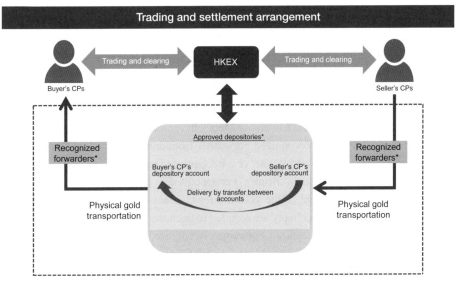

Note: *List of Approved Depositories, Recognised Forwarders will be published on the HKEX website and updated from time to time.

Source: HKEX.

---

11   All first-letter capitalized terms are defined in the trading and clearing rules amendments for the gold products, accessible via the following links:
http://www.hkex.com.hk/eng/rulesreg/traderules/traderuleupdate-hkfe/Documents/49-17-HKFE-Star_e.pdf
http://www.hkex.com.hk/eng/rulesreg/clearrules/clrruleupdate_hkcc/Documents/50-17-HKCC-Star_e.pdf

2. Trading and settlement requirements

An HKFE Clearing Corporation Limited ("HKCC") Participant who wants to take/make physical delivery needs to have: USD and/or CNH settlement account(s) with Settlement Banks; and opened accounts with each of the Approved Depositories or signed Delivery Agreement with another Clearing Participants with delivery capability.

Appendix 3

Chain of integrity[12]

In order to build and maintain a robust and sound mechanism for the quality of the gold bars delivered into the HKEX market, a chain of integrity which is vital in the bullion market globally is formed as below to protect the Exchange and its members.

HKEX requires all Deliverable Metals to be certified by a Recognized Refiner and must be accompanied by documentation issued by the HKCC Participant or its Recognized Forwarder, evidencing that the Deliverable Metals have been shipped or transported to an Approved Depository by a Recognized Forwarder from another Approved Depository, a Recognized Assayer, Recognized Refiner or a Recognized Depository.

Brink's Hong Kong Limited ("Brink's") is appointed as the first Approved Depository to support physical settlement of the gold futures contracts. All HKCC Participant who wants to take/make physical delivery are required to open storage accounts with Brink's.

The latest lists of the Recognized Forwarder, Recognized Assayer, Recognized Refiner and Recognized Depository will be published on HKEX Website.

---

12 All first-letter capitalized terms are defined in the trading and clearing rules amendments for the gold products, accessible via the following links:
http://www.hkex.com.hk/eng/rulesreg/traderules/traderuleupdate-hkfe/Documents/49-17-HKFE-Star_e.pdf
http://www.hkex.com.hk/eng/rulesreg/clearrules/clrruleupdate_hkcc/Documents/50-17-HKCC-Star_e.pdf

Chapter 12

The opportunities for Hong Kong to develop an iron ore derivatives market

13 November 2017

Summary

Hong Kong Exchanges and Clearing Limited (HKEX) launched USD-denominated cash-settled TSI Iron Ore Fines 62% Fe CFR China Futures on its subsidiary, the Hong Kong Futures Exchange (HKFE), on 13 November 2017. It is the first product in the ferrous suite offered by HKEX. Leveraged on HKFE's integrated electronic trading and clearing system, this product aims to enhance price transparency and improve price discovery efficiency in the iron ore derivatives market outside Mainland China.

Iron ore is the key raw material used for making steel and also the second-largest global commodity by trade value after crude oil[1]. China is the world's single largest importer and consumer of iron ore. Given China's high reliance on imported iron ore and its rapid economic growth, the China factor, including China's major economic policies and national development strategies such as the supply-side reform and the strategic initiatives of One Belt One Road (OBOR), would have significant implications on the potential demand and pricing of this commodity.

With rapid developments in recent years, the global iron ore derivatives market, including the Mainland market, continues to gain record turnover. Notwithstanding this, there is still potential for further development. The Mainland domestic market is very active but lacks overseas participation. As for overseas markets, the majority of liquidity resides in the over-the-counter (OTC) swap market which relies on voice broking for trade execution, while the on-exchange markets, which offer more efficient screen-based trading execution, have insufficient liquidity and market depth. Hong Kong sees opportunities in complementing the existing markets and improving the iron ore price discovery on screen through the introduction of iron ore futures on HKEX:

- HKEX's iron ore futures contract is an exchange-traded product with screen-based trading. Compared with the OTC market, it is expected to improve trading convenience, transparency and price discovery efficiency by building up liquidity on screen.

- The iron ore derivatives market has a relatively short development history and its growth potential is yet to be fully realised. The launch of iron ore futures on HKEX's screen-based derivatives market would help further expand the capacity of this market by attracting a broader investor base and enhancing market access through a convenient and transparent infrastructure.

---

1 Source: "The lore of ore", *The Economist*, 13 October 2012 (http://www.economist.com).

- Given the transformation of the pricing model of physical iron ore in the past decade, whether the current index-linked pricing model last or evolve again in the future is to be observed. A new path to this might emerge upon the establishment of a liquid and transparent futures market facilitated by HKEX's iron ore futures.

While the China factor in the iron ore market is significant, the accessibility between the iron ore derivatives market in the Mainland and overseas is yet to be enhanced. Given Hong Kong's strategic position as a global financial centre located at the gateway to China, building a transparent and liquid offshore iron ore futures market in Hong Kong would not only help satisfy the commodity price risk management needs for users of the physical commodity and the trading community, but also offer an attractive investment product of China relevance for institutional and retail investors.

1   China's significance in the iron ore market

1.1   The largest destination market of iron ore

Iron ore is the key raw material to make steel and the second-largest global commodity by trade value after crude oil[2].

Steel is heavily used in many downstream industries, such as real estate, transportation, car manufacturing, energy supply networks, machineries, ship building and home appliances. With the rapid development in the Chinese economy and the growing demand for steel over the past 20 years, China's crude steel production has seen 8-fold increase and reached 808 million tonnes in 2016, making up half of the world's total crude steel production[3]. Being the primary ingredient for steelmaking, iron ore's consumption in China has grown more than 20 folds in the past 20 years to 1.3 billion tonnes in 2016[4].

China is the world's largest iron ore importer in the world, importing 1,024 million tonnes in 2016, and constituting 70% of the world's seaborne trades[5]. With the domestic iron ore reserves being low in grade[6] and high in impurity, to fulfil China's huge demand on medium to high grade iron ore, it has to rely heavily on imports (about 84%[7]) from Australia, Brazil, South Africa, and India, etc.

---

2   Source: "The lore of ore", *The Economist*, 13 October 2012 (http://www.economist.com).
3   Source: Wind, data as of 2016.
4   Source: China Iron and Steel Association, *Bloomberg*, as of 2016.
5   Source: General Administration of Customs of People's Republic of China, as of 2016.
6   China domestic iron ore is around 30% in iron (Fe) content.
7   China has a high reliance on imported iron ore — in every 100 tonnes of iron ore it consumes, 84 tonnes are imported. Source: China Iron & Steel Association, as of 2015.

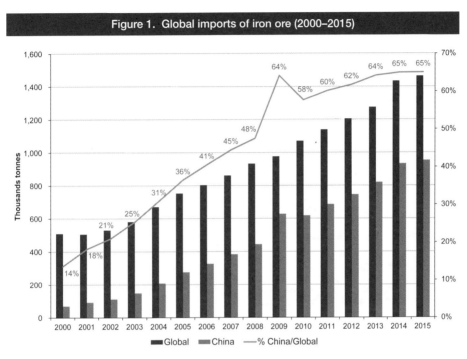

Figure 1. Global imports of iron ore (2000–2015)

Source: Wind.

Due to its heavy industrial reliance on this commodity, China is an active participant in both iron ore physical and derivatives trading. Many Chinese state-owned and private steel mills and trading companies have established a presence overseas. Many of them have trading and financing operations set up in Hong Kong, Singapore or other offshore tax harbours, and some owns overseas mine investment in iron ore reserve-rich regions such as Australia, West Africa, South America and North America.

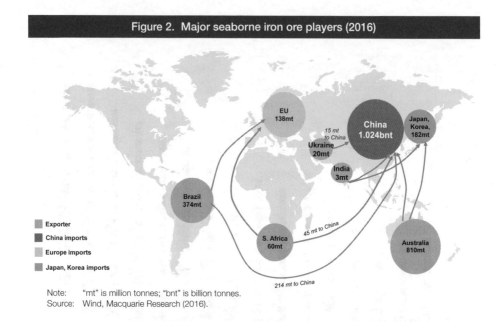

Figure 2. Major seaborne iron ore players (2016)

Note: "mt" is million tonnes; "bnt" is billion tonnes.
Source: Wind, Macquarie Research (2016).

1.2 China's strategic policies impacting the steel industry

With its economic growth adjusted to around 6.5% in 2017 (compared with the growth rate of 6.7% in 2016), China is firmly undertaking a structural reform with a target of eliminating excess capacity and upgrading production efficiency. There are two overarching policies that impact China's steel industry, in particular.

(1) One-Belt-One-Road ("OBOR")

OBOR consists of the Silk Road Economic Belt and the 21st Century Maritime Silk Road. The initiative includes exporting excess capital and capacity to promote trade and building infrastructure networks that connect China with Asia, Europe and Africa along the ancient trade routes. The policy will lead to over 60 bilateral cooperation deals that worth US$100 billion and encompass some 65 countries[8]. The initiative also brings historic opportunities to the steel industry to support exports and relieve overcapacities in China, as per target objectives laid down in the 2017 Government Report. (see Table 1.)

---

8 Source: National Development and Reform Commission (NDRC), as of 2016.

Table 1. 2017 Government Report — Steel industry-related achievements and objectives	
2016 Achievements	2017 Target objectives
• Cut overcapacity and excess inventory, deleverage, reduce costs, and strengthen weaknesses • Exceeded the annual target by successfully cutting overcapacity in the steel industry by 66 million tonnes and the production of coal (a raw material for producing steel) by 290 million tonnes • Achieved initial results in supply-side structural reforms • The OBOR initiative made significant progress and successfully launched a number of major international industrial cooperative projects, created synergy and strengthened ties with participating countries	• Continue to focus on supply-side structural reforms, reduce excess supply and increase vital supply in order to efficiently meet industry demand • Continue to cut overcapacity and excess inventory, deleverage, reduce costs, and strengthen weaknesses: further cut steel production by 50 million tonnes and lower coal production by 150 million tonnes • Strictly enforce environmental, consumption, health and safety regulations; promote corporate mergers and acquisitions; eliminate inferior and excess production capacity • Raise domestic consumption and enhance efficiency to create synergy between supply-side and demand-side reforms, so to unleash the country's full potential

(2) Supply-Side Reform

The 13th Five-Year Plan stated the focus of the steel industry, which is to be on the consolidation of steel mills, reduction of excess capacity, and production efficiency upgrades. The Opinions of the State Council on Reducing Overcapacity in the Iron and Steel Industry restricts new production capacity from being registered and enforces the environmental quality standards to be strictly in compliance with relevant rules and regulations[9]. As of 2016, the industry has cut down 85 million tonnes of capacity successfully, with targeted permanent capacity reduction of 100-150 million metric tonnes by 2020[10].

The OBOR initiative and the Supply-Side Reform target to resolve overcapacity, to stimulate demand and to improve profit margin for the steel-making sector. They also have profound implications on steel and steel-making raw materials prices (such as iron ore, coking coal and coke). Therefore, the importance of risk management by industrial players and other market participants against the price volatility of these commodities is ever-growing.

---

9   Source: Ministry of Industry and Information Technology.
10 Source: The State Council's "Government Work Report 2017".

2 Evolution of the physical market and development of the derivatives market

2.1 Transformation of the iron ore physical market

Since 1960s, iron ore physical trades followed a pattern of annual benchmark pricing, where the sale and purchase price was fixed once a year between global miners (representing the supply side) and leading steel makers (representing the demand side). The price fixed will then become the benchmark to be followed by the rest of the industry. This traditional pricing model was inflexible and lost sight of the changing spot market conditions during the year, leading to increasing defaults in contract performance once the market price deviated from the benchmark price.

The turning point was in 2010, when China rejected the price set by Vale, BHP Billiton and Japanese steel mills. This marked the end of the decades-old annual benchmark pricing and the industry moved on to a quarterly and eventually monthly index-based pricing model. Iron Ore is not the only commodity that underwent such transformation. In fact, similar evolution had happened to thermal coal (in early 2000s), aluminium (in early 1980s), and crude oil (in late 1970s), where the annual benchmark pricing model was abolished and replaced by shorter-term and more flexible pricing models.

For index-based pricing, the physical iron ore price is negotiated based on the monthly average of one or more market-recognised spot price indices, published by price reporting agencies like Platts, The Steel Index (TSI), Metals Bulletin (MB) or some Chinese index providers. This creates flexibility and ensures that the price aligns with the spot market and reflects current market supply and demand. Since then, the spot market in iron ore, which had been constrained by the inflexible annual benchmark pricing model, has started to grow.

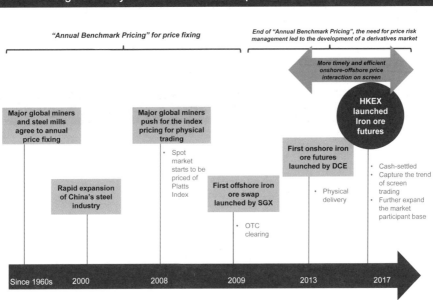

Figure 3. Key milestones in iron ore spot and derivatives markets

Note: DCE — Dalian Commodity Exchange; SGX — Singapore Exchange.

Source: Thomson Reuters, China Iron and Steel Association.

2.2   Emergence of the offshore iron ore derivatives market

Following the rise of indexation and the growth of the iron ore spot market, prices have been increasingly volatile and the need to manage price risks became essential for all users along the value chain — producers, consumers, shippers, traders and financers (banks). The burgeoning needs marked the beginning of the iron ore derivatives market. The first iron ore swap was launched in 2009 by the Singapore Exchange Limited (SGX).

Over the past 9 years, the global iron ore derivatives market (excluding Mainland China) has grown exponentially at an annual rate of 89%[11]. The annual trading volume and year-end open interest in cleared iron ore derivatives outside China reached approximately 1.42 billion tonnes and 72 million tonnes respectively in 2016[12]. To date, a handful of overseas exchanges including SGX, Chicago Mercantile Exchange (CME), Intercontinental

---

11 Source: SGX and CME websites.

12 Source: Futures Industry Association (FIA), as of 2016.

Exchange (ICE), LCH Clearnet[13] and Nasdaq Clearing[14] have offered trading and/or clearing services for the products in the offshore market outside Mainland China. Amongst the underlyings of the offshore derivatives, the TSI Iron Ore Fines 62% CFR China Price Index (TSI 62 Index) is the most commonly referenced benchmark index, representing spot physical iron ore price in USD delivered to North China[15].

Figure 4. Historical iron ore spot prices — TSI 62% Fe CFR China Price Index (USD/tonne)

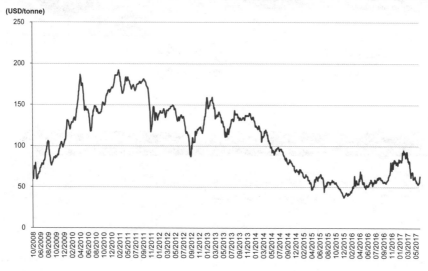

Source: TSI.

After development for a decade, the global iron ore derivatives market outside Mainland China has successfully attracted participation from steel mills, traders, producers and banks. However, as understood from market participants, the market is still predominantly over-the-counter (OTC) with trades matched through voice broking via inter-dealer brokers, with relatively thin liquidity and market depth on screen (transactions are only about 10% of the OTC volume on-exchange[16]). The OTC market has its merit for negotiating large trade or

---

13 LCH Clearnet is part of the LCH Group, which is a leading multi-asset class clearing house, serving a broad number of major exchanges and platforms as well as a range of OTC markets.
14 Nasdaq Clearing is a leading, European Market Infrastructure Regulation (EMIR)-authorised, clearing house providing central counterparty clearing for a broad range of markets and asset classes.
15 The TSI 62% CFR China index refers to the delivery price into North China excluding Qingdao port.
16 An informal estimation by clearing brokers for this research exercise.

tailor-made transactions. However, it is not an efficient and cost-effective way for trading vanilla products where there are naturally plenty of buyers and sellers. Lacking an active screen-based trading platform prevents a diverse investor base from participating in this market. This limits, to some extent, the growth of the iron ore derivatives market outside Mainland China.

2.3   Rapid development of the Mainland's iron ore derivatives market

Dalian Commodity Exchange (DCE) launched a RMB-denominated physically deliverable iron ore futures contract in October 2013, the first iron ore derivatives in Mainland China. Given China's huge risk management and speculative demands, DCE has quickly grown into the world's largest iron ore derivatives market, with annual trading volume and year-end open interest reaching 34 billion tonnes (more than 24 times the entire offshore market trading volume in the period) and 55 million tonnes respectively in 2016[17], surpassing all its counterparts globally.

Although the iron ore derivatives market on DCE continues to attract retail investors, financial institutions and physical users, it is still a domestic market and is not yet opened for direct access to international investors. There are also certain key challenges to be addressed, such as building liquidity across contract months and encourage a higher participation from industrial users.

The establishment and fast development of the onshore derivatives market, nevertheless, has facilitated the liquidity build-up in the offshore market and improved the overall price discovery mechanism for the iron ore market. After DCE introduced its own futures contracts, the trading volume of the iron ore futures on overseas market such as SGX doubled and trading was found to be the most active during the DCE trading hours[18]. According to market observation, cross-market price interaction also became more timely. This proves that a healthy development of the Mainland's onshore futures market adds significant value to the overall price discovery efficiency and also the growth of the offshore market.

---

17 Source: FIA.
18 Source: SGX website.

3 Opportunities for Hong Kong to develop an iron ore derivatives market

3.1 The offshore market calls for higher price transparency

While the overseas iron ore derivatives market was established in 2009, the Mainland market did not come into existence until late 2013. Yet the current size ratio of the Mainland's onshore market to the offshore market is about 24:1[19]. The exponential growth seen in the onshore market could be attributed to a number of reasons such as the abundance of investment money and increased speculative trading. Nevertheless, the screen-based trading model of the onshore market undoubtedly has contributed to the market's rapid growth by facilitating price transparency, thereby attracting diverse participants. In contrast, as understood from market participants, trading in the offshore market (mainly in iron ore swaps) still remains largely OTC via voice broking and the liquidity on screen is still thin. Will a more liquid and transparent offshore screen-based trading benefit the market? The answer is definitely yes.

Table 2. Comparison between exchange-traded derivatives products and OTC derivatives products		
Feature	Exchange-traded derivatives	OTC derivatives
Price transparency	High transparency: Transparent bid/ask spread	Low transparency: Obscure bid/ask spread
Trading efficiency	High efficiency: Centrally matched on electronic platform; ability to match a large amount of orders timely and fairly	Low efficiency: Bilaterally agreed via voice-brokers
Counterparty risk	Centrally cleared, instantly novated at the time of transaction to minimise counterparty risk	Minimised counterparty risk if centrally cleared; but there is often a time gap between the time when the trade is verbally confirmed and the time when the trade is actually novated

---

19 Reference is made to the trading volume data in tonnes of reported exchanges for 2016 in FIA statistics.

(continued)

Table 2. Comparison between exchange-traded derivatives products and OTC derivatives products		
Feature	Exchange-traded derivatives	OTC derivatives
Credit and collateral	Margin-based and cash collateral is accepted	Need to negotiate credit lines and collateral arrangements with banks
Documentation	Only account opening documents are needed	Bilateral documents such as International Swap and Derivatives Agreement (ISDA) are required

An OTC swap market has its merits too, which include (1) being bilateral and off-the-screen (i.e. without market impact) for big trade volumes, and (2) the flexibility to negotiate on bespoke structures. However, voice broking for execution in the OTC market is a very traditional way to discover price and conclude trades. There is no comparison in its speed, accuracy and efficiency with modern electronic trading platforms that offer centralised auto-matching and clearing. In addition, the market price lacks transparency as it is not widely accessible by market participants. This leads to higher cost of trading and asymmetric information. The advantages of screen-based trading (generally adopted for exchange-traded derivatives) are apparent as stated in Table 2 above. Given these comparative advantages, a rising trend in both volume and open interest in exchange-traded derivatives is observed in recent years.

Given the benefits and the increasing market preference for electronic trading, the launch of an exchange-traded iron ore futures contract on HKEX, with integrated price discovery, trading and clearing on one platform, is expected to bring higher transparency, a smoother price discovery process and lower transaction cost to the iron ore market.

3.2   The significant growth potential for iron ore derivatives market

In the past several years, the iron ore derivatives market has grown from scratch to a size of 36 billion metric tonnes globally (including Mainland China) in 2016[20]. Despite this rapid growth, the global derivatives-to-physical trading volume ratio for iron ore is only around 25 times in 2016, compared to the ratio of around 80-100 times for more mature commodities such as gold and copper[21]. The ratio for iron ore is even down to 1.25 times if only the USD-denominated iron ore derivatives volume outside Mainland China was

---

20 In terms of notional trading volume of reported exchanges in FIA statistics.
21 The ratio for a specific commodity is the global derivatives trading volume (including Mainland China) in tonnes on that commodity divided by the global trading volume (including Mainland China) of the physical commodity in tonnes. (Source: FIA, World Gold Council, *Bloomberg.*)

counted[22]. This is due to the shorter history of the iron ore derivatives market. Compared to the derivatives markets of base metals, energy and precious metals, which have a trading history of decades or over a century, iron ore derivatives are still in its early stage of development and the growth potential is yet to be fully realised. (See Figure 5.)

Given their hedging needs, the trading community of the physical commodity are often the first and the core participants of the commodity's derivatives. As the market evolves, more diverse participants including various types of financial institution, investment funds and retail investors will join the market for hedging or speculation. As the investor base becomes more diversified, the market will become more sophisticated, driving up the capacity and the liquidity of the market. This is exactly what is happening for iron ore derivatives now. The transparency and convenience offered by screen-based trading will greatly improve the price discovery efficiency and enables the market to expand.

No one could have correctly envisioned the growth trajectory of the iron ore derivatives market when it was first launched in 2009. Looking forward, it is certain that the iron ore derivatives market structure and dynamics, the composition of market participants and the product offerings will all evolve and improve over time.

Figure 5. Unrealised potential of the iron ore derivatives market

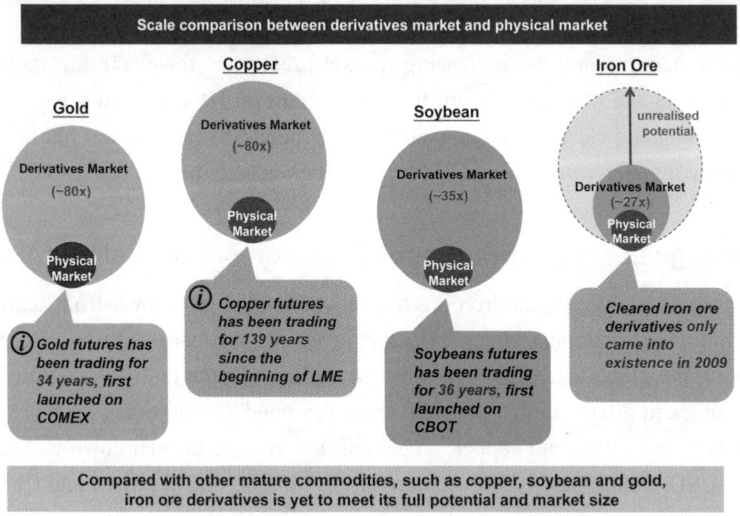

Note: CBOT — Chicago Board of Trade; COMEX — Commodity Exchange, Inc.; LME — The London Metal Exchange. Source of market volumes: FIA, as of 2016.

22 The Mainland iron ore derivatives market offers only futures contracts while the offshore derivatives markets offer swaps, futures and options contracts.

3.3   What is the future for iron ore pricing?

Tremendous changes have taken place in the physical iron ore market since 2010, when the "annual benchmark pricing" model ceased to function and was succeeded by index pricing. The index pricing model, along with the emergence of index-linked derivatives, have fundamentally changed the physical pricing and how market manages its price risk. However, as the market continues to evolve, will new pricing model emerge in the future that better suits the market needs?

Referencing to the trajectory of some commodities with a longer derivatives market history, such as soybeans, copper and crude oil, a "pricing" mechanism is widely adopted in their physical spot trading. This is a spot pricing mechanism to determine a spot commodity price between a buyer and a seller by agreeing upon a basis (premium or discount) over a futures market price for a particular month. So when the spot trade is negotiated, the terms agreed is not a fixed sales price, but an agreed basis (the price difference between the physical spot price and the futures price) plus a futures price. The buyer will have the right to fix the price based on the prices of futures contracts traded on a specific commodity futures market during the agreed Quotational Period ("QP"). This pricing model is considered an effective and market-reflective price determination mechanism. It minimizes the default risk as the buyer has some flexibility in determining the timing of "pricing" and therefore the price. Moreover, by fixing the physical price with reference to a futures market, it facilitates seamless hedging as it eliminates the basis risk between physical and futures market at the time of pricing.

One of the core functions of the futures market is price discovery. By using the futures price to form the basis of physical spot trade, the pricing model based on futures prices is an ultimate demonstration of the price discovery role of the futures market.

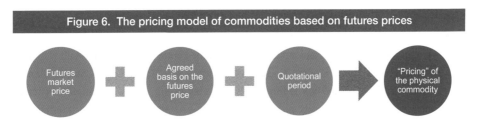

Figure 6. The pricing model of commodities based on futures prices

For iron ore, an interesting trial application of the pricing model was seen in November 2016. At that time, there was a physical deal of 10,000 metric tonnes transacted on Beijing Iron Ore Trading Center Corporation (COREX) which was priced on the iron ore futures price on DCE with a basis.

Usually, for this pricing mechanism to work, certain key conditions have to be fulfilled:

(1) Having a fully functional **liquid and transparent futures market** is the key. The futures price at any point of time must be visible and accessible to all market participants so that the pricing can be done at any given time during the day.

(2) The buyers and sellers in the market need to **trust and recognise the price** of the futures market as indicative and representative of the underlying physical market.

Change will take time. The ferrous market will itself determine what the best for the industry is and whether to evolve out a new pricing model or craft something of its own. This is worth paying close attention to.

3.4　The strategic positioning of Hong Kong and HKEX

As a global financial centre located at the gateway of China, which is the world's second largest economy and the biggest iron ore importer and dominant consumer, Hong Kong has been acting as a "super-connector" between Mainland China and the rest of the world. It is well positioned to build up an iron ore derivatives market to better serve the risk management needs from Chinese enterprises, regional commodity trading firms as well as their business partners across the globe.

HKEX, after its acquisition of the London Metal Exchange (LME) in 2012, has also geared up to better serve the real economy in commodities. The launch of the HKEX Iron Ore Futures contracts on 13 November 2017 is considered a constructive move.

(1) Hong Kong's strategic positioning: The key gateway to China and the "Super-Connector" between China and the world

Hong Kong, strategically positioned as the gateway of China, is a "super-connector" between China and the world. Hong Kong has achieved many breakthroughs and innovations by connecting the two in respect of the financial market. These include the launch of the Stock Connect scheme[23] and various initiatives that have driven itself to become the world's largest offshore RMB centre.

As a major shipping centre and an international trade centre in Asia, around half (US$454 billion in 2016) of Hong Kong's trade was re-exports[24]. Around 20% of Mainland's international trade is routed via Hong Kong. Busy shipping routes and highly efficient ports and logistics make this city prosper. Hong Kong is also a world-class international

---

23　Stock Connect is a pilot programme established upon the Mutual Market Access model. For the first time ever, investors in the Mainland China and Hong Kong markets are able to have direct access to each other's stock markets.

24　Source: Hong Kong Trade Development Council (HKTDC) Research website. Subsequent quotations of Hong Kong's economic data in this paragraph are also from this source.

financial centre (IFC) offering a comprehensive range of financial products and services and a business centre with a large number of international and Mainland enterprises setting up offices in it. Thanks to its free market economy and rule of law, Hong Kong establishes itself as the single largest destination market for Mainland outbound investments, providing a one-stop solution with a wide range of financial services. It also accumulates the world's largest offshore RMB liquidity pool. On the other hand, Hong Kong is the market-of-choice for international investors to access Mainland China. The HKEX securities market ranked first among global market in terms of funds raised by initial public offerings in 2015 and 2016. More than half of the companies listed on HKEX are Mainland enterprises[25] and there are over 150 listed companies in the natural resources related sector as of 31 December 2016[26].

Hong Kong provides the necessary financial infrastructure to serve the business needs of enterprises in all dimensions of import/export activities, fund raising, trade finance, asset management and financial risk management. Among these, financial risk management, in particular the management of asset price risk, is especially important for enterprises engaging in commodities trading. Commodities price risk management is therefore considered a key area of development in the Hong Kong financial market. On one hand, such value-added service would be essential for the commodities trading community to prosper in Hong Kong. On the other hand, it would facilitate China in gaining international pricing power in commodities.

(2) Strategic initiatives like OBOR has given Hong Kong a historic opportunity to facilitate Chinese enterprises to expand their market and services abroad

As mentioned in Section 1.2 above, China's policy initiative OBOR brings historic opportunities to the steel industry to support exports and relieve overcapacities in China. This stimulates risk management and investment needs in the iron ore industry as Chinese enterprises in the industry to expand their market and services abroad. As a result, Hong Kong is provided with a historic opportunity to contribute by leveraging on its strengths in finance, trade and logistics, and its wealth of professionals in a wide range of services. Hong Kong's expertise and unique position as the gateway of China will allow it to continue to play an important role in raising capital, managing risks, leading projects and exporting professional services. Hong Kong is therefore well positioned to develop an offshore iron ore derivatives market for Chinese enterprises to manage their offshore risks.

---

25 As of 31 December 2016, out of a total of 1,973 companies listed on HKEX Main Board and GEM, 1,002 were Mainland enterprises. (Source: HKEX)
26 Source: HKEX.

(3) HKEX, as the financial market operator in Hong Kong, shall enhance Hong Kong's status as an IFC by building a steady and liquid commodity derivatives market to better serve the asset price risk management needs of Mainland, local and international enterprises

Given the unique position of the Hong Kong financial market in connecting Mainland China and the world and the historic opportunities offered by China's strategic development initiatives, HKEX has the comparative advantages, as the financial market infrastructure operator in Hong Kong, to excel in its role to better serve the needs in the commodity derivatives market for China and the world. These advantages include:

- Running one of the most robust securities and derivatives trading, clearing and settlement systems in the world;
- Offering a wide range of products and services covering the segments of equities, equity derivatives, fixed-income and currency (FIC) products and commodity products;
- Operating in a sound regulatory regime, with market rules and regulations aligned with the highest international standards and with emphasis put on investor protection.

In recent years, HKEX has spearheaded a series of innovations and strategic moves to offer support to the gradual opening up of the Mainland financial market. In the equities segment, it launched the Stock Connect scheme jointly with the Shanghai Stock Exchange in October 2014 and the Shenzhen Stock Exchange in December 2016, which essentially led to the establishment of a cross-border mutual market[27]. In the FIC segment, it launched the world's first exchange-traded deliverable RMB currency futures product in September 2012 — the US dollar / offshore RMB (USD/CNH) currency futures, and Bond Connect Northbound trading[28] in July 2017. In the commodities segment, it acquired in 2012 the world's largest base metals marketplace, LME, and launched its first physically delivered dual-currency gold futures on its own Hong Kong platform in July 2017. Further product offering and expansion are being made into the ferrous product line, with the launch of iron ore futures as the first product. This new product aims to serve the needs of the region's physical trading community and financial institutions for the commodity's price risk management. (See Appendix 1 on the product's key features and Appendix 2 on the product's contract specifications.)

---

27 See Chapter 2 of this book, "Shanghai and Shenzhen Stock Connect — A 'Mutual Market' for Mainland and global investors".

28 Bond Connect is a mutual market access programme that enables overseas investors to trade bonds on the China Interbank Bond Market in the Mainland (Northbound trading), and Mainland investors to trade bonds in the Hong Kong market (Southbound trading) through the connectivity links established between the institutional financial infrastructure in the Mainland and Hong Kong. The initial launch is confined to Northbound trading.

Appendix 1

Key features of HKEX's iron ore futures

(1) Exchange-traded futures

The Iron Ore futures contract provides a high level of price transparency and an efficient price discovery process.

(2) Quarterly contracts

The product makes available the first quarterly contracts in the global on-exchange market. This offers market participants a more transparent and convenient platform than OTC swaps to execute order and hedge their positions in quarterly trades, and to facilitate price discovery in forward price curve.

(3) Day trading session and after-hours trading (AHT) session

Trading hours span, from 9:00 am to 1:00 am the next morning (Day trading session: 9:00 am to 4:30 pm; AHT session: 5:15 pm to 1:00 am the next morning). This would serve market participants from across the globe, covering business hours of Mainland China and major offshore markets.

(4) Block trades

Block trade facility is available to facilitate the ease of reporting OTC volume, for them to be cleared on the exchange, reducing counterparty risks.

(5) Tracks the most recognised derivatives benchmark as underlying

The contract settles against the TSI Iron Ore Fines 62% Fe CFR China Index, which is widely referenced in the physical trade of iron ore and is also chosen as the settlement price for the majority of USD-denominated iron ore derivatives contracts.

Appendix 2

Contract specifications of HKEX iron ore futures

TSI Iron Ore Fines 62% Fe CFR China Futures		
Feature	Monthly contracts	Quarterly contracts
Trading code	FEM	FEQ
Contract size	100 tonnes	
Minimum fluctuation	US$0.01 per tonne	
Underlying index	TSI Iron Ore Fines 62% Fe CFR China Index (TSI 62 Index)[29]	
Settlement method	Cash settled	
Contract months	Spot month and the next 23 calendar months	Spot quarter and the next seven calendar quarters (i.e. calendar quarters are January to March, April to June, July to September and October to December)
Trading hours[30] (Hong Kong time)	9:00 a.m. to 4:30 p.m. (day trading session) and 5:15 p.m. to 1:00 a.m. (after-hours trading session)	
Trading hours on last trading day[31] (Hong Kong time)	9:00 a.m. — 4:30 p.m. (day trading session) and 5:15 p.m. — 6:30 p.m. (after-hours trading session)	
Last trading day (LTD)	The last Hong Kong business day of a calendar month that is not a Singapore public holiday	The LTD of the last monthly contract in the calendar quarter
Final settlement price (FSP)	Arithmetic average of all TSI Iron Ore Fines 62% Fe CFR China Index values published in that contract month, rounded to 2 decimal places	Arithmetic average of the FSP of the three corresponding Monthly Contracts in that contract quarter, rounded to 2 decimal places
Final settlement day	The second Hong Kong business day after the LTD[32]	
Exchange fee	Trading fee: US$1.00 per contract per side; [33, 34] Settlement fee: US$1.00 per contract per side[35]	
Levies	USD 0.07 per contract per side[36, 37]	

(continued)

TSI Iron Ore Fines 62% Fe CFR China Futures		
Feature	Monthly contracts	Quarterly contracts
Block trade threshold	Minimum 50 lots	
Holiday schedule	Follow HKFE holiday schedule	

---

29 According to an announcement from Platts on 6 July 2017, TSI 62 Index will merge with the Platts IODEX index starting from 2 January 2018. For details, please refer to Platts Subscriber Notes and Methodology and Specifications Guide

30 There is no trading after 12:30 p.m. on the eves of Christmas, New Year and Lunar New Year. The trading hours on those three days shall be 9:00 a.m. — 12:30 p.m.

31 There is no trading after 12:30 p.m. on the Last Trading Day that is the last Hong Kong Business Day before New Year's Day or the Lunar New Year, and which is also the last day before New Year's Day or the Lunar New Year on which the TSI Iron Ore Fines 62% Fe CFR China Index is published. The trading hours on those two days shall be 9:00 a.m. — 12: 30 p.m.

32 Final Settlement Day shall be the first Hong Kong Business Day after the Last Trading Day if (i) the Last Trading Day is on the last Hong Kong Business Day before New Year's Day or the Lunar New Year, (ii) the Trading Hours of the Spot Month Contract and the Spot Quarter Contract end at 12:30 pm., and (iii) the day trading session of other Contract Months ends at 4:30 pm. For further details please refer to the related Regulations and Contract Specifications on the HKEX website.

33 The amount indicated above is subject to change by the Exchange from time to time.

34 Waived from 13 November 2017 to 11 May 2018, both dates inclusive, excluding the After-Hours Futures Trading Session on 11 May 2018.

35 The amount indicated above is subject to change by the Exchange from time to time.

36 The current rate is set at HK$ 0.54 per contract, for which the USD equivalent will be determined by the Exchange from time to time.

37 Waived from 13 November 2017 to 11 May 2018, both dates inclusive, excluding the After-Hours Futures Trading Session on 11 May 2018.

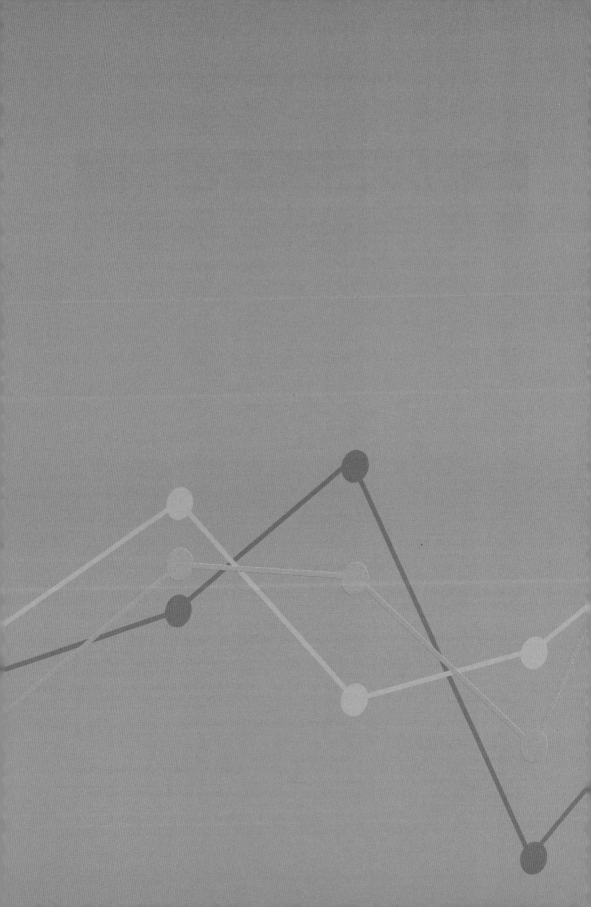

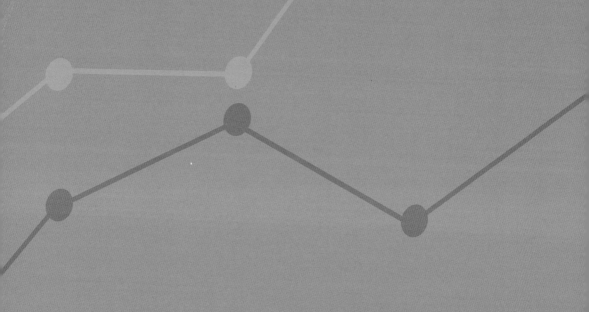

Part 4

Offshore RMB Products Centre

Chapter 13

The liquidity provision mechanism for offshore RMB market

Current status, impact and possible improvements

4 January 2017

Summary

The liquidity provision mechanism for offshore RMB market can be basically stratified into two layers: Long-term liquidity comes mainly from the onshore market through real economic activities (via cross-border trade settlement); while short-term liquidity is mainly provided under the scheme of currency swap agreement among monetary authorities and market financing.

Current status and structural issues of the offshore RMB liquidity provision mechanism: The provision of long-term liquidity for the offshore RMB market largely relies on the channel of cross-border trade settlement which is easily susceptible to the fluctuation of the RMB exchange rate. The short-term liquidity provision lags behind the rapid market development in terms of efficiency, scale and operation time. Also, there is still room for improvement in reducing the imbalance of offshore RMB capital allocation.

Sharp volatility in the short-term interest rate of offshore RMB market would most likely pressurize on the steady expansion of offshore bond market, increase the hedging costs of offshore entities holding RMB assets and may even trigger short-term speculative cross-border capital flows.

Given the inclusion of RMB into the Special Drawing Rights (SDR) basket of the International Monetary Fund (IMF), the demand for RMB assets continues to grow. The sufficient offshore RMB liquidity is essential to enhance market depth and underpin the growth of cross-border trade, offshore RMB investment and financing, foreign exchange transactions and other economic activities. Continuous widening of two-way cross-border capital flows and the improvement of existing market mechanism will effectively pave the way for RMB as a fully convertible international currency and to be widely used in global investment and foreign reserves.

1 Two layers of offshore RMB liquidity provision mechanism

Given the unique development path of offshore RMB market, the liquidity provision for offshore RMB market can be basically classified into long-term and short-term levels.

Long-term liquidity comes mainly from the onshore market through real economic activities (via cross-border trade settlement). Since the launch of cross-border trade settlement in July 2009, the payments settled in RMB from onshore to offshore market have always exceeded the receipts, even the ratio of RMB payment over receipt has been slowly declining from 1:5 in Q1 of 2011 to 1:0.96 at the end of 2015, it was still a net RMB capital outflow. Consequently, cross-border trade settlement has become the main channel to export RMB liquidity to the offshore market. At the end of 2014, the total offshore RMB liquidity pool reached a historical high of RMB 1.6 trillion, including RMB 1.15 trillion (offshore RMB deposits and certificates of deposit) in Hong Kong, RMB 302.2 billion deposits in Taiwan, more than RMB 200 billion deposits in Singapore, most of which are channelled through cross-border trade settlement.

Within the trade-settlement process, Mainland importers and exporters facilitated the expansion of the offshore RMB liquidity pool by settling trades in RMB or USD depending on the relative exchange rate of RMB across markets. The arbitrage mechanism is that when the RMB's exchange rate was expected to appreciate, the RMB exchange rate against USD in the Hong Kong offshore market (CNH) will be much appreciated relative to the onshore RMB exchange rate against USD (CNY). The premium of CNH means that RMB is more valuable in the offshore market, offering an opportunity for enterprises to earn extra benefit if settling imports in RMB. Therefore, the market has the incentive to pay RMB instead of USD for importing, thereby causing an outflow of RMB liquidity from onshore to offshore market[1].

---

[1] See "A Review of RMB Internationalisation from Exchange Rate Perspective" (《從當前的人民幣匯率波動看人民幣國際化》 *International Economic Review*《國際經濟評論》), Issue 1, 2012.

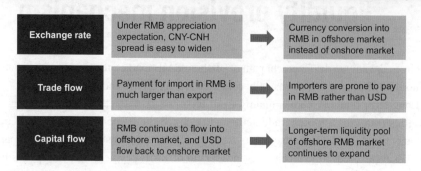

Figure 1. Expansion of offshore RMB liquidity pool under cross-border arbitrage — settle cross-border trade in different locations based on the deviations between CNH and CNY

The short-term RMB liquidity is mainly provided under the scheme of currency swap agreements between monetary authorities and market financing. From the official channels, short-term liquidity provision includes one-day and one-week liquidity arrangements (T+1 settlement in both cases) and overnight liquidity arrangement (T+0 settlement) provided by the Hong Kong Monetary Authority (HKMA). In 2014, the HKMA offered an additional short-term RMB funding up to RMB 10 billion per day to welcome the launch of the Shanghai-Hong Kong Stock Connect Programme, and also designated a number of banks active in the CNH market as Primary Liquidity Providers (PLPs), offering a repurchase (repo) line of RMB 2 billion to each of the PLPs so as to facilitate more market-making and other business activities in the CNH market[2].

---

2   On 27 October 2016, the HKMA announced to expand the scheme from seven PLPs to nine PLPs with the total PLP facility increased from RMB 14 billion to RMB 18 billion.

Table 1. Short-term liquidity provision mechanism via the official channels (as of October 2016)		
Term structure	Interest rate	Scale and funding source
One-week liquidity arrangement (T+1 settlement)	By reference to prevailing market interest rates	Funding from currency swap agreements between monetary authorities
One-day liquidity arrangement (T+1 settlement)	By reference to prevailing market interest rates	Funding from currency swap agreements between monetary authorities
Overnight funding (T+0 settlement)	Average of the three most recent Treasury Markets Association (TMA) overnight CNH HIBOR fixings (inclusive of the fixing on the same day of the repo), plus 50 bps, subject to a minimum at 0.50%	Estimated not to exceed RMB 10 billion in the inception of this arrangement
Intraday funding (T+0 settlement)	Average of the three most recent TMA overnight CNH HIBOR fixings (inclusive of the fixing on the same day of the repo), subject to a minimum of 0%, and to be charged based on the actual time used during the day on a per-minute basis	Not to exceed RMB 10 billion
Primary Liquidity Providers (PLPs)	By reference to prevailing market interest rates	RMB 18 billion in total

Source: HKMA.

In addition, the shortfall of short-term RMB liquidity in the offshore market can also be substituted via foreign exchange (FX) swap market. FX swaps refer to the purchasing of a spot FX contract and simultaneously selling a forward FX contract of the same currency, or selling a spot FX contract and simultaneously purchasing a forward FX contract of the same currency. Currently, the offshore FX swap is available to cover the tenor from intraday, overnight to one-year funding.

Apart from the above, participants in the offshore market can obtain short-term liquidity via the offshore interbank lending market and the repo facilities offered by the offshore clearing bank, which together with the funding from official channels constitute a comprehensive mechanism to provide short-term RMB liquidity to the offshore market.

2 Current status and structural issues

2.1 The long-term liquidity provision has shrank significantly (mainly relying on cross-border trade settlement) since the exchange rate reform on 11 August 2015

As mentioned above, cross-border RMB settlement is the main channel to obtain long-term RMB liquidity. However, it also causes that the flow and scale of offshore long-term liquidity is so susceptible to the fluctuations of the RMB exchange rate. Since the exchange-rate reform on 11 August 2015[3], CNH rate in the offshore market shows more depreciation than CNY rate, leading to a reversal of the arbitrage mechanism and the flowing back of RMB capital to the onshore market. Specifically, when CNH depreciation exceeds that of CNY to a certain basis point, it implies that RMB is valued more onshore, stimulating traders to buy RMB in the offshore market and sell RMB on the onshore market, and then to repatriate RMB capital to the onshore market through cross-border trade channel so as to get the extra benefit from the exchange rate arbitrage. Meanwhile, offshore investors became less confident to hold RMB assets under the RMB depreciation expectation, thereby converting their RMB deposits back into USD or HKD assets. These effects combined led to a drop of RMB deposits in the Hong Kong market from the peak of RMB 1,003.5 billion to RMB 652.9 billion[4], a decline of about 23% compared to the end of 2015.

2.2 The current mechanism for short-term liquidity provision lags behind market development in terms of efficiency, scale and operation time

First, the intra-day RMB liquidity is quite limited in the offshore market. By contrast, offshore transactions have significantly increased, reaching an average daily turnover of RMB 770 billion which, at times, even exceeds that of the HKD settlement volume (see Figures 2 and 3 below). According to the Triennial Survey from Bank for International Settlements (BIS) in 2016, the average daily turnover of over-the-counter (OTC)

---

3   On 11 August 2015, the Mainland Central Bank started to reform its daily USD/CNY mid-price fixing mechanism, which is generally regarded as an important step in RMB exchange rate liberalisation.
4   Source: HKMA, as of end-August 2016.

transactions in the offshore RMB market, including spot, forward and FX swap, totalled US$202 billion, implying a strong market demand for intraday short-term liquidity.

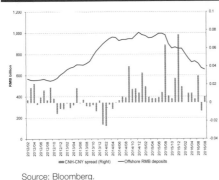

Figure 2. Hong Kong's offshore RMB liquidity pool shrunk since exchange rate reform on 11 August 2015

Source: Bloomberg.

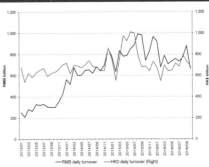

Figure 3. RMB's average daily turnover exceeded that of HKD during some periods after 2015 2H

Source: Hong Kong Interbank Clearing Ltd.

Second, a considerable portion of RMB short-term funding is provided under the official scheme of currency swap agreement with the Mainland central bank (the PBOC), however, the usage of which is subject to the operating hours of the Mainland's interbank market and its clearing systems as well. As a result, the offshore market is most likely to face short-term funding pressures during Mainland long holidays when the onshore clearing systems and funding facilities are suspended.

Third, the offshore RMB swap market, a major funding source for offshore short-term RMB liquidity, easily becomes volatile under global financial shocks, especially under the upcoming US dollar rate hike cycle.

The fourth issue is a lack of effective cross-border funding channel between onshore and offshore money markets. As mentioned above, the existing cross-border RMB liquidity provided to the offshore market is mainly accomplished under the current account and the mid- to long-term capital accounts, including cross-border RMB trade settlement, foreign direct investments (FDI), three types of eligible institutions in the Mainland's interbank bond market and RMB Qualified Foreign Institutional Investors (RQFII), etc., while the onshore money market is still largely closed to non-residents as the RMB is not yet fully convertible[5]. There is, so far, no direct channel linking onshore and offshore markets for

5   For details, see the report by Survey and Statistics Department of People's Bank of China in 2012 —— "China is Basically Ready for Further Opening Up Capital Account" (《中國人民銀行調查統計司課題組報告（2012）》——〈我國加快資本賬戶開放的條件基本成熟〉).

short-term liquidity flows, in particular, for overnight and one-week short-term liquidity, except the currency swap arrangement.

2.3 The offshore RMB funding allocation is concentrated in domestic long-term assets, indicating a certain degree of allocation imbalance

Currently, offshore RMB funds are mainly allocated in: RQFII with RMB 270 billion quota; about RMB 500 billion in Dim Sum Bonds; and RMB 281.6 billion in cross-border RMB loans[6]. Although the RMB loans can facilitate the expansion of the offshore RMB liquidity pool due to the multiplier effect, the above allocations have mostly used up offshore RMB capital. The allocation of offshore RMB funding is largely concentrated on long-term assets in the domestic market, which are neither actively traded nor convenient for repurchasing, making it difficult for offshore financial institutions to adjust their RMB portfolios once a sharp demand for short-term RMB money arises in the offshore market, and thereby resulting in short-term liquidity strains.

Figure 4. Offshore RMB funds are mainly allocated in the segments of Dim Sum Bonds and offshore RMB loans

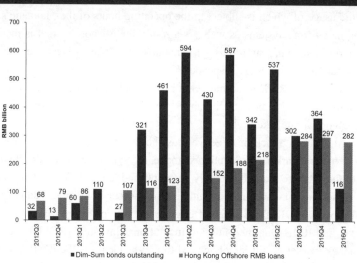

Source: Bloomberg.

---

6   Source: HKMA, as of end of March 2016.

The allocation imbalance of offshore RMB capital might be augmented owing to current contraction of overall offshore RMB liquidity pool, particularly before Mainland holidays (such as, Mid-Autumn Festival and the National Holiday, etc.) when the seasonal factor leads one-way capital flow back to onshore market ahead of the closing of the Mainland interbank market during holidays. In this context, the short-term interest rate in the offshore RMB market has experienced a sharp volatility at the end of September and early October of 2016.

3 Impact of short-term interest rates volatility on the offshore RMB market

3.1 Imposing pressures on the steady expansion of the offshore bond market

Hong Kong has been the world's largest offshore RMB bond market and, also, the major market for overseas institutions investing in RMB bonds. However, since the exchange rate reform on 11 August 2015, the contraction of overall offshore liquidity pool pushed up the offshore RMB financing costs. One-year deposit rates rose to an average of above 4% and the financing costs for three-year Dim Sum Bonds notably jumped almost 200 basis points, while the onshore market adopted a more loosening monetary policy with ample liquidity available. The gradual widening of the interest rate spreads between the onshore and offshore bond markets urged many Dim Sum Bond issuers to move back to the onshore market, leading to a significant drop in offshore bond issuance. For example, nearly 60 percent of property enterprises returned onshore to issue bonds in 2015, most of which have been active in the offshore bond market, causing a fall of bond issuance in the real estate sector from USD24.8 billion in 2014 to US$9.6 billion in the offshore market[7].

---

7 See "Recent Developments of Offshore RMB Bond Market: Attributes and Trends" (〈近期離岸人民幣債券市場的發展態勢、原因及趨勢〉), *China Money* (《中國貨幣市場》), Issue 1, 2016.

3.2 Increasing hedging difficulty for offshore entities in holding RMB assets

Given the inclusion of RMB into the SDR basket, the demand for RMB assets continues to grow when central banks and global investors consider to reallocate into RMB-denominated assets. SDR inclusion itself will not directly spur significant investment need, as SDR assets only accounts for 2.4% of international reserves. However, the achievement of SDR status increases the global acceptance of RMB as a global investment and reserve currency. If the holding share of RMB assets by global institutions or individuals could reach to a level comparable to that of the Japanese yen in global FX reserve assets, it could be expected that over RMB 2 trillion would flow into relevant RMB assets.

Increasing investments result in a surging demand in risk management. Sharp volatility in offshore short-term RMB liquidity will weaken the ability of offshore entities in developing effective interest rate benchmarks for offshore RMB floating loans and appropriate pricing models for RMB assets, or introducing RMB risk management products, thereby imposing hedging difficulties on international investors. It is necessary to develop more instruments to eliminate the effect of interest rate volatility and direct market entities to adjust their FX trading strategies in order to facilitate more foreign capital to participate in the offshore market.

Figure 5. Rebound in foreign holdings of onshore RMB assets by institutions and individuals

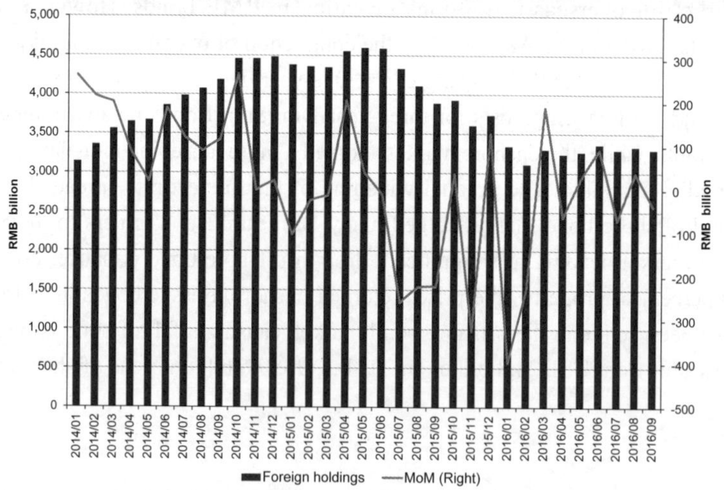

Source: Wind.

3.3   The widening interest rate spreads between onshore and off-shore markets may trigger short-term speculative cross-border capital flows

With successive cuts of onshore interest rates and the reserve requirement ratio, the decline in onshore RMB yield drives domestic RMB capital to look for more profitable opportunities. The widening of interest rate spreads between the onshore and offshore markets may trigger interest rate arbitrage, which has not been obvious in cross-border capital flow at present, and spur potential capital outflow via unofficial channels, aggravating liquidity pressures domestically.

Figure 6. Continuing net payments in RMB from onshore market may aggravate capital outflow pressure

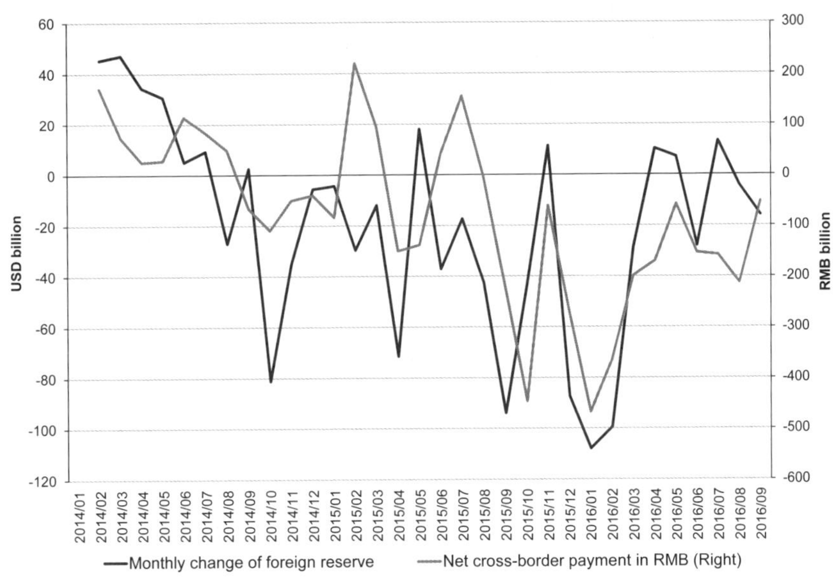

Source: Wind.

4   Possible improvements

Given the increasing RMB transaction and investment activities in the offshore market in terms of RQFII, Shanghai Connect and Shenzhen Connect, both financial transactions and long-term financing need adequate RMB liquidity as a fundamental support. Sufficient liquidity is critical in enhancing market depth and developing of cross-border trades, investments, FX transactions and other economic activities.

It is noteworthy that although international investors can now more directly participate in domestic RMB businesses and RMB transactions have gradually shifted to the onshore market, the development experience of the US dollar and other international currencies demonstrates that currency internationalisation should be accompanied hand in hand by the development of an offshore market, which plays a key role in facilitating the RMB's circulation outside the domestic market and it being widely used in the global economy.

To further improve the offshore RMB liquidity conditions, the following potential solutions may be considered.

4.1   Steadily promoting RMB internationalisation and further opening up two-way channels for cross-border capital flows

As mentioned above, the RMB exchange rate is a major determinant in the expansion of the overall offshore RMB liquidity pool as well as its long-term liquidity provision. Previously, to support the RMB exchange rate reform, onshore market focused more on the stabilisation of the RMB exchange rate. When the market has got used to the new exchange rate pricing mechanism and the policy effects have been gradually showing up, an appropriate facilitation of onshore RMB capital flows to the offshore market and further opening up two-way cross-border flows could be considered for expanding offshore RMB liquidity pool and enhancing the development of the offshore market.

With regard to the offshore circulation channel, a bottleneck has emerged in driving the global use of the RMB via the current account and trade settlement, due to the sluggish global economy and the decline in China's foreign trade. Relying more on direct investment under the capital account, particularly through overseas direct investment (ODI), Mainland enterprises' going overseas and Belt and Road initiatives, will enhance global acceptance of the RMB and address the stagnated development of the offshore RMB market[8].

4.2   Further utilising existing policies to link up onshore and off-shore repo markets

In 2015, the PBOC introduced a new policy related to bond repo agreements[9], allowing offshore institutions to conduct bond repos in the Mainland interbank market in order to channel liquidity from onshore to offshore. Such a move could, to some extent, connect the onshore and offshore capital markets and ease the offshore liquidity issue.

Furthermore, setting up a Bond Connect, a cross-border platform linking the onshore and offshore bond markets could be another potential solution to further improve the convenience and efficiency. The above bond repo policy, which permits overseas institutions to acquire liquidity through bond repo transactions, only applies to onshore bonds, i.e. offshore RMB bonds cannot be used as collateral in the onshore repo market. In addition, neither can onshore RMB bonds held by RQFIIs nor by the three types of eligible institutions be repurchased in the offshore market. Notably, the average outstanding of offshore RMB bonds has amounted to about RMB 500 billion[10] at present, close to the level of bonds holding by foreign participants onshore. Setting up a cross-border Bond Connect platform could facilitate foreign institutions to conduct repo transactions on offshore bond holdings and obtain RMB funding from the onshore market, which will not only increase the offshore RMB liquidity, but also improve the tradability and usability of offshore RMB assets and sustain the stability of the offshore RMB market.

4.3   Developing a market benchmark for interest rate swaps and other derivative products, thereby strengthening the pricing efficiency and risk management capabilities in the offshore RMB market

To promote interest rates marketisation, the PBOC puts greater emphasis on developing market-oriented mechanism in determining interest rates, and gradually reduces the gaps between different yield curves and maturities of interest rate, leading to an increasing

---

8   "Moving forward amid the restructuring of the Hong Kong offshore RMB market" (〈香港離岸市場在調整中前行〉), *China Forex* (《中國外匯》), Issue 15, 2016.

9   For details, see the *PBOC Circular on Offshore RMB Clearing Banks, Offshore Participating Banks Transacting in Bond Repurchase Transactions on Inter-bank Bond Market 2015* (《中國人民銀行關於境外人民幣業務清算行、境外參加銀行開展銀行間債券市場債券回購交易的通知》(2015)).

10   Source: Bloomberg.

correlation between onshore and offshore rates and strengthening the pricing efficiency of CNH Hibor fixing. Increased stability and efficiency of the CNH Hibor fixing will facilitate the introduction of RMB products related to both bond repos and interest rate swaps, strengthen the hedging capability of the offshore RMB market, and ultimately develop a deep market environment for offshore RMB transactions.

Figure 7. Divergence between CNH Hibor fixing and Shibor since 2015 2H

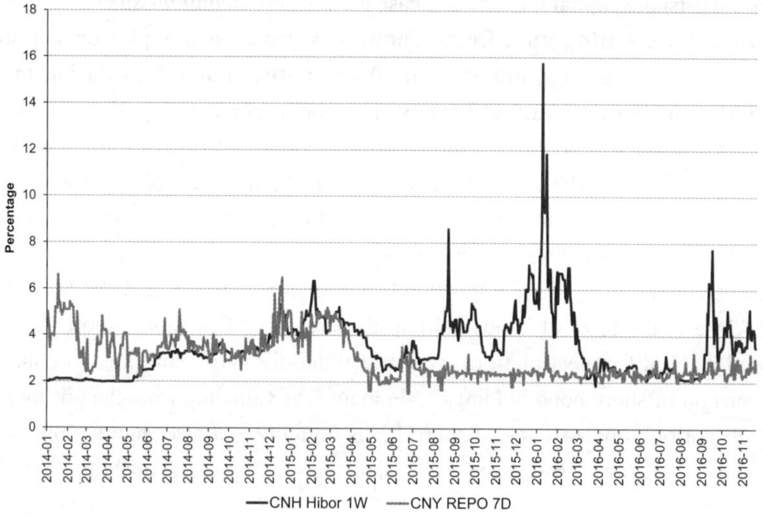

Source: Bloomberg.

4.4 Widening the scale and types of offshore RMB products and further expanding offshore RMB liquidity pool

Removing the aggregate quota for the Shanghai Connect and Shenzhen Connect Programmes and the expansion of Qualified Foreign Institutional Investors (QFII) and RQFII scheme will continue widening the range of investment channels of the offshore RMB market and encouraging large RMB circulation in the offshore markets. Furthermore, along with RMB internationalisation and the opening up of the capital account, the Mutual Market Connectivity Model could be further expanded to more segments, including bonds and commodities. This will attract more overseas investors to tap into the Mainland market through Hong Kong, inducing more RMB capital to be accumulated in the offshore market as a greater variety of instruments are available.

Chapter 14

HKEX towards an offshore RMB product trading and risk management centre

19 April 2017

Summary

Hong Kong is the first market in the world to start offshore Renminbi (RMB) business upon authorisation by the Mainland government in 2003. Following subsequent Mainland policy liberalisation and facilitation by central policy support, RMB financial products began to prosper in Hong Kong, both off-exchange and on-exchange. RMB products on HKEX now comprise bonds, exchange traded funds (ETFs), real estate investment trust (REIT), equities, RMB fixed income and currency (FIC) derivatives and commodity derivatives. RMB ETFs have been the most actively traded RMB product on the HKEX securities market while RMB currency futures are the most popular product on the HKEX derivatives market and achieved record volume in 2016.

HKEX takes the lead in the listing and trading of RMB products, both securities and derivatives, among world exchanges. Only several RMB securities products are found to be offered by a few other exchanges, with low trading in them. On the other hand, RMB currency futures and options are rather popular products offered by a number of exchanges around the globe. Nevertheless, trading is concentrated on the Asian exchanges and among which HKEX has been the most active venue. In fact, trading statistics showed that the RMB futures contracts on HKEX well exhibited their functionality as RMB currency risk management tools at times of high volatility in RMB exchange rate.

The comparative advantages of HKEX for offshore RMB product trading and risk management lie in a number of factors, including geographically in the centre of the One-Belt-One-Road strategic initiative, the active RMB businesses in Hong Kong supported by a large RMB pool, the international investor base of the HKEX markets and the infrastructural efficiency that the markets offer.

An offshore market with a rich supply of RMB products and risk management instruments is fundamental to support the internationalisation of RMB and at the same time maintaining a steady exchange rate level. Towards this end, HKEX's RMB product suite will continuously be enriched, both in the securities and derivatives markets, to serve the growing investor needs as the RMB steadily progresses on its internationalisation. In addition to the recently launched USD/CNH options and Mainland treasury bond futures, other RMB risk management tools would possibly be introduced in the future. HKEX is well positioned to be the offshore RMB product trading and risk management centre for global investors.

1   Background

Renminbi (RMB) business in Hong Kong began in 2004 after China's central bank, the People's Bank of China (PBOC), and the Hong Kong Monetary Authority (HKMA) signed a Memorandum of Understanding in November 2003 which allowed Hong Kong banks to conduct RMB business for individuals. The scope of services was initially limited to remittance, exchange and RMB credit cards. RMB investment products in Hong Kong started with the issuance of RMB bonds, commonly referred to as "dim-sum" bonds, after the State Council of China gave consent to the expansion of RMB business in Hong Kong in January 2007 — allowing Mainland financial institutions to issue RMB financial bonds in Hong Kong. Related rules were issued in June 2007 for implementation of this state policy[1] and late in the same month the first RMB bond was offered in Hong Kong by a Mainland state-owned policy bank[2].

Subsequent policy relaxations have driven the rapid development of the RMB bond market in Hong Kong. In February 2010, according to policy clarification[3], the range of eligible issuers, issue arrangement and target investors of RMB bonds in Hong Kong can be determined in accordance with the applicable regulations and market conditions in Hong Kong. In the same month, the PBOC gave its permission to allow financial institutions to open RMB accounts in Hong Kong that are related to debt financing, which enables the launch of RMB bond funds in Hong Kong. In October 2011, new rules were introduced to allow overseas RMB obtained through legitimate channels, e.g. by overseas issuance of RMB bonds and stocks, to be engaged in direct investment in the Mainland[4].

Fuelled by the increasingly extensive scope of eligible RMB businesses in the Hong

---

1   "Provisional Measures for the Administration of the Issuance of RMB Bonds in the Hong Kong Special Administrative Region by Domestic Financial Institutions"《境內金融機構赴香港特別行政區發行人民幣債券管理暫行辦法》issued jointly by the PBOC and the National Development and Reform Commission (NDRC) on 8 June 2007.

2   The RMB bond was offered by China Development Bank which had an offer size of RMB 5 billion, a coupon rate of 3% and a maturity of 2 years. At least 20% of the issue was offered to retail investors.

3   HKMA's letter of elucidation of supervisory principles and operational arrangements regarding RMB business in Hong Kong, 11 February 2010.

4   "Administrative Measures for the Clearing and Settlement of Foreign Direct Investment in Renminbi"《外商直接投資人民幣結算業務管理辦法》issued by the PBOC; "Notification About Issues Relating to Cross-Border Direct Investment in Renminbi"《關於跨境人民幣直接投資有關問題的通知》issued by the Ministry of Commerce (MoC).

Kong financial sector and the Mainland government's central policy support[5], RMB financial products began to prosper in Hong Kong, both off-exchange and on-exchange, to beyond RMB bonds. In the exchange securities market, RMB bonds have the most listings in number terms while RMB-traded exchange traded funds (ETFs) have become the most actively traded RMB products. In the exchange derivatives market, the RMB futures contract has exhibited its role in RMB exchange rate risk management at times of high exchange rate volatility. (See section 2 below.)

This report gives an account of HKEX's RMB product development in comparison with other exchanges in the world, showing that HKEX takes the lead in exchange-traded RMB products among world markets and drives itself towards a RMB product trading and risk management centre.

2 HKEX's RMB product development

2.1 Securities products

HKEX saw the first listing of RMB bonds on its securities market on 22 October 2010, three years after the issuance of the first RMB bond outside China. This was a ten-year bond issued by an international financial institution, the Asian Development Bank. The listing of the first RMB bond of a Mainland entity was seen in January 2012, issued by the Agricultural Bank of China. The first listings of two RMB Mainland government bonds took place in July 2012, about three years after the issuance of the first offshore RMB Mainland government bond in Hong Kong in September 2009.

Listed RMB security types shortly extended beyond RMB bonds. The first RMB real

---

5   In August 2011, the then Vice Premier Li Keqiang disclosed a series of central policies about Hong Kong development during his visit to Hong Kong. Specifically, policy support would be offered for Hong Kong's development into an offshore RMB business centre; this would include the development of innovative offshore RMB financial products in Hong Kong, an increase in the number of eligible institutions to issue RMB bonds in Hong Kong with an enlarged issue scale. In June 2012, the Mainland Government formally announced a set of policy measures to strengthen cooperation between the Mainland and Hong Kong, among which is the policy support in developing Hong Kong into an offshore RMB business centre.

estate investment trust (REIT) was listed in April 2011; the first RMB ETF (on gold) was listed in February 2012; the first RMB equity was listed in October 2012; and the first RMB warrant was listed in December 2012. By the end of 2016, the total number of listed RMB securities has grown to 179. Figure 1 shows the historical average daily turnover, and the growth in number, of RMB securities on HKEX.

Figure 1. Turnover and number of RMB securities listed on HKEX (Oct 2010–Dec 2016)

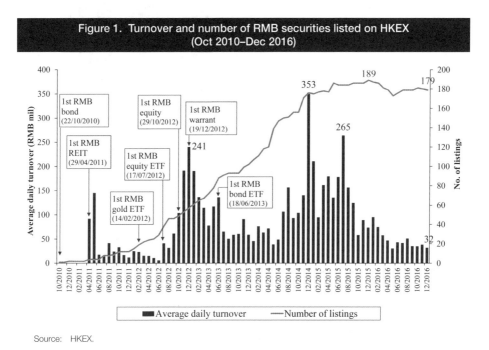

Source: HKEX.

The percentage share of RMB securities in number terms rose to 2% of all listed Main Board securities by the end of 2016. At end-2016, RMB securities consisted mainly of RMB bonds (75%) and ETFs (23%). Among RMB ETFs, stock index ETFs constituted the most (20% of all RMB securities). The proportion of RMB securities in number, though small in respect of all security types on the Main Board, was rather significant for ETFs (31%) and to a considerable extent for debt securities (15%). (See Figures 2 to 4.)

Figure 2. Year-end number of RMB securities listed on HKEX by type (2010–2016)

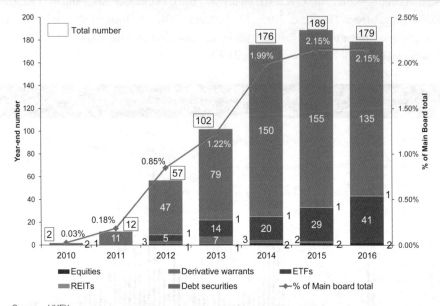

Source: HKEX.

Figure 3. Number of RMB securities listed on HKEX by type (End–2016)

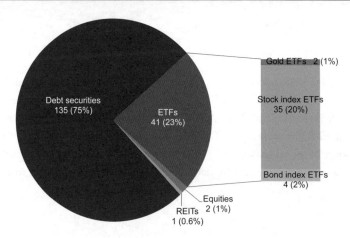

Source: HKEX.

Figure 4. Number of RMB securities as percentage of total number of securities listed on HKEX by type (End–2016)

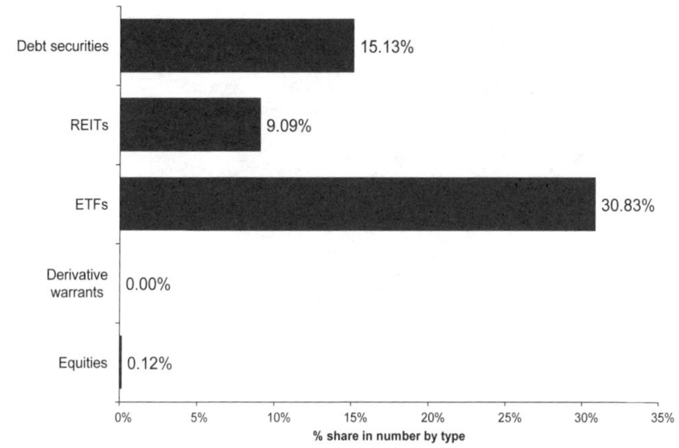

Debt securities — 15.13%

REITs — 9.09%

ETFs — 30.83%

Derivative warrants — 0.00%

Equities — 0.12%

% share in number by type

Source: HKEX.

Trading in RMB securities grew for 4 consecutive years since 2011 before a contraction in 2016. Owing to the small number of listings, securities traded in RMB had only a negligible share of the Main Board market total turnover. Among them, RMB ETFs had the biggest share in each year since their launch year of 2012 — 77% in 2016, mainly from stock index ETFs. The only one REIT came second (20% in 2016). Although RMB bonds had the most listings, they had only a small share by turnover (3%). (See Figures 5 and 6.)

Figure 5. Annual RMB turnover of RMB securities listed on HKEX by type (2010–2016)

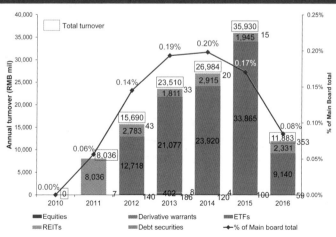

Note: RMB turnover refers to the turnover of RMB trading counters of the RMB securities.
Source: HKEX.

677

Figure 6. Share in RMB turnover of RMB securities listed on HKEX by type (2016)

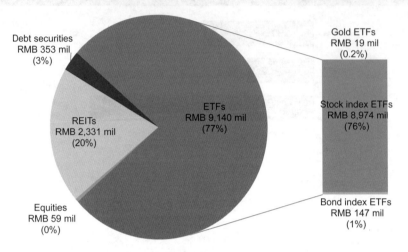

Note: RMB turnover refers to the turnover of RMB trading counters of the RMB securities.
Source: HKEX.

Figure 7. Turnover of all counters of RMB securities listed on HKEX by counter (2016)

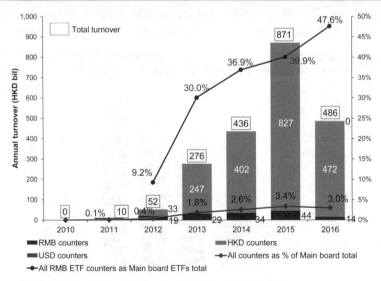

Note: The first dual counters of a RMB security were listed on 12 October 2012.
Source: HKEX.

Starting from October 2012, dual counter trading of RMB securities in other currencies, mainly Hong Kong dollar (HKD) and to a lesser extent US dollar (USD), is available. As at the end of 2016, there are 41 HKD dual counters and 9 USD dual counters of RMB securities on HKEX — one of the two RMB equity securities has a HKD counter; all RMB ETFs except one gold ETF had HKD counters and 9 of them also had USD counters. Compared to RMB counters, trading in the HKD counters of RMB securities were more active. In fact, most of the trading of RMB securities across all counters was concentrated in the HKD counters (over 97% in 2016). The combined turnover of all counters of RMB securities rose to about 3% of Main Board total turnover in 2015 and 2016. In particular, the combined turnover of all counters of RMB ETFs (almost all were stock index ETFs) had an ever increasing share of the total turnover of all ETFs on the Main Board since their launch in 2012, reaching the all-time high of 48% in 2016 (compared to a 31% share of all ETFs in number terms). (See Figure 7.)

In summary, RMB securities on HKEX experienced a steady development. The introduction of dual-counter trading in HKD provides trading convenience to investors and attracted considerable degree of trading. RMB ETFs, mainly stock index ETFs, are particularly attractive to investors, albeit most of the trading was concentrated in the HKD counters.

2.2   Derivative products

The first RMB derivative product traded on HKEX was the US dollar to offshore RMB (USD/CNH) futures launched in September 2012. The product was introduced to provide the market with the currency risk management tool and investment tool in the course of gradual RMB internationalisation. After a modest start, active trading in the product was ignited by the policy move of the PBOC on 11 August 2015 on the formation mechanism of the central parity rate of RMB against US dollar in the Interbank foreign exchange (FX) market to make the currency rate more market-driven. The trading momentum further picked up in 2016 along with increased RMB exchange rate volatility in the year. To serve anticipated increasing demand for more RMB currency derivatives in view of increasing global economic activities conducted in RMB, HKEX introduced three new RMB-traded currency futures of offshore RMB against euro, Japanese yen and Australian dollar — EUR/CNH, JPY/CNH and AUD/CNH, and the USD-traded CNH/USD futures on 30 May 2016.

Another product initiative to support the international use of the RMB and RMB pricing in the real economy is the introduction of RMB-traded commodity futures contracts in December 2014. The first batch of products launched were London metal mini futures contracts on aluminium, copper and zinc. The underlying three metals are those which

China has significant shares in global consumption[6] and which have the most liquid futures contracts traded on the London Metal Exchange (LME)[7], a subsidiary of HKEX. A year later, three more metal mini futures were launched — lead, nickel and tin. These six RMB-traded metal contracts are cash-settled mini contracts of the corresponding physically-settled contracts traded on LME. They are the first metal products outside China which provides for RMB exposure in the underlying assets, supporting RMB benchmarking for metals in the Asian time zone.

Figure 8 shows the historical average daily volume and open interest of RMB derivatives on HKEX since launch.

Figure 8. Volume and open interest of RMB derivatives on HKEX (Sep 2012–Dec 2016)

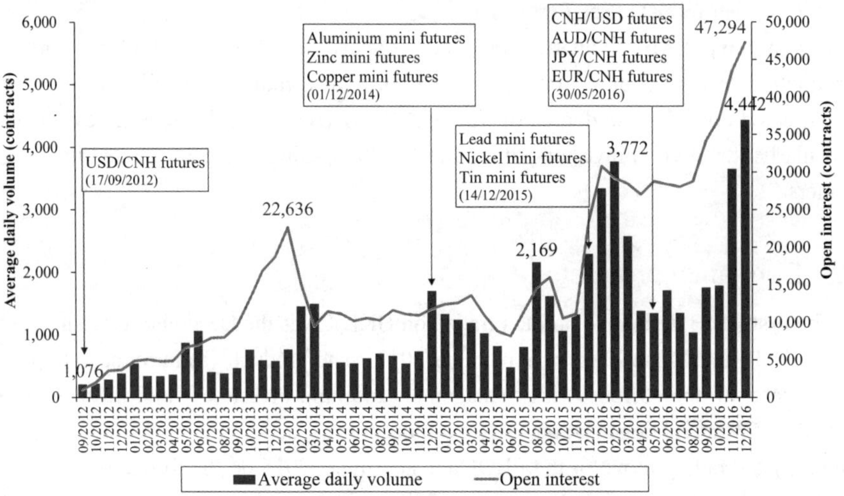

Source: HKEX.

Among the RMB derivatives, the two futures contracts on the CNH rate against the USD are by far the most active ones. The **USD/CNH futures** on HKEX achieved remarkable performance in 2016, recording historical highs of annual contract volume and year-

---

6   China's share in global consumption was 36% for aluminium in 2015 (24,960 kilo tonnes out of 69,374 kilo tonnes, source: World Aluminium, http://www.world-aluminium.org), 46% for copper in 2015 (9,942 kilo tonnes out of 21.8 mil tonnes, source: The Statistics Portal, https://www.statista.com) and 45% for zinc in 2014 (about 6.25 mil tonnes out of 13.75 mil tonnes, source: Metal Bulletin, The Statistics Portal).

7   In 2016, trading volume in futures contracts of aluminium, copper and zinc on LME constituted 35.5%, 24.7% and 18.0% of the total futures trading volume on LME (source: LME).

end open interest. The record-breaking total trading volume of the product in 2016 was 538,594 contracts, an annual increase of 105%; and the record high year-end open interest was 45,635 contracts, a year-on-year increase of 98%. Its average daily volume climbed to 4,325 contracts in December 2016. The newly introduced **CNH/USD futures** also showed a growing contract volume in 2016H2 and its open interest has been continuously building up since launch. Its average daily volume achieved 95 contracts in December 2016; and its open interest reached the highest of 1,494 contracts at year-end. (See Figure 9.)

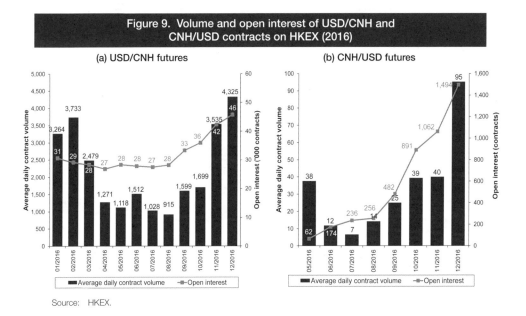

Figure 9. Volume and open interest of USD/CNH and CNH/USD contracts on HKEX (2016)

Source: HKEX.

In fact, the increased volume and open interest in the two futures contracts on HKEX reflects that their functionality as RMB currency hedging tools is being realised. High trading volumes were observed in the contracts at times of high volatility in offshore RMB (CNH) to USD exchange rate (see Figures 10 and 11 below). On 5 January 2017 when CNH rate saw big fluctuations and the overnight (O/N) CNH Hibor fixing rate surged to 33.335%, the full-day trading volume and open interest in USD/CNH futures reached all-time highs of 20,338 contracts and 46,711 contacts respectively. On the day before (4 January 2017), the after-hours futures trading (AHFT) in the USD/CNH contract also reached an all-time high of 3,642 contracts. The risk management functionality of the RMB currency products is also observed in the fact that a moderate but statistically significant degree of correlation (a correlation coefficient of about 0.4 to 0.5) was found between the daily contract volume and open interest of the two products with the CNH Hibor O/N rate during 2016. That is, the higher the liquidity problem of offshore RMB, trading in the futures products tends to be more active and the open interest tends to be higher.

Figure 10. Daily volume and open interest of USD/CNH futures on HKEX and USD/CNH exchange rate (4 Jan 2016–6 Jan 2017)

Source: HKEX for futures data; Thomson Reuters for USD/CNH rate.

Figure 11. Daily volume and open interest of CNH/USD futures on HKEX and USD/CNH exchange rate (30 May 2016–6 Jan 2017)

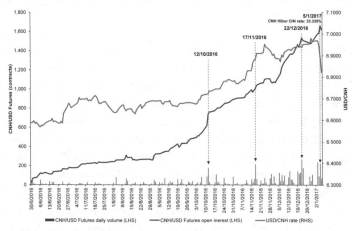

Source: HKEX for futures data; Thomson Reuters for USD/CNH rate.

RMB currency risk management tools on HKEX are further enriched upon the introduction on 20 March 2017 of the first RMB currency options contract, **USD/CNH Options**, which had a first-day trading volume of 109 contracts and an average daily volume of 122 contracts up to the end of March 2017. The increased product variety would also provide opportunities for investors to adopt different investment strategies for their RMB exposure.

On the backdrop of the increasing internationalisation of the RMB and liberalisation of the RMB market, the demand from global investors for hedging against their RMB exposure is ever increasing. To better serve investor needs, HKEX continues to pursue new product initiatives. In June 2016, it launched the **TR/HKEX RMB Currency Index series** (RXY Indices or RXY Index series) which is jointly developed with Thomson Reuters[8]. The RXY Index series is the first tradable index series on RMB outside Mainland China. **Related index futures** with RXY as the underlying could then be launched when conditions are mature.

Furthermore, futures contracts on treasury bonds issued by China's Ministry of Finance (**MOF T-bond futures**) had just been launched on 10 April 2017. RMB derivatives would be useful tools for interest rate hedging, especially under the **Bond Connect scheme** between the bond markets in Mainland and Hong Kong which was publicly addressed by Premier Li Keqiang on 15 March 2017 at the press conference after the close of the National People's Congress meeting.

In summary, RMB currency products are in high demand by global investors in the course of RMB internationalisation, and the RMB currency derivatives on HKEX have been well received by investors to meet their needs. HKEX's RMB derivative product suite is continuously being enriched with more FIC products to serve the growing investor demand.

3   RMB products offered by world exchanges

Not many major exchanges in the world are found to have RMB-traded securities or

---

8   See Chapter 6 of this book, "TR/HKEX RMB Currency indices (RXY)".

derivative products listed on their markets[9]. These findings are described in sub-sections below.

3.1 Securities products

From the official websites of major exchanges, those found to have listed RMB-traded securities include Deutsche Börse (DB), Japan Exchange Group (JPX), London Stock Exchange (LSE), Singapore Exchange (SGX) and Taiwan Stock Exchange (TWSE)[10].

DB formed a joint venture — **China Europe International Exchange (CEINEX)** — with the Shanghai Stock Exchange (SSE) and the China Financial Futures Exchange (CFFEX), which commenced business on 18 November 2015. CEINEX is positioned as the trading and pricing centre for offshore RMB assets in Europe. At the initial stage, product development would be focused on cash securities products traded and settled in RMB and later on derivative products when conditions become mature. As at the end of 2016, RMB-traded securities on CEINEX consist of two ETFs and three debt securities. There are some 14 other ETFs with Chinese assets as underlyings but these are traded in euro. CEINEX introduced the first ETF derivatives (ETF futures with the Mainland stock index, CSI300, as the underlying index) on 20 February 2017; but the product is traded in euro. Following the ETF futures, the corresponding ETF options, also traded in euro, would be launched.

JPX had two RMB-traded bonds on its Pro-Bond Market as of end-2016, the first one was listed in July 2015. There were also a few China-related ETFs, but all were traded in JPY.

LSE offered RMB-trading in two ETFs and over 100 bonds as of end-2016. The first RMB ETF was listed in March 2015, followed by the second one later in September 2016. There were other China-related ETFs but these are traded in GBP or USD.

SGX had one dual-counter stock — a Chinese company (Yangzijiang Shipping Holdings Ltd) — traded in both Singapore dollar (SGD) and RMB, and 96 RMB-traded bonds as of end-2016. There were no RMB-traded ETFs but 6 China-related ETFs, five traded in USD and one in SGD.

TWSE implemented a Dual-Currency Trading Mechanism for Exchange Traded Funds (ETFs), known as "DC-ETF(s)", on 8 August 2016. This is the first time that TWSE opens up foreign currency counters for its securities products. By the end of 2016, there were two

---

9   Information search was done on selected world exchanges' official websites on a best-efforts basis and comprehensiveness and accuracy are not guaranteed.

10   See Appendix 1 for the list of identified RMB-traded securities on HKFX and overseas exchanges.

RMB-traded counters of DC-ETFs.

There are some other major exchanges found to have some China-related ETFs, but these are traded in their domestic currencies. They include Australian Securities Exchange (ASX), Korea Exchange (KRX), New York Stock Exchange (NYSE) and Nasdaq.

As for RMB bonds, it was found that over 400 offshore products were traded on other exchanges. These include Frankfurt Stock Exchange, MarketAxess, Luxembourg Stock Exchange, Taipei Exchange (Gretai Securities Market)[11].

3.2   Derivative products

The RMB derivatives found to be traded on exchanges other than HKEX are confined to RMB currency futures and options only. These include CME Group and BM&FBovespa (BMFB) in Americas; SGX, ICE Futures Singapore (ICE SGP), Taiwan Futures Exchange (TAIFEX) and Moscow Exchange (MOEX) in Asia; Johannesburg Stock Exchange (JSE) in Africa; and Dubai Gold and Commodities Exchange (DGCX) in the Middle East[12]. No exchanges other than HKEX were found to have RMB-traded commodity contracts.

CME Group offers the largest number of RMB currency products among the identified exchanges — 8 futures and 2 options contracts on onshore RMB (CNY) or offshore RMB (CNH) as of end-2016. These are traded on the Group's two sister exchanges:

- **Chicago Mercantile Exchange (CME)** — 4 futures and 2 options. Two CNY futures (one standard and one mini contract) were delisted in May 2016.
- **CME Europe Exchange (CMED)** — 4 futures.

As of end-2016, **SGX** had 5 RMB currency futures on CNY or CNH paired with currencies USD, SGD and euro, and one options contract on RMB currency futures; **ICE SGP** had two RMB currency mini futures on CNY or CNH paired with USD; and TAIFEX had two futures and two options on offshore RMB paired with USD — one standard contract and one mini contract respectively for futures and options.

The other exchanges — **BM&FBovespa, DGCX, JSE** and **MOEX** — each had a RMB currency futures contract as of end-2016.

---

11 Source: Thomson Reuters, 6 January 2017. The number would include multiple counting as the same RMB bond may be traded on multiple exchanges. Note that the list cannot be verified with the official sources of the exchanges.

12 See Appendix 2 for the list of identified RMB derivatives on HKEX and overseas exchanges.

4 Comparison of RMB products on HKEX with world exchanges

4.1 Securities products

HKEX offers by far the largest number of RMB-traded securities, ahead of any other exchanges in the world[13]. **ETF is the most popular on-exchange RMB-traded product type in markets outside Mainland China.** Although there are also a considerable number of listings of RMB-denominated bonds, on-exchange trading is negligible, if any[14].

Table 1. RMB-traded listed securities on HKEX and selected exchanges (Dec 2016)					
Exchange	Equity	ETF	REIT	Debt	Total
HKEX	2	41	1	135	179
CEINEX	0	2	0	3	5
JPX	0	0	0	2	2
LSE	0	2	0	101	103
SGX	1	0	0	96	97
TWSE	0	2	0	0	2

Note: Compiled on a best-efforts basis.
Source: HKEX and the respective exchanges' websites.

Table 1 above gives a comparison of the number of RMB-traded securities products on HKEX with exchanges in the world found to offer RMB-traded securities; and Table 2 below gives a comparison of their trading in RMB ETFs. The average daily turnover (ADT) of RMB ETFs on HKEX in 2016 was RMB 37 million, higher than the other exchanges, even on a per-security basis.

---

13 As far as known from available data and information.
14 Bond trading is often done over-the-counter (OTC) rather than on exchanges. Bond listings on exchanges may be pursued by issuers to enable trading by institutional investors and fund managers who are required in their mandate to invest in securities that are listed on a recognised stock exchange.

Table 2. Total and average daily turnover of RMB ETFs (2016)		
Exchange	Total (RMB mil)	ADT (RMB mil)
HKEX	9,140	37.0
CEINEX*	74	0.3
LSE	0.5	0.0
TWSE	141	1.4

* RMB products are traded on DB platforms.

4.2   Derivative products

RMB currency futures have become the most popular RMB-traded derivatives in the world, with at least 8 other exchanges offering them in addition to HKEX. Contracts in currency pair of USD/CNH receive the greatest investor interest, reflected by their relatively high trading volume. Contracts on RMB against another international currency, Euro, and other domestic currencies such as SGD had negligible or no trading in the past two years (according to the official sources of the exchanges examined). Table 3 below gives the number of RMB derivatives on HKEX and exchanges in the world found to offer RMB derivatives.

Table 3. RMB derivatives on HKEX and selected exchanges (Dec 2016)							
Exchange	Currency		Commodity		Total		Grand total
	Futures	Options	Futures	Options	Futures	Options	
HKEX	5	0	6	0	11	0	11
BMFB	1	0	0	0	1	0	1
CME Group[1]	8	2	0	0	8	2	10
DGCX	1	0	0	0	1	0	1
ICE SGX	2	0	0	0	2	0	2
JSE	1	0	0	0	1	0	1
MOEX	1	0	0	0	1	0	1
SGX	5	1	0	0	5	1	6
TAIFEX	2	2	0	0	2	2	4
Total	26	5	6	0	32	5	37

(1)   Offered by Chicago Mercantile Exchange (CME) — 4 futures and 2 options, and by CME Europe Exchange (CMED) — 4 futures.

Note: Compiled on a best-efforts basis.

Source: HKEX and the respective exchanges' websites.

HKEX was the top among world exchanges in respect of the average daily notional trading value (US$219 million) and year-end open interest of RMB currency derivatives (47,294 contracts) in 2016. SGX and TAIFEX had considerable trading volume in their RMB currency derivatives, probably in relation to the degree of economic activities of Singapore and Taiwan with China[15]. On the face, TAIFEX had a higher contract volume of its RMB derivatives in 2016 than HKEX. However, this was concentrated on the mini contracts and the relatively high notional trading value achieved by TAIFEX in 2015 did not sustain in 2016. As for SGX, the contract volume and notional trading value were relatively high in 2016, while the open interest as at end-2016 was less than 40% of that on HKEX. (See Figure 12.)

Figure 12. Trading and open interest of RMB derivatives on HKEX and selected exchanges (2016)

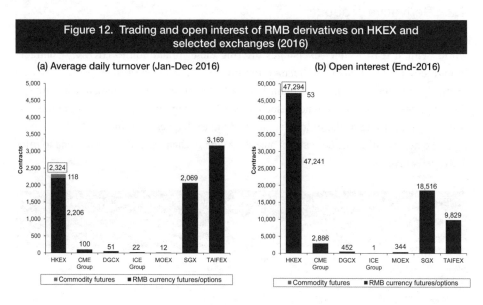

---

15 According to People's Bank of China's RMB Internationalisation Report 2015 (June 2015), Singapore and Taiwan had
    the biggest RMB trade settlement amount following Hong Kong.

(continued)

Figure 12. Trading and open interest of RMB derivatives on HKEX and selected exchanges (2016)

(c) Average daily notional trading value of RMB currency derivatives (2016 vs 2015)

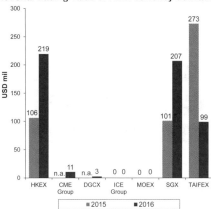

n.a.:   Not available.

Source: HKEX and the respective exchanges' websites.

Among RMB currency futures, four other exchanges offer similar products with the same contract size as the HKEX's standard futures, USD/CNH futures. HKEX also took the lead in trading of the standard contract. (See Figure 13.)

Figure 13. Trading and open interest of the standard USD/RMB contracts on HKEX and selected exchanges (2016 vs 2015)

(a) Average daily turnover (2016 vs 2015)

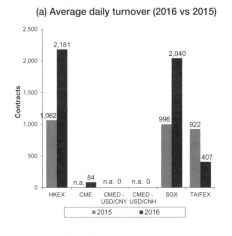

(b) Year-end open interest (2016 vs 2015)

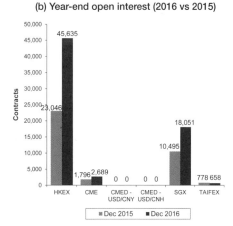

n.a.:   Not available.

Note: Standard contract size is USD100,000. Unless otherwise stated, currency pair is USD/Offshore RMB (CNH).

Source: HKEX and the respective exchanges' websites.

(See Appendix 3 on the average daily volume in 2016 of each RMB currency product on the different exchanges.)

5   HKEX as an offshore RMB product trading and risk management centre

HKEX has achieved a leading position in the listing and trading of RMB products among world exchanges outside Mainland China. RMB ETFs are by far the most well received offshore RMB securities product. While offshore RMB ETFs have achieved modest trading, offshore RMB currency derivatives have gained momentum — a global notional trading value of about US$134 billion (~RMB 935 billion), over 60% growth compared to about US$83 billion (RMB 539 billion) in 2015[16].

Upon the RMB exchange rate mechanism reform in August 2016 and the increasing marketisation of the exchange rate system in China, there is increasing market acceptance of an increased volatility in the RMB exchange rate. According to the Society for Worldwide Interbank Financial Telecommunication (SWIFT) statistics, the RMB has ranked the top 5th or 6th most active currency for international payments by value — a share of 1.68% in January 2017 (6th position) compared to a share of 2.31% in December 2015 (5th position). In the offshore FX market, RMB is said to have the potential to be in the top five most traded currencies[17] and RMB FX risk management tools such as swaps and options have become increasingly popular. In addition, there is a growing diversity in the client base of the RMB currency — all types of bank, institutional investors like Qualified Foreign Institutional Investors (QFIIs), speculators like hedge funds and retail investors in addition to the pioneer group of commercial corporations doing business with China. RMB currency futures and options add to the suite of hedging tools for these investors in RMB FX trading.

---

16 Source: HKEX and respective exchanges' websites, as per analysis given in section 4.
17 By Executive Director of FX products at CME Group, quoted in "China's offshore RMB endgame, Part III: From shadow banking to cyberspace", *Global Capital Euroweek*, 24 April 2015.

Exchanges around the world rush in to offer various risk management instruments for the RMB, to a certain extent, to support the increasing economic activities in RMB with China. HKEX stands out as the leading exchange in RMB products, especially in RMB derivatives, owing to a number of reasons:

(1) Geography

According to the Bank for International Settlements (BIS) Triennial Survey, Hong Kong contributed the largest proportion (39%) of total over-the-counter (OTC) trading in RMB FX instruments outside Mainland China[18]. This was followed by Singapore (22%) and the US (12%). Given China's One-Belt-One-Road (OBOR) initiative put forward by the Chinese leaders in 2013, China's current and near-future economic development with the rest of the world would focus on the Asian area. OBOR consists of the Silk Road Economic Belt (SREB) and 21st Century Maritime Silk Road (MSR) initiatives. The SREB runs through Central Asia, West Asia, the Middle East to Europe, with extension to South Asia and Southeast Asia. The MSR runs through Southeast Asia, Oceania and North Africa.

To support infrastructural projects and economic activities in OBOR, further internationalisation and the use of RMB is expected. It is therefore expected that trading needs of RMB futures for risk management purposes would be the highest in Asia. As Hong Kong is the international financial centre at the centre of OBOR, RMB products and services, both on-exchange and off-exchange, are expected to further prosper.

(2) RMB liquidity pool and RMB businesses in Hong Kong

Since offshore RMB business was offered firstly in Hong Kong back in 2004, Hong Kong has become the global hub for RMB trade settlement, financing and asset management. As noted in section 4.2 above, Hong Kong served the biggest amount of RMB trade settlement in 2015. The dim-sum bond market in Hong Kong is now the largest outside Mainland China[19].

Offshore RMB services in Hong Kong now embrace RMB retail and corporate banking, RMB capital markets, RMB money and FX market and RMB insurance. These services are supported by the formation of a large RMB liquidity pool, which together are conducive to the development of a wide range of RMB products. The expanded variety and usage of RMB products and services in need by customers or investors in turn would boost or help

---

18 In terms of daily average notional turnover value in USD. Source: Triennial Survey statistics on turnover, BIS website.

19 Quoted in "Hong Kong — The global offshore Renminbi business hub", Hong Kong Monetary Authority, January 2016.

sustain the RMB liquidity pool.

Despite the shrinkage of global offshore RMB liquidity pool in the recent wave of RMB depreciation, Hong Kong maintains the largest offshore RMB deposits in the world: RMB 522.5 billion (end-January 2017), compared to RMB 310.7 billion in Taiwan (end-January 2017) and RMB 126 billion in Singapore (end-December 2016)[20]. A large RMB pool supports all RMB activities and active RMB securities and derivatives markets.

(3) International investor base

Hong Kong is a well-known international financial centre and participants in HKEX's markets come from all over the world. According to HKEX's surveys, international investors contributed 39% of the securities market trading (same as local investors) and 28% of the derivatives market trading (larger than the 21% from local investors) in value terms[21]. Active international investor participation contributes to the international pricing of RMB products.

(4) Infrastructural efficiency

The efficient and robust market infrastructure in Hong Kong is fundamental in supporting the activity and continuous development of RMB business and services in Hong Kong. This includes the HKMA's SWIFT-based **RMB Real Time Gross Settlement (RTGS) system** that facilitates market participants from all over the world to handle RMB transactions both with Mainland China and among the offshore markets. While governed by Hong Kong laws[22], the RMB RTGS system in Hong Kong is directly linked up with Mainland China's National Advanced Payment System (CNAPS), enabling the system to handle RMB transactions with Mainland China. There are also linkages with HKD, USD and euro RTGS systems. The RMB RTGS system not only processes RMB interbank payments on an RTGS basis, but also handles RMB bulk clearing and settlement of payment items similar to those handled by the Hong Kong dollar RTGS system. Statistics from SWIFT show that the value of RMB settlement handled by banks in Hong Kong accounted for some 70% of the total offshore RMB payments conducted vis-à-vis Mainland China and within the offshore market globally[23].

RMB RTGS system is accompanied by the **RMB Liquidity Facility** provided by the HKMA to enhance short-term liquidity in the offshore RMB market, which may be affected

---

20 Source: HKMA, Bank of China (Hong Kong) Offshore RMB Express 2017 No. 3, Monetary Authority of Singapore website.
21 Source: HKEX's Cash Market Transaction Survey 2014/15 and Derivatives Market Transaction Survey 2014/15.
22 Settlement finality is protected by the Clearing and Settlement Systems Ordinance in Hong Kong.
23 Quoted in HKMA's "Hong Kong — The Global Offshore Renminbi Business Hub", January 2016.

due to seasonal factors or capital market activities. The facility provides short-term funding need (intraday, overnight, 1-day, 1-week) to banks. In addition, from 27 October 2016, nine banks are appointed as Primary Liquidity Providers (PLPs) for the offshore RMB market.

Apart from the strong support from Hong Kong's overall financial system for the offshore RMB market, the **exchange's infrastructure** also offers efficient support. HKEX's derivatives clearing house, HKFE Clearing Corporation Ltd (HKCC), offers central clearing for exchange-traded derivatives. HKCC is CPSS-IOSCO[24] compliant and operates under an internationally recognised regulatory regime and the protection of Hong Kong laws. International participants are subject to lower capital charges come with HKCC's "Qualifying Central Counterparty (CCP)" status under Basel III. In late September 2016, HKCC relaxed its cash collateral policy, allowing Clearing Participants (CPs) to satisfy their RMB margin requirement of up to RMB 1 billion[25] by any acceptable cash and/or non-cash collateral. This policy relaxation helps reduce investors' funding costs when trading RMB-denominated derivative products. In addition, Hong Kong has an equivalent infrastructure in the OTC derivatives with the launch of **OTC Clear** by HKEX in 2013. The services of OTC Clear now cover certain interest rate swaps (IRS), non-deliverable IRS (NDIRS), cross-currency IRS (CCS) and non-deliverable currency forwards (NDF).

The above platforms form a solid foundation for further development of RMB derivatives in Hong Kong.

As stated in the Chinese Government's work report delivered on 5 March 2017, the RMB would be a significant currency in the international monetary system and RMB exchange rate would be at a basically steady level. An offshore market with a rich supply of RMB products and risk management instruments is fundamental to support the internationalisation of RMB and at the same time maintaining a steady exchange rate level. Towards this end, HKEX's RMB product suite will continuously be enriched, both in the securities and derivatives markets, to serve the growing investor needs as the RMB steadily progresses on its internationalisation. In addition to the recently launched USD/CNH options and T-bond futures, other RMB risk management tools would possibly be introduced in the future. HKEX is well positioned to be the offshore RMB product trading and risk management centre for global investors.

---

24 Committee on Payment and Settlement Systems (CPSS) and the Technical Committee of the International Organization of Securities Commissions (IOSCO).

25 Beyond which HKCC CPs must satisfy their RMB margin requirement by RMB cash.

Appendix 1

List of RMB-traded equities, ETFs and REITs on HKEX and overseas exchanges

(End-2016)

HKEX		
Type	Stock code	Product
Equity	80737	Hopewell Highway Infrastructure Ltd.
Equity	84602	ICBC RMB 6.00% Non-Cum, Non-Part, Perpetual Offshore PrefShs
ETF	82808	E Fund Citi Chinese Government Bond 5-10 Years Index ETF
ETF	82811	Haitong CSI300 Index ETF
ETF	82822	CSOP FTSE China A50 ETF
ETF	82828	Hang Seng H-Share Index ETF
ETF	82832	Bosera FTSE China A50 Index ETF
ETF	82833	Hang Seng Index ETF
ETF	82834	iShares NASDAQ 100 Index ETF
ETF	82836	iShares Core S&P BSE SENSEX India Index ETF
ETF	82843	Amundi FTSE China A50 Index ETF
ETF	82847	iShares FTSE 100 Index ETF
ETF	83008	C-Shares CSI 300 Index ETF
ETF	83010	iShares Core MSCI AC Asia ex Japan Index ETF
ETF	83012	AMUNDI Hang Seng HK 35 Index ETF
ETF	83074	iShares Core MSCI Taiwan Index ETF
ETF	83081	Value Gold ETF
ETF	83095	Value China A-Share ETF
ETF	83100	E Fund CSI 100 A-Share Index ETF

(continued)

HKEX		
Type	Stock code	Product
ETF	83107	C-Shares CSI Consumer Staples Index ETF
ETF	83115	iShares Core Hang Seng Index ETF
ETF	83118	Harvest MSCI China A Index ETF
ETF	83120	E Fund CES China 120 Index ETF
ETF	83122	CSOP China Ultra Short-Term Bond ETF
ETF	83127	Horizons CSI 300 ETF
ETF	83128	Hang Seng China A Industry Top Index ETF
ETF	83129	CSOP China CSI 300 Smart ETF
ETF	83132	C-Shares CSI Healthcare Index ETF
ETF	83136	Harvest MSCI China A 50 Index ETF
ETF	83137	CSOP CES China A80 ETF
ETF	83139	iShares RMB Bond Index ETF
ETF	83146	iShares DAX Index ETF
ETF	83147	CSOP SZSE ChiNext ETF
ETF	83149	CSOP MSCI China A International ETF
ETF	83150	Harvest CSI Smallcap 500 Index ETF
ETF	83155	iShares EURO STOXX 50 Index ETF
ETF	83156	GFI MSCI China A International ETF
ETF	83162	iShares MSCI China A International Index ETF
ETF	83168	Hang Seng RMB Gold ETF
ETF	83170	iShares Core KOSPI 200 Index ETF
ETF	83180	ChinaAMC CES China A80 Index ETF
ETF	83188	ChinaAMC CSI 300 Index ETF
ETF	83199	CSOP China 5-Year Treasury Bond ETF
REIT	87001	Hui Xian Real Estate Investment Trust

(continued)

Overseas exchange	Type	Product
China Europe International Exchange (CEINEX) [Products traded on platforms of Deutsche Börse (DB)]	ETF	BOCI Commerzbank SSE 50 A Share Index UCITS ETF
	ETF	Commerzbank CCBI RQFII Money Market UCITS ETF
London Stock Exchange (LSE)	ETF	Commerzbank CCBI RQFII Money Market UCITS ETF
	ETF	ICBC Credit Suisse UCITS ETF SICAV
Singapore Exchange (SGX)	Equity	Yangzijiang Shipbulding Holdings Ltd
Taiwan Stock Exchange (TWSE)	ETF	Fubon SSE180 ETF
	ETF	Capital SZSE SME Price Index ETF

Sources: HKEX for HKEX products; the respective exchanges' websites for their RMB products.

Appendix 2

List of RMB currency futures/options on HKEX and overseas exchanges

(End-2016)

Exchange	Product	Contract size	Trading currency*	Settlement
HKEX	RMB Currency Futures - USD/CNH Futures	USD100,000	CNH	Deliverable
	RMB Currency Futures - EUR/CNH Futures	EUR 50,000	CNH	Cash settled
	RMB Currency Futures - JPY/CNH Futures	JPY 6,000,000	CNH	Cash settled
	RMB Currency Futures - AUD/CNH Futures	AUD 80,000	CNH	Cash settled
	RMB Currency Futures - CNH/USD Futures	RMB 300,000	USD	Cash settled
	London Aluminium Mini Futures	5 tonnes	CNH	Cash settled
	London Zinc Mini Futures	5 tonnes	CNH	Cash settled
	London Copper Mini Futures	5 tonnes	CNH	Cash settled
	London Lead Mini Futures	5 tonnes	CNH	Cash settled
	London Nickel Mini Futures	1 tonne	CNH	Cash settled
	London Tin Mini Futures	1 tonne	CNH	Cash settled
BMFB	Chinese Yuan Futures	CNY 350,000	BRL	Cash settled
CME	Standard-Size USD/Offshore RMB (CNH) Futures	USD 100,000	CNH	Deliverable
	E-micro Size USD/Offshore RMB (CNH) Futures	USD 10,000	CNH	Deliverable
	Chinese Renminbi/USD Futures	CNY 1,000,000	USD	Cash settled
	Chinese Renminbi/Euro Futures	CNY 1,000,000	EUR	Cash settled
	Chinese Renminbi/USD Options on futures	CNY 1,000,000	USD	Deliverable
	Chinese Renminbi/Euro Options on futures	CNY 1,000,000	EUR	Deliverable

(continued)

Exchange	Product	Contract size	Trading currency*	Settlement
CMED	Euro/Chinese Offshore Renminbi (EUR/CNH) Physically Deliverable Futures	EUR 100,000	CNH	Deliverable
	U.S. Dollar / Chinese Renminbi (USD/CNY) Cash Settled Futures	USD 100,000	CNY	Cash settled
	U.S. Dollar / Chinese Offshore Renminbi (USD/CNH) Physically Deliverable Futures	USD 100,000	CNH	Deliverable
	Euro/Chinese Renminbi (EUR/CNY) Cash Settled Futures	EUR 100,000	CNY	Cash settled
DGCX	US Dollar / Chinese Yuan Futures	USD 50,000	CNH	Cash settled
ICE	Mini Offshore Renminbi Futures	USD 10,000	CNH	Deliverable
	Mini Onshore Renminbi Futures	CNY 100,000	USD	Cash settled
JSE	Chinese Renminbi/Rand Currency Futures	CNY 10,000	ZAR	Cash settled
MOEX	CNY/RUB Exchange Rate Futures	CNY 10,000	RUB	Cash settled
SGX	CNY/SGD FX Futures	CNY 500,000	SGD	Cash settled
	CNY/USD FX Futures	CNY 500,000	USD	Cash settled
	EUR/CNH FX Futures	EUR 100,000	CNH	Cash settled
	SGD/CNH FX Futures	SGD 100,000	CNH	Cash settled
	USD/CNH FX Futures	USD 100,000	CNH	Cash settled
	USD/CNH FX Options on Futures	USD 100,000	CNH	Cash settled
TAIFEX	USD/CNH FX Futures	USD 100,000	CNH	Cash settled
	USD/CNT FX Futures	USD 20,000	CNH	Cash settled
	USD/CNH FX Options	USD 100,000	CNH	Cash settled
	USD/CNT FX Options	USD 20,000	CNH	Cash settled

* CNH = Offshore RMB; CNY = Onshore RMB

Sources: HKEX for HKEX products; the respective exchanges' websites for their RMB products.

Appendix 3

Average daily trading volume and year-end open interest of RMB currency products on HKEX and key overseas exchanges

(2016 vs 2015)

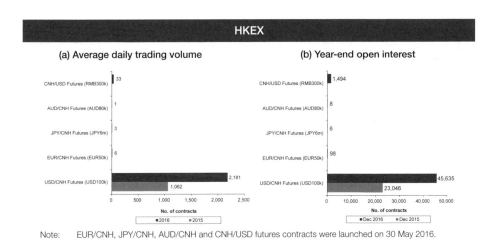

Note: EUR/CNH, JPY/CNH, AUD/CNH and CNH/USD futures contracts were launched on 30 May 2016.

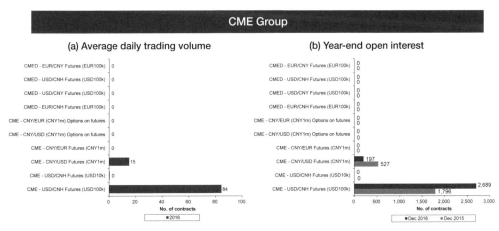

Note: 2015 data is not available.

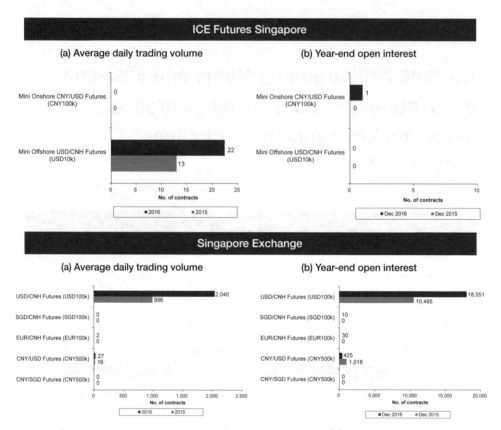

ICE Futures Singapore

(a) Average daily trading volume

Mini Onshore CNY/USD Futures (CNY100k): 0, 0

Mini Offshore USD/CNH Futures (USD10k): 22 (2016), 13 (2015)

No. of contracts

■ 2016 ■ 2015

(b) Year-end open interest

Mini Onshore CNY/USD Futures (CNY100k): 1, 0

Mini Offshore USD/CNH Futures (USD10k): 0, 0

No. of contracts

■ Dec 2016 ■ Dec 2015

Singapore Exchange

(a) Average daily trading volume

USD/CNH Futures (USD100k): 2,040 / 996
SGD/CNH Futures (SGD100k): 0 / 0
EUR/CNH Futures (EUR100k): 2 / 0
CNY/USD Futures (CNY500k): 27 / 16
CNY/SGD Futures (CNY500k): 0 / 0

No. of contracts

■ 2016 ■ 2015

(b) Year-end open interest

USD/CNH Futures (USD100k): 18,051 / 10,495
SGD/CNH Futures (SGD100k): 10 / 0
EUR/CNH Futures (EUR100k): 30 / 0
CNY/USD Futures (CNY500k): 425 / 1,018
CNY/SGD Futures (CNY500k): 0 / 0

No. of contracts

■ Dec 2016 ■ Dec 2015

Note: The above charts exclude the USD/CNH FX options on Futures introduced in December 2016.

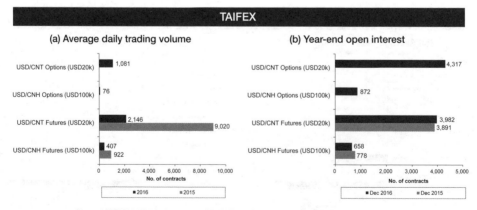

TAIFEX

(a) Average daily trading volume

USD/CNT Options (USD20k): 1,081
USD/CNH Options (USD100k): 76
USD/CNT Futures (USD20k): 2,146 / 9,020
USD/CNH Futures (USD100k): 407 / 922

No. of contracts

■ 2016 ■ 2015

(b) Year-end open interest

USD/CNT Options (USD20k): 4,317
USD/CNH Options (USD100k): 872
USD/CNT Futures (USD20k): 3,982 / 3,891
USD/CNH Futures (USD100k): 658 / 778

No. of contracts

■ Dec 2016 ■ Dec 2015

Note: USD/CNH FX Options and USD/CNT FX Options were launched on 27 June 2016.

Sources: HKEX for HKEX products; the respective exchanges' websites for their RMB products.

Chapter 15

OTC clearing solution for Mainland China's increasing cross-border derivatives trading

14 November 2017

Summary

In the course of Mainland China's market economy reforms, there has been increasing trading activities in over-the-counter (OTC) derivatives by Mainland financial institutions for risk management purposes. Domestically, trading value in interest rate swaps on the interbank market surged as they are increasingly used for hedging against the increased volatility in interest rates along with the market-based interest rate reform. Rapid growth was also seen in the trading of foreign exchange (FX) derivatives on the domestic interbank market, which are used for hedging the increasing FX risk along with the market-based FX rate reform.

Alongside, Mainland financial institutions are increasingly engaged in international businesses with overseas parties in the course of China's market opening and economic development. As a result of providing the funding needs for international trade, infrastructure projects under the One-Belt-One-Road initiative and Mainland enterprises' international business developments, Mainland financial institutions would have an increasing proportion of their balance sheet in foreign currency, the majority of which would be US dollar, the mostly used global currency in international trade and finance. Interest income in foreign currency assets is susceptible to interest rate risk as well as to FX rate risk. Therefore, in addition to the increasing utilisation of risk management instruments in the domestic OTC market, there is also an increasing demand by Mainland financial institutions for hedging their growing foreign asset positions by engaging in OTC interest rate and FX derivatives transactions with foreign institutions.

Under the tightened global regulatory framework for risk management of OTC derivatives after the 2008 Global Financial Crisis, financial institutions would either have to do mandatory central clearing for standardised OTC derivatives or to be subject to higher capital and margin requirements for bilaterally cleared OTC derivatives. In the latter case, they may voluntarily opt for central clearing in order to reduce the transaction costs. International participants domiciled in Europe or the US generally use major clearing houses recognised by their home regulators, such as LCH or Eurex Clearing in Europe, or CME Clearing in the US. Mainland banks which execute OTC derivatives transactions with foreign institutions would have to centrally clear the trades via a recognised central counterparty (CCP). Very often, Mainland banks would not be able to become direct clearing members of the foreign CCPs owing to the uncertainty in the status of the People's Republic of China (PRC) as a netting jurisdiction. As a result, clearing and settlement is usually done via a clearing broker which is a clearing member of an overseas CCP and which acts as the agent for the Mainland banks.

In the Asian time zone, the subsidiary of Hong Kong Exchanges and Clearing Ltd (HKEX) — OTC Clearing Hong Kong Ltd (HKEX OTC Clear) — is a recognised clearing house for OTC derivatives in Hong Kong and is also a qualified central counterparty for US and European institutions to treat risk exposure facing CCP under preferential capital treatment. It is also the only overseas CCP that can accept PRC banks with a Hong Kong branch to be direct clearing members. This practice is unparalleled in the US and European clearing houses. Given that many of the Mainland banks have set up branches or subsidiaries in Hong Kong, becoming a direct member of HKEX OTC Clear would be more cost-effective than appointing a clearing broker for clearing their derivatives transactions in an overseas CCP. Under the solution offered by HKEX OTC Clear, the Mainland banks can do direct clearing via their Hong Kong subsidiaries or branches. Hence, compared to other clearing houses, HKEX OTC Clear offers a more convenient and cost-effective solution to Mainland banks for central clearing of their OTC derivatives transactions. Moreover, in addition to serving OTC transactions in USD and other major currencies, HKEX OTC Clear is more capable, compared to its overseas counterparts, to support Mainland and global financial institutions' OTC transactions in offshore RMB, which are believed to have a high growth potential.

1 The increasing OTC derivatives activities of Mainland financial institutions

Over-the-counter (OTC) derivatives were introduced in Mainland China only a little more than a decade ago. There are virtually no OTC derivatives markets other than the one operated by the **China Foreign Exchange Trade System (CFETS)** which is under the supervision of the People's Bank of China (PBOC) and the State Administration of Foreign Exchange (SAFE). The first derivative product introduced on CFETS was bond forwards in June 2005. The product range is now extended to include foreign exchange (FX), interest rate and credit derivatives.

1.1   CFETS and its products

CFETS, also known as the National Interbank Funding Center, is a sub-institution (直 屬事業單位) of the PBOC established on 18 April 1994. Its main functions are to provide services for interbank FX trading, Renminbi (RMB) borrowing and lending, trading in bonds (including commercial papers) and derivatives, and the associated clearing, informa- tion, risk management and supervision services.

In the money market, CFETS operates the FX market and the RMB market. In the FX market, CFETS is responsible for calculating and disseminating the central rates of the RMB against major currencies including the US dollar (USD), the Euro (EUR), Japanese yen (JPY), British pound (GBP) and Hong Kong dollar (HKD). The RMB market consists of the interbank borrowing and lending market, certificates of deposit, loan transfers, the bond market (including asset-backed securities) and the RMB derivatives market.

The first derivative product formally launched in the interbank market was **bond forwards** in June 2005, nine years after the establishment of the interbank bond market in 1996 and the introduction of treasury bond repurchases (repos) in the same year. Given the acceleration of the market-based interest rate reform, Mainland financial institutions were exposed to increased interest rate risk but were facing increasing difficulties in hedging that risk through existing tools like bond repos and forwards. To provide more interest rate risk management tools, **RMB interest rate swaps (IRS)** were introduced in the interbank bond market on a pilot basis in February 2006. Under the pilot, certain qualified institutions were allowed to conduct RMB IRS within certain limits. The pilot programme ended in a full launch of the product in February 2008. Adding to the suite of hedging and risk management tools, **RMB forward rate agreements (FRA)** were introduced in November 2007, followed by **standardised interest rate derivatives**[1] in November 2014, and credit risk mitigation (CRM) instruments, including **CRM warranty (CRMW)** in 2010 and **credit default swaps (CDS)** in September 2016.

In the interbank FX market, **currency forwards** (RMB against foreign currencies) were introduced in August 2005. **Currency swaps** (RMB against foreign currencies) were introduced as a pilot in April 2006[2] and formally in August 2007 upon release of the related rules. **Currency options** were subsequently introduced in April 2011, followed by **standardised currency swaps** in February 2015 and **standardised currency forwards** in May 2016. These products provide the hedging tools for banks, allowing banks greater flexibility in managing their foreign currency positions.

Participants in CFETS include banks and non-bank financial institutions such as securities companies, insurance companies, trust investment companies, funds and fund management companies, asset management companies and social security funds. The interbank bond market was the first in the interbank market to open to foreign participation (in 2010) — initial qualified participants included central bank-type institutions, RMB clearing banks and participating banks, Qualified Foreign Institutional Investors (QFIIs) and Renminbi QFIIs (RQFIIs). Authorised foreign participants were later expanded to include all legitimately registered financial institutions and their investment products, pension funds and charity funds. Along with subsequent policy relaxations, certain foreign institutional investors including central bank-type institutions, international financial institutions and sovereign funds can now access a wide range of products on the interbank market, including the spot bond and FX market, bond derivatives and interest rate derivatives.

Clearing and settlement services for OTC derivatives transactions on CFETS are provided by three institutions — CFETS itself, the China Central Depository & Clearing Co., Ltd. (CCDC) and the Shanghai Clearing House (SCH). CFETS provides trade confirmation and straight-through processing (STP) to support the clearing and settlement of transactions in its FX market and RMB market via the PBOC's payment and settlement system. It also provides trade offsetting/compression services for IRS and FX swaps. CCDC provides clearing and settlement services for bond and bond derivatives transactions on CFETS. SCH provides central clearing services for a variety of derivatives transactions.

---

1   Standardised interest rate derivatives are IRS and FRA products with standardised expiry and interest duration.
2   RMB swaps against foreign currencies between banks and their customers were introduced earlier in August 2005.

In particular, it was designated by the PBOC in February 2014 to be the central counterparty (CCP) for mandatory central clearing of RMB IRS traded on CFETS[3].

1.2 Derivatives trading activities in the domestic OTC market

As shown in Figure 1, IRS has dominated the trading in bond/interest rate derivatives since 2010. Trading in bond forwards diminished after 2009 and was almost completely taken over by trading in IRS. The trading value of IRS reached RMB 9,920 billion (~US$1.4 trillion) in 2016, with a compound annual growth rate (CAGR) of 76% during the period from 2006 to 2016. However, the trading level is still low in comparison with major international markets — the average daily turnover value in OTC single currency interest rate derivatives in China was about US$4 billion in April 2016, which was 0.3% as much as the daily average of US$1,241 billion in the US and US$1,180 billion in the UK[4]. In terms of currency kind, the OTC single currency interest rate derivatives in RMB had an average daily trading value of US$10 billion in April 2016, which was about 0.7% that in USD, 1.6% that in EUR and 4% that in GBP[5].

As for FX derivatives, RMB/FX swaps (including cross currency swaps)[6] are the dominant product type in terms of trading in nominal principal amount. They reached a total trading value of US$10 trillion in 2016, with a CAGR of 121% during 2006 to 2016 (see Figure 2). In comparison with major international markets, the average daily turnover of OTC FX derivatives (forwards, swaps and options) in RMB was US$134 billion in April 2016, which was about 4% that in USD, 12% in EUR, 19% that in JPY and 30% in GBP[7].

---

3   PBOC announcement《中國人民銀行關於建立場外金融衍生產品集中清算機制及開展人民幣利率互換集中清算業務有關事宜的通知》, 21 February 2014.

4   Source: Bank for International Settlements (BIS) Triennial Survey statistics on OTC derivatives (April 2016), BIS website; daily averages are on "net-gross" basis.

5   The average daily turnover of OTC single currency interest rate derivatives in USD, EUR and GBP was respectively US$1,357 billion, US$641 billion and US$237 billion in April 2016, on "net-net" basis. Source: BIS Triennial Survey statistics on OTC derivatives (April 2016), BIS website.

6   RMB/FX swap involves the actual exchange of two currencies (RMB against FX) on a specific date at a rate agreed in the contract, and a reverse exchange of the same currencies on a specific date further in the future at another rate. RMB/FX cross currency swap involves the exchange of interest payments in two currencies (RMB and FX) for an agreed period of time and may involve also the exchange of principal amounts of the two currencies at a pre-agreed exchange rate at an agreed time in the future.

7   The average daily turnover in OTC FX derivatives in USD, EUR, JPY and GBP in April 2016 were respectively US$3,053 billion, US$1,072 billion, US$701 billion and US$438 billion. Source: BIS Triennial Survey statistics on OTC derivatives (April 2016), BIS website; daily averages are on "net-net" basis.

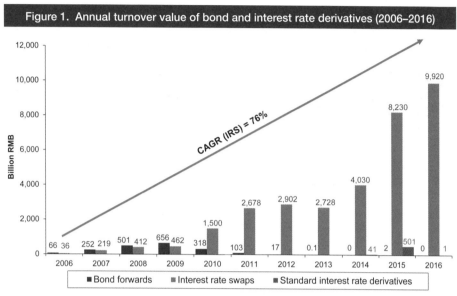

Figure 1. Annual turnover value of bond and interest rate derivatives (2006–2016)

Source: PBOC Annual Reports 2006-2016.

Figure 2. Annual turnover value of RMB/FX derivatives (2006–2016)

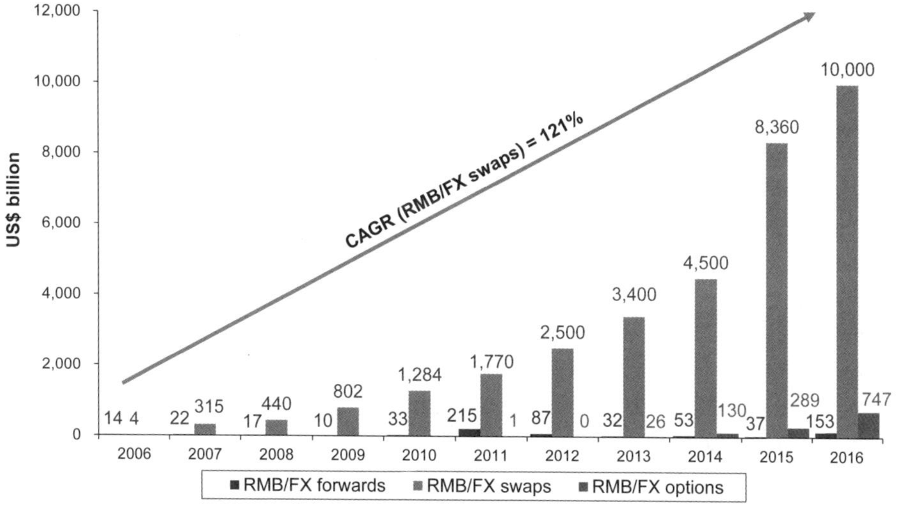

Note: Interbank transactions only, excluding transactions between banks and their clients.

Source: PBOC Annual Reports 2006-2016.

With the rapid growth in FX derivatives trading, especially in RMB/FX swaps instruments, the trading value of FX derivatives has exceeded that in the FX spot market — total FX derivatives trading in 2016 was 1.3 times the trading in the FX spot market. (see Table 1.)

Table 1. RMB/FX market trading value (2016)			
(US$ trillion)	Bank-Client	Interbank	Total
Overall RMB/FX market	3.4	16.8	20.3
Spot FX	2.9	5.9	8.8
FX derivatives	0.5	10.9	11.5
RMB/FX Forward	0.2	0.2	0.4
RMB/FX swaps & cross-currency swaps	0.1	10.0	10.1
RMB/FX options	0.2	0.7	1.0

Note: Numbers may not add up to total due to rounding.
Source: PBOC Annual Report 2016.

Apart from the specified derivative products for trading on the regulated OTC market on CFETS, there would also be interbank trading of other OTC derivatives pursuant to the business needs of the banks. Same as the case in markets around the world, these OTC derivatives would be products created by the buyer/seller to meet their specific needs with customised terms, bilateral trading and settlement. For example, **banks in the Mainland which have corporate clients doing business with foreign partners may need currency or interest rate hedging tools in foreign currencies like USD**. However, no official statistics are available for these trading activities.

1.3 Big growth potential in Mainland OTC derivatives trading

The rapid growth in the trading of IRS could be attributed to the continuous efforts of market-based interest rate reform in the Mainland in the past decade. Along with market economy reform, certain moves of interest rate liberalisation began in early 2000s. The floating range of lending rates were relaxed in January 2004 and was further broadened in October 2004 along with allowing financial institutions to lower the RMB deposit rates below the benchmark rates. Market-based interest rate reform was ascertained by the State Council Standing Committee in June 2013 to be a key financial policy to support economic restructuring. On 24 September 2013, the Market Interest Rate Pricing Mechanism was established as a self-regulatory and coordinative mechanism among financial institutions in the Mainland. It is responsible for self-regulatory management of interest rates in the financial market in accordance with the state's related regulations on interest rates.

In March 2014, the China (Shanghai) Pilot Free Trade Zone became the pioneer in implementing fully market-based foreign currency deposit interest rates in the Mainland. The deposit insurance regulation, which became effective on 1 May 2015, helps pave the way towards full liberalisation of interest rates in the Mainland. The launch of negotiable certificate of deposit (NCD)[8] in June 2016, of which the interest rates are market-driven, signified a further step towards full liberalisation of interest rates.

Market-driven interest rates mean that the borrowing and lending interest rates will fluctuate in consideration of changes in market and economic conditions. **In the light of increasing volatility in interest rates, financial institutions would have increasing demand for interest rate hedging tools such as IRS in the OTC market.** The RMB IRS traded on the interbank market mainly use 7-day repo fixing rate and the Shanghai Interbank Offer Rate (SHIBOR) as the reference rate for the floating end. Figure 3 shows the daily movements of the 1-week SHIBOR, which exhibited a fluctuating rising trend during January 2016 to July 2017.

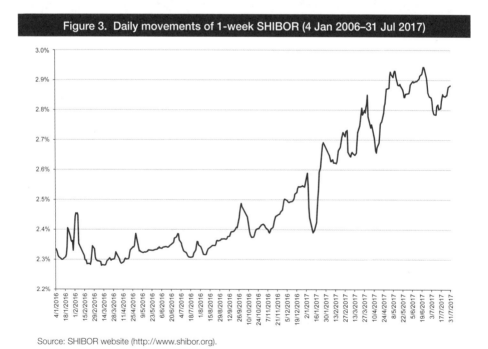

Figure 3. Daily movements of 1-week SHIBOR (4 Jan 2006–31 Jul 2017)

Source: SHIBOR website (http://www.shibor.org).

---

8   NCDs are RMB-denominated book-entry form of deposit instruments issued by banks to investors who are non-financial institutions.

In addition to domestic RMB borrowing and lending activities that would require RMB interest rate risk management, there are increasing overseas financial activities carried out in foreign currencies by Mainland financial institutions. Firstly, along with interest rate liberalisation, the Mainland economy is also increasingly opened resulting in more and more business and financial activities with the world. As a result, **Mainland financial institutions are increasingly engaged in international businesses with overseas parties**, resulting in increasing foreign assets and liabilities. Statistics from SAFE showed that foreign bond assets of Mainland banks (excluding the Central Bank, i.e. the PBOC) had almost doubled from US$48.4 billion at the end of 2015 in a year's time to US$95.2 billion at the end of 2016, rising from 7% to 11% of total foreign financial assets (see Figure 4). Data from the PBOC showed that overseas loans of Mainland financial institutions increased by 50% from RMB 2,328 billion (~US$371 billion) in January 2015 to RMB 3,500 billion (~US$507 billion) in April 2017. (see Figure 5.)

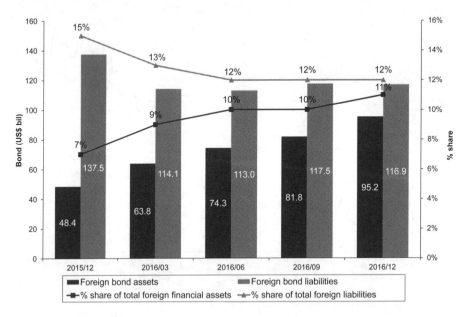

Figure 4. Foreign bond assets and liabilities of Mainland banks (excluding Central Bank) (Dec 2015–Dec 2016)

Source: SAFE data from Wind.

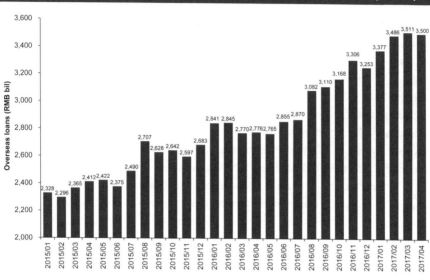

Figure 5. Overseas loans of Mainland financial institutions (Jan 2015–Apr 2017)

Source: PBOC website.

Secondly, given China's **One-Belt-One-Road (OBOR) initiative**[9] put forward by the Chinese leaders in 2013, there will be substantial infrastructure projects to be run along the OBOR countries and China is expected to play a significant role in funding these projects. According to an estimate made by PwC, total new announced project value along the OBOR countries rose 2.1% in 2016 year-on-year to roughly US$400 billion; the final value could rise by as much as 10%[10]. According to a report by the Asian Development Bank (ADB)[11], developing Asia will need to invest US$26 trillion in total or US$1.7 trillion per year in infrastructure until 2030 to maintain its growth momentum, tackle poverty and respond to climate change (or US$1.5 trillion without climate change mitigation and adaptation costs). The infrastructure investment gap — the difference between investment needs and current investment levels — was estimated to be 2.4% of projected gross domestic products (GDP) for the 5-year period from 2016 to 2020. ADB expected the gap

---

9   OBOR consists of the Silk Road Economic Belt (SREB) and 21st Century Maritime Silk Road (MSR) initiatives. The SREB runs through Central Asia, West Asia, the Middle East to Europe, with extension to South Asia and Southeast Asia. The MSR runs through Southeast Asia, Oceania and North Africa. The report, *Industrial cooperation between countries along the Belt and Road* (《「一帶一路」沿線國家產業合作報告》), released by the China International Trade Institute in August 2015 identified 65 countries along OBOR that will be participating in the initiative.

10   *China and Belt & Road Infrastructure, 2016 review and outlook*, PwC, February 2017.

11   *Meeting Asia's infrastructure needs*, ADB, February 2017.

could be bridged to 40% by fiscal reforms and 60% by the private sector.

Mainland financial institutions including policy banks and commercial banks, together with specialised investment funds[12] **and multilateral financial institutions promoted by China**[13] **would play a role in funding the OBOR projects.** In 2016, China had a trade volume with OBOR countries of about US$953.59 billion and a total project contract value of about US$126.03 billion[14]. Funding means for OBOR projects include preferential loans, bank syndicate loans, exports credit insurance, industry funds, bond investments, entrusted asset management, equity holdings, etc.

Thirdly, the Mainland government has been encouraging the "going-out" of domestic enterprises as part of its enterprise reform, especially for the state-owned enterprises. **Enterprises' international business development** including mergers and acquisitions (M&A) is therefore a key development direction. **Mainland banks would provide funding to these enterprises in foreign currencies by means of bond issuance.**

As a result of the above funding needs for international trade, OBOR infrastructure projects and enterprise international business developments, Mainland financial institutions would have an increasing proportion of their balance sheet in foreign currencies, the majority of which would be USD, the mostly used global currency in international trade and finance. **Interest income in foreign currency assets is susceptible to interest rate risk as well as to FX rate risk.** The latter has become more prominent after the exchange rate system reform in August 2015 as a result of which the formation mechanism of the central parity rate of RMB against USD in the interbank FX market has become more market-driven. Figure 6 shows the increased volatility of the offshore RMB (CNH) to USD exchange rate (USD/CNH rate) after the reform.

---

12 Including the Silk Road Fund, China-ASEAN Investment Cooperation Fund and China-Africa Development Fund.
13 Including the Asian Infrastructure Investment Bank (AIIB) and BRICS Development Bank.
14 Source:〈一帶一路資金持再盤點〉, China Merchants Securities, 14 May 2017.

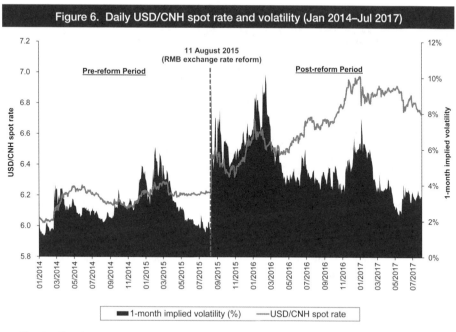

Figure 6. Daily USD/CNH spot rate and volatility (Jan 2014–Jul 2017)

Source: Bloomberg.

Mainland financial institutions therefore would have an increasing demand for hedging their foreign currency assets against interest rate and exchange rate risks through the use of risk management instruments. While the standardised on-exchange tools like bond futures may not be suitable for their customised needs, they would often go for OTC products like IRS and FX swaps with customised terms to match the payment terms of their assets denominated in foreign currencies[15]. These OTC transactions in foreign currencies undertaken by Mainland financial institutions had been mostly in USD — 21% of China's OTC single currency interest rate derivatives turnover in April 2016 was in USD, compared to 79% in RMB and negligible proportions in Australian dollar (AUD) and euro (EUR)[16].

---

15 Domestically, available on-exchange tools are the RMB bond futures — 5-Year and 10-Year Treasury Bond Futures — traded on the China Financial Futures Exchange (CFFEX), the only exchange in Mainland China for trading financial derivatives. Mainland domestic banks may also deal in foreign currency bond derivatives on overseas exchanges upon authorisation obtained from the Mainland authorities.

16 Source: BIS Triennial Survey statistics on OTC derivatives (April 2016), BIS website; the percentages are based on daily averages on "net-gross" basis.

2 Global requirements for risk management of OTC derivatives

2.1 Mandatory clearing and margin requirements

The 2008 Global Financial Crisis exposed significant weaknesses in the resiliency of banks and other market participants to financial and economic shocks. In particular, the lack of transparency in the OTC derivatives markets together with the increasing complexity of derivative instruments and their nexus within the financial sector are considered a major systemic risk factor. The absence of regulation and the bilateral nature of OTC derivatives transactions are causes of the market opaqueness. In response, the Group of Twenty (G20) initiated a reform programme in 2009 to reduce the systemic risk from OTC derivatives. The reform programme initially comprised the following[17]:

- All standardised OTC derivatives should be traded on exchanges or electronic platforms, where appropriate;
- All standardised OTC derivatives should be **cleared through CCPs**;
- OTC derivatives contracts should be reported to trade repositories;
- **Non-centrally cleared derivatives contracts should be subject to higher capital requirements**.

In 2011, the G20 agreed to **add margin requirements on non-centrally cleared derivatives** to the reform programme. Universal adoption of the G20 margin requirements is expected since, if otherwise, the effectiveness of margin requirements could be undermined (i.e. regulatory arbitrage) and it will not be a level-playing field as financial institutions in the low-margin locations could gain a competitive advantage[18].

One of the key principles is that all financial firms and systemically important non-financial entities that engage in non-centrally cleared derivatives must exchange initial and variation margin as appropriate to the counterparty risks. Initial margin should be exchanged without netting. Such bilateral margin would be higher than the initial margin that would be charged by a CCP if these trades are cleared through it. The standardised initial margins required would depend on the asset class, ranging from 1% to 15% of

---

17 Source: *Margin requirements for non-centrally cleared derivatives*, Basel Committee on Banking Supervision and Board of the International Organisation of Securities Commissions (IOSCO), BIS, March 2015.
18 Ditto.

the notional amount of the trade. As required by regulators in the US, Hong Kong and European Union (EU), variation margin requirements would become generally applicable in March 2017. By that date, certain regulated derivatives entities in these regions, including swap dealers in the US and authorised institutions in Hong Kong, are required to collect and post variation margin with covered counterparties.

2.2   Basel III capital requirements

Global financial institutions are now subject to more stringent capital requirements upon the implementation of Basel III[19]. The Basel III framework aims at strengthening the regulation of both capital and liquidity, and to improve the stability and resilience of individual banks and the banking sector as a whole. Banks are required to maintain **a minimum ratio of capital to risk-weighted assets (RWA)**. RWA is calculated by applying a weighting factor (the risk weight) to the asset values, such that "safer" assets are discounted and can therefore be backed by less capital. OTC derivatives are generally considered risky assets and the risk weights would vary according to the type of counterparties. For OTC transactions that are cleared through CCPs which are qualifying (QCCPs[20]), the risk weight could be as low as 2%-4%; otherwise it could be at least 20% on a bilateral basis[21].

2.3   Implementation of global requirements in the US, EU and Hong Kong

In the US, the Commodities Futures Trading Commission (CFTC) took a phased-in approach depending on the category of the trading entity, to implement mandatory central clearing of OTC derivatives transactions, initially for certain classes of IRS and CDS from December 2012 to complete by 9 September 2013[22]. "End-user" clearing exception applies only to counterparties that are not financial entities. On 2 July 2013, the Board of Governors of the Federal Reserve System approved a final rule to establish a new comprehensive regulatory capital framework for all US banking organisations to implement the Basel III

---

19 Pursuant to the report, *Basel III: A Global Regulatory Framework for More Resilient Banks and Banking Systems,* published by the Basel Committee on Banking Supervision in December 2010, worldwide adoption of the Basel III accord is being phased in from 2013 to 2019.

20 Basel III defines a QCCP as an entity that is licensed to operate as a CCP by the relevant regulator and that is based and prudentially supervised in a jurisdiction where the relevant regulator has implemented domestic rules for financial market infrastructures that are consistent with the Principles for Financial Market Infrastructures by CPMI-IOSCO — The Committee on Payments and Market Infrastructures (CPMI) of the IOSCO.

21 Reference is made to the "Final Rule" approved by the US Board of Governors of the Federal Reserve System for implementing Basel III.

22 "OTC derivatives central clearing in the US", Risk Advisors Inc., February 2013.

capital regime (the Basel III US Final Rule).

In the EU, the European Commission adopted new rules in the European Market Infra-structure Regulation (EMIR) in August 2015 to make it mandatory for certain OTC interest rate derivatives transactions to be cleared through CCPs and subsequently in March 2016 to make the same for certain OTC credit derivatives transactions. EMIR's risk mitigation requirements apply to all non-centrally cleared OTC derivatives transactions, which would require the exchange of collateral and bilateral margining. The clearing obligation under Article 4 of EMIR took effect from 21 June 2016 subject to phase-ins that are based on firms' categorisation and derivatives volumes. The clearing obligation applies to contracts between any combination of financial counterparties and non-financial counterparties who exceed the clearing threshold (in terms of gross notional value). The final timeline for all combination of firm categories in OTC transactions to comply with the clearing obligation is 21 December 2018[23].

In Hong Kong, in line with global efforts, the Hong Kong Monetary Authority (HKMA) and the Securities and Futures Commission (SFC), jointly with the Hong Kong Government and stakeholders, have been developing a regulatory regime for the OTC derivatives market in Hong Kong. A series of market consultation has been going on and conclusion actions are and being implemented. Towards establishing an OTC derivatives regulatory regime, the HKMA established the **OTC Derivatives Trade Repository**, which launched its reporting service in July 2013; and the Hong Kong Exchanges and Clearing Limited (HKEX) established the **OTC Clearing Hong Kong Ltd (HKEX OTC Clear)**, which commenced business in November 2013. Phase 1 mandatory central clearing of OTC derivatives transactions commenced in September 2016[24], which covers certain standardised IRS entered into between major dealers.

According to the Basel Committee on Banking Supervision's progress report on adoption of the Basel regulatory framework in April 2017, the US margin requirements for non-centrally cleared derivatives are phased-in beginning on 1 September 2016 and would be fully effective on 1 September 2020. For the EU, the initial margin requirements are being phased in depending on the type of counterparty from 4 February 2017 and the variable margin requirements began to apply from 1 March 2017. For Hong Kong, the margin requirements are in force from 1 March 2017 (subject to a 6-month transitional period).

---

23 Source: UK Financial Conduct Authority website (https://www.fca.org.uk).
24 Applicable in effect to in-scope transactions entered into on or after 1 July 2017.

2.4 Implications to Mainland financial institutions

As a result of the strengthened regulatory requirements discussed above, Mainland financial institutions participating in the OTC derivatives markets, especially when dealing with foreign counterparties, are obliged to follow the overseas mandatory clearing and reporting requirements, and need to take into consideration the cost impacts of non-centrally cleared OTC transactions. Table 2 below summarises the implications.

Table 2. Implications of global requirements on OTC derivatives transactions		
Regulatory requirement	Global basis	Implications
Capital ratio to risk-weighted assets	Basel III	• Bilateral OTC derivatives transactions would have a much higher risk weight than CCP derivatives transactions, implying a higher pricing of non-centrally-cleared OTC contracts to the Mainland counterparty [25].
Mandatory central clearing	G20	• Financial institutions trading with counterparties in jurisdictions that require mandatory central clearing of their OTC derivatives trades are obliged to follow suit; otherwise, they would not be able to trade with these counterparties. • Utilising clearing houses recognised by the home regulator of the overseas counterparties will need to establish connectivity, e.g. via a clearing broker. This may be costly as the clearing broker will charge a commission fee in addition to the clearing fee that is charged by the CCP.
Margining for non-centrally cleared derivatives	G20	• Initial margin and variation margin, on a non-netting basis, will be required for bilateral OTC derivatives trades with foreign counterparties where the jurisdiction follows the G20 requirements, which implies high funding and operating costs than central clearing. • In bilateral clearing, as there are still uncertainties in PRC being a netting jurisdiction [26], Mainland banks would face challenges in executing ISDA collateral agreement [27] with foreign counterparties and therefore may not be accepted by foreign institutions as counterparties in executing OTC derivatives transactions, or even being accepted, may be charged higher margin requirements [28].

---

25 The counterparty will price-in the capital charge when quoting the price to the Mainland bank. The Mainland bank would therefore be subject to a higher pricing, especially when there is no International Swaps and Derivatives Association (ISDA) collateral agreement in place for credit risk protection for the transaction.

26 The status as a "netting jurisdiction" provides for the enforceability of close-out netting for OTC derivatives transactions under the ISDA master agreement in case of bankruptcy of a counterparty. (See "China — The New Netting Jurisdiction", *Derivatives Week*, Vol. XXIII, No.5, 10 February 2014, and "ISDA publishes updated memoranda on China close-out netting", King&Wood Mallesons, 3 April 2017.)

27 This refers to the Credit Support Annex (CSA) of the ISDA Master Agreement, which defines the terms or rules under which collateral is posted or transferred between swap counterparties to mitigate the credit risk arising from "in-the-money" derivatives positions.

28 To clarify on the enforceability of close-out netting provisions under master derivatives agreements with Mainland financial institutions, the China Banking Regulatory Commission (CBRC) issued a reply document dated 4 July 2017 in response to related questions to the Financial and Economic Affairs Committee of the National People's Congress. The reply document states that China's Enterprise Bankruptcy Law in principle does not conflict with the close-out netting provisions of ISDA's related regulations (the Master Agreement) and yet the PRC judiciary ultimately has the power to determine the validity of close-out netting provisions.

In conclusion, under the latest global regulatory requirements for OTC derivatives, **Mainland financial institutions dealing with foreign counterparties in OTC derivatives would be subject to high costs for not using central clearing services for their transactions**. Moreover, the multilateral netting process in central clearing would substantially reduce the margin requirements to a level much lower than the total margins required in the case of bilateral clearing with multiple counterparties. **It would therefore be more preferable for Mainland financial institutions to opt for central clearing of their increasing OTC derivatives transactions with foreign counterparties. The key consideration for the Mainland financial institutions is the practical choice of a relatively low cost central clearer.**

3 OTC clearing services for Mainland banks

Mainland banks which execute OTC derivatives transactions with foreign counterparties may be obliged to adopt the practice of the foreign counterparties to centrally clear the transactions via an international clearing house. Prominent ones in the developed markets include LCH and Eurex Clearing (a company of Deutsche Börse AG) in Europe and CME Clearing in the US. Compared to these clearing houses in the western markets, the OTC clearing arm of HKEX in Hong Kong — HKEX OTC Clear — may offer a preferable alternative.

3.1 LCH[29]

LCH operates a number of clearing arms under LCH.Clearnet Ltd for OTC derivatives clearing — among others, **SwapClear** for IRS and **ForexClear** for FX derivatives which the Mainland banks would be active in.

SwapClear services cover different swap classes (IRS / zero coupon / basis / inflation / FRA), indexes (overnight index swaps (OIS)) and maturities (variable notional swaps). In 2017 up to 28 July, the notional value of OTC transactions cleared by SwapClear amounted to US$526.44 trillion, of which 54% was in USD; and US$122.69 trillion was client

---

29 Relevant data and information are obtained from the LCH website (http://www.lch.com).

clearing, of which 62% was in USD. During the same period, 29% of total cleared volume in notional terms was in IRS (i.e. US$154.28 trillion, of which US$58.69 trillion or 38% was in USD); 30% of client clearing volume in notional terms was in IRS (i.e. US$37.04 trillion, of which 46% was in USD). According to SwapClear, it clears more than 50% of all OTC IRSs and more than 95% of the overall cleared OTC IRS market in the world.

ForexClear serves the non-deliverable currency forwards (NDF) market, covering currency pairs of USD against 12 currencies as of the end of July 2017, including Chinese yuan (CNY). In November 2013, client clearing was launched on ForexClear, allowing clients to access through Futures Commission Merchants (FCMs)[30]. In 2017 up to 22 July 2017, the total volume cleared by ForexClear amounted to US$5.45 trillion, of which 13% (US$712 billion) was in USD/CNY.

OTC transactions must be submitted to LCH for clearing by SwapClear or ForexClear via an Approved Trade Source System (e.g. Bloomberg, MarkitWire) as defined in the LCH rulebook. Membership of SwapClear consists of two types — SwapClear Clearing Members (SCMs) and FCM Clearing Members (FCMs). SCMs are direct Clearing Members that can clear proprietary business and non-US domiciled client business. FCMs are direct Clearing Members that can clear proprietary business, US domiciled client business and non-US domiciled client business. As for ForexClear, participants include FX Clearing Members (FXCCMs), FCM Clearing Members, dealers or clients. Dealers register trades in ForexClear via a clearing agreement with an FXCCM. A client can clear through a FXCCM or FCM.

Applicants for membership of LCH.Clearnet need to satisfy a full range of requirements including minimum net capital requirements and operational requirements such as proper system setups, and certain additional criteria for clearing OTC trades in SwapClear or ForexClear. If the applicant is a bank, it must at all times be appropriately authorised by the banking supervisors of its home country and additionally meet any notification or authorisation requirements set by banking supervisors in the UK.

LCH meets the Basel III criteria of a QCCP in a number of regulatory regimes, including the US, the EU and Hong Kong. In other words, OTC derivatives exposure of banks using LCH's central clearing service would subject to lower capital charges.

Becoming a clearing member of LCH would probably not practical for Mainland banks since the PRC may not be accepted as a netting jurisdiction (see Table 2 and associated footnotes). In the usual case, Mainland banks which use LCH service for clearing their cross-border OTC transactions would have to select the Clearing Member of SwapClear or

---

30 A FCM is an entity that solicits or accepts orders to buy or sell futures contracts, options on futures, retail off-exchange forex contracts or swaps, and accepts money or other assets from customers to support such orders.

ForexClear as the case may be to be their clearing agent. Legal documents such as a Client Clearing Agreement and a Security Deed[31] with the clearing member that are in compliance with the related rules and regulations governing the clearer would be required. SwapClear charges booking fee and maintenance fee for client clearing, on a per-million notional basis. Blended rate and multilateral compression fee schedule are applied to compression service, which enables market participants to reduce the overall notional and number of line items in their portfolios by netting trades. ForexClear charges a per-million notional fee for client clearing and client compression fee is also chargeable for using compression service.

3.2 CME Group OTC Clearing[32]

In the IRS market, CME Group OTC Clearing (CME OTC Clear) covers similar product types as the LCH's SwapClear (fixed/float swaps, OIS, basis swaps, FRAs, etc.) in 21 currencies. In the FX derivatives market, it covers NDFs in 12 currencies including CNY, and cash-settled forwards (CSFs). In the first half of 2017, it had a total clearing volume of US$16.38 trillion in IRS and US$21 million in FX derivatives.

OTC clearing members that will clear OTC derivatives for customers must be registered with the US CFTC as a FCM. OTC clearing members that are incorporated/domiciled in non-US jurisdictions must be subject to a legal and insolvency regime acceptable to the clearing house.

The clearing arms of CME Group — CME Clearing and CME Clearing Europe — meet the criteria established for a QCCP at its US and European clearing houses respectively, covering the OTC derivatives products. Additionally, the operating company of CME Clearing received recognition from the European Securities and Markets Authority (ESMA) to qualify as a QCCP. Therefore, European clients are able to treat CME Clearing as a QCCP.

Similar to the case of clearing at LCH, Mainland banks that clear through CME OTC Clear has to select a clearing firm at CME OTC Clear and complete a Futures Account Agreement with a Cleared OTC Derivatives Addendum. Clearing fee and maintenance fee are applicable on a per-million notional basis. Multilateral compression service is available at a fee.

---

31 A Security Deed is required by the clearing house where no exempt client clearing rule is available for the jurisdiction in question. This acts as a form of protective mechanism that entitles the clearing house to deal with the client assets in case the clearing member defaults, for protection of the client's interests.

32 Relevant data and information are obtained from the CME Group website (http://www.cmegroup.com/clearing.html).

3.3   Eurex Clearing[33]

Eurex Clearing's OTC clearing services (EurexOTC Clear) covers IRS, basis swaps, OIS, zero coupon inflation and FRAs. In December 2016, Eurex Clearing announced that EurexOTC Clear planned to introduce the clearing of OTC FX swaps, OTC FX spots and OTC FX forwards in currency pairs EUR/USD and GBP/USD. In the first half of 2017, a total notional value of €739,625 million (~US$845 billion) in IRS was cleared at EurexOTC Clear.

There are three types of membership at Eurex Clearing for OTC clearing — General Clearing Member (GCM) which can clear their own business and those of all clients; Direct Clearing Member (DCM) which can clear their own business; Basic Clearing Member (BCM) which combines elements of direct clearing membership and traditional service relationship in client clearing. BCMs have principal relationship with the CCP, but require the support of a Clearing Agent for client clearing. GCMs can act as Clearing Agents. Clients in OTC clearing must be disclosed and known to Eurex Clearing as Registered Customers and as such need to enter a tripartite agreement with Eurex Clearing and their Clearing Member.

Eurex Clearing received a conditional registration from the US CFTC as a derivatives clearing organisation under the Commodity Exchange Act, which will become effective after Eurex Clearing meets CFTC's "straight-through processing" requirements. In this situation, Eurex Clearing may clear proprietary positions in IRS for US clearing members but not yet for FCM customer positions[34].

Applicable fees include booking fee, maintenance fee, other administrative charges and fees for additional services like trade netting or multilateral compression.

3.4   HKEX OTC Clear[35]

HKEX OTC Clear was established in 2013 in response to G20's reform programme in 2009, to provide clearing services for OTC derivatives as a CCP. It commenced business on 25 November 2013 after being recognised by the SFC as Recognised Clearing House and was designated by the SFC in August 2016 as a CCP for mandatory clearing for OTC derivatives transactions specified under the Hong Kong OTC derivatives regulatory regime. Client clearing service was subsequently launched in March 2017.

---

33 Relevant data and information are obtained from the Eurex Clearing website (http://www.eurexclearing.com/clearing-en/).

34 Source: CFTC Release pr7316-16, 1 February 2016.

35 Source: HKEX.

Both proprietary clearing and client clearing services are offered for IRS, basis swaps, cross-currency swaps (CCS), non-deliverable IRS (NDIRS) and NDFs (see Appendix 1 for the list of products covered). The clearing volume in notional terms experienced a CAGR of 343% since launch to 2017 and an even higher CAGR of 484% since the full-year operation in 2014 (see Figure 7).

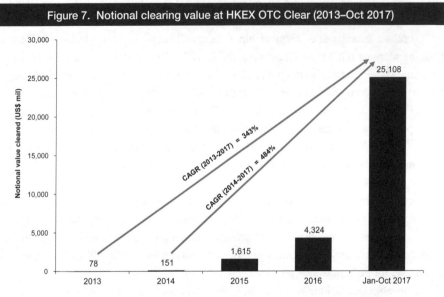

Figure 7. Notional clearing value at HKEX OTC Clear (2013–Oct 2017)

Note: CAGR is calculated using the 12-month pro-rata figure of 2017 based on Jan-Oct 2017 figure.
Source: HKEX.

HKEX OTC Clear accepts Clearing Members which are either Authorised Institutions (AIs) licensed by the HKMA or corporations licensed by the SFC (Licensed Corporations). The OTC clearing and settlement system accepts OTC derivatives transactions from two Approved Trade Registration Systems (ATRS) — MarkitWire, and DSMatch.

In April 2015, HKEX OTC Clear was recognised as a "Third Country CCP" by the ESMA in Europe for OTC derivatives clearing. In September 2015, it received a prescribed CCP status facility under Australia's mandatory clearing regime by the Australia Securities and Investments Commission. In December 2015, it obtained exemption from the US CFTC, allowing it to offer clearing services to US persons for their proprietary transactions without being registered as a Derivatives Clearing Organisation. With these international recognised OTC CCP clearing statuses, HKEX OTC Clear is well positioned to offer central clearing services for Mainland banks' OTC transactions in USD, EUR and RMB with foreign counterparties. The product coverage would be expanded to cover deliverable FX currency forwards and swaps, and FX options in the future, subject to regulatory approval. Trade compression service is also scheduled to be launched in late 2017 to cater for customer needs.

Applicable fees for client clearing services include registration fee and maintenance fee for IRS and registration fee for NDF, in addition to other fees charged on services used by the Clearing Member.

3.5   A choice for Mainland banks

HKEX OTC Clear is the only Recognised Clearing House for OTC derivatives in Hong Kong and is also a QCCP for US and European clients. It operates in the Asian time zone in the Asian international financial centre, Hong Kong, in which the major Mainland banks have set up branches and have been in business operations for years. Moreover, Hong Kong is a major trading market in OTC FX and interest rate instruments in the world[36] and also a prominent offshore RMB centre[37].

In Hong Kong, the HKMA's SWIFT-based **RMB Real Time Gross Settlement (RTGS) system** facilitates market participants from all over the world to handle RMB transactions both with Mainland China and among the offshore markets. The RMB RTGS system has Bank of China (Hong Kong) Limited as its Clearing Bank, which maintains a settlement account with the PBOC and is a member of the China National Advanced Payment System (CNAPS). The RMB RTGS system in Hong Kong is directly linked up with CNAPS and can be regarded as a technical extension of CNAPS in Mainland China, but governed by Hong Kong laws. There are also inter-linkages between RMB, HKD, USD and euro RTGS systems in Hong Kong, enabling payment-versus-payment (PvP) settlement to improve settlement efficiency and to eliminate settlement risk arising from time lags in transactions or time-zone differences. Statistics from SWIFT show that the value of RMB settlement handled by banks in Hong Kong accounted for some 70% of the total offshore RMB payments conducted vis-à-vis Mainland China and within the offshore market globally[38].

---

36 Hong Kong ranked 4th in the turnover of OTC FX instruments and 3rd in the turnover of OTC single currency interest rate derivatives in the world. Source: BIS Triennial Survey statistics on OTC derivatives (April 2016), BIS website; daily averages are on "net-gross" basis.

37 Total RMB deposits in Hong Kong was RMB 534.73 billion as at the end-July 2017 (source: HKMA website), compared to RMB 138 billion as of end-June 2017 for Singapore (preliminary data, source: Monetary Authority of Singapore website), GBP 9 billion (~RMB 77 billion) as of end-March 2017 for the UK (source: Bank of England website) and RMB 307.8 billion as of end-April 2017 for Taiwan (source: Offshore RMB Express No. 6, 2017, Bank of China (Hong Kong)).

38 Quoted in HKMA's report, *Hong Kong — The Global Offshore Renminbi Business Hub*, January 2016.

In view of the progress of RMB internationalisation, offshore RMB derivatives trading is believed to have a high growth potential for the purpose of RMB risk management[39]. In this respect, the participation from Mainland institutions is believed to increase, offering RMB liquidity to meet the demand from global institutions for their increasing RMB assets. With access to the RMB RTGS payment system in Hong Kong, HKEX OTC Clear is able to support Mainland and global financial institutions' OTC transactions in offshore RMB[40], in addition to serving OTC transactions in USD and other major currencies. On the contrary, overseas clearing houses using FX settlement services offered by overseas platforms like CLS would not be able to offer RMB settlement services[41].

With the strategic position of being an OTC clearing house in the Asian region serving Mainland market participants, shareholders of HKEX OTC Clear include five Mainland financial institutions — Agricultural Bank of China Ltd, Bank of China (Hong Kong) Ltd, Bank of Communications Co., Ltd Hong Kong Branch, CCB International Securities Ltd and the Industrial and Commercial Bank of China (Asia) Ltd.

Mainland-incorporated banks' Hong Kong subsidiaries which are AIs or Licensed Corporations can become members of HKEX OTC Clear upon satisfying the membership requirements. Given the extensive business operations of Mainland banks in Hong Kong, this would be relatively more cost-effective than the direct application of Mainland-incorporated banks for membership in the US or Europe clearing houses. For Mainland banks which are not members of major clearing houses of LCH in Europe or the CME Group in the US, client clearing through a clearing broker will be needed in executing OTC derivatives transactions with foreign parties. This would be more costly due to higher transaction fee and commission fee to pay to the clearing broker to compensate for netting and capital cost, and would also be subject to the default risk of the clearing broker. Besides, the issues about operational efficiency and hitches for client clearing via a clearing agent may also be concerns to the Mainland banks. These would include the long onboarding process in engaging a clearing agent whenever the need arises, the dependence on the system infrastructure of the clearing agent and the dependence on the agent's gaining approval for acceptance by the CCP for central clearing of the transaction.

Moreover, HKEX OTC Clear has developed a special solution for admitting Mainland-incorporated licensed banks to be its Clearing Members, within the governance of the Hong

---

39 See also Chapter 14 of this book, "HKEX towards an offshore RMB product trading and risk management centre".

40 Hong Kong Interbank Clearing Ltd (HKICL), the operator of the payment systems, has a total of 141 local RMB clearing members and 68 overseas RMB clearing members as of 25 September 2017 (source: HKMA website).

41 CLS currently provides FX settlement services covering 18 sovereign currencies through accounts established with each of the central banks of the respective currencies. These do not include the RMB and the PBOC. (Source: CLS website.)

Kong regulatory framework. This practice and the resultant clearing solution for Mainland banks are unparalleled in the US and European clearing houses due to the uncertainty in the status of the PRC as a netting jurisdiction. Details are presented in Section 4 below.

4 HKEX OTC Clear's solution for Mainland banks

4.1   PRC Clearing Members

PRC-incorporated banks with Hong Kong branches which are Authorized Institutions regulated by HKMA can be admitted as direct clearing members (referred to as "PRC Clearing Members") of HKEX OTC Clear.

Since HKEX OTC Clear has obtained legal opinions confirming the enforceability of its clearing rules under Hong Kong and PRC law (including provisions relating to close out and set off), it is able to margin PRC Clearing Members on the basis of their net exposure across their entire cleared portfolio. In addition, HKEX OTC Clear enjoys finality protection for payments made in respect of cleared contracts and actions taken against PRC Clearing Members under the default provisions of its clearing rules under Hong Kong law (which is the law governing the clearing rules and collateral posted by PRC Clearing Members and held by HKEX OTC Clear).[42]

Currently, four PRC-incorporated banks — Agricultural Bank of China Ltd, Bank of Communications Co., Ltd, China Minsheng Banking Corporation, Ltd and Shanghai Pudong Development Bank Co., Ltd — have been admitted as direct Clearing Members of HKEX OTC Clear via their Hong Kong branches. Direct members also include the Hong Kong subsidiaries of three other PRC-incorporated banks. (See Appendix 2 for the Clearing Members of HKEX OTC Clear.)

---

42 Source: HKEX.

4.2 Clearing models for cross-border clearing with international counterparties

With direct membership, Mainland banks are able to conduct direct clearing with HKEX OTC Clear without the need to use a clearing broker, thereby avoiding the default risk of the clearing broker and saving commission costs and transaction fees. Direct clearing may be done through the Mainland bank's Hong Kong subsidiary which is a Clearing Member of HKEX OTC Clear, or through its Hong Kong branch if itself is a Clearing Member of HKEX OTC Clear via its Hong Kong branch. The respective operations are illustrated in Figures 8 and 9 below.

Figure 8. Operating model for clearing via the Mainland bank's Hong Kong subsidiary

(a) Trade registration

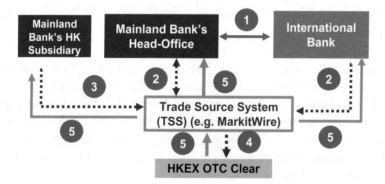

1. Bi-lateral trades done between Mainland Bank's Head-Office and International Bank.
2. Clearing request submitted by both the Mainland Bank's Head-Office (act as a client) and International Bank via TSS.
3. Mainland Bank's HK Subsidiary (act as clearing broker) checks the credit limits before accepting the request.
4. Once accepted, TSS sends the matched record to HKEX OTC Clear for product, margin and credit checks.
5. After the trade is accepted by registration, HKEX OTC Clear will inform Mainland Bank's HK Subsidiary, Mainland Bank's Head-Office and International Bank about the trade clearing status through TSS, and trade will be novated:
   • HKEX OTC Clear vs International Bank
   • HKEX OTC Clear vs Clearing Mainland Bank's HK Subsidiary
   • Mainland Bank's HK Subsidiary vs Mainland Bank's Head-Office

(continued)

Figure 8. Operating model for clearing via the Mainland bank's Hong Kong subsidiary

(b) Margin settlement and collateral management

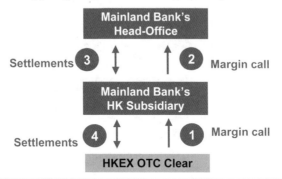

1. HKEX OTC Clear will issue initial margin and variation margin call to the Mainland Bank's HK Subsidiary.
2. Mainland Bank's HK Subsidiary will inform Mainland Bank's Head-Office for the relevant initial and variation margin required by HKEX OTC Clear.
3. Mainland Bank's Head-Office will settle the initial margin and variation margin to Mainland Bank's HK Subsidiary.
4. Mainland Bank's HK Subsidiary will then settle the initial margin and variation margin of Mainland Bank's Head-Office to HKEX OTC Clear.

Source: HKEX OTC Clear.

Figure 9. Operating model for clearing via the Mainland bank's Hong Kong branch

(a) Trade registration

1. Bi-lateral trades done between Mainland Bank's Head-Office and International Bank.
2. Clearing request submitted by Mainland Bank's Head-Office and International Bank via TSS.
3. Once accepted, TSS sends the matched record to HKEX OTC Clear for product, margin and credit checks.
4. After the trade is accepted by registration, HKEX OTC Clear will inform Mainland Bank's HK Branch, Mainland Bank's Head-Office and International Bank about the trade clearing status through TSS, and trade will be novated:
    * HKEX OTC Clear vs International Bank
    * HKEX OTC Clear vs Mainland Bank's Head-Office (under membership of Mainland Bank via its HK Branch)

(continued)

Figure 9. Operating model for clearing via the Mainland bank's Hong Kong branch

(b) Margin settlement and collateral management

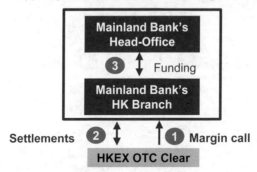

1. HKEX OTC Clear will issue initial margin and variation margin call to the Mainland Bank's HK Branch. The initial and variation margin will be calculated based on all trades cleared by Head-Office and HK Branch.
2. Mainland Bank's HK Branch settles the initial margin and variation margin to HKEX OTC Clear.
3. Internal funding arrangement is required between Mainland Bank's Head-Office and its HK Branch.

Source: HKEX OTC Clear.

The above clearing models would provide a more convenient solution and at lower costs than the model for clearing through a clearing broker, as illustrated in Figure 10.

Figure 10. Operating model for clearing by Mainland banks via a clearing broker at LCH

(a) Trade registration

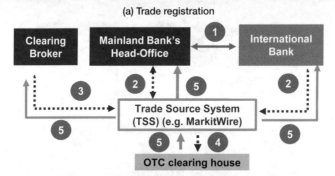

1. Bi-lateral trades done between Mainland Bank's Head-Office and international bank.
2. Clearing request submitted by both the Mainland Bank's Head-Office (act as a client) and International Bank via TSS.
3. Clearing Broker checks the credit limits before accepting the request.
4. Once accepted, TSS sends the matched record to OTC clearing house for product, margin and credit checks.
5. After the trade is accepted by registration, OTC clearing house will inform Clearing Broker, Mainland Bank's Head-Office and International Bank about the trade clearing status through TSS, and trade will be novated:
   - OTC clearing house vs International Bank
   - OTC clearing house vs Clearing Broker
   - Clearing Broker vs Mainland Bank's Head-Office

(continued)

Figure 10. Operating model for clearing by Mainland banks via a clearing broker at LCH

(b) Margin settlement and collateral management

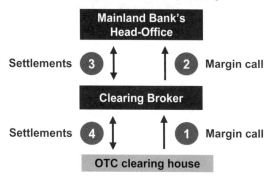

1.   OTC clearing house will issue initial margin and variation margin call to the Clearing Broker.
2.   Clearing Broker will issue statement to Mainland Bank's Head-Office for the relevant initial and variation margin required by OTC clearing house.
3.   Mainland Bank's Head-Office will settle the initial margin and variation margin to Clearing Broker.
4.   Clearing Broker will then settle the initial margin and variation margin of Mainland Bank's Head-Office to OTC clearing house.

Source: HKEX OTC Clear.

4.3   Comparative advantages vis-à-vis other CCPs

As an alternative to indirect clearing through clearing brokers at clearing houses in the US and Europe, the comparative advantages for Mainland banks to do direct clearing with HKEX OTC Clear would include time zone convenience, lower risk exposure and lower costs (see Table 3).

Table 3. Brief comparison of HKEX OTC Clear with other major OTC clearing houses

Attribute	HKEX OTC Clear	CME OTC Clear	LCH (SwapClear/ForexClear)	EurexOTC Clear
Service hours	HK time 08:30 — 19:00 07:30 — 23:00 (Web Portal)	23 hours and 45 minutes per day	SwapClear 07:30 — 24:00 (GMT) 14:30 — 07:00 (HK time) ForexClear 24 hours weekdays 20:00 (GMT) Sunday — 01:00 (GMT) Saturday	8:00 — 22:00 CET
Diversity of membership	EU, US, PRC, HK and other Asian banks		Mainly international banks	

(continued)

Table 3.	Brief comparison of HKEX OTC Clear with other major OTC clearing houses			
Attribute	HKEX OTC Clear	CME OTC Clear	LCH (SwapClear/ForexClear)	EurexOTC Clear
Product coverage	Various types of IRS and NDF	Various types of IRS, NDF, CSF	Various types of IRS, NDF	Various types of IRS
Currency coverage	EUR, USD, RMB and other Asian currencies	Major international currencies		
Membership for Mainland banks	• Direct membership of HK subsidiary • Direct membership of Mainland-incorporated bank via HK branch	The bank (not the branch) may apply for direct membership if satisfying additional regulatory requirements of the jurisdiction		
Clearing mode for Mainland banks	Direct clearing via the Hong Kong subsidiary which is a clearing member; or via its Hong Kong branch if itself is a clearing member via its HK branch	Indirect clearing via a Clearing Broker (another international bank)		
Risk	Lower risk as direct clearing is done via intra-group entities	Subject to default risk of Clearing Broker		
Cost	Lower transaction cost and no commission fee	Higher transaction fee and commission fee for using Clearing Broker to compensate for netting and capital cost		

Note: GMT — Greenwich Mean Time; CET — Central European Time.
Source: HKEX OTC Clear, websites of the respective clearing houses.

Operating in Hong Kong which is a financial hub for Mainland and international users and an offshore RMB centre, HKEX OTC Clear provides a platform for connecting Mainland banks with their international counterparts in clearing their OTC derivatives transactions in both global currencies and RMB. This is as illustrated in Figure 11.

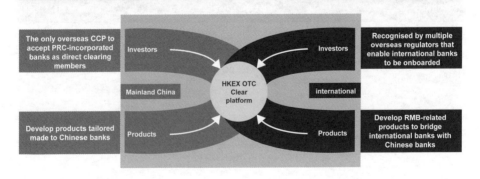

Figure 11. The connectivity platform of HKEX OTC Clear for Mainland and international banks

5   Conclusion

Given the increasing exposure to interest rate risk in foreign currencies and FX risk due to increasing foreign assets held by Mainland financial institutions in USD and other foreign currencies, there is an increasing demand for them to engage in cross-border OTC transactions with foreign counterparties in risk management instruments like IRS and FX derivatives in foreign currencies for hedging their exposure.

Under the tightened global regulatory framework for risk management of OTC derivatives after the 2008 Global Financial Crisis, financial institutions would either have to do mandatory central clearing for standardised OTC derivatives or to be subject to higher capital and margin requirements for bilaterally cleared OTC derivatives. In the latter case, they may voluntarily opt for central clearing in order to reduce the transaction costs.

Unparalleled in the US and European clearing houses, HKEX OTC Clear accepts PRC-incorporated Mainland banks as direct clearing members via their Hong Kong branches so that the Mainland banks can do direct clearing via their Hong Kong branches. In this way, HKEX OTC Clear offers a more convenient and cost-effective solution to Mainland banks for central clearing of their OTC derivatives transactions.

Moreover, there is likely a high growth potential in offshore RMB derivatives trading for RMB risk management in the course of RMB internationalisation. Compared to its overseas counterparts, HKEX OTC Clear is more capable to support Mainland and global financial institutions' OTC transactions in offshore RMB, in addition to serving OTC transactions in USD and other major currencies.

Appendix 1

Products served by HKEX OTC Clear

Product	Currency	Maximum residual term
Single currency interest rate swap (IRS)	CNH (offshore Renminbi)	10 years
	USD	
	EUR	
	HKD	
Single currency basis swap	USD	10 years
	EUR	
	HKD	
Non-deliverable IRS	CNY (onshore Renminbi)	5 years
	INR (Indian Rupee)	10 years
	MYR (Malaysian Ringgit)	
	KRW (Korean Won)	
	THB (Thai Baht)	
	TWD (New Taiwan Dollar)	
Cross currency swap (CCS)	USD vs CNH (offshore Renminbi)	10 years
Non-deliverable currency forward (NDF)	USD / CNY (onshore Renminbi)	2 years
	USD / INR (Indian Rupee)	
	USD / KRW (Korean Won)	
	USD / TWD (New Taiwan Dollar)	

Appendix 2

Clearing members of HKEX OTC Clear

(as of September 2017)

Hong Kong

1. The Bank of East Asia, Limited
2. Hang Seng Bank, Limited
3. The Hongkong and Shanghai Banking Corporation Limited

Mainland China

4. Agricultural Bank of China Limited
5. Bank of China (Hong Kong) Limited
6. Bank of Communications Co., Ltd
7. CCB International Securities Limited
8. China Minsheng Banking Corporation, Ltd
9. Industrial and Commercial Bank of China (Asia) Limited
10. Shanghai Pudong Development Bank Co., Ltd

Europe

11. BNP Paribas
12. Deutsche Bank Aktiengesellschaft
13. Standard Chartered Bank

US

14. Citibank N.A.
15. JP Morgan Bank, National Association

Asia Pacific

16. DBS Bank Ltd.
17. Australia and New Zealand Banking Group Limited (ANZ)

Abbreviations

AI	Authorised Institutions in Hong Kong
BIS	Bank for International Settlements
CCDC	China Central Depository & Clearing Co., Ltd
CCP	Central counterparty
CDS	Credit default swap
CFETS	China Foreign Exchange Trade System
CFTC	Commodity Futures Trading Commission in the US
CNAPS	China National Advanced Payment System
CRM	Credit risk mitigation
CRMW	Credit risk mitigation warrant
CSF	Cash-settled forward
EMIR	European Market Infrastructure Regulation
ESMA	European Securities and Markets Authority
EU	European Union
FRA	Forward rate agreement
FX	Foreign exchange
IRS	Interest rate swap
ISDA	International Swaps and Derivatives Association
NDF	Non-deliverable currency forwards
OIS	Overnight index swap
OTC	Over-the-counter
PBOC	People's Bank of China
QCCP	Qualified central counterparty
QFII	Qualified Foreign Institutional Investor
RQFII	Renminbi Qualified Foreign Institutional Investor
RTGS	Real time gross settlement
SAFE	State Administration of Foreign Exchange in China
SCH	Shanghai Clearing House
SFC	Securities and Futures Commission in Hong Kong
SHIBOR	Shanghai Interbank Offered Rate

Risk statements and disclaimer

Risks of securities trading

Trading in securities carries risks. The prices of securities fluctuate, sometimes dramatically. The price of a security may move up or down, and may become valueless. It is as likely that losses will be incurred rather than profit made as a result of buying and selling securities.

Risks of trading futures and options

Futures and options involve a high degree of risk. Losses from futures and options trading can exceed initial margin funds and investors may be required to pay additional margin funds on short notice. Failure to do so may result in the position being liquidated and the investor being liable for any resulting deficit. Investors must therefore understand the risks of trading in futures and options and should assess whether they are right for them. Investors are encouraged to consult a broker or financial adviser on their suitability for futures and options trading in light of their financial position and investment objectives before trading.

Disclaimer

All information and views contained in this book are for informational purposes only and does not constitute an offer, solicitation, invitation or recommendation to buy or sell any securities, futures contracts or other products or to provide any advice or service of any kind. The views expressed in this book do not necessarily represent the position of Hong Kong Exchanges and Clearing Limited ("HKEX"). Nothing in this book constitutes or should be regarded as investment or professional advice. While information contained in this book is obtained or compiled from sources believed to be reliable, neither HKEX nor any of its subsidiaries, directors or employees guarantee its accuracy, timeliness or completeness for any particular purpose. Neither HKEX nor any of its subsidiaries, directors or employees shall be responsible for any loss or damage arising from the use of, or reliance upon, any information contained in this book.

Afterword

Playing the Exchange's unique role in the development of the Renminbi offshore market

In the rapidly changing world of global finance, exchanges are financial infrastructure with special functions. They are important trading platforms as well as front-line market organisers and innovation promoters. Major international exchanges are increasingly breaking away from the traditional business model and step into more diversified and innovative fields that cover a greater variety of products across multiple markets. HKEX, as an active international exchange in the global avenue, has notably taken such a path in recent years. In particular, with new strategic positioning, it has launched a series of innovative products across multiple asset classes which, to a large extent, have enriched and diversified its product lines.

The launch of each of HKEX's new products, from design, internal and external system adjustments, testing and communication, to post-launch exchanges with the market, requires tremendous time and efforts from different institutions and professionals. In these processes, HKEX Chief China Economist's Office has contributed in various ways and in different areas. To increase Mainland and offshore markets' understanding of renminbi offshore product innovations in Hong Kong, we have worked with various business departments to make a comprehensive overview of HKEX's recent progress on the connectivity programme and renminbi products. This book of collection of research reports, entitled *New Progress in RMB Internationisation: Innovations in HKEX's Offshore Financial Products* is the result of such efforts.

In our opinion, in order for market participants to know a new product well, they must understand the background for the product's launch from a macro and industry level, the development trends in related industries and products in international market as well as the objectives of HKEX's product design. They must also know the risks and returns of such product type from a micro and technical perspective, and its potential application in practical investment. This would require the collaboration of a specialised research department with various business departments, legal and regulatory compliance departments. Each report in this book is a product of cooperation between different HKEX departments and certain HKEX subsidiaries. In producing the reports, the Chief China Economist's Office, as the research organiser, worked well in cooperation with various departments and teams within HKEX, including the equity product development team, the fixed income and currency product team, the commodities team, OTC Clearing Hong Kong, and China Exchanges Services Company Ltd. We jointly produced the drafts for these reports and had addressed the views of many departments in the process of report revision and publishing.

We learned from our interaction with market participants that there is increasing market dissatisfaction with fragmented information about individual products as our product lines continue to expand. Market participants now want more systematic and comprehensive introduction on the basket of products. We hope that through this book which gives a systematic picture on a range of products, financial institutions and HKEX's related business teams can have a reference to rely on when introducing a certain product to the market. Making reference to the book will help understand HKEX's latest product innovations in multiple dimensions.

HKEX Chief China Economist's Office is currently positioned as an in-house think tank, an external thought leader, and a provider of value-added support to HKEX's strategic projects and product innovations. These functions are, in varying degrees, closely related to product innovation, and let us acquire better background knowledge of the course of the product innovations.

I would like to express my special gratitude to HKEX Chief Executive Charles Li for his continuous support and encouragement during the production of this book. I must also thank the regulatory compliance team, the legal services team and the corporate communications team of HKEX for their strong assistance. Without their cooperation and recommendations, this book would hardly be published.

HKEX's growth and development has, in a sense, exemplified the development of the Hong Kong financial industry. It also, from a specific angle, reflects international and Mainland China's economic and financial transformation. Different financial products had been active at different stages of HKEX's development. In other words, these representative, specific product innovations have become symbols of the specific stages of development of HKEX and Hong Kong financial market. From this perspective, the HKEX product innovations discussed in this book manifest the financial innovation currently under way in Hong Kong's financial market and, to a certain extent, reflect the new trend of changes in global finance.

This book may have imperfection given the fast-changing environment in the Mainland and in the international financial market. We welcome any comments you may have for future improvements.

Professor Ba Shusong

Chief China Economist, Hong Kong Exchanges and Clearing Limited

Chief Economist, China Banking Association